W0111302

Periodic Table of the Elements with the Gmelin System Numbers

1 H 2																1 H 2	2 He 1
3 Li 20	4 Be 26											5 B 13	6 C 14	7 N 4	8 O 3	9 F 5	10 Ne 1
11 Na 21	12 Mg 27											13 Al 35	14 Si 15	15 P 16	16 S 9	17 Cl 6	18 Ar 1
19 * K 22	20 Ca 28	21 Sc 39	22 Ti 41	23 V 48	24 Cr 52	25 Mn 56	26 Fe 59	27 Co 58	28 Ni 57	29 Cu 60	30 Zn 32	31 Ga 36	32 Ge 45	33 As 17	34 Se 10	35 Br 7	36 Kr 1
37 Rb 24	38 Sr 29	39 Y 39	40 Zr 42	41 Nb 49	42 Mo 53	43 Tc 69	44 Ru 63	45 Rh 64	46 Pd 65	47 Ag 61	48 Cd 33	49 In 37	50 Sn 46	51 Sb 18	52 Te 11	53 I 8	54 Xe 1
55 Cs 25	56 Ba 30	57** La 39	72 Hf 43	73 Ta 50	74 W 54	75 Re 70	76 Os 66	77 Ir 67	78 Pt 68	79 Au 62	80 Hg 34	81 Tl 38	82 Pb 47	83 Bi 19	84 Po 12	85 At 8a	86 Rn 1
87 Fr 25a	88 Ra 31	89*** Ac 40	104 71	105 71													

* NH_4 23

**Lanthanides 39	58 Ce	59 Pr	60 Nd	61 Pm	62 Sm	63 Eu	64 Gd	65 Tb	66 Dy	67 Ho	68 Er	69 Tm	70 Yb	71 Lu
***Actinides	90 Th 44	91 Pa 51	92 U 55	93 Np 71	94 Pu 71	95 Am 71	96 Cm 71	97 Bk 71	98 Cf 71	99 Es 71	100 Fm 71	101 Md 71	102 No 71	103 Lr 71

A Key to the Gmelin System is given on the Inside Back Cover

Gmelin Handbook of Inorganic Chemistry

8th Edition

Gmelin Handbook of Inorganic Chemistry

8th Edition

Gmelin Handbuch der Anorganischen Chemie

Achte, völlig neu bearbeitete Auflage

Prepared and issued by

Gmelin-Institut für Anorganische Chemie
der Max-Planck-Gesellschaft
zur Förderung der Wissenschaften
Director: Ekkehard Fluck

Founded by — Leopold Gmelin

8th Edition — 8th Edition begun under the auspices of the Deutsche Chemische Gesellschaft by R. J. Meyer

Continued by — E. H. E. Pietsch and A. Kotowski, and by Margot Becke-Goehring

Springer-Verlag Berlin Heidelberg GmbH 1988

STAFF GMELIN HANDBOOK

Dr. G. Bär, D. Barthel, Dr. N. Baumann, Dr. B. Becker, Dr. W. Behrendt, D. Benzaid, Dr. L. Berg, Dipl.-Chem. E. Best, Dr. R. Bohrer, K. D. Bonn, P. Born-Heck, B. Bost, Dr. R. Braun, Dipl.-Ing. V. A. Chavizon, E. Cloos, Dipl.-Phys. G. Czack, I. Deim, Dipl.-Chem. H. Demmer, U. Dietz, Dipl.-Geol. R. Ditz, R. Dowideit, Dipl.-Chem. M. Drößmar, Dr. D. Düster, U. Duggen, R. Durban, M. Engels, Dr. H.-J. Fachmann, Dr. J. Faust, B. Fischer, G. Funk, Dipl.-Ing. N. Gagel, Dr. U. W. Gerwarth, Dipl.-Phys. D. Gras, Dipl.-Bibl. W. Grieser, Dr. I. Haas, Dr. R. Haubold, B. Heibel, Dipl.-Min. H. Hein, A. Heise-Schuster, H.-P. Hente, H. W. Herold, U. Hettwer, Dr. R. Heumüller, Dr. G. Hönes, Dr. W. Hoffmann, Dr. W. Huisl, R. Jonuschat, Dr. R. Jotter, Dr. J. von Jouanne, Dr. B. Kalbskopf, Dipl.-Chem. W. Karl, H.-G. Karrenberg, Dipl.-Phys. H. Keller-Rudek, B. Kirchner, P. Klauck, A. Klusch, Dipl.-Chem. C. Koeppel, Dipl.-Chem. H. Köttelwesch, R. Kolb, E. Kranz, Dipl.-Chem. I. Kreuzbichler, Dr. A. Kubny, Dr. M. Kunz, Dr. W. Kurtz, M. Langer, Dr. U. Lanzendörfer, Dr. B. Ledüc, Dr. A. Leonard, H. Mathis, E. Meinhard, C. Metz, K. Meyer, Dr. M. Mirbach, Dipl.-Chem. B. Mohsin, Dr. U. Neu-Becker, K. Nöring, Dipl.-Chem. R. Nohl, Dipl.-Min. U. Nohl, Dr. U. Ohms-Bredemann, Dr. W. Petz, I. Rangnow, Dipl.-Phys. H.-J. Richter-Ditten, Dipl.-Chem. H. Rieger, E. Rieth, A. Rosenberger, E. Rudolph, G. Rudolph, Dipl.-Chem. S. Ruprecht, Dr. B. Sarbas, Dr. H. Schäfer, Dr. D. Schiöberg, V. Schlicht, B. Schmitt, Dipl.-Chem. D. Schneider, Dipl.-Min. P. Schubert, Dr. K. Schücke, A. Schwärzel, Dr. B. Schwager, Dipl.-Ing. H. M. Somer, Dr. C. Strametz, G. Strauss, Dr. G. Swoboda, Dr. U. Taubald, M. Teichmann, Dr. D. Tille, Dipl.-Ing. H. Vanecek, Dipl.-Chem. P. Velić, Dipl.-Ing. U. Vetter, H.-M. Wagner, Dipl.-Phys. J. Wagner, R. Wagner, Dr. E. Warkentin, Dr. C. Weber, Dr. A. Wietelmann, Dr. M. Winter, Dr. B. Wöbke, K. Wolff

STAFF GMELIN ONLINE DATA SYSTEM

Dr. P. Kuhn, Dr. G. Olbrich

Gmelin Handbook of Inorganic Chemistry

8th Edition

B
Boron Compounds

3rd Supplement Volume 4

Boron and Cl, Br, I, S, Se, Te. Carboranes

With 19 illustrations

AUTHORS

Gert Heller, Institut für Anorganische und Analytische Chemie, Freie Universität Berlin
Berlin, Federal Republic of Germany

Anton Meller, Anorganisch-Chemisches Institut, Universität Göttingen
Göttingen, Federal Republic of Germany

Thomas Onak, Department of Chemistry, California State University
Los Angeles, California, USA

EDITORS

Karl-Christian Buschbeck, Gmelin-Institut, Frankfurt/Main

Kurt Niedenzu, Department of Chemistry, University of Kentucky, Lexington, Kentucky, USA

System Number 13

Springer-Verlag Berlin Heidelberg GmbH 1988

LITERATURE CLOSING DATE: END OF 1984
IN SOME CASES MORE RECENT DATA HAVE BEEN CONSIDERED

Library of Congress Catalog Card Number: Agr 25-1383

ISBN 978-3-662-06140-4 ISBN 978-3-662-06138-1 (eBook)
DOI 10.1007/978-3-662-06138-1

Originally published by Springer-Verlag Berlin Heidelberg New York in 1988.
Softcover reprint of the hardcover 8th edition 1988

Preface

The present Volume 4 of "Boron Compounds" 3rd Supplement concludes the 3rd supplement issue on boron compounds, but will be augmented by a separate index volume for the boron compounds discussed in the four volumes of this specific supplement. This 3rd supplement adds on to the previous volumes dealing with boron compounds beginning (in parentheses: literature closing dates) with the Main Volume (end of 1925), Supplement Volume 1 (end of 1949), Supplement Volume 2 (end of 1975 for boron carbides, mid-1980 for elemental boron), "Boron Compounds" Volumes 1 to 20 and Formula Index, "Boron Compounds" 1st Supplement (3 volumes, end of 1977), and "Boron Compounds" 2nd Supplement (2 volumes, through 1980). In the 3rd supplement the literature is covered uniformly through 1984.

The present volume contains the description of boron compounds with Cl, Br, and I, as well as the systems of boron with S, Se, Te, and Po. The final chapter describes the carboranes; it contains the description of carboranes as well as brief discussions of metallacarboranes and a listing of carborane-containing polymers.

The IUPAC nomenclature is mostly adhered to; occasional abbreviations for compounds are explained in the text. Unless otherwise noted, a positive sign for the chemical shifts on the NMR signals indicates downfield from the references: $(CH_3)_4Si$ for δ^1H and $\delta^{13}C$, $(C_2H_5)_2O{-}BF_3$ for $\delta^{11}B$, aqueous $NaNO_3$ for $\delta^{14}N$, and $CFCl_3$ for $\delta^{19}F$.

Lexington, Kentucky (USA)
Frankfurt am Main
July 1988

Kurt Niedenzu
Karl-Christian Buschbeck

Boron and Boron Compounds in the Gmelin Handbook (Syst. No. 13)

"Bor" (Main Volume)	Historical. Occurrence. The Element. Compounds of B with H, O, N, the Halogens, S, Se, and Te. Literature closing date: end of 1925.
"Bor" (Supplement Volume 1)	Occurrence. The Element. Compounds of B with H, O, N, the Halogens, S, and C. Literature closing date: end of 1949.
"Borverbindungen" 1	Boron Nitride. B–N–C Heterocycles. Polymeric B–N Compounds. Literature coverage from 1950 up to 1972.
"Borverbindungen" 2	Carboranes, Part 1. Nomenclature and Types of Carboranes. Carboranes (without Hetero- and Metallocarboranes, and Higher Carboranes). Literature coverage from 1950 up to 1973 or 1970, respectively.
"Borverbindungen" 3	Compounds of B Containing Bonds to S, Se, Te, P, As, Sb, Si, and Metals. Literature coverage from 1950 to the end of 1973.
"Borverbindungen" 4	Compounds with Isolated Trigonal Boron Atoms and Covalent Boron-Nitrogen Bonding (Aminoboranes and B–N Heterocycles). Literature coverage from 1950 to the end of 1973.
"Borverbindungen" 5	Boron-Pyrazole. Derivatives and Spectroscopic Studies on Trigonal B–N Compounds. Literature coverage from 1950 to the end of 1973.
"Borverbindungen" 6	Carboranes, Part 2. Hetero- and Metallocarboranes. Polymeric Carborane Derivatives. Electronic Properties. Literature coverage from 1950 up to 1974 or 1971, respectively.
"Borverbindungen" 7	Boron Oxides. Boric Acids. Borates. Literature coverage from 1950 to the end of 1973.
"Borverbindungen" 8	The Tetrahydroborate Ion and Its Derivatives. Literature coverage from 1950 to the end of 1974.
"Borverbindungen" 9	Boron-Halogen Compounds, Part 1. Literature coverage from 1950 to the end of 1974.
"Borverbindungen" 10	Boron Compounds with Coordination Number 4. Literature coverage from 1950 to the end of 1975.
"Borverbindungen" 11	Carboranes, Part 3. Dicarba-*closo*-dodecaboranes. Literature coverage from 1950 to the end of 1975.
"Borverbindungen" 12	Carboranes, Part 4. Dicarba-*closo*-dodecaboranes. Literature coverage from 1950 to the end of 1975.
"Borverbindungen" 13	Boron-Oxygen Compounds, Part 1. Literature coverage from 1950 to the end of 1975.

"Borverbindungen" 14	Boron-Hydrogen Compounds, Part 1. Literature coverage from 1950 to the end of 1975.
"Borverbindungen" 15	Amine-boranes. Literature coverage from 1950 to the end of 1975.
"Borverbindungen" 16	Boron-Oxygen Compounds, Part 2. Literature coverage from 1950 to the end of 1975.
"Borverbindungen" 17	Borazine and Its Derivatives. Literature coverage from 1950 to the end of 1976.
"Borverbindungen" 18	Boron-Hydrogen Compounds, Part 2. Literature coverage from 1950 to the end of 1976.
"Borverbindungen" 19	Boron-Halogen Compounds, Part 2. Literature coverage from 1950 to the end of 1976.
"Borverbindungen" 20	Boron-Hydrogen Compounds, Part 3. Literature coverage from 1950 to the end of 1976.
"Boron Compounds"	Formula Index (for the volumes "Borverbindungen" 1 to 20).
"Boron Compounds" 1st Suppl. Vol. 1	Boron and Rare Gases. Boron and Hydrogen. Boron and Oxygen Literature coverage through 1977.
"Boron Compounds" 1st Suppl. Vol. 2	Boron and Nitrogen, Boron and Halogens. Literature coverage through 1977.
"Boron Compounds" 1st Suppl. Vol. 3	Boron and Chalcogens. Carboranes. Formula Index for 1st Suppl. Vol. 1 to 3. Literature coverage through 1977.
"Boron Compounds" 2nd Suppl. Vol. 1	Boron and Noble Gases. Boron and Hydrogen. Boron and Oxygen. Boron and Nitrogen. Formula Index. Literature coverage through 1980.
"Boron Compounds" 2nd Suppl. Vol. 2	Boron and Halogens. Boron and Chalcogens. Carboranes. Formula Index. Literature coverage through 1980.
"Boron Compounds" 3rd Suppl. Vol. 1	Boron and Hydrogen Literature coverage through 1984.
"Boron Compounds" 3rd Suppl. Vol. 2	Boron and Oxygen Literature coverage through 1984.
"Boron Compounds" 3rd Suppl. Vol. 3	Boron and Nitrogen, Boron and Fluorine – 1988 Literature coverage through 1984.
"Boron Compounds" 3rd Suppl. Vol. 4	Boron and Cl, Br, I, S, Se, Te. Carboranes – 1988 (present volume) Literature coverage through 1984.

Table of Contents

6 The System Boron-Chlorine

Anton Meller
Anorganisch-Chemisches Institut, Universität Göttingen
Göttingen, Federal Republic of Germany

Previous coverage, see "Boron Compounds" 2nd Suppl. Vol. 2, 1982, pp. 77/124.

6.1 Binary Species

6.1.1 BCl and BCl_2

BCl. Thermodynamic data obtained from a combination of tungsten transport measurements and theoretical considerations are: $\Delta H_{298} = 151$ kJ/mol; $S_{298} = 213\ J \cdot mol^{-1} \cdot K^{-1}$; $C_{p,298} = 32\ J \cdot mol \cdot K$ [1]. The bond order-dependent enthalpy term E(BCl) is obtained from the relationship $E(BCl) = A[n(BF)]^m$ to be 531.8 kJ/mol (bond order n = 3), and the standard heat of formation $\Delta H_f^{\ominus}(g) = 149.5$ kJ/mol (in good agreement with the value given above) [2].

Generating BCl in a 3.7 m absorption cell by a d.c. glow discharge through BCl_3 led to observation of the pure rotational spectrum of BCl. Rotational transitions of $^{11}B^{35}Cl$ and $^{11}B^{37}Cl$ in the vibrational ground state (split into several components by nuclear electric quadrupole interactions) have been analyzed to yield molecular constants such as the nuclear quadrupole coupling constant eQq; eQq(Cl) for $^{11}B^{35}Cl = -16.737$ MHz and eQq(B) = −3.70 MHz (the latter datum was obtained from a line shape simulation), while eQq(Cl) for $^{11}B^{37}Cl$ was determined to be −13.191 MHz. The corrected nuclear quadrupole coupling constant (taking into account a 0.12 electron delocalization from Cl to B) is −23 MHz. For a tabulation of the observed rotational transition frequencies, additional data and a discussion of the results, see [3].

The infrared spectrum of the $\Delta v = 1$ band of BCl generated in the optical path from BCl_3 by either a microwave or a d.c. discharge has been measured with a tunable diode laser from 828 to 870 cm^{-1}. Spectra of all natural abundant species were present and the v = 1-0, 2-1, 3-2, and 4-3 transitions have been observed. The band center for v = 1-0 transition of $^{11}B^{35}Cl$ is at $v_0 = 829.4087\ cm^{-1}$. A set of eight Dunham coefficients was determined by fitting the data of all isotopic species with the appropriate reduced mass factors. The B–Cl distance was calculated to be r(B–Cl) = 171.52 pm and the dipole moment was estimated to be $\mu = \pm 0.1$ D [4].

The $A^1\Pi$-$X^1\Sigma^+$ transition of BCl has been analyzed up to $v'' = 9$ and $v' = 8$ in a microwave discharge through helium containing traces of BCl_3. Precise molecular constants have been obtained for the two cited electronic states. The excited $A^1\Pi$ state seems to encounter undefined perturbations. The equilibrium constants determined for $^{11}B^{35}Cl$ are compiled in Table 6/1, p. 2 [5]. For the tabulation of rotational constants, origins of the individual bands, vibrational analyses, and equilibrium constants for the less abundant isotopic species, see [5].

Table 6/1
Equilibrium Constants for $^{11}B^{35}Cl$ (in cm^{-1}; standard deviations in parentheses) [5].

	$X^1\Sigma^+$	$A^1\Pi$
B_e	0.684392(73)	0.70792(51)
a_e	0.006764(34) (calcd: 0.0061)	0.01088(58) (calcd: 0.0115)
γ_e	−0.0000277(33)	−0.00034(15)
ε_e	—	0.000077(11)
$D_3\times10^6$	1.919(13) (calcd: 1.82)	2.233(77) (calcd: 1.95)
$\beta_e\times10^6$	—	−0.128(47) (calcd: −0.073)
$\delta_e\times10^6$	—	0.0470(57)
T_e	0	36750.24
ω_e	840.01(17)	852.25(59)
ω_eX_e	5.426(38)	12.91(26)
ω_eY_e	−0.0253(25)	−0.160(41)
ω_eZ_e	—	0.0418(22)

The $A^1\Pi$-$X^1\Sigma^+$ transition of BCl produced by pulsed CO_2 laser irradiation from BCl_3 has also been analyzed [6]. The results presented in Table 6/2 are in fair agreement with the later ones [5]. The dissociation energy $D(BCl^1\Sigma^+)$ was calculated to be 5.6 eV [6]. For the application of molecular calculations in the effective potential approximation using a basis of Slater-type orbitals to BCl, see [7].

Table 6/2
Vibrational Constants for $^{11}B^{35}Cl$ and $^{10}B^{35}Cl$ (in cm^{-1}) [6].

	$X^1\Sigma^+$		$A^1\Pi$	
	$^{11}B^{35}Cl$	$^{10}B^{35}Cl$	$^{11}B^{35}Cl$	$^{10}B^{35}Cl$
ω_e	840.14 ±0.4	870.9 ±0.9	850.09 ±0.4	881.8 ±0.9
ω_eX_e	5.09 ±0.27	5.22 ±0.4	10.9 ±0.2	12.3 ±0.4
ω_eY_e	−0.05 ±0.03	−0.13 ±0.09	−0.26 ±0.03	−0.14 ±0.06
ω_eZ_e	−0.004 ±0.0016	−0.001 ±0.004	0.015 ±0.0016	0.025 ±0.003

The luminescence generated in BCl_3 by radiation from a pulsed CO_2 laser is due to the radiative recombination of BCl molecules with chlorine atoms. The rate constant of this reaction between 2300 and 3100 K is represented by: $k_{eff}=4.38\times10^8$ exp (19300/RT) $cm^3\cdot mol^{-1}\cdot s^{-1}$ [8]. BCl is an intermediate in the discharge synthesis of B_2Cl_4 and can be detected spectroscopically in the plasma from BCl_3 in the radiofrequency discharge [10].

Errors in the calculation of parameters of multi-component systems behind a shock-wave front have been evaluated using an algorithm and computer program. The system originating from BCl_3 and H_2 contains 17 components including BCl and BCl_2. For details, see [9].

BCl_2. Thermochemical data for BCl_2 obtained by a combination of tungsten transport measurements and theoretical estimates are: $\Delta H_{298}=-117$ kJ/mol, $S_{298}=272$ $J\cdot mol^{-1}\cdot K^{-1}$, and $C_{p,298}=46$ $J\cdot mol^{-1}\cdot K^{-1}$ [1].

The structure of the BCl_2 radical has been deduced from an electron spin resonance (ESR) study of an X-ray-irradiated crystal of BCl_3 at 77 K. Two radicals are formed, but brief exposure to UV radiation leaves only the BCl_2 radical. The electronic structure of the radical was obtained by analysis of the determined data (g tensor, ^{11}B magnetic hyperfine tensor, ^{35}Cl magnetic and quadrupole hyperfine tensors). Orbitals containing the unpaired electron are depicted in **Fig. 6-1**. The radical is bent with a Cl–B–Cl angle of about 122°; the unpaired electron is mainly localized at the boron atom [11].

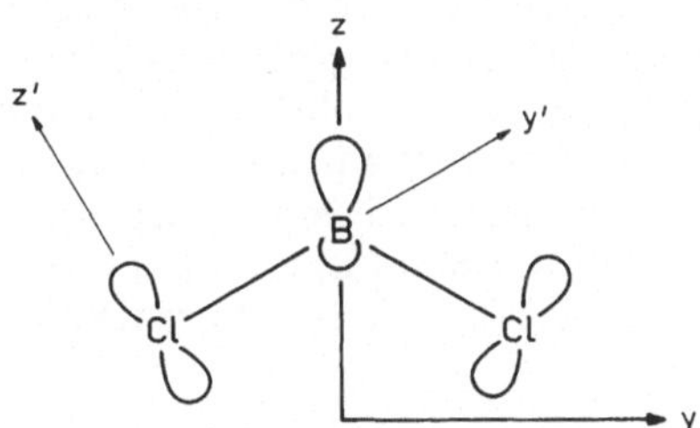

Fig. 6-1. Orbitals containing the unpaired electron in BCl_2 [11].

A kinetic study using ESR and mass spectrometric measurements at a fast flow discharge reactor has been performed. The rate constant for the reaction $H + BCl_3 \rightarrow HCl + BCl_2$ was determined to be $k_{(298)} \simeq 1 \times 10^{-15}$ with an activation energy of 2700 ± 500 cal/mol. For a detailed discussion, see [12]. For errors in the calculation of parameters of multicomponent systems (containing BCl_2) behind a shock-wave, see [9].

BCl_2^+. The semiempirical MNDO SCF–MO method has been used to calculate the following data for BCl_2^+: Heat of formation $\Delta H = 180.1$ kcal/mol; ionization potential = 20 eV [13]. Electrical properties of silicon heavily implanted with BCl_2 show that the effective carrier surface density for BCl_2^+ implants is much lower than for BF_2^+ implants [14].

For column densities of binary boron-chlorine molecules in red giant stars, see [15].

6.1.2 Trichloroborane, BCl_3

6.1.2.1 Preparation, Purification, Isotope Separation

The production of BCl_3 can be performed by (free radical) liquid phase chlorination of $B(OCH_3)_3$ in the presence of CCl_4 over a temperature range of 0 to 100°C (preferably between 40 and 90°C). Cooling is achieved by condensation of the vaporized solvent and an external heat exchanger. The reaction is initiated by radical starters such as light, organic peroxides, or azoisobutyronitrile [16, 17]. Another process uses the direct chlorination of B_2O_3 with Cl_2 in a N_2 atmosphere at 800°C. The resultant BCl_3 is purified by distillation and adsorption (on activated carbon, Al_2O_3, and B_2O_3) to give a product of high purity [18]. The chlorination can also be performed using $B(OH)_3$ absorbed from its aqueous solution on activated carbon, dried in a stream of inert gas at 160 and 300°C, and chlorination at temperatures up to 800°C [19, 20]. Pellets made from borax, activated charcoal, and starch may also be used in the chlorination process [21].

References for 6.1.1 to 6.1.2.1 on pp. 4/5

A model for the combustion of boron powder in chlorine has been proposed; for details, see [22]. The gas composition during the synthesis of BCl_3 can be automatically monitored using gas-liquid chromatography [23]. Directions for the safe handling of BCl_3 are contained in a review (in Japanese) [24].

Purification of BCl_3 can be accomplished by distillation in the presence of KCl and sorbents such as activated charcoal or polydiphenylene oxide [25]. Selective UV photolysis using a KrF eximer laser can be used to remove $COCl_2$ admixed to BCl_3 [26 to 28]. Likewise, admixed organic chlorides can be decomposed by irradiation with a pulsed CO_2 laser to give C, C_2Cl_6, C_6Cl_6, and tar [29]. A continuous CO_2 laser has been used for that purpose, and the energy consumption was 0.8 kWh per hour for 1 kg of BCl_3 [30]. (For a spectroscopic study of the dissociation of BCl_3 in the radiation field of a pulsed CO_2 laser, see [31].) $COCl_2$ can also be removed from BCl_3 by a catalytic process using excess H_2, $COCl_2$, and activated carbon granules at 385 to 400°C [32]. For a method used to purify volatile chlorides by controlled crystallization, see [33]. BCl_3 is removed from $SiCl_4$ by reaction with hexamethylcyclotrisiloxane [34].

Separation or enrichment of $^{11}BCl_3$ and $^{10}BCl_3$ can be achieved by selectively exciting the ν_3 vibrational mode with multi-photon absorption using a TEA CO_2 laser. Enrichment coefficients, kinetics of the process, the influence of low temperatures as well as the admixture of reactants (such as C_nF_{2n+2}) have been studied [35 to 41].

References for 6.1.1 to 6.1.2.1:

[1] Dittmer, G.; Niemann, U. (Philips J. Res. **37** [1982] 1/30).
[2] Holbrook, J. B.; Smith, B. C.; Housecroft, C. E.; Wade, K. (Polyhedron **1** [1982] 701/6).
[3] Endo, Y.; Saito, S.; Hirota, E. (Bull. Chem. Soc. Japan **56** [1983] 3410/4).
[4] Maki, A. G.; Lovas, J. J.; Suenram, R. D. (J. Mol. Spectrosc. **91** [1982] 424/9).
[5] Bredohl, H.; Dubois, I.; Houbrechts, Y.; Nzohabonayo, P. (J. Phys. B **17** [1984] 209/14).
[6] Bogomolova, E. A.; Moskvitina, E. N.; Nikonorov, A. P. (Vestn. Mosk. Univ. Khim. **36** [1981] 531/5; Moscow Univ. Chem. Bull. **36** No. 6 [1981] 9/13).
[7] Zaitsevskii, A. V.; Zviadadze, G. N.; Dzevitskii, B. E. (Deposited Doc. VINITI-2844-83 [1983] 1/8 from C.A. **101** [1984] No. 158052).
[8] Nikonorov, A. P.; Moskvitina, E. N.; Kuzyakov, Yu. Ya.; Stepanov, P. I. (Zh. Fiz. Khim. **57** [1983] 1510/4; Russ. J. Phys. Chem. **57** [1983] 911/4).
[9] Khramova, O. D.; Kuznetsova, L. A.; Kuzyakov, Yu. Ya. (Vestn. Mosk. Univ. Ser. 2 Khim. **25** [1984] 23/7; C.A. **100** [1984] No. 215887).
[10] Briggs, A. G.; Massey, A. G.; Reason, M. S.; Portal, P. J. (Polyhedron **3** [1984] 369/71).

[11] Franzi, R.; Geoffroy, M.; Lucken, E. A. C.; Leray, N. (J. Chem. Phys. **78** [1983] 708/11).
[12] Jourdain, J.-L.; Laverdet, G.; Le Bras, G.; Canbourien, J. (J. Chim. Phys. **78** [1981] 253/7).
[13] Dewar, M. J. S.; Rzepa, H. S. (J. Computat. Chem. **4** [1983] 158/69).
[14] Fuse, G.; Hirao, T.; Inoue, K.; Takayanagi, S.; Yaegashi, Y. (J. Appl. Phys. **53** [1982] 3650/3 from C.A. **97** [1982] No. 15227).
[15] Johnson, H. R.; Sauval, A. J. (Astron. Astrophys. Suppl. Ser. **49** [1982] 77/87 from C.A. **97** [1982] No. 136174).
[16] Richardson, K. W. (U.S. 4239738 [1980] from C.A. **94** [1981] No. 86579).
[17] Delue, N. R.; Crano, J. C. (Ger. Offen. 3011246 [1980] from C.A. **94** [1981] No. 86571).
[18] Yuan, Q.; Zhang, K.; Wang, J. (Huaxue Shijie **23** [1982] 194/6 from C.A. **100** [1984] No. 93302).
[19] Ube Industries, Ltd. (Japan. Kokai Tokkyo Koho 83-20715 [1983] from C.A. **99** [1983] No. 7813).
[20] Iwai, T.; Mizuno, H.; Miura, M. (Eur. Appl. 34897 [1981] from C.A. **95** [1981] No. 222290).

[21] Pai, B. C.; Ramachandran, B. E.; Velpari, V.; Balasubramanian, N. (Indian J. Technol. **22** [1984] 233/5 from C.A. **101** [1984] No. 173912).
[22] Golovko, V. V.; Vovchuk, Ya. I.; Polishchuk, D. I. (Fiz. Goreniya Vzryva **17** No. 5 [1981] 46/50 from C.A. **96** [1982] No. 167032).
[23] Ivanova, N. T.; Malakhovskii, Yu. V.; Alekseeva, Z. I.; Loskutnikova, G. G.; Prigozhina, L. D.; Frangulyan, L. A. (Deposited Doc. SPSTL-637 Khp-D81 [1981] 1/10 from C.A. **98** [1983] No. 118711).
[24] Shirai, T. (Oyo Butsuri **52** [1983] 597/602 from C.A. **100** [1984] No. 108492).
[25] Plesek, J.; Hermanek, S.; Mostecky, J. (Czech. 223494 [1984] from C.A. **102** [1985] No. 81125).
[26] Hyer, R. C.; Hartford, A., Jr.; Atencio, J. H. (Proc. Intern. Conf. Lasers **1980/81** 293/5 from C.A. **96** [1982] No. 145368).
[27] Freund, S. M. (U.S. 4405423 [1983] from C.A. **99** [1983] No. 178355).
[28] Hyer, R. C.; Freund, S. M.; Hartford, A., Jr.; Atencio, J. H. (J. Appl. Phys. **52** [1981] 6944/8 from C.A. **95** [1981] No. 229201).
[29] Farkacova, M.; Vitek, J.; Pola, J.; Dolansky, J.; Plzak, Z. (Chem. Prumysl **34** [1984] 636/8 from C.A. **102** [1985] No. 142250).
[30] Adamova, Yu. A.; Pimenov, V. P.; Skachkov, A. N.; Shmerling, G. V. (Khim. Vysokikh Energ. **18** [1984] 68/71 from C.A. **100** [1984] No. 88038).

[31] Nikonorov, A. P. (Deposited Doc. VINITI-3783 [1979] 357/9 from C.A. **94** [1981] No. 93021).
[32] Chun, D. S. (U.S. 4238465 [1980] from C.A. **94** [1981] No. 86578).
[33] Agliulov, N. Kh.; Polikarpov, Yu. S. (Ref. Zh. Khim. **1982** No. 10L107 from C.A. **97** [1982] No. 103179).
[34] Kray, W. D.; Razzano, J. S. (PCT Intern. Appl. WO 83-03244 [1983] from C.A. **99** [1983] No. 214975).
[35] Hu, C. L.; Chang, C. H.; Liu, Yu. K.; Mi, C.; Ma, H. H.; Yu, S. C.; Kung, F. H.; Chang, Y. W. (Kexue Tongbao **25** [1980] 786/7 from C.A. **94** [1981] No. 164395).
[36] Tang, F.; Yin, Y.; Lu, X. (Zhongguo Jiguang **11** [1984] 416/18 from C.A. **101** [1984] No. 238011).
[37] Academia Sinica (Jiguang **6** No. 11 [1979] 11/16 from C.A. **94** [1981] No. 21915).
[38] Chu, W. S.; Chu, P. I.; Chao. Yu. Y. (Jiguang **7** No. 11 [1980] 61/3 from C.A. **95** [1981] No. 69621).
[39] Zhu, P.; Zhu, W.; Zhao, Y. (Jiguang **8** No. 10 [1981] 20/3, 25 from C.A. **97** [1982] No. 136535).
[40] Tang, F.; Lu, X.; Yin, Y.; Fu, K. (Jiguang **8** No. 3 [1981] 32/4 from C.A. **96** [1982] No. 133030).

[41] Tuccio, S. A.; Heller, D. F. (U.S. 4461686 [1984] from C.A. **101** [1984] No. 159916).

6.1.2.2 Physical Data

Thermodynamic data for gaseous BCl_3 have been obtained by a combination of tungsten transport measurements and theoretical estimates: $\Delta H_{298} = -421$ kJ/mol, $S_{298} = 290\ J \cdot mol^{-1} \cdot K^{-1}$, $C_{p,298} = 64\ J \cdot mol^{-1} \cdot K^{-1}$ [1]. The bond order-dependent enthalpy term $E(BCl) = A[n(BCl)]^m$ (with A being the enthalpy term for the single bond, A = 419 kJ/mol and m = 0.21) was found to be 442.3 kJ/mol (bond order n = 1.33) [1]; the standard enthalpy of formation is given with $\Delta H_f^{\ominus}(gas) = -403$ kJ/mol [2]. MNDO calculations (point group D_{3h}) lead to an enthalpy of formation $\Delta H_f = -87.7$ kcal/mol (= −367 kJ/mol) [5], which differs considerably from the value

References for 6.1.2.2 on pp. 11/3

given elsewhere (403.8 kJ/mol) [3]. The calculated ionization potential of 13.10 eV also seems to be quite high (compare [4]) [5].

Molecular structures and energies have been calculated by the MNDO approximation for BCl_3, $[BCl_3]^+$, all singly and doubly charged ions as well as the corresponding neutral fragments in a study of the mass spectral fragmentation of BCl_3 (and also of B_2Cl_4 and B_4Cl_4). A compilation of relevant data is presented in Tables 6/3 to 6/7 [6].

Table 6/3
MNDO Results of BCl_3 and its Mass Spectrometric Fragments.
Point groups and molecular energies (experimental enthalpies of formation in parenthesis) [6].

	point group	molecular state	$\Delta H_f^\ominus$ in kJ/mol
BCl_3	D_{3h}	$^1A_1'$	−366.7 (−403.8) ([3])
BCl_3^+	C_{2v}	$^2B_2$	+806.3 (+718) ([7])
BCl^+	$C_{\infty v}$	$^2\Sigma^+$	+1150.5 (+1130) ([7])
BCl_2^+	$D_{\infty h}$	$^1\Sigma_g^+$	+753.6 (+644) ([7])
BCl^{2+}	$C_{\infty v}$	$^1\Sigma^+$	+3092.6
BCl_2^{2+}	$D_{\infty h}$	$^2\Pi_g$	+2673.9
BCl_3^{2+}	dissociates to BCl^+ and Cl_2^+		
BCl	$C_{\infty v}$	$^1\Sigma^+$	+37.9
		$^1\Pi$	+669.1
		$^3\Pi$	+493.1
BCl_2	C_{2v}	$^2A_1$	−114.6

Table 6/4
Energy Levels in BCl_3 from MNDO Calculations [6].

−I in eV	symmetry	other assignments [8]	[9]
−13.105	$\tilde{X}A_2'$	A_2'	E'
−13.265	$\tilde{A}E'$	E''	A_2'
−13.437	$\tilde{B}E''$	E'	E''
−15.520	$\tilde{C}A_2''$	A_2''	A_2''
−17.488	$\tilde{D}E'$	E'	E'
−22.493	$\tilde{E}A_1'$	E'	
−26.502	$\tilde{F}E'$	A_1'	A_1'
−27.972	$\tilde{G}A_1'$		

The method of partial retention of diatomic differential overlap (PRDDO) has been used to calculate total energy, HOMO and LUMO, Mulliken charges, and overlap populations for BCl_3, and the resultant data have been compared with those obtained by ab initio wave functions with the same basis set. The results are comparable; however, only a fraction of computing time is required using the former method [11].

Table 6/5

Observed [7] and Calculated [6] Appearance Potentials in eV of Mass-Spectroscopic Fragments from BCl_3.

ion	source	other products	AP_{calc} [6]	AP_{obs} [7]
BCl^+	BCl_3	$\{2Cl^+$ / Cl_2	18.23 / 15.72 $\}$	18.37
BCl_2^+	BCl_3	Cl^+	12.86	12.30
	B_2Cl_4	BCl_2^+	11.68	11.32
BCl_3^+	BCl_3	—	12.16	11.60
BCl_2^{2+}	BCl_3	Cl^+	32.76	33.77 [10]

Table 6/6

Bond Dissociation Energies in kJ/mol, from MNDO Calculations [6].

$D(Cl_2B-Cl)$	373.1	$D(ClB-Cl)$	173.5	$D(ClB^{2+}-Cl)$	539.7
$D(Cl_2B^+-Cl)$	68.3	$D(ClB^+-Cl)$	517.9	$D(ClB^+-Cl^+)$	−146.5
$D(Cl_2B-Cl^+)$	456.0	$D(ClB-Cl^+)$	661.2	$D(B^+-Cl^+)$	−344.6

Table 6/7

Symmetry Constraints on Fragmentation of BCl_3 [6].

fragment	symmetry of transition state	available states of M^+
BCl_3^+	C_{2v}	$\tilde{X}$, $\tilde{A}$, $\tilde{D}$, $\tilde{F}$
$BCl_2^+ + Cl^•$	C_{2v}	all
$BCl^+ + Cl_2$	C_{2v}	$\tilde{A}$, $\tilde{D}$, $\tilde{E}$, $\tilde{F}$, $\tilde{G}$

The surface tension of highly purified BCl_3 was measured by the capillary rise method. The influence of temperature on the surface tension is described by the equation $\sigma = A - BT + CT^2$ and the parameters for BCl_3 from 166.2 to 285.6 K are: A = 67.21, B = 0.2164 and 10^4 C = 1.527 [12, 13]. The solubility of HCl in BCl_3 obeys Henry's law. The temperature dependence of the Henry constant is given by $\ln K = \Delta\bar{H}/RT - \Delta\bar{S}/R$ with $\Delta\bar{H} = -11.0$ kJ/mol and $\Delta\bar{S} = -70.6$ kJ $\cdot mol^{-1} \cdot K^{-1}$ (T from 243.2 to 273.2 K) [14].

The field gradient asymmetry parameter $\eta = 0.557 \pm 0.0011$ has been obtained from a Zeeman study of a BCl_3 single crystal by pulsed Fourier transform nuclear quadrupole resonance spectroscopy. A fine structure has been observed which is due to $^{11}B^{35}Cl$ dipole-dipole interaction [15].

The photoabsorption cross-section of BCl_3 was recorded between 190 to 280 eV using synchrotron radiation. Assignments of the BK spectrum are supported by ab initio calculations. Experimental and calculated term values and oscillator strengths of the BK band of BCl_3 are compiled in Table 6/8, p. 8. The term values of the Cl $L_{2,3}$ spectra and the assignments made are listed in Table 6/9, p. 8. The spectra are depicted in the original paper. For a precise discussion, see [16].

References for 6.1.2.2 on pp. 11/3

Table 6/8

Experimental and Calculated Term Values and Oscillator Strengths of the BK Band of BCl_3 [16].

band	experimental term value in eV	experimental integrated oscillator strength	calculated term value in eV	calculated oscillator strength	assignment
a	7.4	0.022	6.81	0.095	$2p\pi\ (a_2'')$
b	6.7	0.0016			?
c	3.7	0.0036	2.88	forbidden	$3s+\sigma^*\ (a_1')$
d	2.9	0.043	0.64	0.053	$2p\sigma^*\ (e')$
e	2.2	0.0029	2.07	0.00004	$3p\pi\ (a_2'')$
			2.03	0.00105	3p (e′)
f	1.1	0.0046	1.11	0.00001	$4p\pi\ (a_2'')$
g	0.8	0.0019	1.08	0.00815	$4p\sigma\ (e')$

Table 6/9

Experimental and Theoretical Term Values of the L Spectra for BCl_3 (BCl_3 is placed on the (xz) plane and the chlorine atom with a hole is on the z axis. The calculated oscillator strength is shown in parentheses) [16].

experimental in eV	$2px\rightarrow\Phi_s$ in eV	$2py\rightarrow\Phi_s$ in eV	$2pz\rightarrow\Phi_s$ in eV	final orbital Φ_s	$2ps\rightarrow\Phi_s$ in eV
5.9	4.6 (0.0)	4.55 (0098)	4.62 (0.0039)	nonbonded 2p of boron	4.61 (0.0026)
4.3, 4.4	3.11 (0.0002)	3.12 (0.0001)	3.10 (0.0005)	4s	4.05 (0.0003)
2.9	2.13 (0.0)	2.12 (0.0001)	2.13 (0.0001)	4py	2.13 (0.0004)
	2.01 (0.0007)	2.01 (0.0008)	2.01 (0.0016)	4pz	2.00 (0.0000)
	1.98 (0.0001)	1.99 (0.0)	1.99 (0.0001)	4px	1.99 (0.0004)
	1.76 (0.0044)	1.77 (0.0036)	1.60 (0.0097)	σ^*_{BCl}	1.71 (0.0057)
2.0	1.46 (0.0000)	1.47 (0.0003)	1.46 (0.0005)	5s	1.46 (0.0000)
	1.29 (0.0002)	1.29 (0.0001)	1.29 (0.0)	3dxy	1.29 (0.0000)
	1.28 (0.0007)	1.29 (0.0)	1.28 (0.0007)	3dxz	1.28 (0.0000)
	1.28 (0.0001)	1.29 (0.0000)	1.28 (0.0000)	$3dx^2-y^2$	1.28 (0.0000)
	1.26 (0.0)	1.26 (0.0001)	1.26 (0.0001)	3dzy	1.26 (0.0001)
1.3, 1.2	1.08 (0.0)	1.08 (0.0000)	1.08 (0.0000)	5py	1.08 (0.0000)
	1.02 (0.0004)	1.02 (0.0004)	0.97 (0.0000)	5pz	1.01 (0.0018)
	0.99 (0.0001)	1.00 (0.0)	0.99 (0.0007)	5px	0.99 (0.0002)
	0.88 (0.0016)	0.89 (0.0011)	0.80 (0.0074)	$3dz^2$	0.87 (0.0021)

The core binding energies of BCl_3 have been redetermined from the gas phase photoelectron spectra and were evaluated for the boron 1s as 199.98(6) eV with a full width at half maximum (fwhm) of 1.71(20) eV, and for the chlorine $2p_{3/2}$ as 206.84(3) eV with fwhm of

1.34(6) eV (standard deviations in parentheses) [17]. Halogen→boron π-bonding is deduced from the difference between the $1a_2''$ ionization potential and the localized orbital ionization potentials (LOIP) for the halogen valence p orbital. The $1a_2''$ molecular orbital is composed of the pπ-orbitals of all four atoms and the differences measure the stabilization of the halogen pπ-orbitals by π-bonding to the boron atom. The data confirm that there is a steady increase in the π-bonding in going from BI_3 to BF_3. Relevant data for the trihaloboranes are compiled in Table 6/10. For additional details, see [17].

Table 6/10

Valence Ionization Potential Data (in eV) for Trihaloboranes [17].
LOIP = localized orbital ionization potentials; COG = center of gravity ionization potentials.

	LOIP	$1a_2'$	$3e'$	IP		COG	
				$1e''$	$1a_2''$	($1a_2'$ and $3e'$)	$1a_2''$-LOIP
BF_3	16.56	15.95	17.14	16.67	19.13	16.74	2.57
BCl_3	12.34	11.73	12.66	12.39	14.42	12.35	2.08
BBr_3	11.19	10.65	11.71	11.36	13.18	11.36	1.99
BI_3	10.05	9.36	10.42	10.01	11.74	10.06	1.69

Another approach to evaluate the acidity of the trihaloboranes is based on CNDO/2 calculations. Thus, the destabilization ΔE′ of the trihaloboranes by changing from sp^2 to sp^3 hybridization (i.e., adduct formation) is given for BF_3 with 29.00, for BCl_3 with 10.34, and for BBr_3 with 8.49 kcal/mol. For additional details, see [18].

In force field studies by the parametric representation method, experimental data such as isotopic frequencies are used to find the best-fit force field [19]. Valence force constants, mean amplitudes of vibration, shrinkage constants, Coriolis coupling constants, and rotational distortion constants have thus been evaluated and are in fair agreement with experimental results and the data given in "Borverbindungen" 19, pp. 125/32 and "Boron Compounds" 2nd Suppl. Vol. 2, 1982, pp. 81/4. For a tabulation of the calculated data, see [19]; for a valence force field calculation using Wilson's FG matrix method for $^{10}BF_3$ and $^{11}BF_3$ performed with respect to isotope selective excitation, see [20]; also compare [21] and comments concerning these calculations [22]. Harmonic potential functions [22] based on accurate harmonic frequency data [23] have been calculated and the results are presented in Table 6/11. A comparison of calculated harmonic fequencies [22] with the experimental findings [23] is made in Table 6/12, p. 10. A CNDO/2 study of the bending modes of BCl_3 (and BF_3) leads to results which are apparently somewhat less accurate [24].

Table 6/11

Symmetry Potential Constants C and Force Constants F for BCl_3 (C_{11} and C_{33} in Å/mdyn, C_{22} and C_{44} in rad²/Å, and C_{34} in rad/mdyn. The units for the force constants F are the reciprocals of those for the corresponding compliance constants) [22].

$C_{11}=0.2162$	$F_{11}=4.6253$	$C_{34}=0.2373$	$F_{34}=0.5805$
$C_{22}=2.4783$	$F_{22}=0.4035$	$C_{44}=1.4964$	$F_{44}=0.7602$
$C_{33}=0.3108$	$F_{33}=3.6603$		

References for 6.1.2.2 on pp. 11/3

Table 6/12
Harmonic Frequencies for Isotopically Varied BCl_3 (in cm^{-1}).

species		calculated [22]	experimental [23]	difference
$^{10}B^{35}Cl_3$	ω_1	473.542	473.613 ± 0.2	0.071
	ω_2	471.639	471.589 ± 0.05	−0.050
	ω_3	988.144	988.143 ± 2	−0.001
	ω_4	258.80	258.80 ± 2	0.0
$^{11}B^{35}Cl_3$	ω_1	473.542	473.472 ± 0.2	−0.070
	ω_2	451.641	451.689 ± 0.05	0.048
	ω_3	949.465	949.386 ± 2	−0.079
	ω_4	257.93	258.22 ± 2	0.29
$^{10}B^{37}Cl_3$	ω_1	460.57	—	—
	ω_2	470.53	—	—
	ω_3	983.997	984.442 ± 2	0.445
	ω_4	252.18	—	—
$^{11}B^{37}Cl_3$	ω_1	460.57	—	—
	ω_2	450.48	—	—
	ω_3	945.125	945.533 ± 2	0.408
	ω_4	251.36	—	—

The calculated data [22] are confirmed by other calculations [25] using a least-squares interaction method. This latter work [25] contains an excellent review of the vibrational spectroscopy of trihaloboranes combined with careful calculations for the force fields of the E′ modes with the aid of the mass influence on Coriolis constants. The usefulness of Keating coordinates versus valence coordinates as basis of force field approximations is discussed. The resultant force field [25] was used to calculate the mean amplitudes of vibration for $^{11}BCl_3$. The E′ block of the mean-square amplitude matrix for planar symmetrical XY_3 moieties is treated theoretically in terms of the Keating coordinates. New relations which connect the mean-square amplitudes and the force constants are presented and previous work on mean amplitudes and related quantities for trihaloboranes is reviewed [26]. For the final E′ force constants for trihaloboranes in terms of valence, central, and Keating coordinates, see Table 7/2, p. 72 [25]; and for the calculated and observed mean amplitudes of vibration, see Table 7/3, p. 73 [26].

The absorption spectrum of a dilute solution (of about 11 ppm) of $^{11}BCl_3$ in solid argon (84 K) and solid krypton (at 116 K) has been studied in the region of the ν_3 band and temperature effects on the band position have been recorded. The experimental wavenumbers thus obtained for $^{11}B^{35}Cl_3$ are 947 cm^{-1} (Ar, 84 K) and 944.6 cm^{-1} (Kr, 116 K), and for $^{10}B^{35}Cl_3$ 985.6 cm^{-1} (Ar, 84 K) and 983.2 cm^{-1} (Kr, 116 K). Additional weak bands which appear on the low frequency side of the chlorine isotopic features are interpreted as hot bands. For representations of the band profiles, see [27]. The effect of laser irradiation on the sublimation of BCl_3/Xe solid mixtures by selective excitation of the BCl_3 molecules (ν_3, 944.18 cm^{-1}) has also been studied [34].

The absorption of CO_2 laser lines by BCl_3 is reported and extinction coefficients for the 00°1-10°0 band of the laser are reported. A band model calculation summing about 200000 theoretical lines describes the data well if the upper state inertial constants are fit such that $B' = B'' - 0.0013$ and $C' = C'' + 0.005$. The band strength was calculated to be $(5.90 \pm 0.30) \times 10^7$ cm/mol in good agreement with CNDO calculations [28].

The infrared multiphoton decomposition of enriched $^{10}BCl_3$, enriched $^{11}BCl_3$, and natural BCl_3 has been studied in the presence of oxygen. The CO_2 TEA laser was tuned to the P(20) line of the 10.6 μm band (944.2 cm^{-1}), which is in resonance with the ν_3 mode of $^{11}BCl_3$. The dependences of the apparent decomposition rate and luminescence intensity on the laser fluence indicate the initial decomposition of BCl_3 followed by radical scavenging of oxygen. The ratio of the apparent decomposition rates obtained separately with 98% $^{10}BCl_3$ and 97% $^{11}BCl_3$ under identical experimental conditions was 4.7 ± 1.3. With natural BCl_3, however, the corresponding ratio was only 1.3. The large decrease is probably due to the V-V energy transfer between highly excited $^{11}BCl_3$ and unexcited $^{10}BCl_3$ molecules. The isotope selectivity depends only on the pressure of $^{11}BCl_3$ but remains constant with varying pressure of $^{10}BCl_3$. The results are interpreted by a mechanism involving two kinds of vibrationally excited molecules, BCl_3^x and BCl_3^*, where one has a lifetime of a few microseconds while the other decomposes instantaneously [29].

A series of papers (in Chinese) deals with the behavior of BCl_3 in an intense infrared laser field [30 to 33]. The absorption spectrum of vibrationally excited BCl_3 was measured using an IR double resonance technique. A variety of relaxation and energy transfer processes has been observed. The relation $P_{\tau v\text{-}\tilde{v}}(^{11}BCl_3) = 3$ μs·Torr was obtained. A radial sound modulation was detected on the double resonance signal [30]. The visible fluorescence excitation spectrum of BCl_3 induced by a TEA CO_2 infrared laser was measured with respect to the BCl_3 pressure and the laser power and compared with the IR absorption spectrum of BCl_3. The excitation spectra show a red shift of about 20 cm^{-1} [31]. A space- and time-resolved method was used to investigate the dependence of laser-induced pulse visible fluorescence (optoacoustic-optical, OAO, effect) on space and time. There is a notable effect of the geometrical shape of the sample cell on this effect. A pulse acoustic wave, which gives additional excitation to already excited BCl_3 molecules, has been detected and its relationship with the OAO fluorescence was studied [32]. By investigation of the time-resolved absorption spectra of BCl_3 in a high laser energy flux a nonthermal distribution in vibrational states has been observed. Rotational energy transfer plays an important role in generating the distribution. A simplified dynamical collision model is used to explain the development of the vibrational absorption spectrum of BCl_3. The relation $P_{\pi v\text{-}v} \approx c/K'(T_v, T_{o,q})$ is given for the redistribution of the vibrational states and expressions for the equivalent vibrational temperature and the average number of absorbed photons are presented [33].

The dynamic range of reflection under 4-wave interaction in BCl_3 (which is resonant absorber at 10.6 μm) is 10^7 [35]. For the calculation of the reflection efficiency of IR laser irradiation in III/V semiconductors, see [36]. In an atmospheric pressure-active nitrogen (APAN) discharge a brillant blue color was observed upon the introduction of BCl_3 in the after-glow region of the discharge. This is due to the $A^2\Pi \rightarrow X^2\Sigma^+$ transition of BO which forms with O_2 impurities in the N_2. Apparently BO is formed in preference to BN [37].

For ab initio gradient calculations of the π-contribution in the B–Cl bonding of BCl_3, see [38]. For a literature search about swarm parameters and electron impact cross-sections, see [39].

References for 6.1.2.2:

[1] Dittmer, G.; Niemann, U. (Philips J. Res. **37** [1982] 1/30).
[2] Holbrook, J. B.; Smith, B. C.; Housecroft, C. E.; Wade, K. (Polyhedron **1** [1982] 701/6).

[3] Wagman, D. D.; Evans, W. H.; Parker, V. B.; Schumm, R. H.; Halov, I.; Bailey, S. M.; Churney, K. L.; Nuttall, R. L. (J. Phys. Chem. Ref. Data **11** [1982] Suppl. No. 2).
[4] King, G. H.; Krishnamurthy, S. S.; Lappert, M. F.; Pedley, J. B. (Faraday Discussions Chem. Soc. No. 54 [1972] 70/83).
[5] Dewar, M. J. S.; Rzepa, H. S. (J. Computat. Chem. **4** [1983] 158/69).
[6] Bews, J. R.; Glidewell, C. (THEOCHEM **6** [1982] 333/47).
[7] Dibeler, V. H.; Walker, J. A. (Inorg. Chem. **8** [1969] 50/5).
[8] Basset, B. J.; Lloyd, D. R. (J. Chem. Soc. A **1971** 1551/9).
[9] Boyd, R. J.; Frost, D. C. (Chem. Phys. Letters **1** [1968] 649/50).
[10] Mariott, J.; Craggs, J. D. (J. Electron. Control **3** [1957] 194/202).

[11] Marynick, D. S.; Lipscomb, W. N. (Proc. Natl. Acad. Sci. U.S. **79** [1982] 1341/5).
[12] Devyatykh, G. G.; Krasnova, S. G.; Osipova, L. I. (Zh. Fiz. Khim. **55** [1981] 750/1; Russ. J. Phys. Chem. **55** [1981] 419/20).
[13] Osipova, L. I.; Krasnova, S. G. (Deposited Doc. SPSTL-917 Khp-D81 [1981] 61/7 from C.A. **98** [1983] No. 78641).
[14] Krasnova, S. G.; Filimonov, I. V. (Zh. Neorgan. Khim. **26** [1981] 2929/32; Russ. J. Inorg. Chem. **26** [1981] 1566/8).
[15] Litzistorf, G.; Sengupta, S.; Giezendanner, D.; Lucken, E. A. C. (J. Chem. Phys. **75** [1981] 2535/8).
[16] Ishiguro, E.; Iwata, S.; Suzuki, Y.; Mikuni, A.; Sasaki, T. (J. Phys. B **15** [1982] 1841/54).
[17] Beach, D. B.; Jolly, W. L. (J. Phys. Chem. **88** [1984] 4647/9).
[18] Atvars, T. D. Z.; Zaniquelli, M. E. D.; Lin, C. T. (Acta Sud Am. Quim. **3** [1983] 37/48 from C.A. **100** [1984] No. 147579).
[19] Mohan, S.; Rajazaman, S. (Indian J. Pure Appl. Phys. **20** [1982] 230/2).
[20] Arnan, A.; Bertran, J.; Silla, E. (Anales Quim. A **80** [1984] 542/5).

[21] Lin, C. T.; Atvars, T. D. (Spectrosc. Letters **13** [1980] 167/84).
[22] Vicharelli, P. A. (J. Mol. Spectrosc. **92** [1982] 266/8).
[23] Maier II, W. B.; Holland, R. F. (J. Chem. Phys. **72** [1980] 6661/77).
[24] Solomonik, V. G.; Ozerova, V. M.; Sazonova, I. G.; Krasnov, K. S. (Zh. Strukt. Khim. **22** [1981] 155/7; J. Struct. Chem. [USSR] **22** [1981] 124/6).
[25] Cyvin, S. J.; Cyvin, B. N.; Brunvoll, J. (Spectrosc. Letters **17** [1984] 511/23).
[26] Cyvin, S. J.; Cyvin, B. N. (Spectrosc. Letters **17** [1984] 525/35).
[27] Holland, R.; Maier II, W. B.; Freund, S. M.; Beattie, W. H. (J. Chem. Phys. **78** [1983] 6405/14).
[28] Richton, R. E.; Farrow, L. A. (J. Chem. Phys. **76** [1982] 5256/9).
[29] Ishikawa, Y.; Kurihara, O.; Nakane, R.; Arai, Sh. (Chem. Phys. **52** [1980] 143/9).
[30] Xu, J.; Huang, N.; Jiang, Y.; Fu, G.; Wu, Z. (Wuli Xuebao **30** [1981] 1456/63 from C.A. **97** [1982] No. 100891).

[31] Jiang, Y. F.; Huang, N. T.; Zhu, T.; Luan, S. Y.; Lu, Z. Z.; Xu, J. R. (Wuli **10** [1981] 220/2 from C.A. **95** [1981] No. 194592).
[32] Jiang, Y.; Huang, N.; Xu, J.; Lin, G. (Wuli Xuebao **30** [1981] 1464/78 from C.A. **97** [1982] No. 63489).
[33] Xu, J.; Huang, N.; Jiang, Y.; Fu, G.; Wu, Z. (Wuli Xuebao **32** [1983] 854/66 from C.A. **99** [1983] No. 166117).
[34] Wada, K.; Yasui, M.; Uehara, Y.; Yamaguchi, H. (Reza Kenkyu **9** [1981] 676/81 from C.A. **99** [1983] No. 184625).
[35] Basov, N. G.; Kovalev, V. I.; Faizullov, F. S. (Kvantovaya Elektron. [Moscow] **10** [1983] 1276/8 from C.A. **100** [1984] No. 94234).

[36] Basov, N. G.; Kovalev, V. I.; Faizullov, F. S. (Izv. Akad. Nauk SSSR Ser. Fiz. **48** [1984] 1407/14 Bull. Acad. Sci. USSR Phys. Ser. **48** [1984] 158/65 from C.A. **101** [1984] No. 219270).
[37] Rice, G. W.; D'Silva, A. P.; Fassel, V. A. (Appl. Spectrosc. **38** [1984] 149/54).
[38] Kosmus, W.; Kalcher, K. (Monatsh. Chem. **112** [1981] 1123/8).
[39] Davies, D. K. (AD-A122026 [1982] 1/106 from C.A. **98** [1983] No. 226403).

6.1.2.3 Chemical Behavior

Light-Induced Reactions

The formation of $HBCl_2$ from a gas mixture of BCl_3, SiF_4, H_2, and $HCCl_3$ by irradiation with a CO_2 laser is more efficient upon excitation of BCl_3 (at 938 cm^{-1}); the thermal dissociation of $HCCl_3$ is identical in both cases. This observation supports the assumption of a nonthermal mechanism [1]. For comments on this experiment, see [2]. In a study of the kinetics of the IR laser-induced reactions of BCl_3 with H_2, the rate constant of the reaction $BCl_3+H_2\rightarrow HBCl_2+HCl$ was determined to be $k=(8.83\pm0.79)\times10^{-5}$ $m^3\cdot mol^{-1}\cdot s^{-1}$; in the presence of Ar it is $k=(8.16\pm0.73)\times10^{-5}$, and in the presence of CO_2 it is $(2.33\pm0.77)\times10^{-5}$. An anomalous dependence of the reactant consumption on photolysis time in the BCl_3/CH_4 system resulted from quenching of C_2H_6 and C_3H_8 impurities. The vibrational excitation of BCl_3 is the main source of initiation for the reactions in both systems [3], and the same holds true for the corresponding three-component system BCl_3–H_2–CH_4. The reaction is terminated at the stage of the $HBCl_2$ formation [4]. The yield of the reaction $BCl_3+H_2\xrightarrow{h\nu}HBCl_2+HCl$ increases on addition of Xe but is lowered by He [5].

A Rice-Ramsperger-Kassel-Marens (RRKM) calculation (in a Monte-Carlo framework) confirmed experimental results which indicated that absorption of a single photon will cause the dissociation of the van der Waal species Ar–BCl_3 within 1 to 3 picoseconds [326, 349]. It has been observed that the sublimation of Xe from solid BCl_3/Xe mixtures (with low dilution) at 68 K is suppressed by more than 30% by exciting the ν_3 vibration of BCl_3. A dense BCl_3 layer is formed on the surface (due to the higher volatility of Xe), and the diffusion of Xe is inhibited by the vibrational amplitude of the BCl_3 molecules [327, 328]. The diffusion rate of selectively excited $^{10}BCl_3$ or $^{11}BCl_3$ can be used to increase isotope-separation factors in gas diffusion through fine-pored filters [329]. Selective excitation is also used for isotopic enrichment in BCl_3/H_2 and BCl_3/air mixtures by influencing the reaction rates [330]. The multiphoton photolysis of BCl_3 is used in the chemical treatment of semiconductor wafers [331]. The reaction of multiphoton-excited BCl_3 with H_2S leads to the formation of a colorless solid and HCl; the kinetics of this reaction have been shown to be different from that of typical multiphoton dissociation experiments [332]. Pulsed-laser irradiation of BCl_3/SiH_4 mixtures (pulse duration from 10^{-2} to 10^{-6} s, intensity up to 10^7 W/cm^2) results in the formation of $HBCl_2$, B_2H_6, H_2, and $HSiCl_3$ as well as deposition of amorphous Si [333]. The laser-induced reaction between BCl_3 and CH_4 at pressures up to 700 hPa leads mainly to $HBCl_2$ and CH_3Cl (and probably also CH_2Cl_2) [333 to 336]. Photochemical products from the laser-induced reaction between BCl_3 and C_2H_4 have been identified to be HCl, CH_4, C_2H_2, C_2HCl_3, C_2Cl_4, $B_2H_2Cl_4$, $B_2H_3Cl_3$, and $B_2H_4Cl_2$. The results are highly dependent on the total and partial pressures of the reactants. They were compared with the results of discharge experiments and of thermal heating [337]. Laser-promoted dissociation ($\nu=986.6$ cm^{-1}) of ClCH=CHBCl$_2$ gives an isotope enrichment of 80% ^{10}B in the reaction products [338], and addition of BCl_3 as a resonance-buffer gas is claimed to give an additional increase of ^{10}B enrichment in the dissociation product [339]. Apparently, the absorption bands of ClCH=CHBCl$_2$ in the region from 950 to 1150 cm^{-1} are

References for 6.1.2.3 on pp. 19/31

coupled (γ-CH_2) or are shifted upon excitation, since the maximum dissociation of the ^{10}B moiety is observed at a lasing frequency of 949.5 cm^{-1} and corresponds to a long-wavelength shift of over 30 cm^{-1} [340]. The process (in a 14-stage cascade system) is used to obtain 3 kg ^{10}B-enriched (93.5%) product per day [341, 342]. The pressure of BCl_3 promotes the multiphoton dissociation of 1-bromopropane to give HBr, propene, CH_4, and HC≡CH [343]. The BCl_3-sensitized decomposition of hexafluoroacetone induced by a continuous-wave (cw) CO_2 laser gives perfluorinated hydrocarbons, F_3CCOF, CO, and F_2CO [344]. C_6F_5Cl and BCl_2F are the main products from the reaction of BCl_3 with C_6F_6 induced by a 977.21 cm^{-1} cw CO_2 laser; this latter frequency apparently permits a faster V-V energy transfer between the reactants than the irradiation at 949.48 cm^{-1} [345]. Multiphoton decomposition of the CH_3SH–BCl_3 complex can be achieved at low threshold laser fluence (0.5 J/cm^2); HCl is formed in the reaction. However, in a natural isotope mixture of the ^{10}B and ^{11}B species, the isotope selectivity of the reaction is about of the same order as for the use of O_2 or H_2S as scavengers [346, 347].

In TEA CO_2 laser-irradiated BCl_3 decomposition and formation of photoproduced $COCl_2$ occur at a peak power of about 10^8 W/cm^2. No dissociation of $COCl_2$ is observed at 10^6 W/cm^2 [6]. $BClF_2$ was observed as a by-product in the multiphoton photolysis of BCl_3/O_2 mixtures; it originates from the reaction of excited BCl_3 with fluorocarbon lubricant [7]. Laser-induced reactions of BCl_3 with NF_3, N_2F_4, or C_2H_4 lead to gaseous products and the formation of boron nitride and $[BF_4]^-$ [8]. A separation of uranium isotopes uses the reaction of laser-excited $^{235}UF_6$ with BCl_3 [9].

Reactions with Gases in Chemical Vapor Depositions

BCl_3 is the major starting material for the chemical vapor deposition (CVD) of boron (by reduction), boron nitride (by reaction with NH_3), or boron carbide (by reaction with CH_4) on iron and other transition metals or their alloys (from the gas phase) as well as for the CVD synthesis of metal borides. The CVD process can be supported thermally or is achieved by low pressure plasma formation from either a discharge or an arc-produced plasma. The determination of temperature and partial pressure inside a CVD reactor can be performed by Raman scattering [10, 11].

Amorphous boron can be deposited at temperatures between 550 and 750°C by the reaction of BCl_3 downstream of a hydrogen discharge; no BCl-containing material is formed in this process (at the relatively low temperatures employed) [12]. For methods for estimating the maximum deposition rate of boron ($H_2/BCl_3=5:1$, pressure 66.5 kPa), see [13]. The macrokinetics of the CVD of boron (from BCl_3/H_2) by thermal deposition have been studied [14], and a one-dimensional model has been developed to describe the reaction kinetics for CVD deposition of boron using an impinging jet method. It is suggested that the decomposition of $HBCl_2$ might control the deposition rate [15]. For the deposition of thin layers of boron produced by injecting BCl_3 in a blown plasma at low pressures (deposition rate 2×10^{-2} m/s), see [16]. A codeposition of silicon and boron can be achieved from the gas-phase reaction in the system $HSiCl_3/BCl_3/H_2O$ (traces); silanyloxy-boron halides are probably formed in this reaction [19].

The deposition of boron carbide (from the reaction of BCl_3, CH_4, and H_2) has been studied with respect to the mass transfer, thermodynamics, and kinetics of the process. The deposition is kinetically limited at low temperatures; equilibrium conditions are approached at high temperatures [17, 18].

Boron nitride CVD is a technically important process (see also "Boron Compounds" 3rd Suppl. Vol. 3, 1988, Section 4.1.1.10.2, p. 37). Boron nitride is deposited at temperatures between 1200 and 2000°C; its structure is characterized by disordered layers, and the systems used are BCl_3/NH_3, $BCl_3/NH_3/H_2$, $BCl_3/NH_3/N_2$, etc. [20 to 23]. The dependence of the structure on the deposition temperature has been studied [24, 25]. The deposition from glow discharge can be achieved at temperatures of about 600°C [26 to 29]. Deposition from a plasma jet in argon or N_2 by sputtering and similar processes is also described [30 to 32].

TiB_2 layers are produced by CVD from $TiCl_4/BCl_3/H_2$ (with or without admixed Ar) [33 to 36]. The formation of layers infiltrated with boron on the surface of metals can be achieved by solid-state or gas-phase boronizing. Specialty steel is treated in this manner in order to obtain hard and wear-resistant surfaces. If there is sufficient boron available to be implanted, boride layers of Fe_2B and FeB are formed. The latter is brittle and has to be removed by diffusion annealing. The activation energy for the B diffusion was determined to be 25.2 kcal/mol in Fe_2B and 19.7 kcal/mol in FeB [45]. Gas-phase mixtures for the boriding of steel are primarily $BCl_3/H_2/N_2$ mixtures and the composition is controlled in order to avoid steel corrosion (by the HCl formed). Typical boriding temperatures range from 800 to 1000°C. Sometimes other metal halides are added as activators (e.g., $SnCl_4$). The process can be modified by boriding in a discharge plasma [36 to 52]. Such plasma boridation is preferably used for Co/Ni alloys for magnetic recording tapes [53]. Amorphous, hydrogenated silicon for the production of electrophotographic photoconductors can be borided in a CVD process [54, 55].

Reactions with Inorganic Materials

As based on an NMR study, BCl_3 reacts in liquid SO_2 with chloride ions (from tetramethylammonium chloride) to give $[BCl_4]^-$ and $[B_2Cl_7]^-$, both of which exchange rapidly with excess BCl_3 [57]. The reduction of BCl_3 by H_2 (as used in the CVD of boron) has been studied with respect to different substrates [58 to 60].

Reaction of a stream of N_2 and H_2 which is saturated (at 33 to 34°C) with BCl_3 with powdered NaH at 280°C gives B_2H_6 in about 60% yield [61]; $Na[BH_4]$ (solid residue) is prepared by a similar procedure [62]. Treatment of salts of the anions $[BH_4]^-$, $[B_3H_4]^-$, $[B_4H_9]^-$, or $[B_9H_{14}]^-$ with BCl_3 (in most cases without solvent) provides for good yields of the corresponding boranes (B_2H_6, B_4H_{10}, etc.) [63]; $B_{10}H_{14}$ can be obtained by this procedure (originating from B_5H_9) via the reaction of $[N(CH_3)_4][B_9H_{14}]$ with BCl_3 [64]. A complex is formed when B_6H_{16} is reacted with a 10-fold excess of BCl_3 [65]. $[3\text{-}(CH_3)_3N\text{-}closo\text{-}2,4\text{-}C_2B_5H_6][BCl_4]$ is obtained by treatment of the adduct of trimethylamine and 1-Cl-*closo*-2,4-$C_2B_5H_6$ with BCl_3 [66]. In the reaction of BCl_3 with B_5H_9 or 2-Cl–B_5H_8 the latter are substituted with a BCl_2 group in 1-position [350]. BCl_3 and $AlCl_3$ are used in Friedel-Crafts halogenation and deprotonation reactions of 1,2′-$(B_5H_8)_2$ and 2,2′-$(B_5H_8)_2$ [67].

A patent describes the (unlikely!) reaction of BCl_3 with aqueous NH_3 for the manufacture of BN [56].

When BCl_3 is passed through a radiofrequency discharge, BCl is detected spectroscopically; it is concluded that the latter is an intermediate in the formation of B_2Cl_4 [68]. $[(CH_3)_3CBOCH_3]_2$ reacts with BCl_3 to give *t*-butyldichloroborane [69].

Errors in the calculation of parameters of BCl_3/H_2 mixtures behind a shock-wave front have been studied [70]. Thermodynamic studies on the electrochemical synthesis of boron carbide from Li_2CO_3 and BCl_3 have been published [71]. The reaction of BCl_3 with H, O, and OH radicals has been studied; rate constants and the activation energy for $H + BCl_3 \rightarrow BCl_2 + HCl$ with $E_A = 2700 \pm 500$ cal/mol are given [72, 73]. The reaction of a BCl_3–N_2–O_2 system in a fast-flow reactor leads to the formation of BO radicals [74]. Trichlorosilanol reacts with BCl_3 to give $Cl_3SiOBCl_2$ [75, 76].

BCl_3 is used in the carbochlorination of kaolinitic ores [77] and in the removal of SiO_2 from SiC powders [78]. The reaction with zeolites results in the formation of dealuminated zeolites [79]; see also [80].

References for 6.1.2.3 on pp. 19/31

BCl_3 is used in the polymerization of hexachlorocyclotriphosphazene [81 to 84], and reacts with mixed chloro- and trifluoroethoxy-substituted cyclotriphosphazenes to give the corresponding polyphosphazenes [85]. Heteroring systems containing B–N–P–N linkages are obtained on reaction with monomeric phosphazenes [86]. BCl_3 forms 1:1 adducts with $(C_2H_5)_3NO$ or R_3PO ($R=OCH_3$, $N(CH_3)_2$, C_2H_5, etc.), which subsequently can be converted to oligocondensates [87]. BCl_3 reacts with F_5TeOH at −80°C to form $B(OTeF_5)_3$ [88], and with $F_2S{=}NTeF_5$ to give $Cl_2S{=}NTeF_5$ [89]. The reaction of BCl_3 with an excess of perfluoroalkanesulfonic acids gives conjugate superacids such as $[CF_3SO_2OH_2][B(OSO_2CF_3)_4]$ [90]; with $CF_3SO_3Si(CH_3)_3$ (and its homologs) strongly polarized adducts are formed [91]. BCl_3 is used in the conversion of sulfonyl derivatives to disulfides with iodides [92]. 3,4-Dichloro-1,2,5-thiadiazole, $SN_2C_2Cl_2$, gives a 1:1 adduct with BCl_3 [93].

BCl_3 (which is injected into the melt with an Ar stream via a steel lance) removes impurities (such as Si or Mn) from molten Mg [94]. BCl_3 is also used as a critical solvent (to be recycled) in metal (as chloride) recovery from chlorination processes [95].

Submicron TiB_2 powder is prepared in a gas plasma based on $TiCl_4$, BCl_3, and H_2 (or other reactants such as TiH_2, $TiCl_3$, etc.) [96 to 98]; oxide contaminations can be removed from TiB_2 by treatment with BCl_3 at temperatures between 500 and 800°C [99, 100]. ZrB_2 is made by the reaction of zirconium chlorides with BCl_3 and Al or Mg in an NaCl/KCl melt (BCl_3 is fed into the melt by a dispensing tube purged with Ar) [101, 102]. WO_3 and MoO_3 react with BCl_3 to give $(\text{–BCl–O–})_3$ and MoO_2Cl_2 and $MoOCl_4$ or $WOCl_4$ and WO_2Cl_2 [103]; $OsOCl_4$ is obtained from BCl_3 and OsO_4 [104]. Hexafluorides of ruthenium, osmium, and iridium exchange halogen upon reaction with BCl_3, and mixed species such as $Ir_2F_3Cl_6$ are formed [105].

The reaction of $UF_5 \cdot 2SbF_5$ with BCl_3 at low temperature is an effective method for the synthesis of UCl_5 [106]; see also [107]. Liquid BCl_3 converts UO_2F_2 to α- or β-UCl_5 (depending on conditions) while the reaction of gaseous BCl_3 yields UCl_6 [107]. Adhesive or powdered deposits are removed from equipment for the manipulation of UF_6 by treatment with BCl_3 [108]. Reaction of BCl_3 with PuF_6 in perfluoroalkanes occurs according to: $3PuF_6+3BCl_3 \rightarrow 3PuF_4+2BF_3+3Cl_2$ [109].

Trichloroborane in Dry Etching Processes (Plasma Etching, Reactive Ion Etching)

Plasma-assisted dry etching technologies are an indispensible method for high-resolution patterning required in the production of silicon-integrated circuits and devices. The reaction (mechanisms) of BCl_3 in etching of aluminium (alloys) can be described by the equations: $BCl_3+e^- \rightarrow BCl_x+(3-x)Cl+e^-$; $Al+xCl \rightarrow AlCl_x$; $BCl_x+(3-x)Cl \rightarrow BCl_3$; $Cl+Cl \rightarrow Cl_2$. Of these species the BCl_x radicals are most effective in etching of Al_2O_3. Additions of BCl_3 to Cl_2 and $SiCl_4$ (which are the main constituents of the etching gas) are therefore used to remove oxide layers. In such mixtures, atomic Cl etches Al quickly and $SiCl_x$ radicals facilitate the removal of native oxide. The chlorine-based reactive ion etching process of Al/Si produces vertical profiles. In many cases anisotropic profiles are desirable in order to maximize the cross-sectional area of the aluminium lines. Thus, the lateral etch rate und photoresist erosion must be limited. The etch directionality is a function of dc self-bias voltage (and also of the etch time). High etch rates (high Cl_2 content of the gas mixture) are desirable from the standpoint of throughput but can be detrimental in terms of the line profile. Decreasing the amount of Cl_2 in BCl_3 increases the relative concentration of BCl-containing recombinants. The latter species protect the line side walls and decrease the lateral etch rate. The lateral etch rate is increased with respect to the vertical etch rate when the dc bias is made less negative. Tapered profiles result from erosion of the photoresist mask edge during etching. The processes can be followed up by mass spectrometry or optical emission from the plasma; for additional information, see [110, 111]. Information on plasma diagnosis by time-resolved mass spectrometry [112, 113] and on the emission spectroscopy of BCl_3-containing plasma [114 to 116] is available. It seems that the intensity of the 522 nm band (of AlCl origin) is mostly used for the endpoint detection of the etching process. Details on the used equipment are available [117 to 124]. This area is the subject of numerous patents.

CCl_4 is often used in conjunction with the standard [138 to 140] BCl_3/Cl_2 etchant in the dry etching of Al (or its alloys), and PCl_3 or BBr_3 may also be added [125 to 137, 351]. Other gases admixed to the BCl_3 plasma are $FCCl_3$ [141, 142] or $HCCl_3$; the latter is found to promote anisotropic etching [143 to 145], while large amounts of He assist in photoresist preservation. However, anisotropic etching can be achieved also by scanning the plasma [146], by optical control of the luminiscence spectrum and automatic control of the components [147], and (as noted above) by influencing the directionality of the etching process by means of the dc self bias [148, 149]. $SiCl_4$ is often admixed to BCl_3 or BCl_3/Cl_2 [150 to 155], and sometimes SiF_4 or CF_4 is added [156 to 160]. Additional information on the Al_2O_3/Al dry-etching process by a BCl_3-containing plasma is available [161 to 166].

Dry-etching processes with gas mixtures containing BCl_3 are also applied for the etching of boron nitride [167]; gallium arsenide is etched in a BCl_3/Cl_2 plasma [168, 169], and polycrystalline silicon or silicide/polysilicon layers are etched by BCl_3/CCl_4, BCl_3/NF_3, BCl_3/Cl_2, or BCl_3/Br_2 [170 to 176]. Thin layers of SnO_2 are selectively etched by a BCl_3 plasma [177, 178] and additional applications of BCl_3 dry etching in the production of semiconductor devices and integrated circuits are published [179 to 191].

Reactions with Organometallic and Organic Materials

The reaction of BCl_3 with $LiN(t\text{-}C_4H_9)(CH_2C_6H_5)$ leads to the formation of $HB[N(t\text{-}C_4H_9)(CH_2C_6H_5)]_2$ [192]. 1,3-Dimethyl-1,3,2-diazaborolidines form 1:1 adducts with BCl_3 in which the latter is coordinated to one of the annular N atoms [193]. Two BCl_3 molecules coordinate to the diborane(4) derivative $\{[\text{–}N(CH_3)\text{–}(CH_2)_2\text{–}N(CH_3)\text{–}]B\}_2$ (one to each ring) [196]. Reaction of BCl_3 with the lithiated cyclic disilanylamine $[\text{–}Si(CH_3)_2\text{–}(CH_2)_2\text{–}Si(CH_3)_2\text{–}NLi]$ leads to the corresponding bis(amino)chloroborane species [194].

The amino-iminoborane $(tmp)B{=}N(t\text{-}C_4H_9)$ (where tmp = 2,2,6,6-tetramethylpiperidinyl) reacts with BCl_3 to give the bicyclic spiro compound $[\text{–}C(CH_3)_2\text{–}(CH_2)_3\text{–}C(CH_3)_2\text{–}]N(\mu\text{-}BCl_2)(\mu\text{-}BCl)N(t\text{-}C_4H_9)$ [197]. The diamidoboron tetrachloroborate $[(tmp){=}B{=}N(C_2H_5)_2][BCl_4]$ is obtained from $(tmp)(ClB)N(C_2H_5)_2$ and BCl_3 [198]. Reaction of BCl_3 with N,N'-dimethyl thiourea yields six-membered B–N–C heterocycles and $(\text{–}BCl\text{–}NCH_3\text{–})_3$ [199]. For the reaction of BCl_3 with a dilithiated pyridine-2,6-bis(bistrifluoromethylmethanol) derivative to yield a tricyclic chloroborate, see [200]. With $RN[Si(CH_3)_3]CH_2\text{–}C{\equiv}CH$, the corresponding (amino)dichloroborane moiety is obtained by cleavage of the N–Si bond [201].

Fluoroalkylamines such as $(CF_3)_{3-n}N(CHF_2)_n$ react with BCl_3 to give $(CF_3)_{3-n}N(CHCl_2)_n$ (n = 1 to 3) [195]. $RNHCH_2C{\equiv}CH$ reacts with BCl_3 by addition across the C≡C bond and a five-membered B–N–C heterocycle is formed by B–N coordination [201]. The formation of ternary charge-transfer complexes containing BCl_3, NOCl, and aromatic compounds has been mentioned briefly [352].

Various compounds such as $Cl(CH_3)_2Si(CH_2)_3BCl_2$ have been obtained by cleavage of the Si–C bond in silacyclobutanes with BCl_3 [202]. Liquid BCl_3 cleaves Sn–C bonds in $(CH_3)_3SnCl$ and $(CH_3)SnBr$ to give CH_3BCl_2, but there is no reaction with $(CH_3)_2SnX_2$ (X = Cl, Br) [203]. Tetravinyltin reacts with (gas phase) BCl_3 to give $H_2C{=}CH\text{–}BCl_2$ [204]. The formation of $RBCl_2$ and R_2BCl (R = CH_3, C_6H_5) derivatives by the reaction of R_4Sn with BCl_3 in CH_2Cl_2 has also been reported [205]. 1:1 adducts of BCl_3 are obtained with dimethylphosphane [206], $(CH_3)_3As$, and $(C_6H_5)_3As$ [207]. With $(C_6H_5)_4SbCl$, the corresponding tetrachloroborate is formed, and

References for 6.1.2.3 on pp. 19/31

$(C_6H_5)_3SbCl_2$ gives a 1:2 adduct with BCl_3 [208]. It is claimed that the reaction of BCl_3 with $H_2C{=}CH{-}CHO$ gives a 1:1 adduct [209], but an insertion-type reaction seems much more likely. BCl_3 reacts with 2,4-pentanedione to form bis(2,4-pentanedionato)boron(III)hydrogen dichloride, $[B(acac)_2]HCl_2$, while $(acac)BCl_2$ results from the reaction of BCl_3 with $(acac)B(i\text{-}C_3H_7)_2$ [210]. Treatment of 2-mercapto-4-methoxyphenol with BCl_3 gives 2-chloro-5-methoxy-1,3,2-benzoxythioborol [211], and BCl_3 is monoarylated by chloro(3,5-di-*t*-butyl-4-hydroxyphenyl)-mercury(I) to yield the corresponding boroxin (after hydrolysis) [212].

Bis(mesithylthio) complexes of Ge^{II} such as $(CO)_5Cr{-}Ge[SC_6H_2(CH_3)_3]_2$ react with BCl_3 to give the corresponding dihalogermylene species [213]; $(CO)_5W{-}Ge[SC_6H_2(CH_3)_3]_2$ and BCl_3 interact to yield $[(CO)_5W{-}W(CO_5)\text{-}\mu\text{-}GeCl_2]$ containing a bridging germylene ligand [214]. The carbene complex $CH_3C_5H_4(CO)_2MnC(OC_2H_5)N(C_2H_5)_2$ reacts with BCl_3 to give $[CH_3C_5H_4(CO)_2\text{-}MnCN(C_2H_5)_2][BCl_4]$ [215]. From the carbonyl hydride, the complex $(CO)_5MnH{-}BCl_3$ has been prepared [216]. Halogen exchange is observed with perfluoroalkyl complexes of transition metals upon treatment with BCl_3, e.g., $(CO)_5ReC_2F_5 + BCl_3 \rightarrow (CO)_5ReCCl_2CF_3 + BClF_2$ [217, 218]. A cationic dichlorocarbene complex is prepared according to: $[(\eta^5\text{-}C_5H_5)Fe(CO)_2CF_3] + 2BCl_3 \rightarrow [(\eta\text{-}C_5H_5)Fe(CO)_2(CCl_2)][BCl_4] + BF_3$ [219]. Cationic carbyne complexes of the type $[(\eta^5\text{-}C_5H_5)Mn(CO)_2{\equiv}CR][BCl_4]$ were obtained from BCl_3 and the carbene complex $(\eta^5\text{-}C_5H_5)\text{-}Mn(CO){=}C(OCH_3)R$ [220]. The complex $Re(NBCl_3)Cl_2[P(CH_3)_2C_6H_5]_3$ contains a nitrido bridge between Re and B; it is obtained from BCl_3 and $ReNCl_2[P(CH_3)_2C_6H_5]_3$ [221].

The conversion of carbonyl compounds into their corresponding thiocarbonyl analogs is effected by reaction with bis(tricyclohexyl)tin sulfide and BCl_3 [353]. BCl_3 and $C_6H_5\text{-}2\text{-}C_6H_4OH$ interact in the presence of $AlCl_3$ to give boroxarophenanthrenes [222]. 9-Nitroanthracene forms a 1:1 charge transfer complex with BCl_3 [223]. With phenyliodine(III) diacetate and phenyliodine(III) bis(trifluoroacetate), BCl_3 reacts to give $[C_6H_5IOCOR]^+[RCO_2BCl_3]^-$ ($R = CH_3$, CF_3) [224]. Treatment of 2-aryl or 2-diazoarylphenyl azides with BCl_3 produces carbazole and benzocinnoline derivatives [225].

BCl_3 is used to split C–O bonds in boronic esters [226, 227], cyclic acetals [228], methoxyarenes [229], and alkoxybenzaldehydes [230]. It is used in the debenzylation of thionucleosides [231] and in decarbalkoxylation reactions to give malonates [232]. The reaction of BCl_3 with $2\text{-}NO_2{-}C_6H_4OH$ is claimed to give a 1:1 adduct [233]. BCl_3 is used in the addition of trihaloethanols to alkenes [234], in the addition of naphthol with pyruvate esters [235], in reactions with cyclic enamines [236], in the cyclization of alkenyl and alkynyl hexenones [237], in Friedel-Crafts type reactions [238 to 240], and a variety of rearrangement, isomerization, and racemization reactions [241 to 246].

Trichloroborane as Polymerization Catalyst

BCl_3 is used as a component of catalyst systems for the cationic polymerization of α-olefins, of styrene or derivatives thereof, of polar vinyl derivatives (such as acrylonitrile, methacrylates, etc.), isoprene and chloroprene rubbers, phosphazenes, and a variety of other unsaturated compounds. In admixture with (modified) Ziegler-Natta type catalyst systems it is, for example, used in the polymerization of ethylene [247 to 253] and in the copolymerization of ethene/butene mixtures [254, 255]. In some cases the catalyst is supported by inorganic fillers; catalysts containing Zr complexes may be used in such systems [256]. The same catalyst mixtures have been applied for the polymerization of propylene [257 to 263]. Another catalyst system is $BCl_3/C_6H_5C(CH_3)_2Cl$ (or related organic compounds containing a tertiary C–Cl moiety) which is used in the polymerization of isobutylene [264 to 268] and systems containing tertiary alcohols or sterically hindered amines [269 to 273]. Telechelic polymers (of isobutylene), α-, ω-terminated with Cl atoms or $Cl(CH_3)_2C$ groups are obtained by the "inifer" (polyfunctional initiator/transfer agent) technique (for example using 1,4-dicumyl chloride) [274 to 278]. Such techniques (with the aim to obtain perfect terminally functional linear poly(isobutylene)) also make use of effects of the addition of sterically hindered bases (such as 2,6-di-*t*-butylpyridine) [279, 280] and of the $BCl_3/[BCl_4]^-$ co-initiator system [281, 282]; see also [283, 348, 354]. The tertiary alcohol/BCl_3 system is applied to the copolymerization of isobutylene with styrene, indene, etc. [284], and for the polymerization of 2,4,4-trimethyl-1-pentene [285]. 2,4,6-Trimethylstyrene has also been polymerized by the application of BCl_3/cumyl chloride (cumyl = $C_6H_4CH(CH_3)_2$) [286]. The effect of BCl_3 in the copolymerization of $H_2C{=}CHCN$ with cycloolefins has been studied [287] as well as the use of BCl_3 in the alternating copolymerization of methylmethacrylate with styrene [288 to 291]. BCl_3 is used in the production of modified poly(chloroprene) rubber (and the copolymerization of chloroprene with α-methylstyrene or isobutylvinylether) [292 to 294] and for modified isoprene rubber compositions [295, 296]. It serves in the polymerization of aromatic dinitriles [297], and in preparing poly(aryleneether ketones) [298].

Hexachlorocyclotriphosphazenes or hexakis(trifluoroethoxy)phosphazenes give poly(phosphazenes) upon treatment with BCl_3 [299 to 301]; see also [81 to 85] in this section (dealing with reactions with inorganic materials).

Additional Applications of BCl_3

Trichloroborane serves as an ion source in doping silicon or germanium with boron (in the manufacture of transistors and integrated circuits) [302 to 309], and for the doping of electric- or photoconducting polymers [310 to 314, 355]. Furthermore, BCl_3 is used in the preparation of scratch-resistant silicon-based coatings on metallized plastics [315], the preparation of phenolic resin composites used in neutron activation analysis [316], the opacification of gaseous media in the optical and infrared region of the electromagnetic spectrum [317], in glass fiber technology [318 to 320], in the preparation of calcium-intercalated boronated carbon fibers [321], and in lithium/thionyl chloride batteries [322].

Analytical and Environmental Information

A spark-proof ionization analyzer has been developed for the monitoring of BCl_3 concentrations in hydrogen [323]. The equilibrium constants of the reactions of BCl_3 with metal oxides have been calculated in order to evaluate systematic errors caused by impurities in the stationary phase of gas/liquid chromatography. The following lg K_p data at 323 K are given (am = amorphous, cr = crystalline, g = gaseous) [324]:

$$2BCl_3(g) + Fe_2O_3(am) \rightarrow B_2O_3(am) + 2FeCl_3(g) = 38{,}0$$
$$2BCl_3(g) + Al_2O_3(am) \rightarrow B_2O_3(am) + 2AlCl_3(g) = 6{,}5$$
$$2BCl_3(g) + 3CaO(cr) \rightarrow B_2O_3(am) + 3CaCl_2(cr) = 134{,}1$$
$$4BCl_3(g) + 3TiO_2(am) \rightarrow 2B_2O_3(am) + 3TiCl_4(g) = 74{,}1$$
$$4BCl_3(g) + 3SiO_2(am) \rightarrow 2B_2O_3(am) + 3SiCl_4(am) = 1{,}8$$

The environmental pollution caused by BCl_3 emissions in the manufacture and in degradation of photovoltaic cells has been discussed [325].

References for 6.1.2.3:

[1] Pankratov, A. V.; Shmerling, G. V. (Khim. Vysokikh Energ. **17** [1983] 173/5; High Energy Chem. [USSR] **17** [1983] 138/9 from C.A. **98** [1983] No. 188925).

[2] Samsonov, Yu. N.; Petrov, A. K. (Khim. Vysokikh Energ. **17** [1983] 170/3; High Energy Chem. [USSR] **17** [1983] 135/7 from C.A. **98** [1983] No. 207358).

[3] Volkov, S. V.; Gurko, A. F.; Druzheruchenko, A. G.; Lutoshkin, V. I. (Khim. Vysokikh Energ. **17** [1983] 464/7; High Energy Chem. [USSR] **17** [1983] 362/5 from C.A. **99** [1983] No. 184899).
[4] Volkov, S. V.; Gurko, A. F.; Lutoshkin, V. I. (Ukr. Khim. Zh. **48** [1982] 451/3; Soviet Progr. Chem. **48** No. 5 [1982] 1/3; C.A. **97** [1982] No. 48679).
[5] Sazonov, V. N.; Shmerling, G. V. (Kvantovaya Elektron [Moscow] **10** [1983] 10434/5 from C.A. **100** [1984] No. 148367).
[6] Li, L.; Chen, G.; Cai, Z.; Kang, N. (Zhongguo Jiguang **9** [1982] 570/1 from C.A. **100** [1984] No. 59488).
[7] Li, L.; Chen, G.; Zhang, R.; Kang, N. (Zhongguo Jiguang **8** [1981] 33/7 from C.A. **96** [1982] No. 133033).
[8] McDonald, J. K.; Warren, L. C.; Merrit, J. A. (AD-A101035 [1981] 1/20 from C.A. **96** [1982] No. 13556).
[9] Kanter, I. E.; Chen, Ch. L.; Liebermann, R. W. (Fr. Demande 2495493 [1982] from C.A. **97** [1982] No. 152858).
[10] Bouix, J.; Berthet, M. P.; Boubehira, M.; Dazord, J.; Vincent, H. (J. Electrochem. Soc. **129** [1982] 2338/43).

[11] Bouix, J.; Berthet, M. P.; Dazord, J.; Vincent, H. (Rev. Chim. Minerale **18** [1981] 464/74; C.A. **96** [1982] No. 166937).
[12] Vandenbulcke, L. (Proc. Intern. Ion Eng. Congr., Kyoto 1983, Vol. 2, pp. 921/5 from C.A. **101** [1984] No. 162115).
[13] Carlsson, J. O. (Proc. Electrochem. Soc. **79** Pt. 3 [1979] 332/50; C.A. **95** [1981] No. 33583).
[14] Tsirlin, A. M.; Fedorova, T. V.; Florina, E. K. (Fiz. Khim. Obrab. Mater. **1984** No. 3, pp. 99/107 from C.A. **101** [1984] No. 101380).
[15] Michaelidis, M.; Pollard, R. (J. Electrochem. Soc. **131** [1984] 860/8).
[16] Guilly, J.; Pennaneach, M.; Lassau, G. (Symp. Proc. 6th Intern. Symp. Plasma Chem., Montreal 1983, Vol. 1, pp. 91/6 from C.A. **100** [1984] No. 148691).
[17] Vandenbulcke, L. (Proc. Electrochem. Soc. **79** Pt. 3 [1979] 315/31; C.A. **95** [1981] No. 33582).
[18] Zyatkevich, D. P.; Melamed, A. G.; Makarenko, G. N.; Kosolapova, T. Ya.; Popova, O. J. (Dispers. Poroshki i Materialy na ikh Osnove Kiev **1982** 39/41 from C.A. **99** [1983] No. 26885).
[19] Bartsch, K.; Wolf, E. (Cryst. Res. Technol. **17** [1982] 405/9 from C.A. **97** [1982] No. 16163).
[20] Liepins, V.; Kuznetsova, I. G.; Maier, A. A.; Shumilina, O. E.; Bredikhin, V. I. (Tr. Mosk. Khim. Tekhnol. Inst. No. 108 [1979] 89/90 from C.A. **95** [1981] No. 11388).

[21] Wen, T. L.; Li, W. L.; Li, S. Ch. (Guisuanyan Xuebao **8** [1980] 351/6 from C.A. **95** [1981] No. 11448).
[22] Hannache, H.; Naslain, R.; Bernard, C. (J. Less-Common Metals **95** [1983] 221/46).
[23] Shinko, J. S.; Keyes, T. C. (Eur. Appl. 76731 [1983] from C.A. **99** [1983] No. 57854).
[24] Nakae, H.; Matsuda, T.; Uno, N.; Matsunami, Y.; Masumoto, T.; Hirai, T. (Japan. Kokai Tokkyo Koho 84-207811 [1984] from C.A. **102** [1985] No. 118457).
[25] Sano, M.; Aoki, M. (Thin Solid Films **83** [1981] 247/51; C.A. **95** [1981] No. 213137).
[26] Troitskii, V. N.; Grebtsov, B. M.; Domashnev, I. A.; Gurov, S. V. (Proc. 1st Ann. Intern. Conf. Plasma Chem. Technol., San Diego 1982 [1983], pp. 141/58 from C.A. **100** [1984] No. 202318).
[27] Prokoshkin, D. A.; Demin, Yu. N.; Tret'yakov, V. I.; Gornichev, A. A. (Fiz. Khim. Obrab. Mater. **1984** No. 2, pp. 138/9 from C.A. **101** [1984] No. 11116).
[28] Sumitomo Electric Industries, Ltd. (Japan. Kokai Tokkyo Koho 84-227800 [1984] from C.A. **102** [1985] No. 177058).

[29] Lapatovich, W. P.; Proud, J. M.; Riseberg, L. A. (U.S. 4436762 [1984] from C.A. **100** [1984] No. 160999).
[30] Sumitomo Electric Industries, Ltd. (Japan. Kokai Tokkyo Koho 84-16969 [1984] from C.A. **100** [1984] No. 179049).

[31] Nichicon Capacitor, Ltd. (Japan. Kokai Tokkyo Koho 83-147559 [1983] from C.A. **100** [1984] No. 25224).
[32] Pennaneach, M.; Guilly, J.; Membrives, J.; Lassau, G. (Symp. Proc. 5th Intern. Symp. Plasma Chem., Edinburgh 1981, Vol. 2, pp. 609/14 from C.A. **97** [1982] No. 221125).
[33] Motojima, S.; Yamada, M.; Sugiyama, K. (J. Nucl. Mater. **105** [1982] 335/7 from C.A. **97** [1982] No. 62502).
[34] Muhlratzer, A.; Erben, E. (Brit. Appl. 2107742 [1983] from C.A. **99** [1983] No. 9527).
[35] Schintlmeister, W.; Wallgram, W. (Eur. Appl. 83043 [1983] from C.A. **100** [1984] No. 25219).
[36] Newkirk, L. R.; Riley, R. E.; Sheinberg, H.; Valencia, F. A.; Wallace, T. C. (Proc. Electrochem. Soc. **79** Pt. 3 [1979] 515/24).
[37] Hegewaldt, F.; Schmaderer, F.; Singheiser, L.; Wahl, G.; Maag, G. (Ger. Offen. 3139462 [1983] from C.A. **99** [1983] No. 41969).
[38] Hegewaldt, F.; Singheiser, L.; Türk, M. (HTM Härterei-Tech. Mitt. **39** No. 1 [1984] 7/15 from C.A. **100** [1984] No. 178492).
[39] Than, E.; Weber, D.; Luthardt, M.; Marx, G. (Neue Hütte **28** No. 3 [1983] 94/8 from C.A. **98** [1983] No. 202204).
[40] Luthardt, M.; Than, E.; Marx, G. (Tagungsber. Wiss. Tag. T.H. Karl-Marx-Stadt **1983** 58/65 from C.A. **102** [1985] No. 10379).

[41] Than, E.; Luthardt, M.; Finster, J.; Meisel, A.; Marx, G. (SIA Surf. Interface Anal. **5** [1983] 235/8 from C.A. **100** [1984] No. 195833).
[42] Bochmann, G.; Spoerl, T.; Ritzel, B.; Wagner, W. (Wiss. Z. T.H. Karl-Marx-Stadt **24** [1982] 682/9 from C.A. **99** [1983] No. 74721).
[43] Bochmann, G.; Spoerl, T.; Ritzel, B.; Wiesner, S.; Wagner, W. (Neue Hütte **29** No. 1 [1984] 26/8 from C.A. **100** [1984] No. 124867).
[44] Plaenitz, H.; Treffer, G.; Koenig, H.; Marx, G.; Bahr, D. (Ger. [East] 208632 [1984] from C.A. **101** [1984] No. 115091).
[45] Plaenitz, H.; Treffer, G.; Koenig, H.; Marx, G. (Neue Hütte **27** No. 6 [1982] 228/30 from C.A. **97** [1982] No. 186216).
[46] Wierzchon, T.; Michalski, J.; Karpinski, T. (Mechanik **56** [1983] 485/7 from C.A. **100** [1984] No. 143046).
[47] Wierzchon, T.; Pokrasen, S.; Karpinski, T. (HTM Härterei-Tech. Mitt. **38** No. 2 [1983] 57/62 from C.A. **99** [1983] No. 57126).
[48] Filep, E.; Farkas, S. N.; Kolozsvari, Z. C. (Rom. 70576 [1981] from C.A. **96** [1982] No. 108906).
[49] Farkas, S.; Filiep, E.; Kolozsvary, Z. (Metalozn. Obrob. Cieplna No. 51/52 [1981] 66/8 from C.A. **96** [1982] No. 127203).
[50] Wierzchon, T.; Bogacki, J.; Karpinski, T. (Proc. 18th Intern. Conf. Heat Treat. Mater., Detroit 1980, pp. 13/24 from C.A. **94** [1981] No. 125354).

[51] Kemnitz, H.; Morgenstern, B.; Otto, W.; Schoenberg, P.; Gebes, B. (Ger. [East] 159647 [1983] from C.A. **99** [1983] No. 108948).
[52] Xiuyu, M. (Jinshu Rechuli **1983** No. 4, pp. 52/7 from C.A. **99** [1983] No. 198779).
[53] TDK Corp. (Japan. Kokai Tokkyo Koho 84-68821 [1984] from C.A. **101** [1984] No. 221226).

[54] Shimizu, I.; Ogawa, K.; Inoue, E.; Kanbe, J.; Canon K.K. (Ger. Offen. 3209055 [1982] from C.A. **98** [1983] No. 152783).
[55] Shimizu, I.; Ogawa, K.; Inoue, E.; Canon K.K. (Ger. Offen. 3208494 [1982] from C.A. **98** [1983] No. 117102).
[56] Shin-Etsu Chemical Industry Co., Ltd. (Japan. Kokai Tokkyo Koho 83-140142 [1983] from C.A. **100** [1984] No. 16265).
[57] Glavincevski, B.; Brownstein, S. K. (Can. J. Chem. **59** [1981] 3012/5).
[58] Luthardt, M.; Than, E.; Marx, G. (Z. Chem. [Leipzig] **22** [1982] 231/2).
[59] Wagner, W.; Bochmann, G. (Wiss. Z. T.H. Karl-Marx-Stadt **24** [1982] 672/81).
[60] Grange, A.; Corbe, J. (Eur. Appl. 18260 [1980] from C.A. **94** [1981] No. 142008).

[61] Prochazka, V.; Zapletal, V. (Czech. 214026 [1984] from C.A. **101** [1984] No. 25684).
[62] Prochazka, V.; Zapletal, V. (Czech. 212962 [1984] from C.A. **100** [1984] No. 194525).
[63] Leach, J. B.; Toft, M. A.; Himpsl, F. L.; Shore, Sh. G. (J. Am. Chem. Soc. **103** [1981] 988/9).
[64] Shore, S. G.; Toft, M. A.; Himpsl, F. L. (U.S. 4338289 [1982] from C.A. **97** [1982] No. 94883).
[65] Dolan, P. J.; Moody, D. C.; Schaeffer, R. (Inorg. Chem. **20** [1981] 745/8).
[66] Siwapinyoyos, G.; Onak, T. (Inorg. Chem. **21** [1982] 156/63).
[67] Heppert, J. A.; Kulzick, M. A.; Gaines, D. F. (Inorg. Chem. **23** [1984] 14/8).
[68] Briggs, A. G.; Massey, A. G.; Reason, M. S.; Portal, P. J. (Polyhedron **3** [1984] 369/71).
[69] Klusik, H.; Bernott, A. (J. Organometal. Chem. **234** [1982] C17/C19).
[70] Khramova, O. D.; Kuznetsova, L. A.; Kuzyakov, Yu. Ya. (Vestn. Mosk. Univ. Ser. 2 Khim. **25** [1984] 23/7; C.A. **100** [1984] No. 215887).

[71] Shapoval, V. I.; Kushkov, Kh. B.; Novoselova, I. A. (Ukr. Khim. Zh. **48** [1982] 738/42; Soviet Progr. Chem. **48** No. 7 [1982] 70/4).
[72] Jourdain, J.-L.; Laverdet, G.; Le Bras, G.; Combourieu, J. (J. Chim. Phys. Phys. Chim. Biol. **78** [1981] 253/7).
[73] Jourdain, J.-L.; Le Bras, G.; Combourieu, J. (Actes 1st Colloq. Intern. Berthelot-Vieille-Mallard-Le Chatelier, Talence, Fr., 1981, pp. 296/300 from C.A. **98** [1983] No.128732).
[74] Llewellyn, I. P.; Fontijn, A.; Clyne, M. A. A. (Chem. Phys. Letters **84** [1981] 504/8).
[75] Gooden, R.; Mitchell, J. W. (J. Electrochem. Soc. **129** [1982] 1619/23).
[76] Darnell, R. D.; Ingle, W. M. (PCT Intern. Appl. 82-04434 [1982] from C.A. **98** [1983] No. 182038).
[77] Wyndham, R. (Austrian 523570 [1982] from C.A. **98** [1983] No. 111405).
[78] Brynestad, J.; Bamberger, C. E.; Heatherly, D. E.; Land, J. F. (J. Am. Ceram. Soc. **67** [1984] C184/C185 from C.A. **102** [1985] No. 11249).
[79] Fejes, P.; Kiricsi, I.; Hannus, I.; Schobel, G. (Magy. Kem. Folyoirat **89** [1983] 264/9 from C.A. **99** [1983] No. 114989).
[80] Geismar, G.; Westphal, U. (Z. Anorg. Allgem. Chem. **487** [1982] 207/16).

[81] Fieldhouse, J. W.; Graves, D. F. (ACS Sym. Ser. No. 171 [1981] 315/20 from C.A. **96** [1982] No. 52730).
[82] Fieldhouse, J. W.; Graves, D. F. (U.S. 4226840 [1980] from C.A. **94** [1981] No. 16385).
[83] Fieldhouse, J. W.; Fenske, S. L. (U.S. 4327064 [1982] from C.A. **97** [1982] No. 6998).
[84] Fieldhouse, J. W.; Graves, D. F. (Eur. Appl. 46584 [1982] from C.A. **96** [1982] No. 218398).
[85] Horn, H. G.; Kolkmann, F. (Makromol. Chem. **183** [1982] 2427/35).
[86] Cowley, A. H.; Kilduff, J. E.; Wilburn, J. C. (J. Am. Chem. Soc. **103** [1981] 1575/7).
[87] Bravo, R.; Laurent, J. P. (J. Chem. Res. S **1983** 61).
[88] Kropshofer, H.; Leitzke, O.; Peringer, P.; Sladky, F. (Chem. Ber. **114** [1981] 2644/8).

[89] Hartl, H.; Huppmann, P.; Lentz, D.; Seppelt, K. (Inorg. Chem. **22** [1983] 2183/7).
[90] Olah, G. A.; Laali, Kh.; Farooq, O. (J. Org. Chem. **49** [1984] 4591/4).

[91] Olah, G. A.; Laali, Kh.; Farooq, O. (Organometallics **3** [1984] 1337/40).
[92] Olah, G. A.; Narang, S. C.; Field, L. D.; Karpeles, R. (J. Org. Chem. **46** [1981] 2408/10).
[93] Roesky, H. W.; Wehner, E. (Z. Naturforsch. **36b** [1981] 1247/50).
[94] Mena, A.; Charriere, J. M.; Desbrest, J. (Fr. Demande 2516940 [1983] from C.A. **99** [1983] No. 143860).
[95] Rado, T. A. (U.S. 4457812 [1984] from C.A. **101** [1984] No. 95212).
[96] Hoekje, H. H. (U.S. 4282195 [1981] from C.A. **95** [1981] No. 174250).
[97] Hitachi Metals, Ltd. (Japan. Kokai Tokkyo Koho 84-64524 [1984] from C.A. **101** [1984] No. 57252).
[98] Brynestad, J.; Bamberger, C. E. (U.S. Appl. 488870 [1983] from C.A. **101** [1984] No. 135816).
[99] Brynestad, J.; Bamberger, C. E. (U.S. 4452767 [1984] from C.A. **101** [1984] No. 96543).
[100] Brynestad, J.; Bamberger, C. E.; Land, J. F.; Finch, C. B. (J. Am. Ceram. Soc. **66** [1983] C215/C216; C.A. **100** [1984] No. 11556).

[101] Becker, A. J. (U.S. 4414188 [1983] from C.A. **100** [1984] No. 25400).
[102] Becker, A. J. (Proc. Electrochem. Soc. **83** Pt. 7 [1983] 65/9 from C.A. **99** [1983] No. 217346).
[103] Cook, N. D.; Timms, P. L. (J. Chem. Soc. Dalton Trans. **1983** 239/43).
[104] Levason, W.; Ogden, J. S.; Rest, A. J.; Turff, J. W. (J. Chem. Soc. Dalton Trans. **1982** 1877/8).
[105] Burns, R. C.; O'Donnell, T. A. (J. Fluorine Chem. **23** [1983] 1/14).
[106] Rediess, K.; Sawodny, W. (Z. Naturforsch. **37b** [1982] 524/5).
[107] Brown, D.; Berry, J. A.; Holloway, J. H. (J. Chem. Soc. Dalton Trans. **1982** 1385/8).
[108] Jacob, E. (Ger. Offen. 3009933 [1981] from C.A. **96** [1982] No. 59735).
[109] Burns, R. C.; O'Donnell, T. A.; Randall, C. H. (J. Inorg. Nucl. Chem. **43** [1981] 1231/8).
[110] Light, R. W.; Bell, H. B.; Macro, H. (Proc. Electrochem. Soc. **83** Pt. 10 [1983] 530/48).

[111] Park, K. O. (Proc. Electrochem. Soc. **83** Pt. 10 [1983] 300/9).
[112] Lehmann, H. W.; Heeb, E.; Frick, K. (Solid State Technol. **24** No. 10 [1981] 69/74 from C.A. **95** [1981] No. 213888).
[113] Lehmann, H. W.; Heeb, E.; Frick, K. (Proc. Electrochem. Soc. **82** Pt. 6 [1982] 364/78).
[114] Spencer, J. E.; Shu, B. Y. (Proc. Electrochem. Soc. **82** Pt. 7 [1982] 103/7).
[115] Park, K. O.; Rock, F. C. (J. Electrochem. Soc. **131** [1984] 214/5).
[116] Mizutani, T.; Ueki, K.; Iida, S.; Komatsu, H.; Hitachi, Ltd. (Ger. Offen. 3019583 [1980] from C.A. **94** [1981] No. 144061).
[117] Lehmann, H. W.; Frick, K.; Widmer, R. W.; RCA Corp. (U.S. 4387013 [1983] from C.A. **99** [1983] No. 57543).
[118] Nichiden, A. K. K. (Japan. Kokai Tokkyo Koho 82-133633 [1982] from C.A. **98** [1983] No. 26354).
[119] Matsushita Electric Industrial Co., Ltd. (Japan. Kokai Tokkyo Koho 84-87818 [1984] from C.A. **101** [1984] No. 181208).
[120] Straughan, V. E.; Wainer, E. (U.S. 4292384 [1981] from C.A. **96** [1982] No. 13671).

[121] Yamazaki, T., Toshiba Corp. (Eur. Appl. 47002 [1982] from C.A. **96** [1982] No. 191732).
[122] Matsushita Electric Industrial Co., Ltd. (Japan. Kokai Tokkyo Koho 84-06379 [1984] from C.A. **100** [1984] No. 149569).

[123] Hitachi, Ltd. (Japan. Kokai Tokkyo Koho 84-115701 [1984] from C.A. **101** [1984] No. 135603).
[124] Showa Denko K.K. (Japan. Kokai Tokkyo Koho 83-214092 [1983] from C.A. **100** [1984] No. 149568).
[125] Matsushita Electronics Corp. (Japan. Kokai Tokkyo Koho 84-123771 [1984] from C.A. **102** [1985] No. 11087).
[126] Fujitsu, Ltd. (Japan. Kokai Tokkyo Koho 81-139677 [1981] from C.A. **96** [1982] No. 108901).
[127] Fujitsu, Ltd. (Japan. Kokai Tokkyo Koho 80-85670 [1980] from C.A. **94** [1981] No. 23904).
[128] Maa, J. S.; O'Neill, J. J. (J. Vac. Sci. Technol. [2] A **1** [1983] 636/7; C.A. **99** [1983] No. 14670).
[129] Saia, R. J.; Gorowitz, B. (U.S. 4444618 [1984] from C.A. **101** [1984] No. 42306).
[130] Anelva Corp. (Japan. Kokai Tokkyo Koho 83-213877 [1983] from C.A. **100** [1984] No. 214268).

[131] Fujitsu, Ltd. (Japan. Kokai Tokkyo Koho 83-35364 [1983] from C.A. **99** [1983] No. 204696).
[132] Fujitsu, Ltd. (Japan. Kokai Tokkyo Koho 83-220448 [1983] from C.A. **100** [1984] No. 166437).
[133] Tokunaga, K.; Redeker, F. C.; Danner, D. A.; Hess, D. W. (J. Electrochem. Soc. **128** [1981] 851/5).
[134] Takami, S. (JEE J. Electron. Eng. **20** [1983] 82/5 from C.A. **99** [1983] No. 223438).
[135] Müller, W.; Beinvogl, W.; Risch, L.; Sigusch, R. (IEE Proc. Pt. I Solid-State Electron Devices **130** [1983] 136/43 from C.A. **99** [1983] No. 46533).
[136] Halon, B.; Vossen, J. K., Jr. (U.S. 4370195 [1983] from C.A. **98** [1983] No. 130946).
[137] Takada, T.; Tokitomo, K.; Hoshino, H. (Eur. Appl. 23429 [1981] from C.A. **94** [1981] No. 149192).
[138] Schwarzl, S.; Beinvogl, W. (Proc. Electrochem. Soc. **83** Pt. 10 [1983] 310/20 from C.A. **99** [1983] No. 167659).
[139] Halon, B. (U.S. 4375385 [1983] from C.A. **98** [1983] No. 165697).
[140] Hitachi, Ltd. (Japan. Kokai Tokkyo Koho 84-207627 [1984] from C.A. **102** [1985] No. 177616).

[141] Kokusai Electric Co., Ltd. (Japan. Kokai Tokkyo Koho 83-193368 [1983] from C.A. **100** [1984] No. 143630).
[142] Iida, S.; Abe, M. (Rept. Res. Cent. Ion Beam Technol. Hosei Univ. Suppl. No. 2 [1983] 81/4 from C.A. **99** [1983] No. 162508).
[143] Bruce, R. H.; Malafsky, G. P. (J. Electrochem. Soc. **130** [1983] 1369/73).
[144] Bruce, R. H.; Malafsky, G. P. (Proc. Electrochem. Soc. **82** Pt. 6 [1982] 336/47 from C.A. **97** [1982] No. 137163).
[145] Vossen, J. L., Jr.; Halon, B. (U.S. 4370193 [1983] from C.A. **98** [1983] No. 112285).
[146] Horiike, Y.; Yamazaki, T.; Tokura, T.; Okano, H. (Symp. Proc. 6th Intern. Symp. Plasma Chem., Montreal 1983, Vol. 3, pp. 610/5 from C.A. **100** [1984] No. 166353).
[147] Fujitsu, Ltd. (Japan. Kokai Tokkyo Koho 84-129428 [1984] from C.A. **102** [1985] No. 16164).
[148] Light, R. W.; Bell, H. B.; Macro, H. A. (Proc. SPIE Intern. Soc. Opt. Eng. No. 470 [1984] 48/54 from C.A. **101** [1984] No. 82345).
[149] Light, R. W. (J. Electrochem. Soc. **130** [1983] 2225/30).
[150] Reichelderfer, R. F.; Vogel, D. C.; Tang, M. C. (U.S. 4380488 [1983] from C.A. **98** [1983] No. 220517).

[151] Purdes, A. J. (Eur. Appl. 99558 [1984] from C.A. **100** [1984] No. 125594).
[152] Nelson, N. J. (U.S. 4468284 [1984] from C.A. **101** [1984] No. 176081).

[153] Toshiba Corp. Tokuda Seisakusho K.K. (Japan. Kokai Tokkyo Koho 84-220928 [1984] from C.A. **102** [1985] No. 213889).
[154] Tiller, H. J.; Goebel, R.; Voigt, R.; Fendler, R.; Huebner, R. (Ger. [East] 212992 [1984] from C.A. **102** [1985] No. 30490).
[155] Purdes, A. J. (J. Vac. Sci. Technol. [2] A **1** [1983] 712/5).
[156] Jacob, A.; Choe, D. H. (Ger. Offen. 3410023 [1984] from C.A. **101** [1984] No. 238904).
[157] Wang, D. N.; Egitto, F. D.; Maydan, D. (U.S. 4412885 [1983] from C.A. **100** [1984] No. 11422).
[158] Kokusai Electric Co., Ltd. (Japan. Kokai Tokkyo Koho 83-91171 [1983] from C.A. **99** [1983] No. 162704).
[159] Hitachi, Ltd. (Japan. Kokai Tokkyo Koho 82-02585 [1982] from C.A. **96** [1982] No. 134304).
[160] Takada, T.; Shimizu, K. (Eur. Appl. 122776 [1984] from C.A. **102** [1985] No. 196449).

[161] Fujitsu, Ltd. (Japan. Kokai Tokkyo Koho 80-160431 [1980] from C.A. **94** [1981] No. 184424).
[162] Hitachi, Ltd. (Japan. Kokai Tokkyo Koho 80-111164 [1980] from C.A. **94** [1981] No. 56983).
[163] Saia, R. J.; Gorowitz, B. (Solid State Technol. **26** [1983] 247/52 from C.A. **98** [1983] No. 208466).
[164] Toshiba Corp. (Japan. Kokai Tokkyo Koho 84-56743 [1984] from C.A. **101** [1984] No. 102411).
[165] Cho LSI Gijutsu Kenkyu Kumiai (Japan. Kokai Tokkyo Koho 81-137637 [1981] from C.A. **96** [1982] No. 61771).
[166] Reichelderfer, R. F. (Solid State Technol. **25** No. 4 [1982] 160/5).
[167] Hitachi, Ltd. (Japan. Kokai Tokkyo Koho 82-100907 [1982] from C.A. **98** [1983] No. 10518).
[168] Tamura, H.; Kurihara, H. (Japan. J. Appl. Phys. II **23** [1984] 731/3).
[169] Sonek, G. J.; Ballantyne, J. M. (J. Vac. Sci. Technol. [2] B **2** [1984] 653/7).
[170] Fujitsu, Ltd. (Japan. Kokai Tokkyo Koho 83-168233 [1983] from C.A. **100** [1984] No. 16314).

[171] Kravitz, S. H.; Manocha, A. S.; Willenbrock, W. E., Jr. (U.S. 4426246 [1984] from C.A. **100** [1984] No. 130999).
[172] Hitachi, Ltd. (Japan. Kokai Tokkyo Koho 80-107780 [1980] from C.A. **94** [1981] No. 56981).
[173] Beinvogl, W.; Hasler, B. (Ger. Offen. 3315719 [1983] from C.A. **102** [1985] No. 38099).
[174] Toshiba Corp. (Japan. Kokai Tokkyo Koho 81-137636 [1981] from C.A. **96** [1982] No. 78594).
[175] Light, R. W.; Bell, H. B. (J. Electrochem. Soc. **131** [1984] 459/61).
[176] Purdes, A. J. (Eur. Appl. 101828 [1982] from C.A. **100** [1984] No. 184422).
[177] Tokyo Shibaura Electric Co., Ltd. (Japan. Kokai Tokkyo Koho 81-70636 [1981] from C.A. **95** [1981] No. 90001).
[178] Zhang, H.; Ye, M. (Huazhong Gongxueyuan Xuebao **10** [1982] 67/70 from C.A. **98** [1983] No. 44964).
[179] Fujitsu, Ltd. (Japan. Kokai Tokkyo Koho 84-57468 [1984] from C.A. **101** [1984] No. 121479).
[180] Fujitsu, Ltd. (Japan. Kokai Tokkyo Koho 83-168261 [1983] from C.A. **100** [1984] No. 44052).

[181] Fujitsu, Ltd. (Japan. Kokai Tokkyo Koho 84-56741 [1984] from C.A. **101** [1984] No. 102447).

[182] Toshiba Corp. (Japan. Kokai Tokkyo Koho 84-90930 [1984] from C.A. **101** [1984] No. 181213).
[183] Hoya Corp. (Japan. Kokai Tokkyo Koho 84-139033 [1984] from C.A. **102** [1985] No. 15156).
[184] Smith, P. K.; Herndon, T. O.; Burke, R. L.; Day, D. R.; Senturia, S. D. (J. Electrochem. Soc. **130** [1983] 225/7; C.A. **98** [1983] No. 82032).
[185] Fujitsu, Ltd. (Japan. Kokai Tokkyo Koho 84-61144 [1984] from C.A. **101** [1984] No. 102403).
[186] Fujitsu, Ltd. (Japan. Kokai Tokkyo Koho 84-119848 [1984] from C.A. **101** [1984] No. 238935).
[187] Westinghouse Electric Corp. (Japan. Kokai Tokkyo Koho 83-43522 [1983] from C.A. **99** [1983] No. 46756).
[188] Fujitsu, Ltd. (Japan. Kokai Tokkyo Koho 84-56749 [1984] from C.A. **101** [1984] No. 102448).
[189] New Nippon Electric Co., Ltd. (Japan. Kokai Tokkyo Koho 83-196022 [1983] from C.A. **100** [1984] No. 113398).
[190] Toshiba Corp. (Japan. Kokai Tokkyo Koho 83-220447 [1983] from C.A. **100** [1984] No. 166430).

[191] Fujitsu, Ltd. (Japan. Kokai Tokkyo Koho 84-25245 [1984] from C.A. **101** [1984] No. 31916).
[192] Nöth, H.; Weber, S. (Chem. Ber. **117** [1984] 2504/9).
[193] Anton, K.; Nöth, H.; Pommerening, H. (Chem. Ber. **117** [1984] 2479/94).
[194] Beswick, Y. F.; Wisian-Neilson, P.; Neilson, R. N. (J. Inorg. Nucl. Chem. **43** [1981] 2639/43).
[195] Pawelke, G.; Heyder, F.; Buerger, H. (J. Fluorine Chem. **20** [1982] 53/63).
[196] Anton, K.; Nöth, H.; Pommerening, H. (Chem. Ber. **117** [1984] 2495/503).
[197] Nöth, H.; Weber, S. (Z. Naturforsch. **38b** [1983] 1460/5).
[198] Nöth, H.; Rasthofer, B.; Weber, S. (Z. Naturforsch. **39b** [1984] 1058/68).
[199] Maringgele, W. (J. Organometal. Chem. **222** [1981] 17/32).
[200] Lee, D. Y.; Martin, J. C. (J. Am. Chem. Soc. **106** [1984] 5745/6).

[201] Meller, A.; Hirninger, F. J.; Noltemeyer, M.; Maringgele, W. (Chem. Ber. **114** [1981] 2519/35).
[202] Auner, N.; Grobe, J. (Z. Anorg. Allgem. Chem. **500** [1983] 132/60).
[203] Daneshvar, N.; Straughan, B. D.; Gardiner, D. J. (J. Inorg. Nucl. Chem. **43** [1981] 1147/9).
[204] Kranc, D. M. (Diss. Memphis State Univ. 1983, pp. 1/128; Diss. Abstr. Intern. B **44** [1984] 2419).
[205] Dillon, K. B.; Hewitson, G. F. (Polyhedron **3** [1984] 957/62).
[206] Drake, J. E.; Hencher, J. L.; Kashron, L. N. (Can. J. Chem. **59** [1981] 2898/908).
[207] Drake, J. E.; Kashron, L. N.; Majid, A. (Can. J. Chem. **59** [1981] 2417/28).
[208] Sharma, H. K.; Dubby, S. N.; Puri, D. M. (J. Indian Chem. Soc. **59** [1982] 1031/3).
[209] Childs, R. F.; Mulholland, D. L.; Nixon, A. (Can. J. Chem. **60** [1982] 809/12).
[210] Costes, J. P.; Cros, G.; Laurent, J. P. (Syn. React. Inorg. Metal-Org. Chem. **11** [1981] 383/92).

[211] Andrae, K.; Straube, A. (Z. Anorg. Allgem. Chem. **490** [1982] 191/6).
[212] Milaev, A. G.; Pamov, U. B.; Okhlobystin, O. Yu. (Zh. Obshch. Khim. **51** [1981] 2715/20; J. Gen. Chem. [USSR] **51** [1982] 2343/7).
[213] Jutzi, P.; Steiner, W.; Stroppel, K. (Chem. Ber. **113** [1980] 3357/65).
[214] Jutzi, P.; Stroppel, K. (Chem. Ber. **113** [1980] 3366/8).

[215] Fischer, E. O.; Kleine, W.; Schambeck, W.; Schubert, U. (Z. Naturforsch. **36b** [1981] 1575/9).
[216] Richmond, T. G.; Basolo, F.; Shriver, D. F. (Organometallics **1** [1982] 1624/8).
[217] Richmond, T. G.; Shriver, D. F. (Organometallics **2** [1983] 1061/2).
[218] Richmond, T. G.; Shriver, D. F. (Organometallics **3** [1984] 305/14).
[219] Richmond, T. G.; Crespi, A. M.; Shriver, D. F. (Organometallics **3** [1984] 314/9).
[220] Fischer, E. O.; Postnov, U. N.; Kreissl, F. R. (J. Organometal. Chem. **231** [1982] C73/C77).

[221] Dantona, R.; Schweda, E.; Strähle, J. (Z. Naturforsch. **39b** [1984] 733/5).
[222] Bridger, R. F. (S. African 79-06922 [1981] from C.A. **96** [1982] No. 199869).
[223] Snyder, R.; Testa, A. C. (J. Phys. Chem. **85** [1981] 1871/3).
[224] Gallos, J.; Varvoglis, A. (J. Chem. Res. S **1982** 150/1).
[225] Zanirato, P. (J. Chem. Soc. Chem. Commun. **1983** 1065/7).
[226] Matteson, D. S.; Sadhu, K. M. (Organometallics **3** [1984] 614/8).
[227] Matteson, D. S.; Ray, R.; Rocks, R. R.; Tsai, D. J. S. (Organometallics **2** [1983] 1536/43).
[228] Bonner, T. G.; Lewis, D. R. (J. Chem. Soc. Perkin Trans. I **1981** 1807/10).
[229] Carvalho, C. F.; Sargent, M. V. (J. Chem. Soc. Chem. Commun. **1984** 227/9).
[230] Fuji Photo Film Co., Ltd. (Japan. Kokai Tokkyo Koho 84-82333 [1984] from C.A. **101** [1984] No. 130391).

[231] Seela, F.; Menkhoff, S. (Liebigs Ann. Chem. **1982** 813/6).
[232] Padgett, H. C.; Csendes, I. G.; Rapoport, H. (J. Org. Chem. **44** [1979] 3492/6).
[233] Suvorov, B. A.; Orlova, L. N.; Dzhagatspanyan, R. V. (Zh. Obshch. Khim. **52** [1982] 749/54; J. Gen. Chem. [USSR] **52** [1982] 649/53).
[234] Benner, J. P.; Gill. G. B.; Parrott, St. J.; Wallace, B. (J. Chem. Soc. Perkin Trans. I **1984** 291/312).
[235] Piccolo, O.; Valoti, E.; Filippini, L.; Gandolfi, M. (Eur. Appl. 98012 [1984] from C.A. **101** [1984] No. 23137).
[236] Eberson, L.; Malmberg, M.; Nyberg, K. (Acta Chem. Scand. B **38** [1984] 391/6).
[237] Amupitan, J.; Scovell, E. G.; Sutherland, J. K. (J. Chem. Soc. Perkin Trans. I **1983** 755/7).
[238] Mayr, H.; Striepe, W. (J. Org. Chem. **48** [1983] 1159/65).
[239] Toyoda, T.; Sasakura, K.; Sugasawa, T. (J. Org. Chem. **46** [1981] 189/91).
[240] Sugasawa, T.; Toyoda, T.; Sasakura, K. (PCT Intern. Appl. 81-02157 [1981] from C.A. **96** [1982] No. 34864).

[241] Childs, R. F.; Varadarajan, A. (Can. J. Chem. **59** [1981] 3252/5).
[242] von Gentzkow, W. (Eur. Appl. 47423 [1982] from C.A. **96** [1982] No. 217500).
[243] Bach, R. D.; Domagala, J. M. (J. Org. Chem. **49** [1984] 4181/8).
[244] Lewis, D. K.; Bergmann, J.; Manjoney, R.; Paddock, R.; Kalra, B. L. (J. Phys. Chem. **88** [1984] 4112/6).
[245] Lewis, D. K.; Bosch, H. W.; Hossenlopp, J. M. (J. Phys. Chem. **86** [1982] 803/5).
[246] Suzukamo, G.; Fukao, M.; Nagase, T. (Chem. Letters **1984** 1799/802).
[247] Wagner, B. E.; Goeke, G. L.; Karol, F. J.; George, K. F. (Eur. Appl. 55605 [1982] from C.A. **97** [1982] No. 145443).
[248] Sakurai, H.; Katayama, Y.; Ikegami, T.; Furusato, M. (Ger. Offen. 3028479 [1981] from C.A. **95** [1981] No. 8056).
[249] Asahi Chemical Industry Co., Ltd. (Belg. 884739 [1981] from C.A. **95** [1981] No. 8060).
[250] Mitsubishi Petrochemical Co., Ltd. (Japan. Kokai Tokkyo Koho 82-190005 [1982] from C.A. **98** [1983] No. 161311).

[251] Ube Industries, Ltd. (Japan. Kokai Tokkyo Koho 82-141405 [1982] from C.A. **98** [1983] No. 35143).

[252] Ube Industries, Ltd. (Japan. Kokai Tokkyo Koho 82-143307 [1982] from C.A. **98** [1983] No. 72909).
[253] Bottrill, M. (Eur. Appl. 63459 [1982] from C.A. **98** [1983] No. 180484).
[254] Sakurai, H.; Katayama, Y.; Ikegami, T.; Futusato, M. (Eur. Appl. 58549 [1982] from C.A. **98** [1983] No. 17203).
[255] Caunt, A. D.; Gavens, P. D.; McMeeking, J. (Eur. Appl. 32308 [1981] from C.A. **95** [1981] No. 170100).
[256] Zakharov, V. A.; Echevskaya, L. G.; Nesterov, G. A.; Dudchenko, V. K.; Lesnikova, N. P. (Vysokomol. Soedin. Ser. A **26** [1984] 993/7 from C.A. **101** [1984] No. 91488).
[257] Toyo Soda Mfg. Co., Ltd. (Japan. Kokai Tokkyo Koho 84-197408 [1984] from C.A. **102** [1985] No. 96220).
[258] Gavens, P. D.; Caunt, A. D.; Segal, J. A. (Brit. Appl. 2101609 [1983] from C.A. **98** [1983] No. 144013).
[259] Denki Kagaku Kogyo K.K. (Japan. Kokai Tokkyo Koho 81-22302 [1981] from C.A. **95** [1981] No. 62991).
[260] Mitsubishi Chemical Industries Co., Ltd. (Japan. Kokai Tokkyo Koho 82-165408 [1982] from C.A. **98** [1983] No. 126791).

[261] Denki Kagaku Kogyo K.K. (Japan. Kokai Tokkyo Koho 82-23605 [1982] from C.A. **96** [1982] No. 200353).
[262] Ube Industries, Ltd. (Japan. Kokai Tokkyo Koho 82-18706 [1982] from C.A. **96** [1982] No. 163367).
[263] Ube Industries, Ltd. (Japan. Kokai Tokkyo Koho 82-25306 [1982] from C.A. **96** [1982] No. 163377).
[264] Nuyken, O.; Pask, S. D.; Vischer, A.; Walter, M. (Cationic Polym. Relat. Processes Proc. 6th Intern. Symp., Ghent 1983 [1984], pp. 35/42 from C.A. **101** [1984] No. 211775).
[265] Pask, S. D.; Nuyken, O.; Vischer, A.; Walter, M. (Cationic Polym. Relat. Processes Proc. 6th Intern. Symp., Ghent 1983 [1984], pp. 25/33 from C.A. **101** [1984] No. 211774).
[266] Nuyken, O.; Pask, S. D.; Vischer, A. (Makromol. Chem. **184** [1983] 553/62 from C.A. **98** [1983] No. 216064).
[267] Nuyken, O.; Pask, S. D.; Walter, M. (Polym. Bull. [Berlin] **8** [1982] 451/5 from C.A. **98** [1983] No. 35021).
[268] Guhaniyogi, S. C.; Kennedy, J. P.; Ferry, W. M. (J. Macromol. Sci. Chem. A **18** [1982] 25/37 from C.A. **97** [1982] No. 72826).
[269] Nguyen, H. A.; Kennedy, J. P. (Polym. Bull. [Berlin] **9** [1983] 507/14 from C.A. **98** [1983] No. 216054).
[270] Nguyen, H. A.; Kennedy, J. P. (Polym. Bull. [Berlin] **6** [1981] 47/54 from C.A. **96** [1982] No. 86031).

[271] Byrikhin, V. S.; Nesmelov, A. I.; Murachev, V. B.; Belousov, S. I.; Ezhova, E. A.; Shashkina, E. F.; Pravednikov, A. N. (Dokl. Akad. Nauk SSSR **274** [1984] 626/9).
[272] Nguyen, H. A.; Marechal, E. (Polym. Bull. [Berlin] **11** [1984] 99/104 from C.A. **100** [1984] No. 175354).
[273] Kennedy, J. P.; Kelen, T. (J. Macromol. Sci. Chem. A **18** [1982] 129/52 from C.A. **97** [1982] No. 39459).
[274] Fehervari, A.; Kennedy, J. P.; Tudos, F. (Polym. Prepr. Am. Chem. Soc. Div. Polym. Chem. **20** [1979] 320/3 from C.A. **94** [1981] No. 175674).
[275] Kennedy, J. P.; Smith, R. A. (Polym. Prepr. Am. Chem. Soc. Div. Polym. Chem. **20** [1979] 316/9 from C.A. **94** [1981] No. 175673).
[276] Kennedy, J. P.; Ross, L. R.; Lackey, J. E.; Nuyken, O. (Polym. Bull. [Berlin] **4** [1981] 67/74 from C.A. **94** [1981] No. 157336).

[277] Chang, V. S. C.; Kennedy, J. P.; Ivan, B. (Polym. Bull. [Berlin] **3** [1980] 339/46 from C.A. **94** [1981] No. 84583).
[278] Smith, R. A. (Diss. Univ. Akron 1981, pp. 1/179; Diss. Abstr. Intern. B **42** [1981] 1043 from C.A. **95** [1981] No. 204544).
[279] Pask, S. D.; Nuyken, O. (Polym. Bull. [Berlin] **8** [1982] 457/60 from C.A. **98** [1983] No. 54578).
[280] Kennedy, J. P. (J. Macromol. Sci. Chem. A **18** [1982] 3/10 from C.A. **97** [1982] No. 72823).

[281] Wondraczek, R. H.; Kennedy, J. P.; Storey, R. F. (J. Polym. Sci. Polym. Chem. Ed. **20** [1982] 43/51 from C.A. **96** [1982] No. 69504).
[282] Kennedy, J. P.; Chen, F. J. Y. (Polym. Prepr. Am. Chem. Soc. Div. Polym. Chem. **20** [1979] 310/5 from C.A. **94** [1981] No. 175672).
[283] Taha, M.; Rigal, G.; Pietrasanta, Y.; Platzer, N.; Sudres, P.; Raynal, S. (Makromol. Chem. **182** [1981] 2545/55 from C.A. **95** [1981] No. 220420).
[284] Nguyen, H. A.; Kennedy, J. P. (Polym. Bull. [Berlin] **10** [1983] 74/81 from C.A. **99** [1983] No. 88654).
[285] Nguyen, H. A.; Kennedy, J. P. (Polym. Bull. [Berlin] **6** [1981] 55/60 from C.A. **96** [1982] No. 86032).
[286] Gyor, M.; Kennedy, J. P.; Kelen, T.; Tudos, F. (J. Macromol. Sci. Chem. A **21** [1984] 1295/309 from C.A. **101** [1984] No. 152396).
[287] Furukawa, J.; Kobayashi, E.; Wakui, T. (Polym. J. **15** [1983] 435/41 from C.A. **100** [1984] No. 34862).
[288] Hirai, H.; Takenchi, T.; Komiyama, M. (J. Polym. Sci. Polym. Chem. Ed. **19** [1981] 2581/94 from C.A. **95** [1981] No. 187761).
[289] Hirai, H.; Gotoh, Y.; Komiyama, M. (Kenkyu Hokoku Asahi Garasu Kogyo Gijutsu Shoreikai **45** [1984] 1/14 from C.A. **102** [1985] No. 167207).
[290] Hirai, H.; Komiyama, M.; Mori, T. (Kobunshi Ronbushu **41** [1984] 421/4 from C.A. **101** [1984] No. 131197).

[291] Hirai, H.; Takeuchi, K.; Komiyama, M. (J. Polym. Sci. Polym. Chem. Ed. **20** [1982] 159/72 from C.A. **96** [1982] No. 69507).
[292] Kennedy, J. P.; Plamthottam, S. S.; Ivan, B. (J. Macromol. Sci. Chem. A **17** [1982] 637/51 from C.A. **96** [1982] No. 69559).
[293] Kennedy, J. P.; Guhaniyogi, S. C. (J. Macromol. Sci. Chem. A **18** [1982] 103/17 from C.A. **97** [1982] No. 56295).
[294] Kennedy, J. P.; Plamthottam, S. S. (Polym. Prepr. Am. Chem. Soc. Div. Polym. Chem. **20** [1979] 98/101 from C.A. **94** [1981] No. 176395).
[295] Kitahara, S.; Hirokawa, Y.; Kawada, H.; Fujii, T.; Sugi, N.; Hasegawa, H.; Yoshioka, A. (Eur. Appl. 87109 [1983] from C.A. **99** [1983] No. 196443).
[296] Kitahara, S.; Hirokawa, Y.; Kawada, H.; Fujii, T.; Sugi, N.; Hasegawa, H.; Yosioka, A. (Eur. Appl. 87110 [1983] from C.A. **99** [1983] No. 177305).
[297] He, X.; Huang, Z. (Huaxue Xuebao **42** [1984] 576/9 from C.A. **101** [1984] No. 91510).
[298] Janson, V.; Gors, H. C. (PCT Intern. Appl. 84-03891 [1984] from C.A. **102** [1985] No. 204469).
[299] Horn, H. G.; Kolkmann, F. (Makromol. Chem. **183** [1982] 1833/41).
[300] Horn, H. G.; Kolkmann, F. (Makromol. Chem. **183** [1982] 1843/54).

[301] Sennett, M. S.; Hagnauer, G. L.; Singler, R. F. (Polym. Mater. Sci. Eng. **49** [1983] 297/300 from C.A. **100** [1984] No. 175457).
[302] Oki Electric Industry Co., Ltd. (Japan. Kokai Tokkyo Koho 84-175720 [1984] from C.A. **102** [1985] No. 124047).

[303] Nippon Electric Co., Ltd. (Japan. Kokai Tokkyo Koho 83-190052 [1983] from C.A. **100** [1984] No. 113332).
[304] Nippon Electric Co., Ltd. (Japan. Kokai Tokkyo Koho 83-207641 [1983] from C.A. **100** [1984] No. 149561).
[305] Fujitsu, Ltd. (Japan. Kokai Tokkyo Koho 84-04053 [1984] from C.A. **100** [1984] No. 201832).
[306] Fujitsu, Ltd. (Japan. Kokai Tokkyo Koho 84-02320 [1984] from C.A. **100** [1984] No. 201825).
[307] Fujitsu, Ltd. (Japan. Kokai Tokkyo Koho 84-04054 [1984] from C.A. **100** [1984] No. 201834).
[308] Nippon Telegraph and Telephone Public Corp. (Japan. Kokai Tokkyo Koho 83-101455 [1983] from C.A. **99** [1983] No. 150689).
[309] Gavrilov, G. M.; Evdokimov, V. I. (Poluch. Svoistva Tonkikh Plenok **1982** 33/7 from C.A. **99** [1983] No. 14211).
[310] Ikeda, S.; Kobayashi, Y.; Shirakawa, H. (PCT Intern. Appl. 80-02142 [1980] from C.A. **94** [1981] No. 56965).

[311] Lewis, D.; Kalina, D. W.; Tsai, T. E. (U.S. 4456548 [1984] from C.A. **101** [1984] No. 82606).
[312] Ehrlich, D. J.; Tsao, J. Y. (Appl. Phys. Letters **44** [1984] 267/9 from C.A. **100** [1984] No. 77260).
[313] Yata, S. (Eur. Appl. 67444 [1982] from C.A. **98** [1983] No. 82317).
[314] Steigerwald, W. E.; Ambros, P.; Geyer, H. (U.S. 4271045 [1981] from C.A. **95** [1981] No. 53587).
[315] Ichimitsu Kogyo K. K. (Japan. Kokai Tokkyo Koho 80-133466 [1980] from C.A. **94** [1981] No. 85859).
[316] Drynkin, V. I.; Beln'kii, B. V.; Kerzin, A. L. (At. Energiya SSSR **55** [1983] 300/3 from C.A. **100** [1984] No. 60955).
[317] Godefroy, P. (U.S. 4328117 [1982] from C.A. **97** [1982] No. 14677).
[318] Dainichi Nippon Cables, Ltd. (Japan. Kokai Tokkyo Koho 80-144437 [1980] from C.A. **94** [1981] No. 213058).
[319] Shintani, T.; Utsumi, A.; Sukawa, T.; Kobayashi, R. (Ger. Offen. 2947074 [1980] from C.A. **94** [1981] No. 51783).
[320] Takagi, H.; Kokubo, T. (Yogyo Kyokaishi **92** [1984] 444/7 from C.A. **101** [1984] No. 135656).

[321] Sara, R. V. (Eur. Appl. 68752 [1983] from C.A. **98** [1983] No. 162354).
[322] Bressan, J.; Feuillade, G.; Wiart, R. (J. Electrochem. Soc. **129** [1982] 2649/53).
[323] Gol'tsblat, V. L.; Drobiz, A. M.; Semenenko, V. A. (Khim. Prom. Ser. Avtom. Khim. Proizvod. **1981** No. 5, pp. 21/4 from C.A. **96** [1982] No. 125034).
[324] Krylov, V. A.; Salganskii, Yu. M.; Sokolova, G. V. (Zh. Analit. Khim. **39** [1984] 671/7; C.A. **101** [1984] No. 122085).
[325] Moskowitz, P. D.; Perry, P.; Wilenitz, I. (EPA-600-S7-82-066 [1983] 1/4 from C.A. **99** [1983] No. 42729).
[326] Adams, J. E. (J. Chem. Phys. **78** [1983] 1275/80).
[327] Wada, K.; Yabushita, S.; Yamaguchi, H.; Uehara, Y.; Ohmori, F.; Nakazaki, Y. (Reza Kenkyu **11** [1983] 434/8 from C.A. **100** [1984] No. 57110).
[328] Wada, K.; Yasui, M.; Uehara, Y.; Yamaguchi, H.; Yamanaka, M. (Chem. Phys. Letters **94** [1983] 527/30).
[329] Petrov, Yu. N. (Proc. Intern. Conf. Lasers **1980/81** 852/8 from C.A. **96** [1982] No. 111959).
[330] Hu, C. L.; Chang, C. H.; Liu, Y. S.; Mi, C.; Kung, F. A.; Yu, S. C.; Chang, Y. W.; Ma, H. H. (Zhougguo Kexue Jishu Daxue Xuebao **9** No. 2 [1979] 19/22 from C.A. **94** [1981] No. 216272).

[331] Denison, D. R.; Hartsough, L. D. (Ger. Offen. 3013679 [1980] from C.A. **94** [1981] No. 75605).
[332] Takeuchi, K.; Kurihara, O.; Nakane, R. (Chem. Phys. **54** [1981] 383/6).
[333] Adamova, Yu. A.; Akinfiev, N. N.; Boitsov, B. S.; Zubkov, V. I.; Pankratov, A. V.; Skachov, A. N. (Khim. Vysokikh Energ. **15** [1981] 365/9; High Energy Chem. [USSR] **15** [1981] 280/4 from C.A. **95** [1981] No. 106880).
[334] Schramm, B. (Springer Ser. Chem. Phys. **6** [1979] 274/6).
[335] Volkov, S. V.; Gurko, A. F., Lutoshkin, V. I. (Teor. Eksperim. Khim. **17** [1981] 564/7; Theor. Exptl. Chem. [USSR] **17** [1981] 442/5).
[336] Gurko, A. F.; Lutoshkin, V. I.; Volkov, S. V. (Ukr. Khim. Zh. **49** [1983] 376/82; Soviet Progr. Chem. **49** No. 4 [1983] 41/6 from C.A. **98** [1983] No. 225137).
[337] Pessine, F. B. T.; Lin, C. T. (Acta Sud Am. Quim. **1** [1981] 19/33 from C.A. **96** [1982] No. 208244).
[338] Abzianidze, T. G.; Abdushelishvili, G. I.; Bakhtadze, A. B.; Gverdtsiteli, I. G.; Kaminskii, A. V.; Kudziev, A. G.; Tkeshelashvili, G. I.; Tsinadze, T. B. (Pis'ma Zh. Tekhn. Fiz. **8** [1982] 1234/6 from C.A. **98** [1983] No. 42561).
[339] Abdushelishvili, G. I.; Abzianidze, T. G.; Egiazarov, A. S.; Tkeshelashvili, G. I.; Tsinadze, T. B. (At. Energya SSSR **57** [1984] 203/4 from C.A. **102** [1985] No. 36524).
[340] Kuz'menko, V. A. (Zh. Fiz. Khim. **58** [1984] 727/9; Russ. J. Phys. Chem. **58** [1984] 436/8).

[341] Jensen, R. J.; Hayes, J. K.; Cluff, C. L.; Thorne, J. M. (IEEE J. Quantum Electron. **16** [1980] 1352/6 from C.A. **94** [1981] No. 74608).
[342] Jensen, R. J.; Thorne, J. M.; Cluff, C. L. (U.S. Appl. 228036 [1984] from C.A. **100** [1984] No. 93280).
[343] Fu, K.; Lu, X.; Tang, F.; Yin, Y.; Shi, H. (Jiguang **8** No. 9 [1981] 54/6 from C.A. **97** [1982] No. 64035).
[344] Pola, J.; Engst, P.; Horak, M. (Collection Czech. Chem. Commun. **47** [1982] 912/7).
[345] Engst, P.; Horak, M.; Pola, J. (Collection Czech. Chem. Commun. **48** [1983] 1314/22).
[346] Ishikawa, Y. I.; Kurihara, O.; Arai, S.; Nakane, R. (J. Phys. Chem. **85** [1981] 3817/20).
[347] Ishikawa, Y.; Kurihara, O.; Arai, S.; Nakane, R. (Reza Kagaku Kenkyu No. 3 [1981] 69/75 from C.A. **96** [1982] No. 34333).
[348] Ivan, B.; Kennedy, J. P.; Kelen, T.; Tudos, F. (J. Macromol. Sci. Chem. A **16** [1981] 533/42 from C.A. **95** [1981] No. 43783).
[349] Stratt, R. M.; Adams, J. E. (J. Chem. Phys. **78** [1983] 2368/73).
[350] Gaines, D. F.; Heppert, J. A.; Coons, D. E.; Jorgenson, M. W. (Inorg. Chem. **21** [1982] 3662/5).

[351] Reichelderfer, R. F.; Vogel, D. C.; Tang, M. S. K. (Brit. Appl. 2087315 [1982] from C.A. **97** [1982] No. 202216).
[352] Brownstein, S.; Gabe, E.; Lee, F.; Tan, L. (J. Chem. Soc. Chem. Commun. **1984** 1566/8).
[353] Steliou, K.; Mrani, M. (J. Am. Chem. Soc. **104** [1982] 3104/6).
[354] Kennedy, J. P.; Chen, F. J. Y. (Polym. Prepr. Am. Chem. Soc. Div. Polym. Chem. **20** [1979] 310/5 from C.A. **94** [1981] No. 175672).
[355] Ikeda, S.; Kobayashi, Y.; Shirakawa, H. (PCT Intern. Appl. 80-02146 [1980] from C.A. **94** [1981] No. 56965).

6.1.3 Adducts of BCl_3 with Nitrogen Donors

H_3N–BCl_3. Argon-matrix isolation of H_3N–BCl_3 and $H_3{}^{15}N$–BCl_3 has been performed and IR bands (in cm^{-1}) have been assigned as follows: 1352, $\delta_s(NH_3)$; 853, $\nu_{as}(^{10}BCl_2)$; 836, $\nu_{as}(^{11}BCl_2)$; 760, $\nu(BN)$ or $\varrho(NH_3)$; 506, $\nu_s(BCl_2)$ [1]. The most frequent use of BCl_3/NH_3 gas mixtures is in the preparation of boron nitride, especially in chemical vapor deposition. See the CVD section of Chapter 6.1.2 on pp. 14/5. However, BCl_3/NH_3 is also used in boriding thoriated tungsten [2].

$(CH_3)_3N$–BCl_3 is contained in photopolymerizable compositions for relief image formation [3].

$(C_2H_5)_3N$–BCl_3 is formed in the thermal decomposition of $[N(C_2H_5)_4][BCl_4]$ (as shown by differential thermal analysis) [4]. The mass spectrum contains primarily peaks due to the $(C_6H_5)_3N$ and BCl_3 fragments [5].

C_5H_5N–BCl_3 is prepared by addition of pyridine to a solution of BCl_3 in cyclohexane. The X-ray structure shows a monoclinic crystal with a = 6.168(1), b = 15.327(1), and c = 9.741(1) Å; space group $P2_1/c$-C^5_{2h} (No. 14); Z = 4; D_x = 1.569(3) g/cm^3. The B–N distance is 1.592 Å [6].

Adducts have also been prepared from quinoline, i.e., C_9H_7N–BCl_3, and the influence of the complexation on the electronic relaxation of quinoline has been studied [7]; from N-benzylideneaniline, $C_{13}H_{11}N$–BCl_3, including a study of electron transfer and polarization upon complex formation (μ = 9.0 D) [8]; and the species 4-C_6H_5–CO–$C_6H_4NH_2$–BCl_3, where the coordination at the nitrogen atom has been confirmed by IR specroscopy, and conductivity data have been obtained (together with those of AlX_3 (with X = Cl, Br, I) adducts) [9]. Amine complexes of BCl_3 are used for curing the working layers of magnetic tapes [10]. For NMR (^1H) and IR data on the complex C_2H_5CN–BCl_3, see [11].

6.1.4 Adducts of BCl_3 with Oxygen Donors

The IR spectra (in an Ar matrix) of **$(CH_3)_2O$–BCl_3** and **$(CD_3)_2O$–BCl_3** have been recorded. The following assignments have been made (in cm^{-1}): $\nu_{as}(^{10}BCl_3)$, 815; $\nu_{as}(^{11}BCl_3)$, 777; $\nu_{as}(CO)$, 999; $\nu(CO)$, 903; $\nu_s(BCl_3)$, 415; $\nu(BO)$, 608. The assignments are supported by the spectra of the corresponding adducts with BF_3 and BBr_3 [32].

Trimethyl- and triethylsilanyltrifluoromethanesulfonate form strongly polarized donor-acceptor complexes with BCl_3 involving O-coordination; ionization to silicenium ions is not observed. The NMR spectra show strong deshielding of the ^{29}Si nucleus upon complex formation ($\Delta\delta^{29}Si \sim 30$ ppm) and the formation of four-coordinate boron [34].

$(CH_3)_3SiO(BCl_3)SO_2CF_3$ has been prepared by slow addition of liquid BCl_3 to $(CH_3)_3SiOSO_2CF_3$ at −78°C with efficient stirring. The ^{29}Si NMR spectrum of the resultant oil (recorded in liquid BCl_3 at −35°C) consisted of two signals, $\delta^{29}Si = 72.35$ ppm (for the complex, which corresponds to a low field shift of $\Delta\delta^{29}Si = 28.81$ ppm) and a signal at $\delta^{29}Si = 29.02$ ppm (for $(CH_3)_3SiCl$ formed by ligand exchange). Upon standing at −75°C the complex slowly precipitates from BCl_3. A solution of the complex in $ClFSO_2$ exhibits the ^{11}B NMR signal at $\delta = -2.62$ ppm [34].

Analogous complexes, e.g., $R_3SiO(BCl_3)SO_2CF_3$ (R = C_2H_5, n-C_4H_9) and $RR'_2SiO(BCl_3)SO_2CF_3$ (R = i-C_3H_7, R′ = CH_3, R = CH_3, R′ = C_6H_5) have been prepared, as was the adduct $R_2Si(OSO_2CF_3)_2 \cdot BF_3$ (R = C_2H_5). When sterically crowded species such as $(i$-$C_3H_7)_3SiOSO_2CF_3$ were treated with BCl_3, only ligand exchange to give $(i$-$C_3H_7)_3SiCl$ has been observed. For additional data, see [34]; compare also [35].

6.1.5 The Tetrachloroborate Ion, $[BCl_4]^-$

A set of molecular constants, kinetic constants, force constants, mean amplitudes of vibration, and Coriolis coupling constants has been obtained for $[BCl_4]^-$ (and 21 more tetrahedral XY_4 species) [12] by the method of [13]. The following data are given for $[BCl_4]^-$ [12]: Kinetic constants (in 10^{23} g): $k_d=4.4514$, $k_{dd}=0.4515$, $k_\alpha=0.2984$, $-k'_{\alpha\alpha}=-0.0242$, $-k'_{d\alpha}=0.3193$. Force constants (in mdyn/Å): $f_d=1.4768$, $f_{dd}=0.6340$, $f_\alpha=0.0571$, $-f_{\alpha\alpha}=0.0159$, $-f_{d\alpha}=0.0850$. Compliance constants (in Å/mdyn): $c_d=14.8710$, $-c_{dd}=4.8584$, $c_\alpha=29.7886$, $-c'_{\alpha\alpha}=27.1044$, $-c'_{d\alpha}=11.4845$. Valence mean square amplitudes of vibration at 298.16 K (in 10^{-3} Å^2): $\sigma_d=3.6967$, $\sigma_{dd}=0.7054$, $\sigma_\alpha=19.9163$, $\sigma'_{\alpha\alpha}=8.1172$, $\sigma'_{d\alpha}=2.5624$. Mean amplitudes of vibration at 298.16 K (in 10^{-2} Å): $l_{(x-y)}=6.0800$, $l_{(y-y)}=7.6127$. Coriolis coupling constants: $\zeta_{33}=0.8232$, $\zeta_{44}=-0.3232$, $\zeta_{34}=0.4837$, $\zeta_{23}=0.3438$, $-\zeta_{24}=0.9392$.

Calculations on the electronic structure of $[BCl_4]^-$ have been performed by the discrete variational X_α method (DVM X_α). This method uses an LCAO basis of both numerical Hartree-Fock (HF) functions and analytical double-zeta (DZ) Clementi functions. The ionization potentials of the valence levels, the energy of the first electronic transition, atomic charges and overlap populations (by a Mulliken population analysis) are listed in Table 6/13 [14, 15].

Table 6/13
Ionization Potentials, the Energy of the First Electronic Transition, Atomic Charges, and Overlap Populations of $[BCl_4]^-$ (calculated by the DVM X_α procedure) [14, 15].

	HF	DZ		HF	DZ
level			transition		
$2t_1$	4.9	4.5	$2t_1 \rightarrow 7a_1$	8.5	8.5
$7t_2$	5.5	5.0	atomic charges		
2e	6.1	5.7	Q_B	−0.12	0.15
$6t_2$	8.3	7.8	Q_{Cl}	−0.22	−0.29
$6a_1$	10.4	9.9	overlap populations		
$5t_2$	18.2	17.5	$P_{B\text{-}Cl}$	0.26	0.26
$5a_1$	20.2	19.6	$P_{Cl\text{-}Cl}$	−0.07	−0.07

The $[BCl_4]^-$ ion reacts with BCl_3 to give the **$[B_2Cl_7]^-$** ion, rapidly exchanging with excess of BCl_3 as shown by ^{11}B NMR data. However, it is not possible to isolate $[N(CH_3)_4][B_2Cl_7]$ and it decomposes on attempted crystallization [16].

Glasses containing lithium chloroborate ($Li_2O/Li_2Cl_2/B_2O_3$) belong to the group of fast-ion conducting glasses. NMR (^7Li, ^{19}F) studies show that the ionic conduction is produced only by the Li^+ ions [17, 18].

A combination of the matrix-isolation technique with the salt/molecule reaction has been used to study the $Cs[BCl_4]$ contact ion pair in an Ar matrix. The IR spectrum exhibits splitting of the triply degenerate B–Cl stretching band (for undisturbed T_d symmetry) into an A_1 mode (to higher energies) and an E mode (at lower energies) by the polarizing power of the metal cation (monodentate C_{3v} symmetry of $[BCl_4]^-$ in the ion pair). The observed splitting corresponds to a change of the B–Cl valence force constant of $\Delta K=1.6$ [19].

$[Sb(C_6H_5)_4][BCl_4]$ has been obtained from $[Sb(C_6H_5)_4]Cl$ and BCl_3 in benzene solution at ambient temperature. The colorless crystals melt at 245°C. IR data are discussed [20].

References for 6.1.3 to 6.1.6 on pp. 34/5

Among other tetrachloroborates, $[(tmp)B{=}N(C_2H_5)_2][BCl_4]$ (tmp = 2,2,6,6-tetramethylpiperidinyl) exists in equilibrium with $(tmp)ClB{-}N(C_2H_5)_2$ and BCl_3 [21], and the cationic carbyne complexes $[\eta^5\text{-}C_5H_5(CO_2)MCR][BCl_4]$ (M = Mn, Re; R = aryl) are worth mentioning [22]. ^{31}P NMR data have been recorded for tetrachloroborates of phosphonium ions such as $[(C_2H_5)_2PCl_2][BCl_4]$ [23] and various sulfonium salts, e.g., $[H_3C{-}S{-}CH_2{-}S(CH_3)CH_2Cl][BCl_4]$, have also been prepared (available NMR data: ^1H, ^{13}C) [36].

In an estimate of the electron affinity (EA) of $[BCl_4]^-$ (which neglects the difference between vertical and adiabatic ionization potentials) the EA of the **$[BCl_4]^•$** radical is considered to be equal to the first IP of $[BCl_3]^-$. Thus, the EA of $[BCl_4]^•$ is evaluated as 4.9 eV (HF) or 4.5 eV (DZ basis). This leads to the assumption that the $[BCl_4]^•$ radical behaves as a "superhalogen" [14, 15].

6.1.6 Additional Binary Species

B_2Cl_2 is observed in a laser-induced surface reaction of BCl_3 on Pb [24]. As based on an MNDO calculation, the enthalpy of formation $\Delta H_f = -110.5$ kcal/mol [33].

B_2Cl_4 and derivatives thereof are discusssed in Section 2.4.3 of "Boron Compounds" 3rd Suppl. Vol. 1, 1987. Recent reports describe possible intermediates in the discharge synthesis of B_2Cl_4 [25], the adduct formation with B_6H_{10} at −45°C [26], and the reaction with difunctional bases to give heterocyclic species [27].

B_4Cl_4 reacts with lithium alkyls to give, for example, $C_2H_5B_4Cl_3$, $(C_2H_5)_2B_4Cl_2$, and $(t\text{-}C_4H_9)_4B_4$ [28].

$[B_6Cl_6]^{2-}$ has been prepared from $Na_2[B_6H_6]$ and chlorine in aqueous alkaline solution and was characterized by ^{11}B NMR and vibrational spectroscopy [29].

B_8Cl_8 is formed by thermal decomposition of B_2Cl_4 (20% in CCl_4) in nearly 90% yield. With the fivefold amount of $t\text{-}C_4H_9Li$, $(t\text{-}C_4H_9)_9B_9$ is obtained [30]. B_8Cl_8 and B_9Cl_9 react with Al_2Br_6 to give B_8Br_8 and B_9Br_9, respectively [31].

References for 6.1.3 to 6.1.6:

[1] Hunt, R. L.; Ault, B. S. (Spectrosc. Intern. J. **1** [1982] 31/44).
[2] Kuleshov, Yu. S.; Khotin, V. M.; Nefedov, V. G.; Sergienko, R. I. (Zashch. Pokrytiya Met. No. 14 [1980] 88/9 from C.A. **97** [1982] No. 221157).
[3] Green, G. E.; Losert, E.; Paul, J. G. (Eur. Appl. 62611 [1982] from C.A. **98** [1983] No. 170377).
[4] Sheik, S. U. (Therm. Anal. Proc. 6th Intern. Conf. Therm. Anal., Bayreuth, FRG, 1980, Vol. 2, pp. 165/8; C.A. **94** [1981] No. 75999).
[5] Sheik, S. U.; Ghafoor, A. (Pakistan J. Sci. Ind. Res. **23** [1980] 23/4).
[6] Töpel, K. H.; Hensen, K. (Acta Cryst. B **37** [1981] 969/71).
[7] Snyder, R.; Testa, A. C. (J. Phys. Chem. **88** [1984] 5948/50).
[8] Kharabaev, N. N.; Kogan, V. A.; Osipov, O. A. (Zh. Obshch. Khim. **52** [1982] 2617/20; J. Gen. Chem. [USSR] **52** [1982] 2314/6).
[9] Vezzosi, I. M.; Zanoli, A. F.; Peyronel, G. (Spectrochim. Acta A **37** [1981] 593/5).
[10] Timofeev, E. N.; Gmyrko, L. A.; Rubinson, E. B.; Artamonova, T. I. (Neserebryan. Fot. Materialy Magnit. Lenty M. **1982** 89/93 from C.A. **99** [1983] No. 32056).

[11] Tandon, S. (Indian J. Chem. A **23** [1984] 65/6).
[12] Mohan, S.; Mukunthan, A. (Indian J. Pure Appl. Phys. **20** [1982] 315/7).

[13] Thirugnanasarnbandam, P.; Mohan, S. (Indian J. Phys. B **51** [1977] 342; from [12]).
[14] Gutsev, G. L.; Boldyrev, A. I. (Zh. Neorgan. Khim. **27** [1982] 868/72; Russ. J. Inorg. Chem. **27** [1982] 487/9).
[15] Gutsev, G. L.; Boldyrev, A. I. (Chem. Phys. Letters **84** [1981] 352/5).
[16] Glavincevski, B.; Brownstein, S. K. (Can. J. Chem. **59** [1981] 3012/5).
[17] Bray, P. J.; Hintenlang, D. E.; Mulkern, R. V.; Greenbaum, S. G.; Tran, D. C.; Drexhage, M. (J. Non-Cryst. Solids **56** [1983] 27/32).
[18] Bray, P. J.; Lui, M. L.; Hintenlang, D. E. (Wiss. Z. Friedrich Schiller Univ. Jena Math. Naturwiss. Reihe **32** [1983] 409/26 from C.A. **99** [1983] No. 81369).
[19] Hunt, R. L.; Ault, B. S. (Spectrochim. Acta A **37** [1981] 63/9).
[20] Sharma, H. K.; Dubey, S. N.; Puri, D. M. (J. Indian Chem. Soc. **59** [1982] 1031/3).

[21] Nöth, H.; Rasthofer, B.; Weber, S. (Z. Naturforsch. **39b** [1984] 1058/68).
[22] Fischer, E. O.; Chen, J.; Schubert, U. (Z. Naturforsch. **37b** [1982] 1284/8).
[23] Dillon, K. B.; Nisbet, M. P.; Waddington, T. C. (Polyhedron **1** [1982] 123/7).
[24] Lin, C. T. (Springer Ser. Chem. Phys. **39** [1984] 137/42 from C.A. **102** [1985] No. 68024).
[25] Briggs, A. G.; Massey, A. G.; Reason, M. S.; Portal, P. J. (Polyhedron **3** [1984] 369/71).
[26] Dolan, P. J.; Moody, D. C.; Schaeffer, R. (Inorg. Chem. **20** [1981] 744/8).
[27] Haubold, W.; Hrebicek, J.; Sawitzky, G. (Z. Naturforsch. **39b** [1984] 1027/31).
[28] Davan, T.; Morrison, J. A. (J. Chem. Soc. Chem. Commun. **1981** 250/1).
[29] Preetz, W.; Fritze, J. (Z. Naturforsch. **39b** [1984] 1472/7).
[30] Emery, S. L.; Morrison, J. A. (J. Am. Chem. Soc. **104** [1982] 6790/1).

[31] Markwell, A. J.; Massey, A. G.; Portal, P. J. (Polyhedron **1** [1982] 135/7).
[32] Hunt, R. L.; Ault, B. S. (Spectros. Intern. J. **1** [1982] 45/61).
[33] Dewar, M. J. S.; Rzepa, H. S. (J. Computat. Chem. **4** [1983] 158/69).
[34] Olah, G. A.; Laali, K.; Farooq, O. (Organometallics **3** [1984] 1337/40).
[35] Olah, G. A.; Field, L. D. (Organometallics **1** [1982] 1485/7).
[36] Hartge, K.; Akgün, E. (Liebigs Ann. Chem. **1981** 47/51).

6.2 With Hydrogen or Organic Substituents

6.2.1 Dichloroborane, $HBCl_2$, and Its Adducts

$HBCl_2$ has been generated in the gas phase from BCl_3 (at 1 to 3 Torr pressure) by reaction with $Na[BH_4]$ (loosely packed over 15 cm in a 12 mm o. d. Pyrex tube) at 250°C in quantitative yield [1]. It is also formed in a variety of multiphoton-initiated reactions of BCl_3 with hydrogen [2 to 7], methane [7, 8], or silane [9] in the radiation field of CO_2 lasers. These reactions have already been discussed with the photo-induced reactions of BCl_3 in Section 6.1.2.3, pp. 13/4.

Equilibrium states in the BCl_3/H_2 system have been calculated using the EHSYS program, including gas mixtures containing between 1 and 50 vol% of BCl_3 and at temperatures between 800 and 1400 K. The equilibrium conditions are important for the chemical vapor deposition of boron [10]. The standard enthalpy of formation for H_2BCl (see p. 36) has been calculated under the assumption that the bond enthalpy terms E(B–H) and E(B–Cl) are the same as in $HBCl_2$, and that E(B–Cl) is the same for BCl_3, $HBCl_2$, and H_2BCl. This is certainly a rough simplification (compare [1, 11]) considering that results are calculated to hundredth parts of a %. Data of the equilibrium states for 1:9 vol% BCl_3/H_2 and also for the system $BCl_3/H_2/H_2O$ containing between 0.02 to 0.5% H_2O are presented in a table in [10], where a full tabulation of the results is given.

References for 6.2.1 to 6.2.3 on pp. 38/9

Analysis of published thermochemical data leads to a single bond enthalpy term E(B–H) = 371±12 kJ/mol. The bond-order dependent enthalpy terms for the boron-chlorine bonds have been calculated to be E(B–Cl) = 448.7 kJ/mol (assumed bond order 1.5). The enthalpy of formation is calculated to be $\Delta H_f^\ominus(g) = -248.1$ kJ/mol [11].

The He(I) photoelectron spectrum of $HBCl_2$ has been recorded and interpreted with the aid of ab initio calculations on the STO-4-31 G level [1, 12]. Results are presented in Table 6/14 and **Fig. 6-2** [1].

Table 6/14

Cation States, Ionization Potentials (IP), Vibrational Structure, and ab initio Orbital Energies ε_j for $HBCl_2$ ([a] vertical IP's, adiabatic IP's in parentheses; [b] corresponding molecular ground states: $\nu_1 = 2617$ cm^{-1} (ν(BH)); $\nu_2 = 7400$ cm^{-1} (ν(^{11}BCl)); $\nu_3 \cong 290$ cm^{-1} (δ_s(BCl)); [c] valence orbital numbering; [d] STO-31G basis set. Koopmanns' IP's scaled by 0.92. Total energy = −943.27106 a.u.) [1, 12].

cation state	experimental IP in eV [a]	ν^+ in cm^{-1} [b]	calculated molecular orbital [c]	ε_j in eV [d]
$\tilde{X}^2B_2$	(11.91)			
	11.91	860±40	$4a_1$	−11.60
$\tilde{A}^2A_1$	12.35	(320±40)?	$3b_2$	−11.63
$\tilde{B}^2A_2$	12.35		$1a_2$	−11.85
	(13.60)			
$\tilde{C}^2B_1$	13.68	650±40	$1b_1$	−13.27
	(14.65)			
$\tilde{D}^2B_2$	14.95	610±40	$2b_2$	−14.39
$\tilde{E}^2A_1$	15.29		$3a_1$	−15.27
$\tilde{F}^2A_1$	17.71	670±60	$2a_1$	−17.79
		2510±40		
			$1b_2$	−28.01

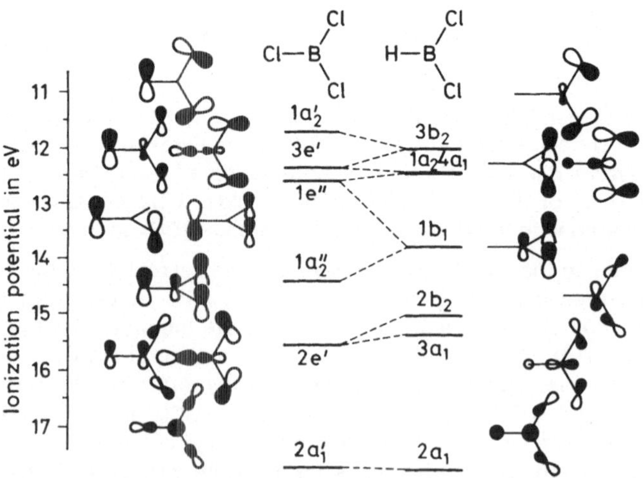

Fig. 6-2. Schematic orbital diagrams and a comparison of the photoelectron spectra of BCl_3 and $HBCl_2$ [1].

Table 6/15 gives the gross overlap populations of BCl_3, $HBCl_2$, and H_2BCl (STO-4-31G basis set). Calculated π-stabilization energies are $\Delta\pi(BCl_3)=1a_2''-1e''$, $\Delta\pi(HBCl_2)=1b_1-1a_2$ [1]; for a conclusive discussion, see the original work [1, 12].

Table 6/15
Gross Overlap Populations of BCl_3, $HBCl_2$, and H_2BCl [1].

mole-	B			Cl			H	LUMO	$\Delta\pi$
cule	total	σ	π	total	σ	π	total	in eV	in eV
BCl_3	4.632	4.199	0.433	17.123	15.244	1.879	—	2.07	2.10
$HBCl_2$	4.666	4.367	0.299	17.181	15.307	1.874	0.971	2.17	1.42
H_2BCl	4.695	4.358	0.157	17.237	15.371	1.866	1.034	2.38	—

$HBCl_2$ is less effective than BCl_3 in opening cyclic acetals to give α-chloro ethers [14]. The chemical vapor deposition of boron has been analyzed using an impinging jet apparatus and a one-dimensional model has been developed to describe the reaction. Apparently, the decomposition of $HBCl_2$ on the surface controls the deposition rate [13]; see also [10].

The determination of $HBCl_2$ in tetrahydrofuran or diethyl ether can be done by a coulometric titration with electrochemically generated bromine in acetic acid, and with IO^- in NaOH solutions at a Pt electrode [15].

$(CH_3)_3N$–$BHCl_2$ is obtained by adding 3.0 g (41 mmol) of $(CH_3)_3N$–BH_3 to 50 mL of 2 M NaOCl and stirring for 7 h at 85°C. The product is extracted with $(C_2H_5)_2O$ and purified by sublimation under vacuum. The melting point is 149 to 150°C. In the IR spectrum, ν(BH) is observed at 2460 cm^{-1} [16]. NMR data are reported with $\delta^{11}B=4.7$ ppm (J(BH) = 147.5 Hz) and $\delta^{1}H=2.94$ ppm (CH_3) [17].

$(CH_3)_2S$–$BHCl_2$ is used (in the presence of BCl_3) for the hydroboration of 1-methylcyclooctene at 0°C to give highly pure *trans*-(2-methylcyclooctyl)dichloroborane [18].

6.2.2 Monochloroborane, H_2BCl, and Its Adducts

H_2BCl. Based on the assumption that the bond energy term E(B–Cl) is the same in H_2BCl as in BCl_3 (which certainly introduces some error), the standard enthalpy of formation has been calculated to be $\Delta H_{f,298}=-22.3$ kcal/mol; for additional data, see [10]. Using the published thermochemical data (B–H single bond enthalpy terms and bond-order dependent B–Cl enthalpy terms), the standard heat of formation in the gaseous state is predicted to be $\Delta H_f^{\ominus}=-109.8$ kJ/mol (−26.24 kcal/mol) [1]. Gross orbital populations for H_2BCl are given in Table 6/15 [1].

Monochloroborane has been reacted with alkynes to give divinylchloroboranes [19].

$(CH_3)_3N$–BH_2Cl is prepared in 75% yield by treatment of $(CH_3)_3N$–BH_3 (dissolved in C_6H_6) with gaseous HCl. The melting point is 83 to 84°C. In the ^1H NMR spectrum, a singlet with $\delta^{1}H=2.8$ ppm is observed for the methyl groups. In the IR spectrum, ν(BH) is observed at 2400 cm^{-1} [16].

$(C_2H_5)_3N$–BH_2Cl, obtained analogously to $(CH_3)_3N$–BH_2Cl in 83% yield, has a melting point of 40 to 41°C. The ^1H NMR spectrum exhibits signals at $\delta=2.9$ ppm (quartet) and $\delta=1.2$ ppm (triplet). In the infrared spectrum, ν(BH) is at 2420 cm^{-1} [16].

References for 6.2.1 to 6.2.3 on pp. 38/9

Similarly, quinuclidine-chloroborane, $C_7H_{13}N–BH_2Cl$, m.p. = 132°C; δ^1H = 1.8 (multiplet), 2.1 (septet), and 3.1 ppm (triplet), ν(BH) = 2400 cm^{-1} (doublet), has been described [16]. N-Methylmorpholine-chloroborane, $C_5H_{11}NO \cdot BH_2Cl$, has a melting point of 56°C [16].

$(C_2H_5)_2O–BH_2Cl$ reacts with 2,3-dimethyl-2-butene to give a complex mixture of thexylborane, thexylchloroborane, and thexyldichloroborane (thexyl = $(CH_3)_2CHC(CH_3)_2$) [17]. Monochloroborane-diethyletherate does not reduce 9-chloro-9-borabicyclo[3.3.1]nonane to the B–H species [20]. However, the reagent reacts with (+)-limonene to give (after subsequent reduction with $Li[AlH_4]$) limonylborane [21].

$(–CH_2–)_4O–BH_2Cl$ reacts with 2,3-dimethyl-2-butene to give the tetrahydrofuran adduct of thexylchloroborane [17]. Reaction with $HSCH_2SH$ yields a compound with adamantane structure [22].

$(CH_3)_2S–BH_2Cl$ is stable at ambient temperature. Reaction with 2,3-dimethyl-2-butene yields the dimethylsulfide adducts of thexylborane [17, 23]. $(CH_3)_2S–BH_2Cl$ does not reduce 9-chloro-9-borabicyclo[3.3.1]nonane [20].

6.2.3 Additional Species

$(CH_3)_2CHC(CH_3)_2BHCl$, thexylchloroborane, can be easily prepared from equimolar amounts of 2,3-dimethyl-2-butene and $(CH_3)_2S–BH_2Cl$ in CH_2Cl_2 at 0°C. The reaction is complete after 1 h stirring at 25°C. NMR data: $\delta^{11}B$ = 6.9 ppm with J(BH) = 128 Hz for the adduct with dimethylsulfide. The dimethylsulfide adduct is used in the regioselective hydroboration of various alkenes [17, 23 to 27]. Reactions of thexylchloroborane with alkynes lead to thexyl(alkenyl)chloroboranes [28]. Treatment of the dimethylsulfide complex with trimethylamine gives $(CH_3)_3N \cdot BH(Cl)C(CH_3)_2CH(CH_3)_2$ [17, 24].

$[H_3BCl]^-$ has been prepared in nearly quantitative yield according to:

$$M[BH_4] + BCl_3 \xrightarrow[\text{ambient temperature}]{CH_2Cl_2} \tfrac{1}{2}B_2H_6 + M[HBCl_3] \quad (M = [N(n\text{-}C_4H_9)_4]^+,\ [CH_3P(C_6H_5)_3]^+),$$

by the abstraction of a hydride ion from $[BH_4]^-$. The solid products are free-flowing materials which are stable at room temperature under N_2; solutions in CH_2Cl_2 are stable for several hours. The ^{11}B NMR spectrum of the $[HBCl_3]^-$ anion consists of a doublet (which gives a single resonance upon proton decoupling) at $\delta^{11}B$ = 3.0 ppm (J(BH) = 158 Hz). In the IR spectrum, ν(BH) is observed at 2480 cm^{-1} [29]. The $[HBCl_3]^-$ ion is also prepared by treating a solution of $[\{(C_6H_5)_3P\}_2N]Cl$ in CH_2Cl_2 with B_2H_6 at −78°C, thus giving $[\{(C_6H_5)_3P\}_2N][H_3BCl]$. The IR spectrum exhibits ν(BH) modes at 2340, 2290, and 2240 cm^{-1}. The ^{11}B NMR spectrum (besides the signal of $[H_3BCl]^-$ at δ = −14.6 ppm, quartet, J(BH) = 134 Hz) contains a triplet at δ = −2.8 ppm (J(BH) = 131 Hz) corresponding to $[H_2BCl_2]^-$, and minor signals for $[B_2H_7]^-$ and $[B_3H_8]^-$. The ^1H NMR spectrum contains the quartet of $[H_3BCl]^-$ at δ = 2.1 ppm, and the triplet for $[H_2BCl_2]^-$ at 3.4 ppm [30].

References for 6.2.1 to 6.2.3:

[1] Frost, D. C.; Kirby, C.; McDowell, C. A.; Westwood, N. P. C. (J. Am. Chem. Soc. **103** [1981] 4428/32).

[2] Pankratov, A. V.; Shmerling, G. V. (Khim. Vysokikh Energ. **17** [1983] 173/5 from C.A. **98** [1983] No. 188925).

[3] Samsonov, Yu. N.; Petrov, A. K. (Khim. Vysokikh Energ. **17** [1983] 170/3 from C.A. **98** [1983] No. 207358).

[4] Sazonov, V. N.; Shmerling, G. V. (Kvantovaya Elektron. [Moscow] **10** [1983] 1043/5 from C.A. **100** [1984] No. 148367).
[5] Pankratov, A. V.; Shmerling, G. V. (Khim. Vysokikh Energ. **16** [1983] 69/72 from C.A. **96** [1982] No. 134852).
[6] Sazonov, V. N.; Shmerling, G. V. (Kvantovaya Elektron. [Moscow] **9** [1982] 370/2 from C.A. **96** [1982] No. 190551).
[7] Volkov, S. V.; Gurko, A. F.; Druzheruchenko, A. G.; Lutoshkin, V. I. (Khim. Vysokikh Energ. **17** [1983] 464/7 from C.A. **99** [1983] No. 184899).
[8] Volkov, S. V.; Gurko, A. F.; Lutoshkin, V. I. (Ukr. Khim. Zh. **48** [1982] 451/3; Soviet Progr. Chem. **48** No. 5 [1982] 1/3 from C.A. **97** [1982] No. 48679).
[9] Adamova, Yu. A.; Akinfiev, N. N.; Boitsov, B. S.; Zubkov, V. I.; Pankratov, A. V.; Skachkov, A. N. (Khim. Vysokikh Energ. **15** [1981] 365/9 from C.A. **95** [1981] No. 106880).
[10] Wagner, W.; Bochmann, G. (Wiss. Z. T.H. Karl-Marx-Stadt **24** [1982] 672/81).

[11] Holbrook, J. B.; Smith, B. C.; Housecroft, C. E.; Wade, K. (Polyhedron **1** [1982] 701/6).
[12] Westwood, N. P. C. (NATO ASI Ser. B **90** [1983] 275/8 from C.A. **99** [1983] No. 45662).
[13] Michaelidis, M.; Pollard, R. (J. Electrochem. Soc. **131** [1984] 860/8).
[14] Bonner, T. G.; Lewis, D.; Rutter, K. (J. Chem. Soc. Perkin Trans. I **1981** 1807/10).
[15] Kuleshova, O. D.; Gorbunov, A. I. (Zh. Analit. Khim. **38** [1983] 2031/5 from C.A. **100** [1984] No. 61150).
[16] Kelly, H. C.; Yasui, S. C.; Twiss-Brooks, A. B. (Inorg. Chem. **23** [1984] 2220/3).
[17] Brown, H. C.; Sikorski, J. A. (Organometallics **1** [1982] 28/37).
[18] Brown, H. C.; Racherla, U. S. (J. Org. Chem. **48** [1983] 1389/91).
[19] Sobieralksi, T. J.; Hancock, K. G. (J. Am. Chem. Soc. **104** [1982] 7533/41).
[20] Brown, H. C.; Kulkarni, S. U. (J. Organometal. Chem. **218** [1981] 299/307).

[21] Jadhav, P. K.; Kulkarni, S. U. (Heterocycles **18** [1982] 169/73).
[22] Binder, H.; Diamantikos, W.; Dermentzis, K.; Hansen, H. D. (Z. Naturforsch. **37b** [1982] 1548/52).
[23] Brown, H. C.; Sikorski, J. A.; Kulkarni, S. U.; Lee, H. D. (J. Org. Chem. **47** [1982] 863/72).
[24] Sikorski, J. A. (Diss. Purdue Univ. 1981, pp. 1/185; Diss. Abstr. Intern. B **42** [1982] 4431).
[25] Brown, H. C.; Basavaiah, D.; Racherla, U. S. (Synthesis **1983** 886/8).
[26] Sikorski, J. A.; Brown, H. C. (J. Org. Chem. **47** [1982] 872/6).
[27] Brown, H. C.; Cha, J. S.; Nazer, B.; Yoon, N. M. (J. Am. Chem. Soc. **106** [1984] 8001/2).
[28] Zweifel, G.; Pearson, N. R. (J. Org. Chem. **46** [1981] 829/30).
[29] Toft, M. A.; Leach, J. B.; Himpsl, F. L.; Shore, S. G. (Inorg. Chem. **21** [1982] 1952/7).
[30] Lawrence, S. H. (Diss. Ohio State Univ. 1984, pp. 1/143; Diss. Abstr. Intern. B **45** [1984] 1773).

6.2.4 (Organyl)chloroboranes

6.2.4.1 (Organyl)dichloroboranes, $RBCl_2$

CH_3BCl_2. The reaction of (excess) liquid BCl_3 with $(CH_3)_3SnCl$ at 20°C gives CH_3BCl_2 and $(CH_3)_2SnCl_2$; $(CH_3)_3SnBr$ can accordingly also be used in the preparation of CH_3BCl_2 [1]. Mixtures of BCl_3, CH_3BCl_2, and $(CH_3)_2SnCl_2$ are obtained from $(CH_3)_4Sn$ or $(CH_3)_3SnCl$ and BCl_3 in CH_2Cl_2, as shown in an NMR study [2]. Analysis of published thermochemical data lead to a B–C single bond enthalpy term of $E(B–C) = 350 \pm 10$ kJ/mol. The standard enthalpy of formation is calculated to be $\Delta H_f^\ominus = -337.4$ kJ/mol (gaseous state), and the standard enthalpy of atomization to be 2510.7 kJ/mol [3].

References for 6.2.4 on pp. 46/8

The microwave spectrum of methyldichloroborane has been reported. The spectrum is complicated by double-halogen quadrupole interaction, the variety of isotopic forms, and by many excited torsional lines arising from the low barrier to internal rotation. The ground state spectrum has been assigned for three isotopic species and the rotational constants are listed in Table 6/16 [4]. The structure of CH_3BCl_2 is depicted in **Fig. 6-3**. The following ^{35}Cl quadrupole coupling constants (in MHz) have been observed: $\chi_{aa} = -21.7$, $\chi_{bb} = 12.5$, $\chi_{cc} = 9.2$ [4].

Table 6/16
Rotational Constants for CH_3BCl_2 [4]. A, B, C in MHz;
Δ in amu·Å².

	CH_3BCl_2	$CH_3{}^{10}BCl_2$	$CH_3B^{35,37}Cl_2$
A	6029.54	6040.69	5996.28
B	3221.10	3220.73	3133.03
C	2096.64	2097.83	2055.02
Δ	0.330	0.329	0.336

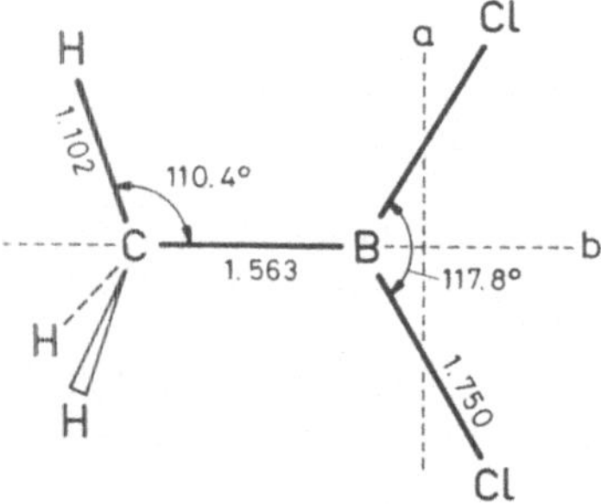

Fig. 6-3. The zero point average structure of methyldichloroborane (r(CB) and r(BCl) ±0.005 Å, ∢(ClBCl) ±0.4°) [4].

$H_2C{=}CHBCl_2$ has been prepared from $(H_2C{=}CH)_4Sn$ and BCl_3 in a vacuum line [5].

$C_2H_5BCl_2$ reacts with octahydroxycyclobutane to yield tetrakis[ethylboranediylbis(oxy)]-cyclobutane [6]. Ethyldichloroborane is widely used as a component of catalysts for the polymerization of ethylene and other olefins [7 to 14], and in the alternating photochemical copolymerization of methacrylate with styrene [15, 16].

$C_6H_5BCl_2$. The preparation according to $(C_6H_5)_4Sn + 4\,BCl_3 \rightarrow SnCl_4 + 4\,C_6H_5BCl_2$ has been described in detail [17]. For the formation of phenyldichloroborane from BCl_3 and $(C_6H_5)_4Sn$, $(C_6H_5)_3SnCl$, $(C_6H_5)_2SnCl_2$, and $C_6H_5SnCl_3$ in CH_2Cl_2 solution as studied by NMR, see [2].

The standard enthalpy of formation of $C_6H_5BCl_2$ (in the gaseous state) has been calculated to be $\Delta H_f^\ominus(g) = -266.0$ kJ/mol. The bond-order dependent enthalpy term E(B–Cl) is given with 462.6 kJ/mol [3].

The reaction of $C_6H_5BCl_2$ with $(CH_3)_3SiN(CH_3)_2$ (1:1 molar ratio) leads to $(CH_3)_2N{-}B(C_6H_5)Cl$ [18]. 1-Phenyl-3-borolene is obtained from $C_6H_5BCl_2$ and $MgC_4H_6 \cdot 2thf$ (butenediyl bis(tetrahydrofuran)magnesium) [19]. 3,5-Dimethyl-2,6-diphenyl-1,3,5-triaza-2,6-diboracyclohexanone-4 is isolated from the reaction of $C_6H_5BCl_2$ with $[CH_3(H)N]_2CO$ [20]. For additional reactions of $C_6H_5BCl_2$ with urea and thiourea derivatives, see [20]. Treating the amino-iminoborane (tmp)B=N(*t*-C_4H_9), (tmp = 2,2,6,6-tetramethylpiperidinyl) with $C_6H_5BCl_2$ yields the spiro compound tmp(-BC_6H_5-)(-BCl_2-)N(*t*-C_4H_9) [21]. With 2-chloro-1,3-dimethyl-1,3,2-diazaborolidine, a 1:1 adduct is formed by coordination of $C_6H_5BCl_2$ to one of the nitrogen atoms [22]. In the reaction with α-, β-unsaturated *o*-aryl azides, a 1,2-shift of the phenyl group is observed and the corresponding N-phenylaminodichloroboranes are obtained after N_2 elimination [23]. With octahydrocyclobutane, the reaction of $C_6H_5BCl_2$ gives tetrakis[phenylboranediylbis(oxyl)]cyclobutane [6]. Cyclic rhena-β-ketoiminato chloro(phenyl)borane has been prepared from the corresponding *cis*-substituted rhenium complex [24]. $C_6H_5BCl_2$ is used as a catalyst in the polymerization of hexachlorocyclotriphosphazene [25] and in *erythro*-selective aldol reactions [26].

$C_6F_5BCl_2$ reacts with mesidine (= 2,4,6-trimethylaniline) or *p*-anisidine (= 4-methoxyaniline) to give, ultimately, compounds such as 2,4,6-$(CH_3)_3C_6H_2$–NHB(Cl)C_6F_5, 2,4,6-$(CH_3)_3$-C_6H_2–NHB(C_6F_5)N[2,4,6-$(CH_3)_3C_6H_2$]B(Cl)C_6F_5, and [2,4,6-$(CH_3)_3C_6H_2NH]_2BC_6F_5$, depending on the duration of the pyrolysis [27].

$Cl_2B[(CH_3)_3Si]C{=}C{=}O$ has been prepared in 67% yield by reacting 5 g of $(CH_3)_3SiC{\equiv}COC_2H_5$ and 4.13 g of BCl_3 in 10 mL of pentane at −50°C. The boiling point is 54 to 56°C/20 Torr. NMR data: $\delta^1H = 0.21$ ppm; $\delta^{13}C = -2.3$ ppm (CH_3Si), 29.0 ppm (⟩C=), and 168.8 ppm (=C=O). In the IR spectrum, ν(C=C=O) appears at 2100 cm^{-1} [78].

***cis*-Cl_2B–C(*t*-C_4H_9)=C(*t*-C_4H_9)BCl_2** is prepared in 83% yield from di-*t*-butylethyne and B_2Cl_4. Colorless crystals, m.p. 99°C; mass spectrum and NMR data are given ($\delta^{11}B = 56.8$ ppm, $\delta^1H = 1.20$ ppm, $\delta^{13}C$ data are given) [43].

A survey of new data on additional (organyl)dichloroboranes is presented in Tables 6/17 and 6/18, p. 42.

Table 6/17
Survey of (Organyl)dichloroboranes, $RBCl_2$.

R	comments	Ref.
ClCH=CH (*trans*)	laser irradiation, isotope selective dissociation	[28 to 32]
n-C_3H_7	reaction with octahydroxycyclobutane	[6]
i-C_3H_7	^1H NMR study from $(t\text{-}C_4H_9)_2BOCH_3 + BCl_3$	[33]
	reaction with $(C_6H_5)_3CCH(CH_2OH)_2$ to give a 1,3,2-dioxaborinane	[34]
	reaction with $[t\text{-}C_4H_9][(CH_3)_3Si]NH$	[35]
	reaction with octahydroxycyclobutane	[6]
cyclo-C_3H_5	IR and Raman spectrum (gas, liquid, solid); complete vibrational assignment; conformation	[36]
n-C_4H_9	$\Delta H_f^\ominus = -393.3$ kJ/mol; thermochemical data	[3]
	reaction with $[CH_3(H)N]_2CO$ and other urea and thiourea compounds	[20]
t-C_4H_9	preparation	[72]
	reaction with $(C_6H_5)_3CCH(CH_2OH)_2$	[34]
cyclo-C_5H_5	MNDO calculation about the 1,5-sigmatropic shift of the BCl_2 group	[79]

References for 6.2.4 on pp. 46/8

Table 6/17 (continued)

R	comments	Ref.
$(CH_3)_2CHC(CH_3)_2$	preparation, b.p. 75°C; NMR ($\delta^{11}B = 65.2$ ppm) reaction with thexylborane (thexyl = $(CH_3)_2CHC(CH_3)_2$)	[41, 42]
$C_6H_5CH_2$	regiospecific addition to 1-alkynes	[37]
$(CH_3)_2SiCl(CH_2)_3$	preparation from silacyclobutane derivative and BCl_3; mass spectrum, NMR (^1H), IR	[38]
$(CH_2{=}CH)_2SiCl(CH_2)_3$	preparation, NMR (^1H), addition reactions	[38]
$CH_3(C_6H_5)SiCl(CH_2)_3$	preparation, NMR (^1H)	[38]
$(C_6H_5)_2SiCl(CH_2)_3$	preparation, NMR (^1H), addition reaction	[38]
$(CH_3)_2SiClCH_2Si(CH_3)_2CH_2$	preparation, mass spectrum, NMR (^1H)	[38]
$(C_6H_5)_2SiClCH_2Si(C_6H_5)_2CH_2$	preparation, NMR (^1H)	[38]
$t\text{-}C_4H_9CH{=}C(CH_2C_6H_5)$	preparation, mass spectrum, NMR (^1H, ^{11}B)	[37]

Table 6/18
Donor Adducts of $RBCl_2$.

R	donor	comments	Ref.
$n\text{-}C_5H_{11}$	$(CH_3)_2S$	reaction to yield $n\text{-}C_5H_{11}(n\text{-}C_6H_{13})BCl$	[39]
$3\text{-}C_6H_{13}$	$(CH_3)_2S$	preparation from *cis*-3-hexene + $HCl_2B{-}SCH_3/BCl_3$; thermal isomerization	[40]
$(CH_3)_2CHC(CH_3)_2$	$(CH_3)_3N$	preparation; NMR: $\delta^{11}B = 14.3$ ppm, $\delta^1H = 2.83$ ppm (CH_3)	[41]

6.2.4.2 (Diorganyl)chloroboranes, R_2BCl

$(CH_3)_2BCl$. For the formation of $(CH_3)_2BCl$ in a 1:1 mixture of $(CH_3)_4Sn$ and BCl_3 in CH_2Cl_2 (NMR study), see [2]. Analysis of published thermochemical data resulted in a single bond enthalpy term E(B–C) = 350 ± 10 kJ/mol. Together with the bond-order dependent term for the B–Cl bond, these data have been used to obtain the following enthalpies (for the gaseous state): Standard enthalpy of formation $\Delta H_f^\ominus = -285.6$ kJ/mol; standard enthalpy of atomization $\Delta H_f^\ominus = 3681.3$ kJ/mol [3].

The reaction of $(CH_3)_2BCl$ with $[(CH_3)_3Si]_2N{-}O[Si(CH_3)_3]$ yields the dimethyl[silanyl(silanyloxy)amino]borane derivative $(CH_3)_2BN[Si(CH_3)_3][OSi(CH_3)_3]$ [44].

$(C_2H_5)_2BCl$ reacts with $[(CH_3)_3Si]_2N{-}O[Si(CH_3)_3]$ to give $(C_2H_5)_2B{-}N[Si(CH_3)_3][OSi(CH_3)_3]$ [44]. With $K[(CO)_6CrCN]$, $(CO)_5CrCNB(C_2H_5)_2$ is obtained [45]. Diethylchloroborane has been tested as alternation regulating catalyst in the photochemical copolymerization of methylmethacrylate with styrene [15], and in catalyst compositions for the polymerization of olefins [73].

$(C_6H_5)_2BCl$. The formation of diphenylchloroborane from BCl_3 and $(C_6H_5)_{4-n}SnCl_n$ (n = 0, 1, 2) in CH_2Cl_2 solution has been studied by ^{11}B NMR spectroscopy [2]. The bond-order dependent bond enthalpy term of the B–Cl bond has been calculated to be E(B–Cl) = 489.7 kJ/mol, and the standard enthalpy of formation is $\Delta H_f^\ominus = -95.3$ kJ/mol [3].

The reaction of $(C_6H_5)_2BCl$ with $(CH_3)_3SiN_3$ gives $(C_6H_5)_2BN_3$ [46], and with $(CH_3)_2NN$-$[B(CH_3)_2]Li$ the species $(CH_3)_2NN[B(C_6H_5)_2][B(CH_3)_2]$ is obtained [47]. Alkynes such as $HC{\equiv}C(n\text{-}C_4H_9)$ and $HC{\equiv}C(t\text{-}C_4H_9)$ are organoborylated upon treatment with $(C_6H_5)_2BCl$ to give, primarily, $C_6H_5(Cl)B{-}C(C_6H_5){=}CH(n\text{-}C_4H_9)$ and $C_6H_5BClC(C_6H_5){=}CH(t\text{-}C_4H_9)$ [37].

Additional (diorganyl)chloroboranes, R_2BCl, are listed in Table 6/19 and unsymmetrical species of the type RR′BCl are compiled in Table 6/20, p. 44.

Table 6/19
Survey of (Diorganyl)chloroboranes, R_2BCl.

compound	comments	Ref.
$(H_2C{=}CH)_2BCl$	reaction with $Ru_3(CO)_{12}$	[77]
$(n\text{-}C_3H_7)_2BCl$	formation from BCl_3 and (2,4-pentanedionato)-dipropylborane	[48]
	reaction with $[(CH_3)_3Si]_2N{-}O[Si(CH_3)_3]$	[44]
	reaction with $(CH_3)_3SiN_3$	[49]
$(i\text{-}C_3H_7)_2BCl$	reaction with $[(CH_3)_3Si]_2N{-}O[Si(CH_3)_3]$	[50]
	reaction with $(CH_3)_3SiN_3$	[49]
$(C_2H_5CH{=}CH)_2BCl$	preparation from $H_2BCl + HC{\equiv}CC_2H_5$	[51, 52]
$[CH_3CH{=}(CH_3)C]_2BCl$	preparation from $H_2BCl + H_3CC{\equiv}CCH_3$	[51, 52]
$(n\text{-}C_4H_9)_2BCl$	thermodynamic data	[3]
	reactions with α-chlorosulfides, thioacetals, and $HC[S(C_2H_5)_3]$	[53]
	reaction with $(C_2H_5OCH_2)_2$	[54]
	reaction with $(CH_3)_3SiN_3$	[49]
$(i\text{-}C_4H_9)_2BCl$	reaction with $(CH_3)_3SiN_3$	[49]
$(s\text{-}C_4H_9)_2BCl$	reaction with $(CH_3)_3SiN_3$	[49]
	reaction with 2-butene	[57, 74]
$(t\text{-}C_4H_9)_2BCl$	reaction with $[(CH_3)_3Si]_3SiLi$	[55]
$(cyclo\text{-}C_5H_9)_2BCl$	reaction with $RCH{=}P(C_6H_5)_3$ ($R = CH_3$, C_2H_5, C_6H_5)	[56]
	reaction with 4-octyne; reactions with terminal alkynes	[57, 74]
$(n\text{-}C_5H_{11})_2BCl$	reaction with $RCH{=}P(C_6H_5)_3$ ($R = CH_3$, C_6H_5)	[56]
$(C_6F_5)_2BCl$	reaction with $(CH_3)_3SiN_3$	[46]
$[C_2H_5CH{=}(C_2H_5)C]_2BCl$	preparation from H_2BCl and $C_2H_5C{\equiv}CC_2H_5$	[51, 52]
$(cyclo\text{-}C_6H_{11})_2BCl$	reaction with $RCH{=}P(C_6H_5)_3$ ($R = C_6H_5$)	[56]
$(n\text{-}C_6H_{13})_2BCl$	reaction with 2-butyne or 1-bromo-1-alkynes; reaction with 1-octyne	[57, 58, 74]
$(s\text{-}C_6H_{13})_2BCl$	reaction with 1-bromo-1-octyne	[58]

References for 6.2.4 on pp. 46/8

Table 6/19 (continued)

compound	comments	Ref.
$(2\text{-}CH_3\text{-}C_6H_4)_2BCl$	reaction with $(CH_3)_3SiN_3$	[46]
$(t\text{-}C_4H_9CH{=}CH)_2BCl$	preparation from H_2BCl and $HC{\equiv}C(t\text{-}C_4H_9)$	[51, 52]
$(n\text{-}C_8H_{17})_2BCl$	reaction with 3-hexyne	[57]
$[\{(CH_3)_3Si\}_2CH]_2BCl$	reaction with electron-rich olefin or pyridine for B-radical formation	[59]
$[(CH_3)_2SiCl(CH_2)_3]_2BCl$	preparation from BCl_3 and silacyclobutane (1:2); mass spectrum, NMR (^1H), IR	[38]
cyclic species		
$(-CH_2-)_5BCl$	additive to aldehyde-carbanion reactions for regio- and stereoselectivity	[60]
	reaction with alkynes	[61]
9-Cl-9-borabicyclo[3.3.1]-nonane	adduct with $(CH_3)_2S$: reaction with hydridic reducing agents	[62]
3-Cl-7-CH_3-3-bora-bicyclo[3.3.1]nonane	preparation from 1-boraadamantane and HCl	[63]
	reaction with $NaNH_2$	[64, 80]
[structure: benzo-fused boracycle with B–Cl and R′]	($R' = H$, $n\text{-}C_4H_9$, $t\text{-}C_4H_9$) preparation; mass spectrum, NMR (^1H) (not for $R' = t\text{-}C_4H_9$)	[37]

Table 6/20
Unsymmetrical (Diorganyl)chloroboranes, RR′BCl.

R	R′	comments	Ref.
$(CH_3)_2CHC(CH_3)_2$	$cyclo\text{-}C_5H_9$	preparation; methanolysis, reaction with olefin and $[HB(O\text{-}i\text{-}C_3H_7)_3]^-$	[64, 76, 80]
$(CH_3)_2CHC(CH_3)_2$	$n\text{-}C_5H_{11}$	reaction with olefins and $[HB(O\text{-}i\text{-}C_3H_7)_3]^-$	[76]
$(CH_3)_2CHC(CH_3)_2$	$n\text{-}C_4H_9CH{=}CH$	preparation; reaction with chloropropargylide	[65]
$(CH_3)_2CHC(CH_3)_2$	$t\text{-}C_4H_9CH{=}CH$	preparation; reaction with chloropropargylide	[65]
$(CH_3)_2CHC(CH_3)_2$	$C_2H_5CH{=}CC_2H_5$	preparation; reaction with chloropropargylide	[65]
$C_6H_5CH_2$	$t\text{-}C_4H_9(Cl)C{=}CH$	preparation; mass spectrum, NMR (^1H, ^{11}B)	[37]

Table 6/20 (continued)

R	R'	comments	Ref.
$(CH_3)_2CHC(CH_3)_2$	n-C_6H_{13}	preparation, reaction with $H_2C{=}CNCH(CH_3)OCOCH_3$	[66]
C_6H_5	$R(C_6H_5)C{=}CH$	R = n-C_4H_9, t-C_4H_9; preparation, oxidation, deuterolysis	[37]
$(CH_3)_2CHC(CH_3)_2$	n-C_8H_{17}	preparation, methanolysis, reaction with olefins and $HB(O$-i-$C_3H_7)_2$	[64, 75, 76, 80]
C_6H_5	$RCH{=}CC_6H_5$	(R = n-C_4H_9, t-C_4H_9) preparation; mass spectrum, NMR (^1H, ^{11}B)	[37]
C_6H_5	n-$C_4H_9[(CH_3)_3Si]$-$C{=}CC_6H_5$	preparation; mass spectrum, NMR (^1H, ^{11}B, ^{13}C)	[37]
CH_3	$(CH_3)_2SiCl$-$CH_2SiCl(CH_3)CH_2$	preparation; mass spectrum, NMR (^1H)	[38]

Species Containing Two BCl Groups

(t-$C_4H_9BCl)_2$ and **(t-$C_4D_9BCl)_2$** have been synthesized by the reaction of (t-$C_4H_9BOCH_3)_2$ and (t-$C_4D_9BOCH_3)_2$ with BCl_3. They were reacted with Na/K alloy to give the $[(-t\text{-}C_4H_9B-)_4]^{\bullet -}$ and $[(-t\text{-}C_4D_9B-)_4]^{\bullet -}$ radical anions [67]. The reaction with alkynes, $RC{\equiv}C$-$Si(CH_3)_3$ (R = CH_3, C_6H_5, CH_2-t-C_4H_9, $Si(CH_3)_3$), leads to $R[(CH_3)_3Si]C{=}C[B(t\text{-}C_4H_9)Cl]_2$ (mass spectrum, NMR (^1H, $\delta^{11}B \approx$ 75 ppm; ^{13}C)) by 1,1-diboronation [68]. With alkynes, $R'C{\equiv}CR''$ (R' = R'' = H, CH_3, C_2H_5; R' = H; R'' = $Si(CH_3)_3$), 1,2-diboronation occurs and compounds of the type $[Cl(t\text{-}C_4H_9)B]C(R'){=}C(R'')[B(t\text{-}C_4H_9)Cl]$ (mass spectrum, NMR (^1H, ^{11}B, ^{13}C)) are obtained [68]. The reaction of $[(CH_3)_3Si]_2C{=}C[B(t\text{-}C_4H_9)Cl]_2$ with Na/K alloy in pentane leads to the borirane species **1** with an exocyclic B=C double bond [69, 70]. On the other hand, the reaction of (t-$C_4H_9BCl)_2$ with $RC{\equiv}CSn(CH_3)_3$ (R = t-C_4H_9, $Sn(CH_3)_3$) leads to the borirene compounds **2** [71]. Treatment of the derivatives $[Cl(t\text{-}C_4H_9)B]C(R){=}C(R)[B(t\text{-}C_4H_9)Cl]$ (R = CH_3, C_2H_5) gives 1,3-dihydro-1,3-diboretes **3** [70].

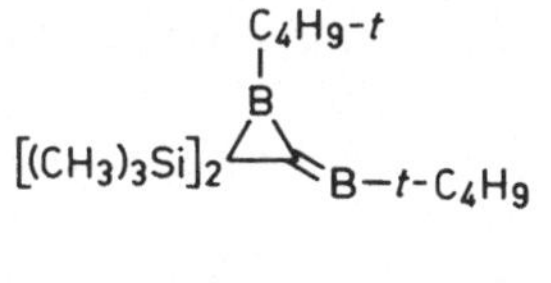

C_4H_9-t | B | $[(CH_3)_3Si]_2$ | B-t-C_4H_9

1

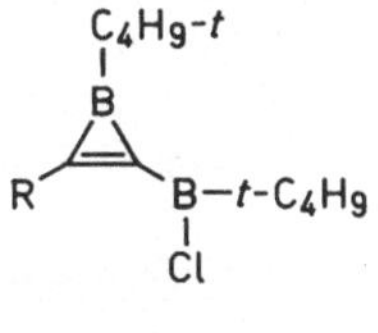

C_4H_9-t | B | R | B-t-C_4H_9 | Cl

2

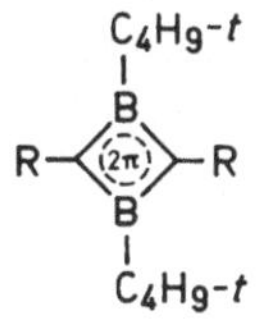

C_4H_9-t | B | R | (2π) | R | B | C_4H_9-t

3

References for 6.2.4 on pp. 46/8

References for 6.2.4:

[1] Daneshvar, N.; Straughan, B. P.; Gardiner, D. J. (J. Inorg. Nucl. Chem. **43** [1981] 1147/9).
[2] Dillon, K. B.; Hewitson, G. F. (Polyhedron **3** [1984] 957/62).
[3] Holbrook, J. B.; Smith, B. C.; Housecroft, C. E.; Wade, K. (Polyhedron **1** [1982] 701/6).
[4] Cox, A. P. (J. Mol. Struct. **97** [1983] 61/76).
[5] Kranc, D. M. (Diss. Memphis State Univ. 1983, pp. 1/128; Diss. Abstr. Intern. B **44** [1984] 2419).
[6] Yalpani, M.; Köster, R.; Wilke, G. (Chem. Ber. **116** [1983] 1336/44).
[7] Karol, F. J.; Goeke, G. L.; Wagner, B. E.; George, K. F. (Eur. Appl. 83-456 [1983] from C.A. **99** [1983] No. 105895).
[8] Asahi Chemical Industry Co., Ltd. (Japan. Kokai Tokkyo Koho 82-05709 [1982] from C.A. **96** [1982] No. 200348).
[9] Asahi Chemical Industry Co., Ltd. (Japan. Kokai Tokkyo Koho 84-56408 [1984] from C.A. **101** [1984] No. 152507).
[10] Speca, A. N. (Brit. Appl. 2087907 [1982] from C.A. **97** [1982] No. 128284).

[11] Gessell, D. E. (Eur. Appl. 105727 [1984] from C.A. **101** [1984] No. 55652).
[12] McDaniel, M. P.; Johnson, M. M. (Eur. Appl. 55863 [1982] from C.A. **101** [1984] No. 38990).
[13] Pullukat, T. J.; Hoff, R. E.; Lynch, M. W. (Ger. Offen. 3328061 [1984] from C.A. **101** [1984] No. 131286).
[14] Asahi Chemical Industry Co., Ltd. (Japan. Kokai Tokkyo Koho 84-58011 [1984] from C.A. **101** [1984] No. 91641).
[15] Hirai, H.; Takeuchi, K.; Komiyama, M. (J. Polym. Sci. Polym. Chem. Ed. **19** [1981] 2581/94 from C.A. **95** [1981] No. 187761).
[16] Hirai, H. (Proc. 28th IUPAC Intern. Union Pure Appl. Chem. Macromol. Symp., Amherst, Ma., 1982, p. 94 from C.A. **99** [1983] No. 176516).
[17] Gerwarth, U. W.; Weber, W. (Inorg. Syn. **22** [1983] 207/8).
[18] Beswick, Y. F.; Wisian-Neilson, P.; Neilson, R. H. (J. Inorg. Nucl. Chem. **43** [1981] 2639/43).
[19] Herberich, G. E.; Hessner, B.; Söhnen, D. (J. Organometal. Chem. **233** [1982] C35/C37).
[20] Maringgele, W. (J. Organometal. Chem. **222** [1981] 17/32).

[21] Nöth, H.; Weber, S. (Angew. Chem. **96** [1984] 998; Angew. Chem. Intern. Ed. Engl. **23** [1984] 994).
[22] Anton, K.; Nöth, H.; Pommerening, H. (Chem. Ber. **117** [1984] 2479/94).
[23] Leardini, R.; Zanirato, P. (J. Chem. Soc. Chem. Commun. **1983** 396/7).
[24] Lukehart, C. M.; Raja, M. (Inorg. Chem. **21** [1982] 2100/1).
[25] Fieldhouse, J. W.; Graves, D. F. (ACS Symp. Ser. No. 171 [1981] 315/20 from C.A. **96** [1982] No. 52730).
[26] Hamana, H.; Sasakura, K.; Sugasawa, T. (Chem. Letters **1984** 1729/32).
[27] Morse, J. G.; Glanville, W. K. (Inorg. Chem. **23** [1984] 11/3).
[28] Kuz'menko, V. A. (Zh. Fiz. Khim. **58** [1984] 727/9; Russ. J. Phys. Chem. **58** [1984] 436/8).
[29] Abdushelishvili, G. I.; Abzianidze, T. G.; Egiazarov, A. S.; Tkeshelashvili, G. I.; Tsinadze, T. B. (At. Energiya SSSR **57** [1984] 203/4 from C.A. **102** [1985] No. 36524).
[30] Jensen, R. J.; Thorne, J. M.; Cluff, C. L. (U.S. Appl. 228036 [1984] from C.A. **100** [1984] No. 93280).

[31] Jensen, R. J.; Hayes, J. K.; Cluff, C. L.; Thorne, J. M. (IEEE J. Quantum Electron. **16** [1980] 1352/6 from C.A. **94** [1981] No. 74608).

[32] Abzianidze, T. G.; Abdushelishvili, G. I.; Bakhtadze, A. B.; Gverdtsiteli, I. G.; Kaminskii, A. V.; Kudziev, A. G.; Tkeshelashvili, G. I.; Tsinadze, T. B. (Pis'ma Zh. Tekhn. Fiz. **8** [1982] 1234/6 from C.A. **98** [1983] No. 42561).
[33] Kuznetsov, V. V.; Zakharov, K. S.; Gren, A. I. (Teor. Eksperim. Khim. **20** [1984] 742/4; Theor. Exptl. Chem. [USSR] **20** [1984] 700/2).
[34] Bogatskii, A. V.; Gren, A. I.; Kuznetsov, V. V.; Bogatskaya, Z. D.; Egorova, S. P.; Vil'chinskaya, N. I.; Luzhetskaya, T. V.; Shakib, B. (Khim. Prom. Ser. Reakt. Osobo Chist. Veshchestva **1980** No. 2, pp. 26/8 from C.A. **94** [1981] No. 121624).
[35] Paetzold, P. I.; von Plotho, C.; Schmid, G.; Boese, R. (Z. Naturforsch. **39b** [1984] 1069/75).
[36] Durig, J. R.; Trowell, P. L.; Szafran, Z.; Johnston, S.A.; Odom, J. D. (J. Mol. Struct. **74** [1981] 85/95).
[37] Binnewirtz, R. J.; Klingenberger, H.; Welte, R.; Paetzold, P. (Chem. Ber. **116** [1983] 1271/84).
[38] Auner, N.; Grobe, J. (Z. Anorg. Allgem. Chem. **500** [1983] 132/60).
[39] Kulkarni, S. U.; Basavaiah, D.; Zaidlewicz, M.; Brown, H. C. (Organometallics **1** [1982] 212/4).
[40] Brown, H. C.; Racherla, U. S. (J. Org. Chem. **48** [1983] 1389/91).

[41] Brown, H. C.; Sikorski, J. A. (Organometallics **1** [1982] 28/37).
[42] Brown, H. C.; Sikorski, J. A.; Kulkarni, S. U.; Lee, H. D. (J. Org. Chem. **47** [1982] 863/72).
[43] Hildenbrand, M.; Pritzkow, H.; Zenneck, U.; Sieber, W. (Angew. Chem. **96** [1984] 371/2; Angew. Chem. Intern. Ed. Engl. **23** [1984] 371).
[44] Paetzold, P.; von Bennigsen-Mackiewicz, T. (Chem. Ber. **114** [1981] 298/305).
[45] Hoefler, M.; Loewenich, H. (J. Organometal. Chem. **226** [1983] 229/37).
[46] Paetzold, P.; Truppat, R. (Chem. Ber. **116** [1983] 1531/9).
[47] Fußstetter, H.; Nöth, H. (Liebigs Ann. Chem. **1981** 633/41).
[48] Costes, J. P.; Cros, G.; Laurent, J.-P. (Syn. React. Inorg. Metal-Org. Chem. **11** [1981] 383/92).
[49] Meier, H.-U.; Paetzold, P.; Schröder, E. (Chem. Ber. **117** [1984] 1954/64).
[50] Paetzold, P.; von Plotho, C.; Schwan, H.; Meier, H.-U. (Z. Naturforsch. **39b** [1984] 610/4).

[51] Sobieralski, T. J.; Hancock, K. G. (J. Am. Chem. Soc. **104** [1982] 7533/41).
[52] Sobieralski, T. J. (Diss. Univ. California, Davis 1980, pp. 1/226; Diss. Abstr. Intern. B **41** [1981] 4527).
[53] Bagdasaryan, G. B.; Badalyan, K. S.; Indzhikyan, M. G. (Arm. Khim. Zh. **37** [1984] 175/81).
[54] Bagdasaryan, G. B.; Airyan, L. S.; Indzhikyan, M. G. (Arm. Khim. Zh. **33** [1980] 744/7).
[55] Biffar, W.; Nöth, H. (Z. Naturforsch. **36b** [1981] 1509/15).
[56] Bestmann, H. J.; Arenz, T. (Angew. Chem. **96** [1984] 363/4; Angew. Chem. Intern. Ed. Engl. **23** [1984] 381).
[57] Brown, H. C.; Basavaiah, D.; Kulkarni, S. U. (J. Org. Chem. **47** [1982] 171/3).
[58] Brown, H. C.; Basavaiah, D. (J. Org. Chem. **47** [1982] 754/6).
[59] Carty, A. J.; Gynane, M. J. S.; Lappert, M. F.; Miles, S. J.; Singh, A.; Taylor, N. J. (Inorg. Chem. **19** [1980] 3637/41).
[60] Yamamoto, Y.; Saito, Y.; Maruyama, K. (J. Chem. Soc. Chem. Commun. **1982** 1326/8).

[61] Basavaiah, D.; Brown, H. C. (J. Org. Chem. **47** [1982] 1792/3).
[62] Brown, H. C.; Kulkarni, S. U. (J. Organometal. Chem. **218** [1981] 299/307).
[63] Lagutin, N. A.; Mitin, N. I.; Zubairov, M. M.; Arkhipova, T. N.; Petracheva, T. K.; Mikhailov, B. M.; Smirnov, V. N.; Baryshnikova, T. K.; Govorov, N. N. (Khim. Farm. Zh. **17** [1983] 1077/80).
[64] Gurskii, M. E.; Mikhailov, B. M. (Izv. Akad. Nauk SSSR Ser. Khim. **1981** 394/8; Bull. Acad. Sci. USSR Div. Chem. Sci. **30** [1981] 312/5).

[65] Zweifel, G.; Pearson, N. R. (J. Org. Chem. **46** [1981] 829/30).
[66] Brown, H. C.; Basavaiah, D.; Racherla, V. S. (Synthesis **1983** 886/8).
[67] Klusik, H.; Berndt, A. (J. Organometal. Chem. **234** [1982] C17/C19).
[68] Klusik, H.; Pues, C.; Berndt, A. (Z. Naturforsch. **39b** [1984] 1042/5).
[69] Wehrmann, R.; Pues, C.; Klusik, H.; Berndt, A. (Angew. Chem. **96** [1984] 372/4; Angew. Chem. Intern. Ed. Engl. **23** [1984] 372).
[70] Klusik, H.; Berndt, A. (Angew. Chem. **95** [1983] 895/6; Angew. Chem. Intern. Ed. Engl. **22** [1983] 877).

[71] Pues, C.; Berndt, A. (Angew. Chem. **96** [1984] 306/7; Angew. Chem. Intern. Ed. Engl. **23** [1984] 313).
[72] Klusik, H.; Berndt, A. (J. Organometal. Chem. **232** [1982] C21/C23).
[73] Berge, C.; Mack, M. P. (U.S. 4458027 [1984] from C.A. **101** [1984] No. 152522).
[74] Kulkarni, S. U.; Basavaiah, D.; Brown, H. C. (J. Organometal. Chem. **225** [1981] C1/C5).
[75] Brown, H. C.; Basavaiah, D.; Bhat, N. G. (Organometallics **2** [1983] 1468/70).
[76] Brown, H. C.; Cha, J. S.; Nazer, B.; Kim, S. C.; Krishnamurthy, S. (J. Org. Chem. **49** [1984] 885/92).
[77] Herberich, G. E.; Pampaloni, G. (J. Organometal. Chem. **240** [1982] 121/7).
[78] Ponomarev, S. V.; Nikolaeva, S. N.; Molchanova, G. N.; Kostyuk, A. S.; Grishin, Yu. K. (Zh. Obshch. Khim. **54** [1984] 1817/21; J. Gen. Chem. [USSR] **54** [1984] 1620/5).
[79] Schoeller, W. W. (J. Chem. Soc. Dalton Trans. **1984** 2233/5).
[80] Brown, H. C.; Sikroski, J. A.; Kulkarni, S. U.; Lee, H. D. (J. Org. Chem. **47** [1982] 863/72).

6.3 With Oxygen

ClBO. The unstable ClBO molecule has been generated by a 60-Hz discharge in a BCl_3/O_2 gas mixture. The pure rotational spectra of eight isotopic species have been recorded by microwave spectroscopy: $^{35}Cl^{11}B^{16}O$, $^{35}Cl^{11}B^{18}O$, $^{35}Cl^{10}B^{16}O$, $^{35}Cl^{10}B^{18}O$, $^{37}Cl^{11}B^{16}O$, $^{37}Cl^{11}B^{18}O$, $^{37}Cl^{10}B^{16}O$, and $^{37}Cl^{10}B^{18}O$. The ν_3 (B–Cl) stretching band with its vibration-rotational structure has been observed by infrared diode laser spectroscopy. From the combined data, the molecular constants were obtained for the eight species in the ground state, for $^{35}Cl^{11}B^{16}O$ and $^{37}Cl^{11}B^{16}O$ in the $\nu_2=1$ state, and for $^{35}Cl^{11}B^{16}O$, $^{37}Cl^{11}B^{16}O$, and $^{35}Cl^{10}B^{16}O$ in the $\nu_3=1$ state. The ν_3 band origin was determined at 676.0368, 668.5912, and 680.7642 cm^{-1} in the last cited isotopic species. The observed ground state rotational constants have been used to calculate the bond lengths r_s(B–Cl) with 1.68274(19) Å and r_s(B–O) = 1.20622(21) Å (range of deviation in parentheses) [1]. The ground state constants of ClBO and its molecular constants in the (010) and (001) states are listed in Tables 6/21 and 6/22, $(eQq)_0(^{35}Cl)$ in $^{35}Cl^{11}B^{16}O$ is −47.7(15) [1].

Table 6/21

Molecular Constants of ClBO in the Ground State (in MHz; values in parentheses denote three standard deviations and apply to the last digits; [a)] H_0 was included in the fit ($-0.68(45) \times 10^{-8}$ MHz)) [1].

isotopic species	B_0	$D_0\times10^3$	isotopic species	B_0	$D_0\times10^3$
$^{35}Cl^{11}B^{16}O$ [a)]	5202.3960(24)	1.2960(75)	$^{37}Cl^{11}B^{16}O$	5091.7410(22)	1.2412(60)
$^{35}Cl^{11}B^{18}O$	4871.2315(21)	1.1246(48)	$^{37}Cl^{11}B^{18}O$	4763.1850(17)	1.0757(39)
$^{35}Cl^{10}B^{16}O$	5224.5791(51)	1.304(14)	$^{37}Cl^{10}B^{16}O$	5115.2149(51)	1.252(14)
$^{35}Cl^{10}B^{18}O$	4887.2858(14)	1.1244(33)	$^{37}Cl^{10}B^{18}O$	4780.3511(30)	1.0833(69)

Table 6/22

Molecular Constants of ClBO in the (010) and (001) States (in MHz; values in parentheses denote three standard deviations and apply to the last digits; [a] obtained from diode laser data; [b] ν_3 band origin) [1].

isotopic species	state	B_v	$D_v \times 10^3$	$H_v \times 10^8$	ν_0 in cm^{-1} [b]
$^{35}Cl^{11}B^{16}O$	$(010)^+$	5220.0624(78)	1.334(21)		
	$(010)^-$	5214.4857(78)	1.315(21)		
	(001)	5189.4573(30)	1.3427(9)	−0.88(47)	676.0368(3)
$^{37}Cl^{11}B^{16}O$	$(010)^+$	5108.9751(72)	1.281(20)		
	$(010)^-$	5103.6225(72)	1.266(20)		
	(001) [a]	5078.88(14)	1.291(48)		668.5912(27)
$^{35}Cl^{10}B^{16}O$	(001) [a]	5211.10(11)	1.314(30)		680.7642(30)

The B–Cl bond length r(BCl) = 1.6827 Å of ClBO is very similar to that of ClBS (1.681 Å), while the nuclear quadrupole coupling constants differ considerably with eQq = −47.7 MHz for ClBO but −42.54 MHz for ClBS. This observation can be interpreted in terms of the π-character (π_c), the ionic character of the σ-bond (i_σ), and the total ionic character, which are evaluated with 0.129, 0.5, and 0.371 in ClBO for the B–Cl bond, and with 0.225, 0.5, and 0.275 for that in ClBS. The B–Cl bond length in ClBO is shorter than in the diatomic molecule BCl (r = 1.7153 Å), while r(BO) is about the same as in the BO molecule (r = 1.2045 Å). For additional details, see [1].

Thermodynamic data of ClBO have been calculated by an MNDO treatment and are as follows: enthalpy of formation $\Delta H_f = -92.4$ kcal/mol; ionization potential IP = 13.35 eV, dipole moment $\mu = 1.54$ D; point group $C_{\infty v}$ [2]. Thermodynamic data of gaseous ClBO and $(-BCl-O-)_3$ have been obtained by a combination of tungsten transport measurements and theoretical estimation. The following enthalpies, entropies, and heat capacities have been evaluated [30]: ClBO: $\Delta H_{298} = -339$ kJ/mol; $S_{298} = 237$ $J \cdot mol^{-1} \cdot K^{-1}$; $C_{p,298} = 46$ $J \cdot mol^{-1} \cdot K^{-1}$. $(-BCl-O-)_3$: $\Delta H = -1632$ kJ/mol; $S_{298} = 382$ $J \cdot mol^{-1} \cdot K^{-1}$; $C_{p,298} = 134$ $J \cdot mol^{-1} \cdot K^{-1}$.

For a computation of equilibrium states in systems B/Cl/H/O between 800 and 1400 K and the calculated content of ClBO, see [3]. The vertical ionization potential of ClBO has been computed by applying Rayleigh-Schrödinger perturbation corrections to Koopmans's theorem (PCKT). The results are compiled in Table 6/23 [4]. The decomposition potential of ClBO has been calculated (in a thermodynamic study) to be about 1.13 to 1.17 eV (between 900 to 1300 K) [5].

Table 6/23

Calculated Vertical Ionization Potentials (VIP's) in eV for ClBO (KT = zeroth order, ΔE(3) = third order, GA = geometric approximation; [a] PCKT fails to converge for this cationic state) [4].

orbital	KT (4-31G)	ΔE(3)	$\Delta(E^{GA})$	$(\Delta E)^{GA}$	scaled	average GA
3π	13.58	12.79	12.58	12.35	12.85	12.47
2π	15.77	14.83	14.68	14.53	—	14.61
9σ	17.04	16.69	15.46	14.79	15.86	15.13
8σ	18.19	17.09	16.92	16.76	—	16.84
7σ	32.41			[a]		

References for 6.3 on pp. 52/3

Cl_2BOH has been observed in the reaction of BCl_3 with OH radicals in a fast flow discharge teflon reactor (coupled to an electron spin resonance and a mass spectrometer). For the reaction $BCl_3 + OH^{\bullet} \rightarrow Cl_2BOH + Cl$, a constant of $k_{298} = (4.6 \pm 0.7) \times 10^{-12}$ ($cm^3 \cdot molecule^{-1} \cdot s^{-1}$) has been obtained [6].

$B(ClO_4)_3$ can be prepared according to $BCl_3 + 3HClO_4 \rightarrow B(ClO_4)_3 + 3HCl$, at −78°C as a colorless, crystalline solid which decomposes slowly at 20°C. However, attempted preparation at −30°C leads to a violent decomposition. A low-temperature vibrational study shows that, due to the bonds between the ClO_4 groups and the boron atom with a BO_3 central arrangement, the symmetry is reduced from T_d to C_{3h}, and for that point group 42 vibrations (8A′ + 6A″ + 9E′ + 5E″; IR: A″, E′, Raman: A′, E′, E″) are to be expected for the molecule. The vibrational spectrum including assignments is listed in Table 6/24 [7]. For a detailed discussion, see [7].

Table 6/24
The Vibrational Spectrum of $B(ClO_4)_3$ (in cm^{-1}; vs = very strong, s = strong, m = medium, w = weak, br = broad, sh = shoulder; ClO_t: free Cl–O bonds; ClO_p: Cl–O bonded to boron atoms) [7].

IR	Raman		assignment
1360w	1321w	E′	$\nu_d BO_3$
1295s		A″	
	1271s	A′	$\nu_{as} ClO_t$
1265s		E′	
	1056vs	A′	$\nu_s ClO_t$
1019 to 995s	1037m	E′	
	950?		
860w	852w		$BCl(ClO_4)_2$?
740m			
	714m	A′	$\nu_s ClO_p$
690s	704m	E′	
675sh		A′	πBO_3

IR	Raman		assignment
	670s	A′	$\nu_s BO_3$
620s	625w	E′	
	584m	A′	$\delta_{as} ClO_3$
	573m	E″	
565m		A″	
515m			
	503vs	A′	$\delta_s ClO_3$
442m	455m	E′	
410sh	417m	E′	$\delta_d BO_3$
308sh	308m	E′	
268sh		A″	$\varrho_r ClO_3$
	250m	A′ + E″	
	165	A′	$\delta BOCl$

$Cl_2B(ClO_4)$ (a colorless liquid at −78°C forms a glassy mass at −180°C) and **$ClB(ClO_4)_2$** (colorless solid at −78°C which decomposes to give $Cl_2B(ClO_4)$ and $B(ClO_4)_3$ at room temperature) are prepared from stoichiometric amounts of BCl_3 and pure anhydrous $HClO_4$ at −78°C. $Cl_2B(ClO_4)$ has a maximal symmetry C_s and $ClB(ClO_4)_2$ a maximal symmetry of C_{2v}, which is, however, reduced to C_s (by steric reasons). Vibrational spectra and assignments for both species are compiled in Table 6/25 [7].

Cl_2BOCH_3 is a component of catalysts used in the polymerization of olefins [8].

$ClB(OCH_3)_2$. The reaction of $ClB(OCH_3)_2$ with CH_3Li leads mainly to $CH_3B(OCH_3)_2$ and $(CH_3)_3B$ [9].

$ClB(C_6H_5)(OCH_3)$ is used in the preparation of $R_2NB(C_6H_5)(OCH_3)$ [10].

$ClB[–O–(CH_2)_2–O–]$, prepared from BCl_3 and ethylene glycol, reacts with carbonyl compounds to give enol boronates [11].

Table 6/25
Vibrational Spectra and Assignments for $Cl_2B(ClO_4)$ and $ClB(ClO_4)_2$ (in cm^{-1}; vs = very strong, s = strong, m = medium, w = weak, sh = shoulder; ClO_t: terminal Cl–O bonds; ClO_p: Cl–O bonded to boron atoms) [7].

$Cl_2B(ClO_4)$				$ClB(ClO_4)_2$	
IR	Raman		assignment	Raman	assignment
1295 s			700 + 595	1298 w	$\nu_{as}BO$
1270 s	1283 m, br	A″	$\nu_{as}ClO_t$	1172 w	$\nu_{as}ClO_t$
1210 m			625 + 582	1057 m	$B(ClO_4)_3$
1178 s	1173 m	A′	$\nu_{as}ClO_t$	1043 vs	ν_sClO_t
1040 sh	1043 m	A′	$\nu_sClO_t + BCl(ClO_4)_2$ (?)	1030 w	$BCl_2(ClO_4)$
1028 s	1026 s	A′	νBO	831 m	νBCl
1003 m	1010 sh	A″	$\nu_{as}BCl$	821 w	νBCl
970 w			BCl_3	746 w	?
830 w	830 w		$BCl(ClO_4)_2$	715 m	νClO_p
700 s	700 s	A′	νClO_p	700 m	$BCl_2(ClO_4)$
660 m	662 w		$BCl(ClO_4)_2$	669 s	ν_sBO
625 s	625 s	A′		643 m	$\delta_{as}ClO_3$
590 s	595 s	A′	$\delta_{as}ClO_3$	626 m	$BCl_2(ClO_4)$
582 s	585 m	A″		597 m	$BCl_2(ClO_4)$
522 s	527 s	A′	ν_sBCl	587 s }	$\delta_{as}ClO_3$
492 m	475 m	A′	δ_sClO_3	571 m }	
	462 m		$BCl(ClO_4)_2$	526 s	$BCl_2(ClO_4)$
	413 w		$BCl(ClO_4)_2$	505 m	$B(ClO_4)_3$
398 s	396 m	A′	πCl_2BO	467 s	δ_sClO_3
378 s	378 vs	A′	ϱClO_3	415 m	πO_2BCl
337 s	337 m	A″	ϱClO_3	398 w	—
273 s	273 s	A′	$\delta ClBO$	398 w	—
	257 m		BCl_3	391 w	—
	225 s	A′	δCl_2B	381 m	$BCl_2(ClO_4)$
	150 s	A′	$\delta BOCl$	341 s	ϱClO_3
				312 m	—
				273 m	$BCl_2(ClO_4)$
				264 s	ϱClO_3
				253 m	$B(ClO_4)_3$
				234 s	$\delta OBCl$
				225 sh	—
				203 w	—
				167 w	—
				161 w	$\delta BOCl$
				158 w	—

$ClB(OC_2H_5)_2$ is a component of catalysts for the polymerization of olefins [12, 13].

 References for 6.3 on pp. 52/3

$ClB[C(Cl)(s\text{-}C_4H_9)(n\text{-}C_5H_{11})]OC_2H_5$ is an intermediate in the synthesis of (R)-(−)-3-methyl-4-nonanone [14].

$ClB[\text{–}O\text{–}CH_2\text{–}CH(CH_3)\text{–}O\text{–}]$ reacts with $(CH_3)_2SO[NSi(CH_3)_3]$ to give the corresponding sulfoimine by cleavage of the N–Si bond [15].

$ClB[\text{–}O\text{–}(CH_2)_3\text{–}O\text{–}]$ has been condensed with 1-bromoadamantane (by treatment with sodium) for the preparation of 2-(1-adamantyl)-1,3,2-dioxaborinane [16].

$ClB(1,2\text{-}O)_2C_6H_4$. NMR data: $\delta^{11}B = 29.0$ ppm; ^{13}C NMR data are also available [17]; $\delta^{17}O = 150.5$ ppm (in C_6D_6 relative to external H_2O) [18]. The compound reacts with $(CH_3)_3SiSLi$ to give the corresponding trimethylsilanylthioborane [19].

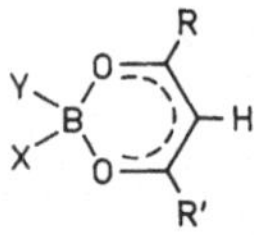

X = Cl; Y = Cl, $n\text{-}C_3H_7$; R = R′ = CH_3, $t\text{-}C_4H_9$ [20]
X = ClO_4, SO_3CF_3; Y = C_6H_5; R = C_6H_5; R′ = CH_3, C_6H_5 [21]
X = Y = Cl; R = R′ = $(CH_3)_2N$ [22]

4

Derivatives of **4** and salts thereof, $[B(acac)_2]Z$ (Z = HCl_2^-, $[BCl_4]^-$), have been synthesized and characterized by mass spectral and NMR (^1H, ^{11}B, ^{13}C) data [20]. Additional compounds based on the skeleton of **4** (R = C_6H_5, R′ = CH_3, C_6H_5; X = ClO_4, SO_3CF_3, Y = C_6H_5) have been prepared and shown to contain four-coordinate boron rather than to be 1,3,2-dioxaborinium(1+) cations. For NMR (^1H, ^{11}B, ^{19}F) data, see [21].

By the reaction of BCl_3 with N,N,N′,N′-tetramethylmalonic diamide, a species of type **4** with X = Y = Cl and R = R′ = $(CH_3)_2N$ has been obtained and was characterized by NMR (^1H, ^{13}C) and IR data [22].

$H_2O\text{–}BCl_3$ in polar solvents (likely to contain solvolysis products) serves as initiator in the preparation of chlorooligo(butylenes) [23], in the polymerization of α-methylstyrene [24], indene [25], and β-pinene [26] to give Cl-terminated polymers.

For the low-temperature IR laser chemical transformation of the epichlorohydrin-BCl_3 system (at 77 K), see [27]. Adducts have been described between Cl_3PO and BCl_3 [28], and between carboxylic esters, RCOOR′ (R = CH_3 R′ = CH_3, C_6H_5, 4-CH_3–C_6H_4, 4-F–C_6H_4; R′ = CH_3, R = $C_6H_5CH_2$, C_6H_5, 4-F–C_6H_4), and BCl_3 and redistribution reactions between different BX_3 (X = F, Cl, Br) species have been studied by NMR (^1H, ^{19}F) [29].

References for 6.3:

[1] Kawaguchi, K.; Endo, Y.; Hirota, E. (J. Mol. Spectrosc. **93** [1982] 381/8).
[2] Dewar, M. J. S.; Rzepa, H. S. (J. Computat. Chem. **4** [1983] 158/69).
[3] Wagner, W.; Bochmann, G. (Wiss. Z. T.H. Karl-Marx-Stadt **24** [1982] 672/81).
[4] Frost, D. C.; Kirby, C.; Lau, W. M.; McDowell, C. A.; Westwood, N. P. C. (J. Mol. Struct. **100** [1983] 87/94).
[5] Shapoval, V. I.; Kushkhov, Kh. B.; Novoselova, I. A. (Ukr. Khim. Zh. **48** [1982] 738/42; Soviet Progr. Chem. **48** No. 7 [1982] 70/4).
[6] Jourdain, J.-L.; Laverdet, G.; Le Bras, G.; Combourieu, J. (J. Chim. Phys. **78** [1981] 253/7).
[7] Chausse, T.; Pascal, J. L.; Potier, A.; Potier, J. (Nouv. J. Chim. **5** [1981] 261/9).
[8] Sakurai, H.; Katayama, Y.; Ikegami, T.; Furusato, M. (U.S. 4471066 [1984] from C.A. **101** [1984] No. 192669).

[9] Brown, H. C.; Cole, T. E. (Organometallics **2** [1983] 1316/9).
[10] Cragg, R. H.; Miller, T. J. (J. Organometal. Chem. **235** [1982] 135/41).

[11] Gennari, C.; Colombo, L.; Poli, G. (Tetrahedron Letters **25** [1984] 2279/82).
[12] Mack, M. P.; Berge, C. T. (Ger. Offen. 3409754 [1984] from C.A. **102** [1985] No. 7265).
[13] Matsuura, K.; Shikatani, Y.; Kamiishi, H.; Kuroda, N.; Miyoshi, M. (Fr. Demande 2543556 [1984] from C.A. **102** [1985] No. 167330).
[14] Brown, H. C.; Jadhav, P. K.; Desai, M. C. (Tetrahedron **40** [1984] 1325/32).
[15] Haubold, W.; Fehlinger, H. G.; Frey, G. (Z. Naturforsch. **36b** [1981] 157/60).
[16] Shegoleva, T. L.; Shaskova, E. M.; Mikhailov, B. M. (Izv. Akad. Nauk SSSR Ser. Khim. **1981** 1098/104; Bull. Acad. Sci. USSR Div. Chem. Sci. **30** [1981] 853/63).
[17] Goetze, R.; Nöth, H.; Pommerening, H.; Sedlak, D.; Wrackmeyer, B. (Chem. Ber. **114** [1981] 1884/93).
[18] Wrackmeyer, B.; Köster, R. (Chem. Ber. **115** [1982] 2022/34).
[19] Hennemuth, K.; Meller, A.; Wojnowska, M. (Z. Anorg. Allgem. Chem. **489** [1982] 47/54).
[20] Costes, J.-P.; Cros, G.; Laurent, J.-P. (Syn. React. Inorg. Metal-Org. Chem. **11** [1981] 383/92).

[21] Narula, C. K.; Nöth, H. (Z. Naturforsch. **38b** [1983] 1161/4).
[22] Hartke, K.; Krug, B.; Hoffmann, R. (Liebigs Ann. Chem. **1984** 370/80).
[23] Tessier, M.; Nguyen, A. H.; Marechal, E. (Polym. Bull. [Berlin] **4** [1981] 111/8 from C.A. **94** [1981] No. 140210).
[24] Kennedy, J. P.; Chou, R. T. (J. Macromol. Sci. Chem. A **18** [1982] 47/76 from C.A. **97** [1982] No. 72828).
[25] Puskas, J.; Kaszas, G.; Kennedy, J. P.; Kelen, T.; Tudos, F. (J. Macromol. Sci. Chem. A **18** [1982] 1263/74 from C.A. **98** [1983] No. 17099).
[26] Kennedy, J. P.; Liao, T. P.; Guhaniyogi, S.; Chang, V. S. C. (J. Polym. Sci. Polym. Chem. Ed. **20** [1982] 3219/27 from C.A. **98** [1983] No. 4842).
[27] Saltybaev, D. K.; Zhubanov, B. A.; Agashkin, O. V.; Kozhabekov, S. S.; Ushanova, V. Zh.; Vladul, A. T. (Izv. Akad. Nauk Kaz.SSR Ser. Khim. **1982** No. 5, pp. 51/4 from C.A. **98** [1983] No. 4841).
[28] Suvorov, A. V.; Sevast'yanova, T. N.; Burlyand, A. I. (Zh. Fiz. Khim. **56** [1982] 1112/6; Russ. J. Phys. Chem. **56** [1982] 678/80).
[29] Hartmann, J. S.; McGarvey, B. D.; Raman, C. V. (Inorg. Chim. Acta **49** [1981] 63/7).
[30] Dittmer, G.; Niemann, U. (Philips J. Res. **37** [1982] 1/30).

6.4 With Nitrogen

6.4.1 (Monoamino)chloroboranes and Related Species

6.4.1.1 Species Containing BCl_2 Groups

$Cl_2BN(CH_3)_2$ is formed in the reaction of BCl_3 with $(CH_3)_2NP(S)Cl_2$ [1]. NMR data: $\delta^{11}B$ = 30.8 ppm, $\delta^{13}C$ = 40.10 ppm [2]. $Cl_2BN(CH_3)_2$ is used in the synthesis of (amino)-1,1-divinylboranes [3, 4]. With $[(CH_3)_3Si]_3SiLi$ it reacts to give $(CH_3)_2NB[Si\{Si(CH_3)_3\}_3]_2$ [5], and with $(CH_3)_3SnLi$, $(CH_3)_2NB(Cl)Sn(CH_3)_3$ and $(CH_3)_2NB[Sn(CH_3)_3]_2$ are obtained depending on the molar ratio of the reactants [6].

$Cl_2BN(C_2H_5)_2$ is formed in the thermal decomposition of $[N(C_2H_5)_4][BCl_4]$ [7]. NMR data: $\delta^{11}B$ = 30.6 ppm [2, 8], $\delta^{13}C$ = 15.2 and 44.9 ppm [2]. With [tmp]Li (tmp = 2,2,6,6-tetramethylpiperidinyl), $ClB[N(C_2H_5)_2][tmp]$ is obtained [9]. The reaction with $(CH_3)_3SnLi$ (molar ratio 1:2) gives $(C_2H_5)_2NB[Sn(CH_3)_3]_2$ [6].

References for 6.4 and 6.5 on pp. 67/70

$Cl_2BN(C_6H_5)_2$. The preparation from BCl_3 and $(C_6H_5)_2NH$ (in CH_2Cl_2) and subsequent treatment with $(C_2H_5)_3N$ in benzene has been reinvestigated. The boiling point is 85°C/10^{-6} Torr, $\delta^{11}B = 32.6$ ppm [8]. $Cl_2BN(C_2H_5)_2$ is used as a starting material in the synthesis of 9-mesityl-9,10-dihydro-9-boraanthrylidene [10].

$(Cl_2B)_3N$ reacts with three molar equivalents of $(CH_3S)_2Pb$ to give $[\{(CH_3)_2S\}_2B]_3N$ [11].

A survey of new data on additional species containing a BCl_2 group is compiled in Table 6/26.

Table 6/26
Additional (Amino)dichloroboranes.

compound	comments	Ref.
$Cl_2BN(i\text{-}C_3H_7)_2$	preparation; NMR ($\delta^{11}B = 30.9$ ppm)	[8]
	reaction with $[(-CH=)_4B-N(i\text{-}C_3H_7)]^{2-}$ to give a classical (C_4B_6) ring compound	[12]
$Cl_2BN(n\text{-}C_4H_9)_2$	formation upon thermal decomposition of $[N(n\text{-}C_4H_9)_4][BCl_4]$	[7, 13]
$Cl_2BN(s\text{-}C_4H_9)_2$	reaction with $[(CH_3)_3Si]_2NLi$	[14]
$Cl_2B(tmp)$	tmp = 2,2,6,6-tetramethylpiperidinyl reaction with $(i\text{-}C_3H_7)NLi-NH(i\text{-}C_3H_7)$ in the preparation of 1,2-diaza-3-boririne and with tmp–B≡N(t-C_4H_9) in the preparation of 1-aza-2,3-diboririne compounds	[15]
$Cl_2BNH[C_6H_2\text{-}2,4,6\text{-}(t\text{-}C_4H_9)_3]$	preparation; mass spectrum, NMR ($\delta^{11}B = 32.8$ ppm)	[16]
$Cl_2BN(CH_3)(n\text{-}C_4H_9)$	preparation; NMR ($\delta^{11}B = 30.6$ ppm), reaction with $[K(t\text{-}C_4H_9)P]_2$ to give diphosphorboriranes	[8]
$Cl_2BN(CH_3)(t\text{-}C_4H_9)$	preparation; NMR ($\delta^{11}B = 30.8$ ppm), reaction with $[K(t\text{-}C_4H_9)P]_2$	[8]
$Cl_2BN(CH_3)(C_6H_5)$	preparation; NMR ($\delta^{11}B = 31.2$ ppm), reaction with $[K(t\text{-}C_4H_9)P]_2$	[8]
	reaction with cycloalkenyllithium	[3]
$Cl_2BN(CH_3)(cyclo\text{-}C_6H_{11})$	preparation; NMR ($\delta^{11}B = 30.4$ ppm), reaction with $[K(t\text{-}C_4H_9)P]_2$	[8]
$Cl_2BN(CH_2C_6H_5)(CH_2C{\equiv}CH)$	preparation; mass spectrum, NMR: (^1H, ^{11}B)	[17]
$Cl_2BN(C_6H_5)(3,3'\text{-dithienyl-4})$	preparation (from $C_6H_5BCl_2$ and 4-azido-3,3'-dithienyl); dehydrochlorination to give a borazaro compound	[18]
$Cl_2BN[Si(CH_3)_3]_2$	reactions with $i\text{-}C_3H_7MgCl$, $t\text{-}C_4H_9Li$, $(CH_3)_3SiCH_2MgCl$, and $(CH_3)_3SiN(CH_3)_2$	[19]

Table 6/26 (continued)

compound	comments	Ref.
$[Cl_2BN(CH_3)P(O)Cl_2]_2$	(8-membered ring) preparation; mass spectrum, NMR (^1H, ^{11}B, ^{13}C, ^{14}N, ^{17}O, ^{31}P), IR	[20]
$[Cl_2BN(C_6H_5)P(O)Cl_2]_2$	(8-membered ring) preparation; NMR (^1H, ^{11}B, ^{13}C), IR	[20]
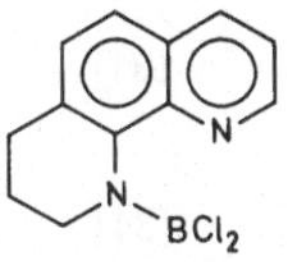	preparation; NMR (^1H, ^{13}C)	[85]
cyclic species		
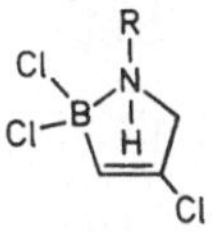	($R = i\text{-}C_3H_7$, $C_6H_5CH_2$) preparation (from $RNHCH_2C{\equiv}N + BCl_3$); mass spectrum, NMR ($^1H$, ^{11}B)	[17]
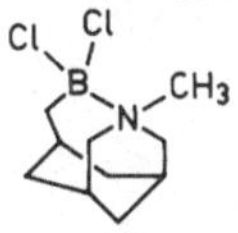	preparation (from 1-boraadamantane and CH_3NCl_2); NMR (^1H, ^{11}B, ^{13}C)	[42]

6.4.1.2 Species Containing BCl Groups

$(CH_3)ClBN(CH_3)_2$ reacts with $(CH_3)_3SiSLi$ to give $(CH_3)_2N(CH_3)B{-}SSi(CH_3)_3$ [21].

$(C_6H_5)ClBN(CH_3)_2$ can be prepared from $C_6H_5BCl_2$ and $(CH_3)_3SiN(CH_3)_2$ in ether (at 0°C) in 80% yield [22], and by ligand exchange between $C_6H_5BCl_2$ and $C_6H_5B[N(CH_3)_2]_2$ [23, 30]. ^{13}C NMR data (in ppm): $\delta = 40.6 + 40.0$ (CH_3), 137.2 (phenyl, C(1)), 132.7 (C(2,6)), 129.0 (C(4)), and 127.5 (C(3,5)) [23]. Reactions to give bis(amino)phenylboranes have been reported with $i\text{-}C_4H_9NH_2$ [24], piperidine derivatives [25], and lithiated silanylamines [22]; $(C_6H_5)(C_2H_5S)$-$BN(CH_3)_2$ results from the reaction with $Pb(SC_2H_5)_2$ [26]. With C_6H_5NCO, insertion into the B–N bond gives $C_6H_5(Cl)BN(C_6H_5)CON(CH_3)_2$ [23, 27, 28]. The corresponding sulfinamidoborane is obtained on treatment with $[(CH_3)_3Si]CH_3NS(O)CH_3$ by cleavage of the Si–N bond [29]. The reaction with sodium pyrazole-1-ylate gives a low yield of 4,8-bis(dimethylamino)-4,8-diphenylpyrazabole [30].

$(C_6H_5)ClBN(C_2H_5)_2$. The preparation by redistribution between $C_6H_5BCl_2$ and C_6H_5B-$[N(C_2H_5)_2]_2$ (3 h) yields 75% of $C_6H_5{-}ClBN(C_2H_5)_2$, boiling point: 50°C/0.1 Torr. NMR data: $\delta^{13}C$ (in ppm) $= 15.3 + 148.8$ (CH_3), $43.6 + 42.7$ (CH_2), 138 (aromatic C(1)), 132.0 (C(2,6)), 128.6 (C(4)), 127.5 (C(3,5)) [23]. Various bis(amino)boranes have been prepared from $(C_6H_5)ClBN(C_2H_5)_2$ by reaction with piperidine derivatives [25] and with $t\text{-}C_4H_9NH_2$ [24]. The reaction with $(CH_3)_3GeSLi$ yields $(C_2H_5)_2N(C_6H_5)B{-}SGe(CH_3)_3$ [21], and the corresponding sulfinamidoborane derivative is obtained with $(CH_3)_3SiN[CH_3][S(O)CH_3]$ [29].

Additional aminoborane-type species containing a BCl group are surveyed in Table 6/27, p. 56.

$[ClBN(CH_3)_2]_2$ reacts with $R{-}C{\equiv}C{-}Si(CH_3)_3$ ($R = C_6H_5$, $(CH_3)_3Si$) to give $[(CH_3)_3Si]RC{=}C$-$[BCl\{N(CH_3)_2\}]_2$ [39].

$H_2C[BCl\{N(CH_3)_2\}]_2$, upon treatment with potassium, yields $H_2C[{-}B\{N(CH_3)_2\}{-}B\{N(CH_3)_2\}{-}]_2$-$CH_2$ [40].

References for 6.4 and 6.5 on pp. 67/70

Table 6/27
Survey of Aminoborane-Type Species Containing a BCl Group.

compound	comments	Ref.
noncyclic species		
$HClBN(s\text{-}C_4H_9)_2$	preparation from $Cl_2BN(s\text{-}C_4H_9)_2$ and $Li[BH_4]$; mass spectrum, NMR (^1H, ^{11}B)	[14]
$(i\text{-}C_3H_7)ClBNH(i\text{-}C_3H_7)$	preparation (from iminoborane and HCl); NMR (^1H, ^{11}B)	[31]
$(n\text{-}C_4H_9)ClBNH(t\text{-}C_4H_9)$	preparation (from iminoborane and HCl); NMR (^1H, ^{11}B)	[31]
$(C_6H_5)ClBNH(t\text{-}C_4H_9)$	preparation; NMR (^1H, ^{13}C)	[23]
$(H_2C{=}CH)ClBNHC_6H_5$	preparation; reaction with cycloalkenyllithium	[3, 4]
$(C_6F_5)ClBNHR$	($R = 2,4,6\text{-}(CH_3)_3C_6H_2$, $4\text{-}CH_3OC_6H_4$) preparation; mass spectrum	[32]
$(C_6H_5)ClBN(CH_2CH{=}CH_2)_2$	preparation; mass spectrum, NMR (^1H, ^{13}C)	[23, 33]
$(C_6H_5)ClBN(pr)_2$	(pr = $n\text{-}C_3H_7$, $i\text{-}C_3H_7$) preparation; NMR (^1H, ^{11}B); reaction with C_6H_5NCO	[23, 28, 33]
$(C_6H_5)ClBN(bu)_2$	(bu = $n\text{-}C_4H_9$, $s\text{-}C_4H_9$) preparation; NMR (^1H, ^{11}B)	[23, 33]
$(C_6H_5)ClBN(i\text{-}C_5H_{11})_2$	preparation; NMR (^1H, ^{13}C)	[23]
$(C_6H_5)ClB(piperid)$	(piperid = piperidinyl, 2-methylpiperidinyl, 3-methylpiperidinyl, 4-methylpiperidinyl) preparation; NMR (^1H, ^{13}C)	[23]
$(C_2H_5CH{=}CH)ClBN(CH_3)C_6H_5$	formation from the $B(C_4H_7)_2$ derivative; mass spectrum, NMR (^1H), IR	[3, 4]
$(C_6H_5)ClBN(C_2H_5)C_6H_5$	preparation; NMR (^1H, ^{13}C)	[23, 33]
$(C_6H_5)ClBN(s\text{-}C_4H_9)C_6H_5$	preparation; NMR (^1H, ^{13}C)	[23]
$(C_6H_5)ClBN(C_6H_5)C(O)N(CH_3)_2$	preparation by insertion (of C_6H_5NCO); IR	[23, 27]
$[(CH_3)_3Si]ClBN(CH_3)_2$	preparation (by $(CH_3)_3SiLi$); NMR (^1H, ^{11}B, ^{13}C, ^{14}N); reaction with K	[34]
$[(CH_3)_3Sn]ClBNR_2$	($R = CH_3$, C_2H_5) preparation (by $(CH_3)_3SnLi$); NMR (^1H, ^{11}B, ^{14}N)	[6]
$(C_2H_5)ClBN(t\text{-}C_4H_9)Si(CH_3)_3$	preparation; mass spectrum, NMR (^1H, ^{13}C); pyrolysis to iminoborane	[35]
$(n\text{-}C_3H_7)ClBN(t\text{-}C_4H_9)Si(CH_3)_3$	preparation; mass spectrum, NMR (^1H, ^{13}C); pyrolysis to $(n\text{-}C_3H_7)B{=}N(t\text{-}C_4H_9)$	[35]

Table 6/27 (continued)

compound	comments	Ref.
$(n\text{-}C_4H_9)ClBN(t\text{-}C_4H_9)Si(CH_3)_3$	preparation; mass spectrum, NMR (^1H, ^{13}C); pyrolysis to $(n\text{-}C_4H_9)B{=}N(t\text{-}C_4H_9)$	[35]
$(i\text{-}C_4H_9)ClBN(t\text{-}C_4H_9)Si(CH_3)_3$	preparation; NMR (^1H, ^{11}B, ^{13}C, ^{14}N); IR; pyrolysis to Dewar-borazine	[36]
$(t\text{-}C_4H_9)ClBN(t\text{-}C_4H_9)Si(CH_3)_3$	preparation; mass spectrum, NMR (^1H, ^{11}B, ^{13}C); pyrolysis to $(t\text{-}C_4H_9)B{\equiv}N(t\text{-}C_4H_9)$	[37]
$RClBN[Si(CH_3)_3]_2$	($R = i\text{-}C_3H_7$, $t\text{-}C_4H_9$, $(CH_3)_3SiCH_2$) preparation; NMR (^1H, ^{13}C)	[19]
$C_6H_5ClBN[\text{-}Si(CH_3)_2\text{-}(CH_2)_2\text{-}(CH_3)_2Si\text{-}]$	preparation; NMR (^1H, ^{13}C)	[22]
species containing two boron atoms		
$[(CH_3)_2N]ClB\text{-}Si(O\text{-}t\text{-}C_4H_9)_2\text{-}B[N(CH_3)_3]_2$	preparation; mass spectrum, NMR (^1H, ^{11}B, ^{13}C)	[38]
$[(CH_3)_3Si]RC{=}C[BCl\text{-}N(CH_3)_2]_2$	($R = C_6H_5$, $(CH_3)_3Si$) preparation; mass spectrum, NMR (^1H, ^{11}B, ^{13}C)	[39]
$[(CH_3)_2N]ClBC(t\text{-}C_4H_9){=}C(t\text{-}C_4H_9)BCl[N(CH_3)_2]$	reaction with K to give the corresponding 1,3-dihydro-1,3-diboret	[41]
cyclic species		
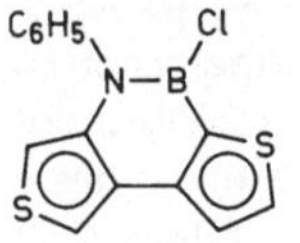	preparation (from α,β-unsaturated *o*-aryl azide and $C_6H_5BCl_2$), alcoholysis	[18]
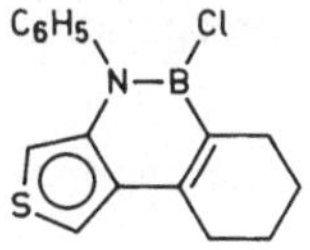	preparation (from α,β-unsaturated *o*-aryl azide and $C_6H_5BCl_2$), alcoholysis	[18]
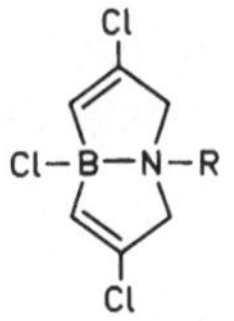	($R = C_2H_5$, $n\text{-}C_3H_7$, $i\text{-}C_3H_7$, $n\text{-}C_4H_9$) preparation from $RN(CH_2\text{-}C{\equiv}CH)_2$ and BCl_3; mass spectrum, NMR (^1H, ^{11}B)	[17]
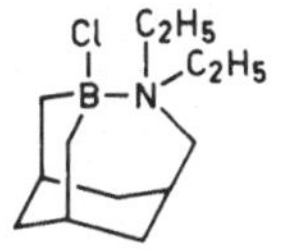	preparation; NMR ($\delta^{11}B = 9.5$ ppm, ^{13}C)	[42]

References for 6.4 and 6.5 on pp. 67/70

6.4.2 Bis(amino)chloroboranes

$ClB[N(CH_3)_2]_2$ forms a 1:1 adduct with $AlCl_3$ in CH_2Cl_2 at −80°C which, however, decomposes upon warming to room temperature to give a mixture of products (as shown by NMR and conductivity measurements) [43]. Reactions with allylic ethers [44], and $CH_3CH{=}CH{-}CH_2MgCl$ [45] have been described; see also [46, 47] for applications in stereoselective reactions. With $(CH_3)_3N{=}S{=}O$, the compound $[(CH_3)_2N]_2B{-}NSO$ is obtained, and reaction with $(CH_3)_3Si{-}N{=}S{=}N{-}Si(CH_3)_3$ yields $[\{(CH_3)_2N\}_2BN]_2S$ [48]. The reaction with $[(CH_3)_3Si]_3SiLi$ gives $[(CH_3)_2N]_2B{-}Si[Si(CH_3)_3]_3$ [5], and with $[(CH_3)_2N]_2BSi(CH_3)_3$ the product is $[(CH_3)_2N]ClBSi(CH_3)_3$ [34]. Treatment of a mixture of $ClB[N(CH_3)_2]_2$ and $(t\text{-}C_4H_9O)_2SiBr_2$ with potassium in hexane gives $[\{(CH_3)_2N\}_2B]_2Si(O\text{-}t\text{-}C_4H_9)_2$ and some $[(CH_3)_2N]ClBSi[B\{N(CH_3)_2\}_2][O\text{-}t\text{-}C_4H_9]_2$ [38]. With $(CH_3)_3SiSLi$ the product is $[(CH_3)_2N]_2BSSi(CH_3)$ [21]. On the other hand Si–N and Si–C bonds can be cleaved by $ClB[N(CH_3)_2]_2$ and $[(CH_3)_2N]_2BN(CH_3)S(O)CH_3$ [29] and $[(CH_3)_2N]_2B{-}C{-}[{=}N{-}CH_2{-}CH(C_6H_5){-}N(CH_3){-}]$ [49] are thus obtained from $(CH_3)_3SiN(CH_3)S(O)CH_3$; $(CH_3)_3SnLi$ leads to $(CH_3)_2NB[Sn(CH_3)_3]_2$ [6].

$ClB[N(C_2H_5)_2]_2$ reacts with $(t\text{-}C_4H_9O)_3SiSNa$ to give $[(C_2H_5)_2N]_2BSSi(O\text{-}t\text{-}C_4H_9)_3$, and with $(CH_3)_3SiLi$ to form $[(C_2H_5)_2N]_2BSSi(CH_3)_3$ [21]; with $(CH_3)_3SnLi$, the product is $[(C_2H_5)_2N]_2B{-}Sn(CH_3)_3$ [6]. The sufinamidoborane $[(C_2H_5)_2N]_2BN(CH_3)S(O)CH_3$ is obtained from $(CH_3)_3SiN{-}[CH_3][S(O)CH_3]$ [29].

$ClB({-}NCH_3{-}CH_2{-}CH_2{-}NCH_3{-})$ is formed upon treatment of $({-}NCH_3{-}CH_2{-}CH_2{-}NCH_3{-})B{-}NCH_3{-}CH_2{-}CH_2{-}NCH_3{-}B({-}NCH_3{-}CH_2{-}CH_2{-}NCH_3{-})$ with BCl_3 (3:1) [50]; the corresponding B-azido derivative is obtained with LiN_3 [51]. $GaCl_3$ and $AlBr_3$ form 1:1 adducts with 2-chloro-1,3-dimethyl-1,3,2-diazaborolidine and are coordinated to one of the nitrogen sites [52]. Analogous adducts have been obtained with BCl_3, BBr_3, $C_6H_5BCl_2$, $C_6H_5BBr_2$ [53], and $AlCl_3$ [54]. With pyridine, the boronium salt 1,3-dimethyl-1-pyridine-1-yl-1,3,2-diazaborolidinium chloride (containing three-coordinate boron) is formed [55]. The product of the reaction with pyrazole, 2-pyrazolyl-1,3-dimethyl-1,3,2-diazaborolidine, shows fluxional behavior [56, 57]. Reaction with 3-*t*-butylamino-5-phenyl-1,2,3-dithiaborole in the presence of triethylamine yields $[{-}NCH_3{-}CH_2{-}CH_2{-}NCH_3{-}]B{-}N(t\text{-}C_4H_9)B[{-}S{-}S{-}CH{=}C(C_6H_5){-}]$ [58, 59]. The reaction with $(CH_3)_3SiSLi$ gives $({-}NCH_3{-}CH_2{-}CH_2{-}NCH_3{-})B{-}SSi(CH_3)_3$, and with $(t\text{-}C_4H_9O)_3SiSNa$ the species $({-}NCH_3{-}CH_2{-}CH_2{-}NCH_3{-})B{-}SSi(O\text{-}t\text{-}C_4H_9)_3$ is formed [21]. $({-}NCH_3{-}CH_2{-}CH_2{-}NCH_3{-})B{-}NSO$, $[\{({-}NCH_3{-}CH_2{-}CH_2{-}NCH_3{-})B\}_2N{=}]_2S$, and related compounds are obtained from the reaction with N-trimethylsilanylated sulfur-nitrogen compounds [48]. Reaction with tris(alkoxy)bromosilanes, $(RO)_3SiBr$ ($R = i\text{-}C_3H_7$, $t\text{-}C_4H_9$) by treatment with potassium in hexane, gives $({-}NCH_3{-}CH_2{-}CH_2{-}NCH_3{-})B{-}Si(OR)_3$; similar but more complex species are obtained if $(t\text{-}C_4H_9O)_2SiBr_2$ is employed as reagent [38]. With $(CH_3)_3Si[CH_3][NS(O)CH_3]$, the corresponding sulfinamidoborane is formed by elimination of $(CH_3)_3SiCl$ [29].

$ClB[{-}NCH_3{-}(CH_2)_3{-}NCH_3{-}]$. The reaction with pyrazole derivatives leads to the corresponding 2-(pyrazol-1′-yl)-1,3,2-diazaborolidines [60].

A survey of new data on additional bis(amino)chloroborane is compiled in Table 6/28.

Table 6/28
Bis(amino)chloroboranes.

compound	comments	Ref.
$ClB[N(i\text{-}C_3H_7)_2]_2$	preparation; NMR (^{13}C), reaction with $AlCl_3$ to give $[(i\text{-}C_3H_7)_2N{=}B{=}N(i\text{-}C_3H_7)_2][AlCl_4]$	[43]
	reaction with LiN_3; reaction with C_6H_5NCO	[28, 51]

Table 6/28 (continued)

compound	comments	Ref.
$ClB(tmp)_2$	(tmp = 2,2,6,6-tetramethylpiperidinyl) preparation; mass spectrum, NMR (^1H, ^{11}B, ^{13}C); reactions with BCl_3, $AlCl_3$, $GaCl_3$ to give [tmp=B=tmp][ECl_4] (E = Al, Ga)	[9]
$ClB[N(C_2H_5)_2][tmp]$	preparation; mass spectrum, NMR (^1H, ^{11}B, ^{13}C); reactions with $AlCl_3$, $GaCl_3$, $TaCl_5$ to give the two-coordinate diamino-boron cation	[9]
$ClB[N(s\text{-}C_4H_9)_2][NHSi(CH_3)_3]$	preparation; mass spectrum, NMR (^1H, ^{11}B); reaction with $Li[BH_4]$	[14]
$ClB[N(s\text{-}C_4H_9)_2][N\{Si(CH_3)_3\}_2]$	preparation; mass spectrum, NMR (^1H, ^{11}B); reaction with $Li[BH_4]$	[14]
$ClB[N\text{–}Si(CH_3)_2\text{–}CH_2\text{–}CH_2\text{–}Si(CH_3)_2\text{–}]_2$	preparation; NMR (^1H, ^{13}C (variable temperature))	[22]
$ClB[N(i\text{-}C_3H_7)\text{–}NH(i\text{-}C_3H_7)][tmp]$	(tmp = 2,2,6,6-tetramethylpiperidinyl) preparation; NMR (^{11}B, ^{13}C); reaction with $t\text{-}C_4H_9Li$ to give diazaboririne	[15]
species containing two boron atoms		
$ClB[N(n\text{-}C_4H_9)_2]\text{–}N(n\text{-}C_4H_9)\text{–}BCl[N(n\text{-}C_4H_9)_2]$	(and similar species) in pyrolysis of $[(n\text{-}C_4H_9)_4N][BCl_4]$	[7]
$ClB(C_6F_5)\text{–}NR\text{–}B(C_6F_5)\text{–}NHR$	R = 2,4,6-$(CH_3)_3C_6H_2$; formation; osmometric mass, NMR (^1H, ^{19}F), IR	[32]
$(tmp)ClBN(t\text{-}C_4H_9)\text{–}BCl(tmp)$	preparation; mass spectrum, NMR (^{11}B, ^{13}C); reaction with Na/K to give 1-aza-2,3-boririne	[15]

The cyclic ureidoboranes **5**, **6**, and **7** (X = O, S) are formed in the reaction of $(CH_3HN)_2CX$ with $C_6H_5BCl_2$ and were identified by mass spectrometry [61].

5 **6** **7**

For cyclic (amino)chloroboranes containing B–B bonds [62], see "Boron Compounds" 3rd Suppl. Vol. 1, 1987, Section 2.4.3.3, p. 69.

6.4.3 Boron-Nitrogen Heterocycles

6.4.3.1 Chloroborazines and Their Derivatives

$(-BCl-NH-)_3$. Vibrational spectra of isotopically substituted B-trichloroborazine have been recorded and the assignments were supplemented (especially for Raman frequencies of species A_1' and E''). Results compiled from the spectra of $(-BCl-NH-)_3$, $(-^{10}BCl-NH-)_3$, $(-BCl-^{15}NH-)_3$, $(-^{10}BCl-^{15}NH-)_3$, $(-BCl-ND-)_3$, $(^{10}BCl-ND-)_3$, $(-BCl-^{15}ND-)_3$, and $(-^{10}BCl-^{15}ND-)_3$ are given in Table 6/29 [63].

Table 6/29

Fundamental Vibrations of Isotopically Substituted 2,4,6-Trichloroborazine [63].

		$(BClNH)_3$	$(^{10}BClNH)_3$	$(BCl^{15}NH)_3$	$(^{10}BCl^{15}NH)_3$	$(BClND)_3$	$(^{10}BClND)_3$	$(BCl^{15}ND)_3$
A_1'	ν_1	3424	3423	3417	3415	2557	2537	2534
	ν_2	1120	1175	1125	1175	1128	1172	1122
	ν_3	863	862.5	835	838	840	834	810.5
	ν_4	360	360	360	360	360	359	360
A_2''	ν_8	704.7	706.5	702.8	703.8	504.1	510.8	510.4
	ν_9	648.3	667.9	643	663.5	655.5	675	647.9
	ν_{10}	104	(104)	(104)	(104)	(104)	(104)	(104)
E'	ν_{11}	3443	3435	3441	3442	2551.5	2541.5	2539
	ν_{12}	1441.5	1478	1423	1457	1405	1443.5	1387
	ν_{13}	1339	1381	1331	1374	1258	1296.5	1249
	ν_{14}	1038	1049	1026	1044	814.4	823.6	817.3
	ν_{15}	752	757.9	734.2	745.8	736.7	743.1	726.7
	ν_{16}	374.8	376.5	369.2	372.1	365.1	367.5	361.7
	ν_{17}	173	172	173	173	173	173	173
E''	ν_{18}	—	—	—	—	—	—	—
	ν_{19}	642	656	640	650	642	658	636
	ν_{20}	156	155	150	150	149	148	145

$(-BCl-NH-)_3$ is used in the chemical vapor deposition of α-BN layers, especially on Mo, W, and Ta substrates between 770 to 1600°C. Apparently, the layers show good adhesion to these substrates due to the formation of boride layers on the surfaces of the substrate metals [64, 65].

$(-BCl-NCl-)_3$ has been applied for the chemical vapor deposition of α-BN layers on silica (at 900°C) [66, 67].

$(-BCl-NCH_3-)_3$ is formed (together with cyclic borylated derivatives of N,N'-dimethylthiourea) in the reaction of $(CH_3NH)_2CS$ with BCl_3 [61]; $\delta^{11}B = 31.8$ ppm, $\delta^{13}C = 35.20$ ppm [2]. Scrambling reactions in the system $(-BCl-NCH_3-)_3 + [-B(n\text{-}C_3H_7)-NCH_3-]_3$ leading to the formation of $B_3Cl_3N_3(CH_3)_{3-n}(n\text{-}C_3H_7)_n$ (with n = 1, 2) have been studied [68].

$[-BCl-N\{Si(CH_3)_3\}-]_3$ eliminates $(CH_3)_3SiCl$ above 260°C in a pyrolytic condensation [69].

The polycyclic species **8** has been reported; $\delta^1H = 0.39$ ppm, $\delta^{11}B = 31.5$ ppm ($h_{1/2} = 320$ Hz), $\delta^{13}C = 4.61$ pm [70].

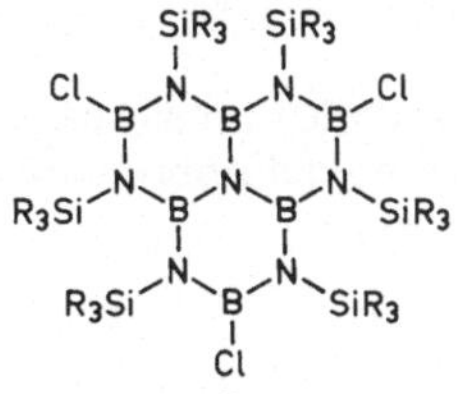

8

$(R = CH_3)$

6.4.3.2 Additional Boron-Nitrogen Heterocycles

The reactions of the amino(imino)borane (tmp)B=N(t-C_4H_9) (tmp = 2,2,6,6-tetramethylpiperidinyl) with BCl_3 [71] and $C_6H_5BCl_2$ [72] yield B–N heterocycles containing three- and four-coordinate boron in a four-membered ring. Thus, the species **9**, **10**, and **11** have been obtained.

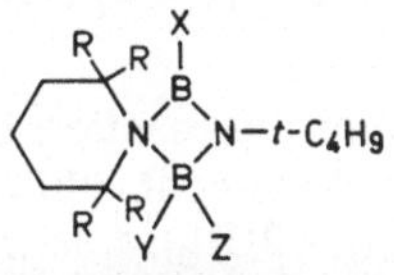

9: X = Y = Z = Cl
10: X = C_6H_5, Y = Z = Cl
11: X = Y = Cl, Z = C_6H_5
$(R = CH_3)$

9 is obtained in pure state by performing the reaction at −40°C in pentane. The melting point is 124°C and the following NMR data are given (in ppm): $\delta^1H = 1.69$ (t-C_4H_9), 1.55 (H(2,3,4)), 1.46 ($4 \times CH_3$); $\delta^{11}B = 27.5$ ($h_{1/2} = 76$ Hz), 9.5 ($h_{1/2} = 25$ Hz); $\delta^{13}C = 62.5$ (C(1,5)), 53.0 ($\underline{C}(CH_3)_3$), 40.3 (C(2,4)), 32.2 ($2 \times CH_3$), 30.4 ($C(\underline{C}H_3)_3$), 26.1 ($2 \times CH_3$), 15.9 (C(3)) [71].

10 and **11** are obtained as a mixture in which $\delta^{11}B$ is given for **10** with 36.8 and 9.6 ppm, and for **11** with 29.0 and 21.9 ppm; for ^{13}C NMR data, see [72].

6.4.4 Chloroboranes Containing B–O and B–N Bonds

The heterocyclic compounds **12** and **13** have been described [62]; see Section 2.4.3.3 in "Boron Compounds" Suppl. Vol. 1, 1987, pp. 67/70.

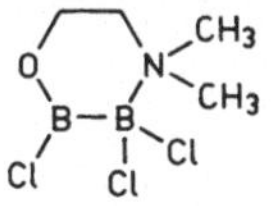

12 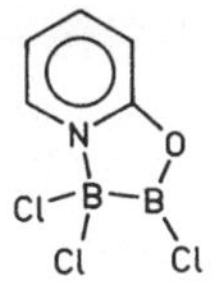 **13** 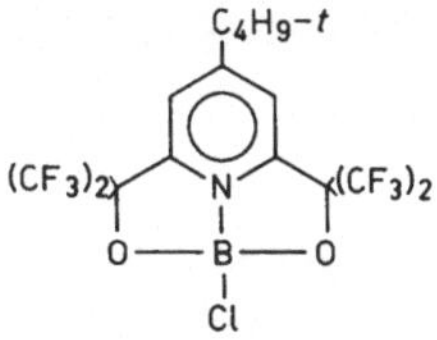 **14**

14 is synthesized starting from the dilithium salt of the corresponding pyridinediol and BCl_3 (4 h, 25°C). It is an intermediate in the preparation of compounds containing penta- and hexacoordinate hypervalent boron [73].

References for 6.4 and 6.5 on pp. 67/70

6.4.5 Ionic Species

Ionic species containing two-coordinate formally cationic boron have been obtained by the reaction of sterically hindered bis(amino)chloroboranes with BCl_3 to give the corresponding tetrachloroborates **15** and **16** [9, 72, 75].

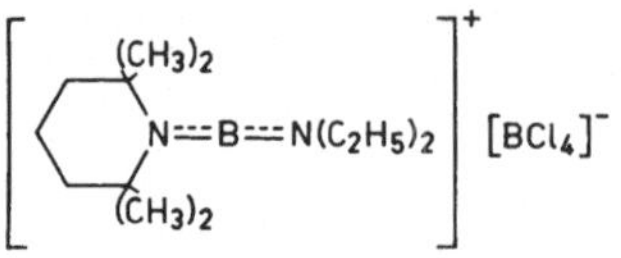

15 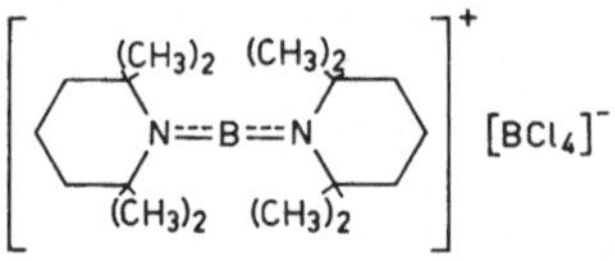 **16**

15 is in equilibrium with (tmp)ClB–N(C_2H_5)$_2$ and BCl_3 (tmp = 2,2,6,6-tetramethylpiperidinyl) as based on NMR data; **16** can be isolated in pure state [9]. For the preparation of the latter, 1.49 g (4.56 mmol) of (tmp)$_2$BCl dissolved in 15 mL of CH_2Cl_2 at −40°C and 0.68 g (5.8 mmol) BCl_3 are condensed under stirring. The reaction mixture is slowly allowed to reach ambient temperature and volatiles are removed under reduced pressure to give a yield of 88% of yellowish **16** with a melting point 111 to 115°C. NMR data (in ppm): $\delta^{11}B = 35.8$ and 7.2; $\delta^{1}H =$ 1.68 (H(2,3,4)), 1.53 (4 × CH_3); $\delta^{13}C = 58.1$ (C(1,5)), 37.4 (C(2,4)), 31.1 (C(6,7)), 16.5 (C(3)) (in CH_2Cl_2 solution) [9].

The salt **17** has been obtained from the reaction of the amino(imino)borane (tmp)B=N-(*t*-C_4H_9) with $C_6H_5BCl_2$ and subsequent treatment with $AlCl_3$. NMR data (in ppm): $\delta^{11}B = 46.4 +$ 37.6; $\delta^{1}H = 1.97$ (H(2,3,4)), 1.72 + 1.56 (2 × 2CH_3), 1.34 (*t*-C_4H_9), 7.52 − 7.70 (C_6H_5); $\delta^{27}Al = 103.7$ [72].

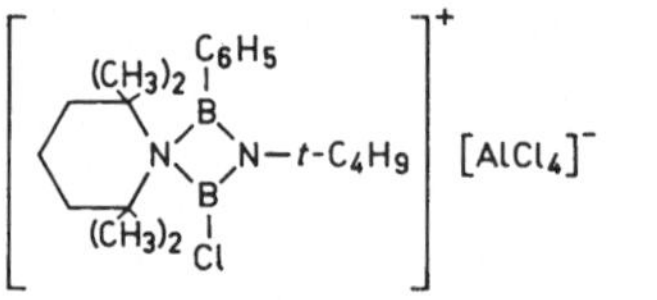

17 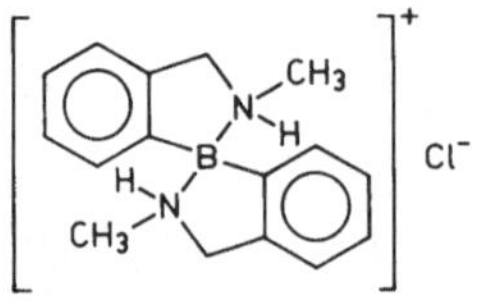 **18**

Compound **18** has been prepared from N-methylbenzylamine by lithiation (in the presence of tetramethylethylenediamine) and reaction with $B(OCH_3)_3$, distillation of the methyl ester of the resultant B–N heterocycle and subsequent hydrolysis. The structure was assigned on the basis of mass spectrum and NMR (^{1}H, ^{11}B = 10.1 ppm, ^{13}C) data, as well as the high stability and its solubility in H_2O and $CHCl_3$ [76].

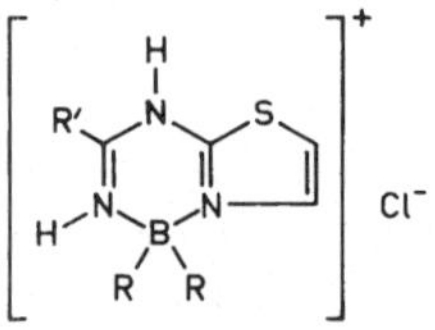

19

R = *n*-C_3H_7

R′ = CH_3, C_6H_5

19 is obtained by addition of HCl (in (C_2H_5)$_2$O) to the corresponding heterocycle; IR group frequencies are reported [77].

The ammonium and nitryl tetraperchloratoborates **$NH_4[B(ClO_4)_4]$** and **$NO_2[B(ClO_4)_4]$**, respectively, have been synthesized by adding a large excess of anhydrous $HClO_4$ to a stoichiometric BCl_3–$MClO_4$ mixture (M = NH_4^+, NO_2^+) at −80°C. IR and Raman spectra show that the ClO_4 groups are monodentate and strongly bonded to the boron atoms. Assignments are made for

$\nu_{as}(ClO_t) \approx 1270\ cm^{-1}$, $\nu_s(ClO_t) \approx 1040\ cm^{-1}$, and for the ClO moiety bonded to the boron atom $\nu(ClO_p) \approx 700\ cm^{-1}$. The NH_4^+ cation produces absorptions for $\nu(NH)$ at $3270 + 3205\ cm^{-1}$, and the following bands due to NO_2 are observed: $\nu_1(R) = 1401\ cm^{-1}$, $\nu_2(IR) = 560\ cm^{-1}$, $\nu_3(IR) = 2330\ cm^{-1}$, $(\nu_2 + \nu_3)(IR) = 3720\ cm^{-1}$. The vibrations due to the $[B(ClO_4)_4]^-$ ions are listed in Table 6/30 [74].

Table 6/30
The Vibrational Spectrum (frequencies in cm^{-1}) of the $[B(ClO_4)_4]^-$ Anion in $NH_4[B(ClO_4)_4]$ and $NO_2[B(ClO_4)_4]$ (vs = very strong, s = strong, m = medium, w = weak, sh = shoulder) [74].

$NH_4[B(ClO_4)_4]$		$NO_2[B(ClO_4)_4]$			assignment
IR	Raman	IR	Raman		
—	1272 m	—	1280 w	—	—
1260 vs	—	1255 s	—	A_1	$\nu_{as}ClO_t$
1225 sh	—	1240 sh	—	B_2+E	
—	1237 m	—	1226 w	B_1	
—	1140 sh	—	—	—	—
—	1110 sh	—	—	—	—
—	1070 vs	—	1065 s	—	ν_sClO_t
—	—	—	—	A_1	
1065 m	—	1075 s	—	—	—
912 sh	937 sh	930	944 s	B_2+E	ν_dBO_4
895 vs	925 s	885	927 s		
—	765 s	—	777 m	A_1	ν_sClO_p
740 s	—	720 s	751	B_2+E	
—	700 s	—	699 s	A_1	ν_sBO_4
—	672 s	—	677 s	E	$\delta_{as}ClO_3$
662 s	—	660 s	—	—	
—	640 s	—	641 m	B_1	
620 s	618 m	620 m	620	E	
—	600 s	—	600 m	A_1	
582 s	585 sh	—	585 m	B_2	
—	—	—	575 m	—	
—	530 s	—	527 s	B_2+E	δ_dBO_4
520 m	—	—	—	E	
490 s	—	—	—	B_2?	δ_sClO_3
462 s	—	466	—	E	
—	462 s	—	472 s	A_1	
418 sh	425 w	420 sh	428 m	—	—
—	320 m	—	314	A_1	δBO_4
—	295 s	—	288 s	A_1	ϱClO_3
—	275 s	—	270 s	E	
—	220 s	—	249 s	B_1+B_2+E	
—	—	—	233 s		
—	—	—	200 m	—	
—	—	—	140 m	A_1+B_2+E	$\delta BOCl$

References for 6.4 and 6.5 on pp. 67/70

The salt $[(CH_3)_2NH_2][(CH_3)_3SnBCl_3]$ has been obtained by reaction of $[(CH_3)_2N]_2B{-}Sn(CH_3)_3$ with four equivalents of HCl in ether at −30°C; $\delta^{11}B = 2.2$ ppm (in dimethylformamide) [6].

6.5 With Fluorine

The formation of **$BClF_2$** has been observed in the reaction of BCl_3 with fluorocarbon grease during the laser-induced photolysis of BCl_3/O_2 mixtures [78].

The force fields of $BClF_2$ and **BCl_2F** have been calculated using standardized computing parameters (and for $BClF_2$ also using the experimental geometry). Valence force constants obtained are compiled in Table 6/31. For a discussion see [80].

Table 6/31

Calculated Valence Force Constants for $BClF_2$ and BCl_2F (in $N \cdot m^{-1}$; set I for $BClF_2$ uses experimental geometry, set II standard geometry) [80].

force constants	$BClF_2$ set I	$BClF_2$ set II	BCl_2F	force constants	$BClF_2$ set I	$BClF_2$ set II	BCl_2F
f_r	723.6	711.6	405.7	$f_{r\alpha}$	11.8	10.9	−6.4
f_R	448.6	438.8	982.0	$f_{r\beta}$	−27.9	−23.1	−33.9
f_α	100.0	100.6	117.8	$f'_{r\beta}$	16.0	12.3	40.3
f_β	65.6	67.2	61.9	$f_{R\alpha}$	−38.6	−36.1	−38.9
f_Θ	47.6	45.8	42.6	$f_{R\beta}$	19.3	18.0	19.5
f_{rr}	68.3	80.6	−30.5	$f_{\alpha\beta}$	−50.0	−50.3	−58.9
f_{rR}	62.9	56.9	284.9	$f_{\beta\beta}$	−15.6	−16.9	−2.9

Mixed Trihaloborane Adducts with O and N Donors

Trihaloborane adducts of acetate and benzoate esters, D, readily exchange halogen to form mixed trihaloborane adducts $D{-}BF_nCl_{3-n}$ (n = 0 to 3). ^1H and ^{19}F NMR shifts for adducts of BCl_2F and $BClF_2$ with carboxylic esters RC(O)OR′ (R = CH_3, R′ = CH_3, C_2H_5, C_6H_5, 4-CH_3–C_6H_4, 4-F–C_6H_4; R′ = CH_3, R = $C_6H_5CH_2$, C_6H_5, 2-Cl–C_6H_4, 4-F–C_6H_4 (except for $BFCl_2$), 4-NO_2–C_6H_4, 4-CH_3O–C_6H_4, 4-CH_3–C_6H_4) have been reported and results are tabulated [83].

The rate of halogen distribution in mixed trihaloborane adducts of tertiary amines is strongly dependent upon the base strength of the amine which, on the other hand, has only little effect on the ^{19}F NMR shifts. These latter are mainly dependent upon steric effects of the amine substituents as estimated from the "cone angle" of the amines [84]. The authors give an extensive tabulation on NMR (^{11}B, ^{13}C, ^{19}F) complexation shifts as well as of ^{11}B-^{19}F coupling constants. Partial data are compiled in Tables 6/32 to 6/35 [84]. Complexation of tertiary amines of the type NRR′R″ stops the inversion and chirality of the donors make the fluorines of an acceptor BF_2X diastereotopic (and hence magnetically nonequivalent). By ^{11}B decoupling the system is reduced to an AB pattern, from which the $\delta^{19}F$ values of the individual fluorines on $^2J(FBF)$ are readily determined. Results are given in Table 6/34, p. 67 [84].

Table 6/32
^{19}F Chemical Shifts of Amine-Trihaloborane Adducts (δ in ppm, low field to $CFCl_3$; [a] estimated; [b] magnetically nonequivalent fluorines; [c] each diastereomer has a distinct chemical shift) [84].

amine	amine		^{19}F chemical shifts							
	pK_b	cone angle (±2°)	$D \cdot BF_3$	$D \cdot BClF_2$	$D \cdot BCl_2F$	$D \cdot BBrF_2$	$D \cdot BBr_2F$	$D \cdot BF_2I$	$D \cdot BFI_2$	$D \cdot BBrClF$
quinuclidine	2.9	134	−163.2	−141.9	−131.7	−133.0	−120.4	—	—	−125.5
$N(CH_3)_3$	4.24	138	−164.1	−143.4	−132.6	−134.0	−120.8	−119.6	—	−126.0
$N(CH_3)_2(C_2H_5)$	4.01	144	−161.4	−139.7	−129.2	−130.8	−117.6	−117.4	—	−123.5
$N(CH_3)(C_2H_5)_2$	3.71	149	−156.0	−133.9	−121.2	−125.0	−109.5	−111.8	−97.2	−114.8
$N(C_2H_5)_3$	3.35	161	−150.5	−127.8	−115.2	−119.6	−104.2	−105.9	−92.5	−108.0
$N(n\text{-}C_3H_7)_3$	3.35	165	−151.5	−128.3	−115.4	−119.5	−104.1	−106.3	−91.8	−109.4
$N(n\text{-}C_4H_9)_3$	3.11	161	−151.0	−128.2	−115.2	−119.3	−103.7	−106.2	−91.9	—
$N(C_2H_5)(i\text{-}C_3H_7)_2$	—	205	−142.4	−119.6	−108.6	−110.3	—	—	—	—
$N(CH_3)_2(C_6H_5)$	8.9	143	−158.3	−137.4	−126.3	−128.7	−115.1	−115.0	−102.0	−120.3
$4\text{-}CH_3C_6H_4N(CH_3)_2$	8.4	—	−158.5	−137.3	−126.3	−128.8	−115.2	−115.6	−102.6	−120.3
$4\text{-}O_2NC_6H_4N(CH_3)_2$	13.2	—	−150.5	—	—	—	—	—	—	—
$N(CH_3)(C_2H_5)(C_6H_5)$	7.99	142	−157.1	−136.5	−125.8	[b]	−114.3	[b]	−102.0	−119.4 −120.3[c]
$N(C_2H_5)_2(C_6H_5)$	7.4	144	−147.4	−127.5	−117.3	−119.3	−106.8	—	—	−111.6
$N(CH_3)(C_2H_5)(CH_2C_6H_5)$	4.3[a]	—	−156.4	−134.0	−121.2	[b]	−109.7	[b]	−97.6	−115.0 −115.4
4-methylpyridine	7.98	—	−142.4	−125.4	−118.8	−117.3	−110.5	—	—	—
uncomplexed BF_nX_{3-n}	—	—	−125.0	−73.5	−26.0	−56.6	5.4	—	—	−10.2

References for 6.4 and 6.5 on pp. 67/70

Table 6/33
^{11}B Chemical Shifts (in ppm) of Tertiary Amine-Trihaloborane Adducts ([a] not detectable) [84].

acceptor	uncomplexed haloborane	donor				
		quinuclidine	$N(C_2H_5)(i\text{-}C_3H_7)_2$	$N(CH_3)_2(C_6H_5)$	$N(CH_3)(C_2H_5)(C_6H_5)$	$N(CH_3)(C_2H_5)(CH_2C_6H_5)$
BF_3	10.0	−0.5	0.2	−0.1	0.1	−0.1
BF_2Cl	19.8	4.1	4.9	4.3	4.3	4.6
$BFCl_2$	32.3	7.7	8.8	8.0	8.0	8.3
BCl_3	46.5	9.2	—	10.0	10.1	9.8
$BBrF_2$	19.5	3.6	3.3	3.6	3.9	3.8
BBr_2F	29.0	3.1	—	3.0	3.3	3.2
BBr_3	38.7	−3.9	—	−3.5	−3.4	−3.7
BF_2I	[a]	—	—	—	2.3	1.3
BFI_2	[a]	—	—	−17.3	−16.9	−16.0
BI_3	−7.9	−53.1	—	−54.0	−54.0	−53.3
$BBrCl_2$	44.7	5.8	—	6.4	6.7	6.3
BBr_2Cl	42.3	1.4	—	1.8	2.3	1.8

Table 6/34

NMR Parameters of Diastereotopic Fluorine Atoms in RR′R″N–BF_2X Adducts ([a] not resolved in the ^{11}B spectra; in the ^{19}F spectra, an additional 0.13 ppm splitting of the 1:1:1:1 quartet peaks is observed; [b] not detected) [84].

adduct	$^2J(FBF)$ in Hz	^{19}F chemical shifts in ppm from $CFCl_3$ F_A	F_B
$D^* = C_6H_5CH_2N(CH_3)C_2H_5$			
$D^* \cdot BF_2I$	39.7	−110.8	−112.8
$D^* \cdot BF_2Br$	48.5	−121.8	−122.5
$D^* \cdot BF_2Cl$ [a]	—	—	—
$D^* = N(CH_3)(C_2H_5)(C_6H_5)$			
$D^* \cdot BF_2I$	33.8	−114.0	−115.8
$D^* \cdot BBrF_2$	44.1	−127.4	−128.1
$D^* \cdot BClF_2$ [b]	—	—	—

Table 6/35

^{11}B-^{19}F Coupling Constants (in Hz) in Tertiary Amine-Trihaloborane Adducts [84].

amine	BF_3	$BClF_2$	BCl_2F	$BBrF_2$	BBr_2F	BF_2I	BFI_2	BBrClF
quinuclidine	15.2	43.5	68.6	53.9	89.6	—	—	78.9
$N(CH_3)_3$	15.1	44.8	69.5	54.2	89.3	65.8	—	78.0
$N(CH_3)_2(C_2H_5)$	15.7	44.1	70.1	54.9	90.8	68.6	—	80.5
$N(CH_3)(C_2H_5)_2$	17.7	45.6	72.1	55.9	93.2	68.6	116.7	82.4
$N(C_2H_5)_3$	18.2	46.9	72.9	56.8	93.6	69.6	118.1	83.3
$N(n\text{-}C_3H_7)_3$	19.1	46.2	72.9	56.2	93.8	68.1	118.6	82.4
$N(n\text{-}C_4H_9)_3$	17.7	46.5	72.6	54.9	92.2	64.7	11.7	—
$N(C_2H_5)(i\text{-}C_3H_7)_2$	20.1	49.5	75.9	60.3	—	—	—	—
$N(CH_3)_2(C_6H_5)$	13.2	44.1	67.5	54.5	91.2	66.3	116.8	79.4
$N(CH_3)_2(4\text{-}CH_3C_6H_4)$	12.8	43.0	67.7	54.9	90.2	67.2	115.7	79.4
$N(CH_3)(C_2H_5)(C_6H_5)$	12.8	43.1	69.1	54.2	90.2	68.7	116.2	79.9
$N(C_2H_5)_2(C_6H_5)$	—	45.1	70.6	55.4	94.1	—	—	84.8
$N(CH_3)(C_2H_5)(CH_2C_6H_5)$	16.7	45.6	72.1	55.6	92.7	66.8	116.6	81.9
4-methylpyridine	11.8	38.3	62.8	49.2	82.2	—	—	—
uncomplexed BF_nX_{3-n}	14.5	33	78	58	121	—	—	95

References for 6.4 and 6.5:

[1] Wade, S. R.; Willey, G. R. (J. Inorg. Nucl. Chem. **43** [1981] 1465/8).
[2] Nöth, H.; Wrackmeyer, B. (Chem. Ber. **114** [1981] 1150/6).
[3] Sobieralski, T. J.; Hancock, K. G. (J. Am. Chem. Soc. **104** [1982] 7533/41).
[4] Sobieralski, T. J. (Diss. Univ. California, Davis 1980, pp. 1/226; Diss. Abstr. Intern. B **41** [1981] 4527).
[5] Biffar, W.; Nöth, H. (Z. Naturforsch. **36b** [1981] 1509/15).

[6] Nöth, H.; Schwerthöffer, R. (Chem. Ber. **114** [1981] 3056/62).
[7] Sheikh, S. U. (Therm. Anal. Proc. 6th Intern. Conf. Therm. Anal., Bayreuth, FRG, 1980, pp. 165/8).
[8] Baudler, M.; Marx, A. (Z. Anorg. Allgem. Chem. **474** [1981] 18/30).
[9] Nöth, H.; Rasthofer, B.; Weber, S. (Z. Naturforsch. **39b** [1984] 1058/68).
[10] Lapin, S. C.; Brauer, B. E.; Schuster, G. B. (J. Am. Chem. Soc. **106** [1984] 2092/100).

[11] Nöth, H.; Staudigl, R.; Storch, W. (Chem. Ber. **114** [1981] 3024/43).
[12] Herberich, G. E.; Ohst, H.; Mayer, H. (Angew. Chem. **96** [1984] 975/6; Angew. Chem. Intern. Ed. Engl. **23** [1984] 969).
[13] Sheikh, S. U.; Ghafoor, A. (Pakistan J. Sci. Ind. Res. **23** [1980] 21/2 from C.A. **94** [1981] No. 113678).
[14] Nutt, W. R.; Wells, R. L. (Inorg. Chem. **21** [1982] 2473/6).
[15] Dirschl, F.; Nöth, H.; Wagner, W. (J. Chem. Soc. Chem. Commun. **1984** 1533/5).
[16] Hitchcock, P. B.; Jasim, H. A.; Lappert, M. F.; Williams, H. D. (J. Chem. Soc. Chem. Commun. **1984** 662/4).
[17] Meller, A.; Hirninger, F. J.; Noltemeyer, M.; Maringgele, W. (Chem. Ber. **114** [1981] 2519/35).
[18] Leardini, R.; Zanirato, P. (J. Chem. Soc. Chem. Commun. **1983** 396/7).
[19] Li, B.-L.; Goodman, M. A.; Neilson, R. H. (Inorg. Chem. **23** [1984] 1368/71).
[20] Nöth, H.; Storch, W. (Chem. Ber. **117** [1984] 2140/56).

[21] Hennemuth, K.; Meller, A.; Wojnowska, M. (Z. Anorg. Allgem. Chem. **489** [1982] 47/54).
[22] Beswick, Y. F.; Wisian-Neilson, P.; Neilson, R. H. (J. Inorg. Nucl. Chem. **43** [1981] 2639/43).
[23] Cragg, R. H.; Miller, T. J. (J. Organometal. Chem. **232** [1982] 201/14).
[24] Cragg, R. H.; Miller, T. J. (J. Organometal. Chem. **217** [1981] 283/90).
[25] Cragg, R. H.; Miller, T. J. (J. Organometal. Chem. **235** [1982] 143/50).
[26] Cragg, R. H.; Miller, T. J. (J. Organometal. Chem. **243** [1983] 387/92).
[27] Cragg, R. H.; Miller, T. J. (J. Organometal. Chem. **255** [1983] 143/52).
[28] Cragg, R. H.; Miller, T. J. (J. Organometal. Chem. **260** [1984] 1/5).
[29] Meller, A.; Maringgele, W.; Armbrecht, M. (Z. Naturforsch. **36b** [1981] 1411/5).
[30] Niedenzu, K.; Seelig, S. S.; Weber, W. (Z. Anorg. Allgem. Chem. **483** [1981] 51/62).

[31] Paetzold, P.; von Plotho, C.; Schwan, H.; Meier, H.-U. (Z. Naturforsch. **39b** [1984] 610/4).
[32] Morse, J. G.; Glanville, W. K. (Inorg. Chem. **23** [1984] 11/3).
[33] Brown, C.; Cragg, R. H.; Miller, T. J.; Smith, D. O'N. (J. Organometal. Chem. **220** [1981] C25/C26).
[34] Biffar, W.; Nöth, H.; Schwerthöffer, R. (Liebigs Ann. Chem. **1981** 2067/80).
[35] Paetzold, P.; von Plotho, C. (Chem. Ber. **115** [1982] 2819/25).
[36] Paetzold, P.; von Plotho, C.; Schmidt, G.; Boese, R. (Z. Naturforsch. **39b** [1984] 1069/75).
[37] Paetzold, P.; von Plotho, C.; Schmid, G.; Boese, R.; Schrader, B.; Bougeard, D.; Pfeiffer, U.; Gleiter, R.; Schäfer, W. (Chem. Ber. **117** [1984] 1089/102).
[38] Pfeiffer, J.; Maringgele, W.; Meller, A. (Z. Anorg. Allgem. Chem. **511** [1984] 185/92).
[39] Klusik, H.; Pues, C.; Berndt, A. (Z. Naturforsch. **39b** [1984] 1042/5).
[40] Fisch, H.; Pritzkow, H.; Siebert, W. (Angew. Chem. **96** [1984] 595; Angew. Chem. Intern. Ed. Engl. **23** [1984] 608).

[41] Hildenbrand, M.; Pritzkow, H.; Zenneck, U.; Siebert, W. (Angew. Chem. **96** [1984] 371/2; Angew. Chem. Intern. Ed. Engl. **23** [1984] 371).
[42] Mikhailov, B. M.; Shagova, E. A.; Etinger, M. Yu. (J. Organometal. Chem. **220** [1981] 1/9).
[43] Hiagashi, J.; Eastman, A. D.; Parry, R. W. (Inorg. Chem. **21** [1982] 716/20).
[44] Hoffmann, R. W.; Kemper, B. (Tetrahedron Letters **23** [1982] 845/8).

[45] Hoffmann, R. W.; Weidmann, U. (J. Organometal. Chem. **195** [1980] 137/46).
[46] Hoffmann, R. W.; Kemper, B. (Tetrahedron Letters **21** [1980] 4883/6).
[47] Hoffmann, R. W.; Kemper, B. (Tetrahedron **40** [1984] 2219/24).
[48] Haubold, W.; Fehlinger, H. G.; Frey, G. (Z. Naturforsch. **36b** [1981] 157/60).
[49] Sicker, U.; Meller, A.; Maringgele, W. (J. Organometal. Chem. **231** [1982] 191/203).
[50] Anton, K.; Fußstetter, H.; Nöth, H. (Chem. Ber. **117** [1984] 2542/6).

[51] Pieper, W.; Schmitz, D.; Paetzold, P. (Chem. Ber. **114** [1981] 3801/12).
[52] Anton, K.; Euringer, C.; Nöth, H. (Chem. Ber. **117** [1984] 1222/34).
[53] Anton, K.; Nöth, H.; Pommerening, H. (Chem. Ber. **117** [1984] 2479/94).
[54] Narula, C. K.; Nöth, H. (Inorg. Chem. **23** [1984] 4147/52).
[55] Narula, C. K.; Nöth, H. (J. Chem. Soc. Chem. Commun. **1984** 1023/4).
[56] Weber, W.; Niedenzu, K. (J. Organometal. Chem. **205** [1981] 147/56).
[57] Alam, F.; Niedenzu, K. (J. Organometal. Chem. **240** [1982] 107/19).
[58] Habben, C.; Maringgele, W.; Meller, A. (Z. Naturforsch. **37b** [1982] 43/53).
[59] Noltemeyer, M.; Sheldrick, G. M.; Habben, C.; Meller, A. (Z. Naturforsch. **38b** [1983] 1182/91).
[60] Alam, F.; Niedenzu, K. (J. Organometal. Chem. **243** [1983] 19/30).

[61] Maringgele, W. (J. Organometal. Chem. **222** [1981] 17/32).
[62] Haubold, W.; Hrebicek, J.; Sawitzky, G. (Z. Naturforsch. **39b** [1984] 1027/31).
[63] Molt, K.; Sawodny, W. (Z. Anorg. Allgem. Chem. **474** [1981] 182/91).
[64] Koide, S.; Nakamura, K.; Yoshimura, K. (Kenkyu Kiyo Nihon Daigaku Bunrigakubu Shizen Kagaku Kenkyusho No. 14 Pt. 5 [1979] 9/14 from C.A. **94** [1981] No. 19465).
[65] Liepins, V.; Kuznetsova, I. G.; Maier, A. A.; Shumilina, O. E.; Bredikhin, V. I. (Tr. Mosk. Khim. Tekhnol. Inst. **108** [1979] 89/90 from C.A. **95** [1981] No. 11388).
[66] Constant, G.; Feurer, R. (J. Less-Common Metals **82** [1981] 113/8).
[67] Feurer, R.; Cros, G.; Constant, G. (Proc. 3rd Eur. Conf. Chem. Vapor Deposition, Neuchâtel, Switz., 1980, pp. 49/55 from C.A. **95** [1981] No. 174092).
[68] Costes, J. P.; Cros, G. (J. Chem. Res. S **1982** 37).
[69] Paciorek, K. J. L.; Kratzer, R. H.; Harris, D. H.; Smythe, M. E. (Polym. Prepr. Am. Chem. Soc. Div. Polym. Chem. **25** [1984] 15/6 from C.A. **101** [1984] No. 55318).
[70] Gasparis-Ebeling, T.; Nöth, H. (Angew. Chem. **96** [1984] 301; Angew. Chem. Intern. Ed. Engl. **23** [1984] 303).

[71] Nöth, H.; Weber, S. (Z. Naturforsch. **38b** [1983] 1460/5).
[72] Nöth, H.; Weber, S. (Angew. Chem. **96** [1984] 998; Angew. Chem. Intern. Ed. Engl. **23** [1984] 994).
[73] Lee, D. Y.; Martin, J. C. (J. Am. Chem. Soc. **106** [1984] 5745/6).
[74] Chausse, T.; Pascal, J. L.; Potier, A.; Potier, J. (Nouv. J. Chim. **5** [1981] 261/9).
[75] Nöth, H.; Weber, S.; Rasthofer, B.; Narula, C.; Konstantinov, A. (Pure Appl. Chem. **55** [1983] 1453/61).
[76] Lauer, M.; Wulff, G. (J. Organometal. Chem. **256** [1983] 1/9).
[77] Dorokhov, V. A.; Zolotarev, B. M.; Mikhailov, B. M. (Izv. Akad.Nauk SSSR Ser. Khim. **1981** 869/74; Bull. Acad. Sci. USSR Div. Chem. Sci. **30** [1981] 649/53).
[78] Li, L.; Chen, G.; Zhang, R.; Kang, N. (Jiguang **8** No. 4 [1981] 33/7 from C.A. **96** [1982] No. 133033).
[79] Mohan, S.; Ravikumar, K. G. (Acta Phys. Polon. A **64** [1983] 671/81).
[80] Aron, J.; Ford, T. A. (S. African J. Chem. **35** [1982] 129/38).

[81] Kamaray, V.; Ramasamy, R.; Venkateswarlu, K. (Bull. Classe Sci. Acad. Roy. Belg. **66** [1980] 381/4).

[82] Hahn, R.; Boehling, H.; Fruwert, J. (Acta Chim. [Budapest] **115** [1984] 33/7).
[83] Hartmann, J. S.; McGarvey, B. D.; Raman, C. V. (Inorg. Chim. Acta **49** [1981] 63/7).
[84] Fox, A.; Hartmann, J. S.; Humphries, R. E. (J. Chem. Soc. Dalton Trans. **1982** 1275/83).
[85] Klebe, G.; Hensen, K.; von Jouanne, J. (J. Organometal. Chem. **258** [1983] 137/46).

7 The System Boron-Bromine

Anton Meller
Institut für Anorganische Chemie, Universität Göttingen
Göttingen, Federal Republic of Germany

7.1 Binary Species

7.1.1 BBr and BBr_2

BBr can be detected spectroscopically when BBr_3 is passed through a radiofrequency discharge in order to prepare B_2Br_4 [2]. For the reaction $BBr_3 + BBr \rightarrow B_2Br_4$, the value of the standard Gibbs potential ΔG_f° of B_2Br_4 is unknown; however, the highly positive value of ΔG_f° of BBr (+195.4 kJ/mol) in conjunction with ΔG_f° for BBr_3 (−232.5 kJ/mol) ensures that ΔG_f° of this reaction will be highly negative for any reasonable value of ΔG_f° (B_2Br_4) [1]. In the discharge, an intensive $A \rightarrow X$ band system of BBr is observed in which the longest wavelength band is the (3,5), corresponding to a vibrational excitation energy of 43.7 kJ/mol [1]. The standard enthalpy of formation has been calculated to be ΔH_f°(gas) = +234.3 kJ/mol and the bond order-dependent enthalpy term of the B–Br bond is given with E(B–Br) = 437.6 kJ/mol [2]. Thermochemical data obtained by a combination of tungsten transport measurements and theoretical estimates for BBr_3 (in the gaseous state) are ΔH_{298} = 228 kJ/mol, S_{298} = 225 kJ/mol, and $C_{p,298}$ = 33 J/mol·K [3]. The dissociation energy has been calculated on the basis of molecular constants (from literature data) to be D = 4.4 eV [4].

BBr_2. Thermodynamic data for gaseous BBr_2 have been obtained by a combination of tungsten transport measurements and theoretical estimates. The following numerical values are given: ΔH_{298} = 30 kJ/mol; S_{298} = 295 $J \cdot mol^{-1} \cdot K^{-1}$; $C_{p,298}$ = 49 $J \cdot mol^{-1} \cdot K^{-1}$ [3].

7.1.2 Tribromoborane, BBr_3

7.1.2.1 Preparation and Physical Data

Highly pure BBr_3 (useful as a dopant for optical fibers) is obtained by the reaction of elemental boron with Br_2 at 750°C in N_2 atmosphere and subsequent fractional distillation. Impurities (in %) from heavy metals range between 3.5×10^{-7} (Co) and 10^{-8} (Cr) [5]. The use of BBr_3 as an auxiliary gas in the isotope separation of $^{235}UF_6/^{238}UF_6$ by laser photoexcitation has been patented [6].

The core binding energies of trihaloboranes have been redetermined (with a probable uncertainty of ±0.005 eV) by gas-phase X-ray photoelectron spectroscopy. For BBr_3, however, the boron 1s peak is hidden under the bromine $3p_{1/2}$ peak. The core binding energies are given with E_B (in eV) = 199.0 (B 1s) and 76.57 (Br $3d_{5/2}$) [7]. For additional information including the valence ionization potential and the stabilization of the bromine $p\pi$ orbitals by π-bonding to the boron atom, see Table 6/10, p. 9, and the original reference [7]. The photoabsorption cross-section of BBr_3 (as well as those of the other trihaloboranes) has been measured between 190 and 280 eV using synchrotron radiation. Assignments for the BK spectrum are supported by ab initio calculations for the BK bands of BF_3 and BCl_3. The experimental results are compiled in Table 7/1, p. 72 [8].

The force field of BBr_3 has been studied using the parametric representation method. For the resulting force constants, mean amplitudes of vibration, shrinkage constants, Coriolis coupling constants, and rotational distortion constant, see tabulation in [9]. An excellent review of experimental vibrational data and normal coordinate analyses for the trihaloboranes

References for 7.1 on pp. 79/83

is reported elsewhere [10]. Harmonic force fields of the E' species are produced by using isotopic frequencies and Coriolis constants as additional data. The usefulness of Keating coordinates versus valence coordinates as basis of force field approximations is discussed. Final force fields are proposed for trihaloboranes with the aid of the mass influence of Coriolis constants (see Table 7/2). For a detailed discussion, see [10].

Table 7/1

Experimental Term Values and Oscillator Strengths of the BK Bands of BBr_3 (the ionization threshold energy is 199.0 eV, see [7]) [8].

band	term value in eV	integrated oscillator strength	assignment
a	7.0	0.021	$2p\pi\ (a_2'')$
b	6.2	0.003	?
c	4.1	0.003	$3s+\sigma^*(a_1')$
d	3.2	0.039	$2p\sigma^*(e')$
e	2.0	0.001	$3p\pi\ (a_2'')$
			$3p\sigma\ (e')$
f	1–3	0.0043	$4p\pi\ (a_2'')$
g	0.6	0.0005	$4p\sigma\ (e')$

Table 7/2

Final E' Force Constants (in mdyn/Å) for Trihaloboranes in Terms of Valence, Central, and Keating Coordinates (F_{ij} based on valence, F_{ij}^c on central, F_{ij}^k on Keating coordinates; [a] frequencies from [50]; [b] frequencies from [51]) [10].

	(ij)	(11)	(12)	(22)
BF_3	F_{ij}	6.698	−0.3849	0.5089
	F_{ij}^c	6.892	0.9931	2.036
	F_{ij}^k	6.423	−0.1051	0.6785
BCl_3[a]	F_{ij}	3.248	−0.1663	0.2343
	F_{ij}^c	3.374	0.4791	0.9373
	F_{ij}^k	3.134	−0.0358	0.3124
BCl_3[b]	F_{ij}	3.297	−0.2001	0.2570
	F_{ij}^c	3.374	0.4900	1.0281
	F_{ij}^k	3.151	−0.0598	0.3428
BBr_3	F_{ij}	2.622	−0.1475	0.1975
	F_{ij}^c	2.704	0.3891	0.7899
	F_{ij}^k	2.518	−0.0387	0.2633
BI_3	F_{ij}	1.953	−0.0909	0.1371
	F_{ij}^c	2.050	0.2933	0.5485
	F_{ij}^k	1.894	−0.0135	0.1828

A theoretical treatment of the E' block of the mean-square amplitude matrix (Σ) for planar, symmetrical XY_3 molecules in terms of Keating coordinates has been performed. Mean amplitudes of vibration (l) and Bastiansen-Morino shrinkage effect (Δ) are considered in a numerical example for BF_3. The force fields as given in Table 7/2 were applied in order to calculate the mean amplitudes of vibration for BCl_3, BBr_3, and BI_3 as presented in Table 7/3. For a comprehensive review of this and related matter, see [11].

Table 7/3
Calculated and Observed Mean Amplitudes of Vibration (in Å) for Trihaloboranes ($^{11}BX_3$; the ω_1 frequencies employed are 471 cm^{-1} for BCl_3 [50], 278 cm^{-1} for BBr_3 [52], and 190 cm^{-1} for BI_3 [52]; [a] frequencies from [50]; [b] temperature not given; [c] frequencies from [51]; [d] calculated value) [11].

molecule	temperature in K	calculated l_{XY}	calculated l_{YY}	observed l_{XY}	observed l_{YY}	Ref.
BF_3	0	0.04250	0.05215			
	289.15	0.04265	0.05550	0.0398 ± 0.0029	0.0541 ± 0.0020	[53]
	293.15	0.04266	0.05562	0.042_0 ± 0.002	0.055_8 ± 0.003	[54]
	298.15	0.04267	0.05578			
BCl_3[a]	0	0.04785	0.05368			
	298.15	0.04926	0.06904	0.045 ± 0.003_5	0.0637 ± 0.004_0	[55][b]
	304.15	0.04934	0.06954	0.0505 ± 0.0023	0.0700 ± 0.0023	[53]
BCl_3[c]	0	0.04785	0.05287			
	298.15	0.04923	0.06691	0.045 ± 0.003_5	0.0637 ± 0.004_0	[55][b]
	304.15	0.04931	0.06738	0.0505 ± 0.0023	0.0700 ± 0.0023	[53]
BBr_3	0	0.04809	0.04611			
	290.15	0.05094	0.07123	0.0514 ± 0.0063	0.0759 ± 0.0031	[56]
	298.15	0.05112	0.07213			
BI_3	0	0.05054	0.04482			
	298.15	0.05567	0.08458			
	333.15	0.05690	0.08915	(0.059)[d]	0.078 ± 0.006	[57]

7.1.2.2 Chemical Behavior

Chemical Vapor Deposition and Related Phenomena

Amorphous boron deposits are produced by laser chemical vapor deposition (LCVD) of BBr_3 (for applications in microelectronics) [12]. A kinetic model has been developed for the interpretation of the experimental growth rate of boron layers by pyrolysis of BBr_3 under reduced pressure. It is assumed that BBr_3 is adsorbed with subsequent dissociation on the boron surface without activation energy. The boron atoms combine with the substrate lattice and Br is chemisorbed in the form of atoms. Bromine may either desorb as atoms or recombine to give Br_2 which can be desorbed in the molecular form or as a bromide. The growth rate depends on the pressure, and the deposition yield depends on the flow rate of the BBr_3 and the pumping speed in the reactor. For details on the calculations, see [13, 14].

References for 7.1 on pp. 79/83

Reduction of BBr_3 with hydrogen can be used for the CVD of elemental boron [15]. Codeposition of boron-doped silicon can be achieved by gas-phase reactions of an $HSiCl_3/BBr_3/H_2O$ mixture (which decreases the boron incorporation into the Si lattice); silanoxy(halo)boranes are probable intermediates in this process [16]. The results of experimental investigations in the $HSiCl_3/H_2$ system with addition of BBr_3 are in agreement with thermodynamic calculations of the dependence of the boron contents on the given parameters (up to a concentration of 5×10^{16} boron atoms per cm^3 in the deposited silicon). However, at low boron concentrations a significant decrease of the incorporation is observed which is not explained by the proposed model [17]. Thermodynamic equilibrium calculations in the B/Si/Br and B/Si/Br/H systems show that the solid-phase composition is strongly dependent on the amount of hydrogen introduced. The Si content of the solid phase increases with the partial pressure of hydrogen. It is possible to describe the entire B/Si phase diagram when a large excess of hydrogen is used. Sometimes, solid two-phase mixtures of SiB_{15} and SiB_6 are observed. If no hydrogen is introduced, only a single solid phase is formed at pressures of about 0.02 Torr. The silicon content of this phase is below 6.5% [18]. The electrical properties (resistivity, thermoelectric power) of such boron-rich phases with some Si content have been studied and a sketch of the density of states in the band gap (especially for B/Si/Fe phases) is given [19].

The system $BBr_3/TiCl_4$ using H_2 as carrier gas is used for boronizing Armco Fe and chromium or chromium/molybdenum alloy steels. Layers of FeB and Fe_2B are formed and the content of Fe_2B increases with the boronizing temperature from 773 to 1373 K. The growth of the boride layers is diffusion-controlled and proportional to the square root of the boriding time. The activation energy of boron diffusion is 25.2 kcal/mol in Fe_2B and 19.7 kcal/mol in FeB. In order to obtain a Fe_2B-rich bridge layer (with superior properties), the FeB phase formed initially by low-temperature boronizing is subsequently transformed to Fe_2B by annealing in argon. In high-temperature boronizing there is a considerable deposition of boron on the reactor walls if BBr_3/H_2 is used as the boriding source. $SnCl_4$ or $TiCl_4$ are sometimes used in the activation of low-temperature BBr_3/H_2 boronizing [20 to 23]. Boronizing can also be achieved by direct reduction of (alloyed) iron oxide in the presence of BBr_3 (or trialkylboranes) [24].

Carbon fibers can be coated with TiC and TiB_2 layers by treatment with $TiCl_4/BBr_3$ (preferably between 1040 and 1120°C) [25]. BBr_3 is employed in order to reduce the carbonization time of polyacrylonitrile fibers (at 1000°C in Ar) in the production of carbon fibers [26]. A model has been developed for the boron deposition in silicon using a BBr_3 source. The profile of deposited layers as a function of the doping gas composition can be modeled by the change of silicon self-interstitial concentration [27]; see also [28, 29]. The use of BBr_3 in optical fiber fabrication using vapor-phase axial deposition (VAD) and flame hydrolysis of the halide has been studied [30 to 32].

Reactions with Inorganic Materials

BBr_3 is oxidized by O_2 at temperatures above 250°C; the reaction yields gaseous B_2O_3 (independently of the stoichiometry of the reactants) [33]. The wet hydrochloric acid oxidation of BBr_3-diffused bases is applied in the manufacture of very large scale integration transistors (VLSI) [34]. The reaction of BBr_3 with H, O, and OH radicals has been studied by electron spin resonance and mass spectrometry via a flow-discharge reactor. The relevant rate constants are ($cm^3\cdot molecule^{-1}\cdot s^{-1}$): $BBr_3+H\rightarrow HBBr_2+Br$: $k_{1(295\,K)}=(2.8\pm0.5)\times10^{-13}$; $BBr_3+O\rightarrow OBBr_2+Br$: $k_{2(295\,K)}=(5.9\pm0.8)\times10^{-12}$; $BBr_3+OH\rightarrow HOBBr_2+Br$: $k_{3(295\,K)}=(1.4\pm0.4)\times10^{-11}$. The reaction of BBr_3 with O_2 and H_2O might be used in the generation of underground explosions for the extraction of bituminous oils from oil shale deposits [37].

By reaction of BBr_3 with the anions $[BH_4]^-$, $[B_3H_8]^-$, $[B_4H_9]^-$, or $[B_9H_{14}]^-$ in the absence of solvent, the corresponding boranes (B_2H_6, B_4H_{10}, B_5H_{11}, or $B_{10}H_{14}$, respectively) are obtained in high yields [38 to 40]. Treatment of $[BrB_3H_7]^-$ with BBr_3 results in a 15% yield of $2\text{-}BrB_4H_9$ [40]. A Friedel-Crafts catalytic halogenation system consisting of $AlBr_3$ and BBr_3 is used in the bromination of 1,2'-$(B_5H_8)_2$ and 2,2'-$(B_5H_8)_2$ [41]. B_6H_{10} forms a complex with BBr_3 which undergoes halogen exchange to give BrB_6H_9 [42].

Treatment of zeolite HY with BBr_3 gives HBr and Si–O–BBr_2 centers [43]. The reaction of $(CF_3)_2NH$ with BBr_3 at 100°C followed by reduction with Mg in tetrahydrofuran affords CF_3NC [44], and selective halogen exchange is observed when fluoroalkylamines $(CF_3)_{3-n}N(CHF_2)_n$ (n=1 to 3) are treated with BBr_3 to give the corresponding $CHBr_2$ moieties [45].

The acidity order of the trihaloboranes, which classifies BBr_3 as a stronger Lewis acid than BCl_3 or BF_3, can be supported by the result of CNDO/2 calculations. Thus, the instabilization of BX_3 upon complexation by the hybridization change from sp^2 to sp^3 is only 8.49 kcal/mol for BBr_3 as compared to 10.34 kcal/mol for BCl_3 and 29.00 kcal/mol for BF_3 [46]. Trialkylsilanyltriflates form strongly polarized donor-acceptor complexes with BBr_3, e.g., $[R_3Si][CF_3SO_2]$-O-BBr_3 (R = CH_3, C_2H_5) [47, 48], and with perfluoroalkanesulfonic acids the conjugate superacids $[(C_nF_{2n+1})SO_2OH_2][B\{OSO_2(C_nF_{2n+1})\}_4]$ are formed [49].

The reaction of BBr_3 with solid WO_3 leads to the formation of $WOBr_2$ and B_2O_3 [58]. Treatment of OsF_6 or IrF_6 with BBr_3 yields $Os_2Br_4F_5$ and $Ir_2Br_4F_5$, respectively [59], and the reaction with UF_5 gives α-UBr_5 but with UCl_6 the β-UBr_5 is obtained (if CH_2Cl_2 is present) [60]. Adhesive deposits in equipment for handling of UF_6 can be removed by reaction with BBr_3 followed by treatment with F_2 [61]. The reaction of PuF_6 with BBr_3 (in perfluoro-alkanes) affords PuF_4, BF_3, and Cl_2 [62].

Polymers obtained from $[(CH_3)_3Si]_2NH$ and BBr_3 are described in a patent [140], and the polymerization of (cyclic)polyhalophosphazenes catalyzed by BBr_3 has been studied [141, 142].

Reactions with Organometallic and Organic Materials

When $(CH_3)_3SnBr$ is dissolved in an excess of liquid BBr_3, the products are CH_3BBr_2 and $(CH_3)_2SnBr_2$; and with $(CH_3)_2SnCl_2$, transhalogenation results in the formation of $(CH_3)_2SnBr_2$ and BCl_3 [63]. With (2,4-pentanedionato)dipropylboron, (2,4-pentanedionato)boron dibromide is obtained [64].

t-Butylimino(2,2,6,6-tetramethylpiperidinyl)borane, (tmp)B=N(*t*-C_4H_9), reacts with BBr_3 to give the four-membered heterocycle [–BBr–N(*t*-C_4H_9)–BBr_2–tmp–] containing both three- and four-coordinate boron and nitrogen [65, 66]. The reaction of $(tmp)_2BF + 2BBr_3$ leads to the boronium(1+)salt $[(tmp)B(tmp)][BBr_4]$ and $BrBF_2$ [67]. With 2-bromo-1,3-dimethyl-1,3,2-diazaborolidine, a 1:1 adduct is formed with BBr_3 coordinated to one of the annular nitrogen atoms [68]. If two of the 1,3-dimethyl-1,3,2-diazaborolidine rings are connected by a B–B bond, BBr_3 can be coordinated to either one or both rings [69]. If 3,5-bis(ethylthio)-1,2,4,3,5-trithiadiborolane is treated with BBr_3, the ethylthio groups are replaced by Br [70]. In the reaction of 3,5-dimethyl-1,2,4,3,5-trithiadiborolane with $^{10}BBr_3$, the ^{10}B atoms are incorporated into the ring and 3,5-dibromo-1,2,4,3,5-trithiadiborolane is formed [71].

N-Borylated propargylamines are obtained from BBr_3 and $RN[Si(CH_3)_3]CH_2C{\equiv}CH$ (R = C_2H_5, $C_6H_5CH_2$) by elimination of $(CH_3)_3SiBr$. With $RNHCH_2C{\equiv}CH$, a 1-azonia-2-borata-3-cyclopentene ring is formed by haloboration of the C≡C triple bond and intramolecular N–B coordination; $RN(CH_2C{\equiv}CH)_2$ and $N(CH_2C{\equiv}CH)$ give bicyclic and tricyclic systems in analogous fashion [72]. Reaction of BBr_3 with $(CH_3NH)_2CS$ yields $(\text{–BBr–NCH}_3)_3$, $BrB(\text{–NCH}_3\text{–CS–})_2NCH_3$, and other products [73]. The reaction of BBr_3 with $(CH_3)_3SiC{\equiv}COC_2H_5$ gives $BrB[(CH_3)_3Si]\text{–C=C=O}$ [74], and with $(C_6H_5)_3SiLi$ the trisilanylborane $[(C_6H_5)_3Si]_3B$ is obtained [75]. Silacyclobutanes are cleaved by BBr_3 to yield $R_2SiBr(CH_2)_3BBr_2$ (R = alkyl, vinyl, C_6H_5) [76].

References for 7.1 on pp. 79/83

BBr_3 is used as a catalyst in the hydroboration reaction of alkynes with thexylchloroboranes (thexyl=1,1,2-trimethylpropyl) [79]. It is also used for the synthesis of alkyldibromoboranes and dialkylbromoboranes via redistribution reactions with trialkylboranes (catalyzed by $(CH_3)_2S–BH_3$) [80]. Bromoboronation of terminal alkynes with BBr_3 gives $Br_2B–CH=CRBr$ [81]. A mixture of mono- and dibromocyclopentadienes is obtained from hexachlorocyclopentadiene [82], and BBr_3 is also applied in selective and regio-specific brominations of polycyclic aromatic compounds [83]. The reaction of BBr_3 with *cis*-3,4-dichlorocyclobutene yields *trans*-3,4-dibromocyclobutene [84].

The bis(mesitylthiogermylene)chromiumpentacarbonyl complex, $\{2,4,6\text{-}(CH_3)_3C_6H_2\}_2Ge\text{-}Cr(CO)_5$, is converted to the corresponding dibromogermylene complex by reaction with BBr_3 [77], while the corresponding tungsten complex gives the three-membered ring compound $[–GeBr_2–W(CO)_5–W(CO)_5–]$ [78].

BBr_3 is used in converting alkylidene complexes of tungsten and chromium to *trans*-halo(tetracarbonyl)metallocarbynes [85, 86]. The species $BrWOR_3 \cdot BBr_3$ (R = mesityl) is obtained from the individual components (in C_6H_6) and used as catalyst in olefin metathesis [87]. Transition-metal perfluoroalkyl complexes such as $(CO)_5MnCF_3$ are readily converted into the corresponding tribromoalkyl derivatives [88]. Regiospecific halogen exchange is observed at the C atom bonded to the metal; thus, $(CO)_5ReC_2F_5$ is converted to $(CO)_5ReCBr_2CF_3$ [89]. Treatment of *cis*-$(CO)_4ReCOCH_3(CH_3C{=}NH_2^+)$ with NaH and then BBr_3 gives the corresponding rhena-β-ketoiminatodibromoborane chelate [90]. $[As(C_6H_5)_4][ReNCl_4]$, upon treatment with an excess of BBr_3, gives $[As(C_6H_5)_4][Br_4Re{=}NBBr_3]$ [91]; and *trans*-$Os[NC(CCl_3)NCCl(CCl_3)]_2Cl_4$ reacts with (excess) BBr_3 to yield $Os[NC(CCl_3)NCBr(CCl_3)]_2Br_4$ [92]. The reaction between *closo*-3,3-$[(C_6H_5)_3P]_2$-3-H-3,1,2-$RhC_2B_9H_{11}$ and BBr_3 leads to the hydrogen-bonded ion pair $[(C_6H_5)_3PH]$*closo*[-3-$(C_6H_5)_3P$]-3,3-Br_2-3,1,2-$RhC_2B_9H_{11}$ [93].

BBr_3 is a potent reagent for cleaving of ethers, lactones, acetals, esters, and glycosidic bonds [94 to 104]. It is used in the determination of polyphenylene glycol bonded to epoxy-activated Sepharose 6B [105], and in the chemical degradation of residual organic matter from laminated cyanobacterial mats [106].

Sulfoxides are deoxygenated by treatment with BBr_3 [107], and CH_3SCH_2X (X = Cl, Br) forms an adduct $XCH_2(CH_3)S–BBr_3$ [108]. Sulfonic acid derivatives are reduced by iodide in the presence of BBr_3 to yield species of lower oxidation state; an intermediate $[BrSO_2OBBr_2]$ is assumed [109]. BBr_3 catalyzes the tritium-labelling of steroids and aryl rings [110, 111], and it is used together with $Os_3(CO)_{12}$ in the catalytic hydrogenation of carbon monooxide [112]. Additional applications of BBr_3 are in the preparation of image-providing materials for photography [113]. Ultratrace impurities in BBr_3 are determined by flameless atomic spectroscopy [114].

7.1.2.3 Selected Adducts of BBr_3 with Lewis Bases

$(CH_3)_2O–BBr_3$ and **$(CD_3)_2O–BBr_3$** have been studied (in an argon matrix) by IR spectroscopy. The following assignments are given (in cm^{-1}): $\nu_{as}(^{10}BBr_3)$ at 722, $\nu_{as}(^{11}BBr_3)$ at 680, $\nu_{as}(CO)$ at 980, $\nu_s(CO)$ at 887; a weak band at 418 cm^{-1} is tentatively assigned to $\nu(BO)$ or $\delta_s(BBr_3)$. For a discussion in context with the spectra of the corresponding adducts with BF_3 and BCl_3, see [114].

It has been claimed (on the basis of conductance and IR studies) that dichloroacetic acid and BBr_3 will form an adduct $Cl_2CHCOOH \cdot BBr_3$ which, in the presence of an excess of dichloroacetic acid, should give $[Cl_2CHCOOH_2][Cl_2CHCOOBBr_3]$ [115]. However, it is unlikely that in the presence of RCOOH a species containing a BBr_3 group will exist.

N,N-Disubstituted amides such as $HCON(C_6H_5)_2$, $CH_3CON(CH_3)_2$, $C_6H_5CON(CH_3)_2$, and $C_6H_5CON(C_6H_5)_2$ form 1:1 adducts with BBr_3; the enthalpies for the complexation range between −139.9 kJ/mol (for $CH_3CON(CH_3)_2$) and −89.7 kJ/mol (for $C_6H_5CON(C_6H_5)_2$). The reported IR data suggest coordination of the BBr_3 to the CO group [116].

Strongly polarized donor-acceptor complexes are formed between trimethyl- and triethylsilanyltriflate and BBr_3 (triflate = trifluoromethane sulfonate, CF_3SO_3H) [47].

$(R_3Si)(CF_3SO_2)O–BBr_3$ (R = CH_3, C_2H_5). The individual components are combined at about −40°C. Upon warming and reaching ambient temperature, two ^{29}Si NMR signals are observed (R = CH_3) with δ = 61.52 ppm for the adduct and δ = 25.74 ppm for $(CH_3)_3SiBr$ (formed by ligand exchange); $\delta^{11}B$ for the complex = −2.45 ppm. If **$(C_2H_5)_2Si(CF_3SO_3)_2$** is reacted with BBr_3 in CH_2Cl_2, again a 1:1 adduct is observed which exhibits $\delta^{29}Si$ = 31.04 ppm and $\delta^{11}B$ = −2.45 ppm. In contrast, $(t\text{-}C_4H_9)(CH_3)_2SiOSO_2CF_3$ reacts only under cleavage of the Si–O bond to yield $(t\text{-}C_4H_9)(CH_3)_2SiBr$; no silicenium ion formation has been observed [47]; see also [48].

The reaction enthalpy of the formation (in kJ/mol at 298 K) of the 1:1 adducts **$CH_3NO_2–BBr_3$** and **$C_6H_5NO_2–BBr_3$** has been determined to be ΔH = −58.2 and −55.6, respectively [143].

$H_3N–BBr_3$. The IR spectrum has been recorded using matrix isolation in Ar and the suggested assignments are supported by the spectra of $H_3^{15}N–BBr_3$ and $D_3N–BBr_3$. The reported data are: $\nu_{as}(BBr_3)$ at 804 + 839 cm^{-1}; $\nu_s(BBr_3)$ at 399 cm^{-1}; $\varrho(NH_3)$ at 685 cm^{-1}; ν(BN) at 600 + 620 cm^{-1} [117].

$C_2H_5CN–BBr_3$ is prepared by condensation of C_2H_5CN on BBr_3 at −196°C, warming to −4°C for 1 h, and removing excess of BBr_3 under reduced pressure and drying the remaining material under vacuum at −15°C. After recrystallization from benzene, the melting point of the adduct is 175 to 180°C. The proton NMR signals of the CH_3-group (triplet) and the CH_2-group (quartet) are shifted downfield by 25 and 41 Hz, respectively (solvent: $CHCl_3$), as compared to the free nitrile. In the IR spectrum, ν(C≡N) appears near 2340 cm^{-1} [118].

$C_5H_5N–BBr_3$. The complexation enthalpy at 298 K has been determined to be −156 kJ/mol [119]. For phosphorescence spectra, lifetimes, and zero-field splitting parameters of the triplet state (also for the isoquinoline adduct), see [120].

$3\text{-}X\text{–}C_5H_4N–BBr_3$ (X = F, Cl, Br, CN). The following ^{11}B-chemical shifts have been observed: X = F: δ = 7.8 ppm; X = Cl: δ = 7.8 ppm; X = Br: δ = 7.9 ppm; X = CN: δ = 7.7 ppm. For **$4\text{-}NC\text{–}C_5H_4N–BBr_3$**, $\delta^{11}B$ = 7.7 ppm. For a tabulation of the ^1H NMR spectra and a comparison with the corresponding BF_3 adducts ($\delta^{11}B \approx -0.3$ ppm), see [121].

$C_6H_5–CO–C_6H_4\text{-}4\text{-}H_2N–BBr_3$. For IR and conductivity data, see [122].

$C_6H_5(C_6H_5CH{=})N–BBr_3$. The electronic distribution has been calculated by the CNDO approximation, and the dipole moment was found to be μ_c = 10.4 D [123].

$H_3P–BBr_3$. The mean amplitudes of vibration for phosphane-tribromoborane (and also for **$D_3P–BBr_3$** and **$F_3P–BBr_3$**) have been calculated on the basis of structural and spectroscopic data from the literature. For a comprehensive tabulation of the results, see [124].

$X_3P–BBr_3$ (X = F, Cl, Br). A CNDO treatment was used to calculate electron transfer, dipole moments, and the complexation energies; results are tabulated [143].

$Br_3P–BBr_3$. The ^{81}Br nuclear magnetic resonance (NQR) spectrum has been recorded and the Zeeman effect in a single crystal (orthorhombic) was studied at 293 K. The ^{81}Br NQR data are listed in Table 7/4, p. 78. The asymmetry parameters η (see Table 7/4) are quite large for the bromine atoms for BBr_3, while those of the PBr_3 group are small. This effect probably is due to

References for 7.1 on pp. 79/83

the B–Brπ-bonding. The bond angles have been determined with ∢(BrBBr) = 107.0° ± 0.1° and ∢(BrPBr) = 109.1° ± 0.3° [125].

Table 7/4
^{81}Br NQR Parameters of $Br_3P–BBr_3$ at 293 K [125] and of BBr_3 at 77 K [126] (η is the asymmetry parameter, f denotes the degree of π-bonding).

compound		frequency in MHz	$e^2q_{zz}Q/h$ in MHz	η	f	assignment
$PBr_3 \cdot BBr_3$	ν_1	150.31	299.48	0.151 ± 0.004	0.05	BBr_3
	ν_2	151.68	302.63	0.120 ± 0.007	0.04	
	ν_3	212.53	425.01	0.025 ± 0.005		PBr_3
	ν_4	213.22	426.41	0.020 ± 0.004		
BBr_3		146.41	283.80	0.45	0.12	

$(CH_3)_2HP–BBr_3$. The IR and Raman spectra of dimethylphospane-tribromoborane and its isotopically labelled species **$(CH_3)_2PD–BBr_3$** and **$(CD_3)_2PH–BBr_3$** have been recorded in the solid state and in solutions at ambient temperature. The fundamentals were assigned for C_s symmetry. Frequencies, potential energy distribution, and force constants have been calculated from a modified Urey-Bradley force field model. The P–B stretching mode is assigned at 765 + 775 cm^{-1}, and $\nu_{as}(BBr_3)$ at 616 cm^{-1}. For a comprehensive tabulation of the spectra, see [127].

$(CH_3)_3P–BBr_3$. The molecular structure of trimethylphosphane-tribromoborane was obtained from gas-phase electron diffraction and vibrational spectroscopic data. The following bond distances (in ppm) and bond angles (in degrees) have been obtained: r(BBr) = 201.0; r(PB) = 194.6; r(CP) = 180.4; r(CH) = 109.8. ∢(BrBBr) = 111.7, ∢(CPC) = 108. The rotational barrier about the P–B bond is estimated to be about 10 kcal/mol [128].

For **$[(C_2H_5)_2N]_2P(CH_2)_4P[N(C_2H_5)_2]_2 \cdot 2BBr_3$**, see [129]; for the use of **$(C_6H_5)_3P–BBr_3$** as cross-linking agent for epoxy resins, see [130].

$(CH_3)_3As–BBr_3$ and **$(C_6H_5)_3As–BBr_3$** have been prepared and the IR, Raman, ^{1}H, and ^{13}C NMR spectra have been recorded and assigned. The assignments are supported by modified Urey-Bradley force field calculations. The As–B stretching modes are assigned at 655 cm^{-1} (for the methyl) and at 625 cm^{-1} (for the phenyl) derivative; $\nu_{as}(BBr)_3$ is at 618 and 543 cm^{-1}, respectively. For a comprehensive tabulation of the spectra, see [131].

7.1.3 The Tetrabromoborate Ion, $[BBr_4]^-$

Using magic angle rotation (MAR) at about 1.2 kHz, sufficiently narrow ^{11}B NMR lines are obtained for solid tetrahaloborates $[PX_4][BX_4]$ so that the signals for the ions $[BBr_4]^-$ and $[BCl_nBr_{4-n}]^-$ can be resolved. For a depiction of the spectra, see [132]. Molecular constants (kinetic constants, force constants, mean amplitudes of vibration, Coriolis coupling constants) have been calculated for 22 molecules of the type XY_4 of T_d point group by application of the kinetic constant method developed earlier by the authors [133]. Data for $[BBr_4]^-$ have been evaluated as follows: Kinetic constants (in 10^{-23} g): $k_d = 9.9278$, $k_{dd} = 1.0583$, $k_\alpha = 0.6471$, $-k'_{\alpha\alpha} = -0.0808$, $-k'_{d\alpha} = 0.748$; force constants (in mdyn/Å): $f_d = 1.2283$, $f_{dd} = 0.5056$, $f_\alpha = 0.0454$,

$-f_{\alpha\alpha}=0.0100$, $-f_{d\alpha}=0.0732$; compliance constants (in Å/mdyn): $c_d=14.858$, $-c_{dd}=4.8312$, $c_\alpha=33.6942$, $-c'_{d\alpha\alpha}=30.5516$, $-c'_{d\alpha}=13.0089$; valence mean square amplitudes of vibration (in 10^{-3} Å^2) at 298.16 K: $\sigma_d=3.8098$, $\sigma_{dd}=0.7169$, $\sigma_\alpha=23.81$, $\sigma'_{\alpha\alpha}=10.4191$, $\sigma'_{d\alpha}=2.9227$; mean amplitudes (in 10^{-1} Å) of vibration at 298.16 K: $l_{(B-Br)}=6.1724$, $l_{(Br-Br)}=8.0927$. Coriolis coupling constants: $\zeta_{33}=0.9131$, $\zeta_{44}=-0.4131$, $\zeta_{34}=0.3504$, $\zeta_{23}=0.2407$, $-\zeta_{24}=0.9706$ [133].

$[N(C_2H_5)_4][BBr_4]$. For an interpretation of mass spectral fragmentation, see [134]. For cationic metalcarbyne complexes, $[(\eta^5\text{-}C_5H_5)(CO)_2M{\equiv}CR][BBr_4]$ (M = Mn, Re; R = C_6H_5, 4-CF_3–C_6H_4), see [135, 136].

Tetrabromide salts of bis(amino)boron cations with two-coordinate boron such as [(tmp)=B=tmp]$[BBr_4]$ (tmp = tetramethylpiperidinyl) have been synthesized from $(tmp)_2BF$ + 2BBr_3 [64]. The decomposition temperature of the species is 195 to 198°C. In the ^{11}B NMR spectrum, $\delta=35.4$ ppm (cation) and −25.1 ppm $[BBr_4]^-$; for ^{1}H and ^{13}C NMR data, see [64].

7.1.4 Additional Binary Species

Species containing B–B bonds are summarized in Chapter 2 of "Boron Compounds" 3rd Suppl. Vol. 1, 1987, pp. 66/7, 145, 151, 182. For a high-yield synthesis of B_2Br_4 from $B_2(OCH_3)_4$ and BBr_3, see [137]; possible intermediates in the discharge synthesis of B_2Br_4 are discussed in [1]. The preparation, ^{11}B NMR, and vibrational spectra of $[B_6Br_6]^{2-}$ are reported by [138], and B_8Br_8 and B_9Br_9 have been prepared from the corresponding chlorine-substituted clusters by reaction with Al_2Br_6 [139].

References for 7.1:

[1] Briggs, A. G.; Massey, A. G.; Reason, M. S.; Portal, P. J. (Polyhedron **3** [1984] 369/71).
[2] Holbrook, J. B.; Smith, B. C.; Housecroft, C. E.; Wade, K. (Polyhedron **1** [1982] 701/6).
[3] Dittmer, G.; Niemann, U. (Philips J. Res. **37** [1982] 1/30).
[4] Bogomolova, E. A.; Moskvitina, E. N.; Nikonorov, A. P. (Vestn. Mosk. Univ. Ser. II Khim. **22** [1981] 531/5).
[5] Yuan, Q.; Zhang, K.; Wang, J. (Huaxue Shijie **23** [1982] 194/6 from C.A. **100** [1984] No. 94302).
[6] Matieu, P. (Belg. 883842 [1980] from C.A. **95** [1981] No. 140669).
[7] Beach, D. B.; Jolly, W. L. (J. Phys. Chem. **88** [1984] 4647/9).
[8] Ishiguro, J. E.; Iwata, S.; Suzuki, Y.; Mikuni, A.; Sasaki, T. (J. Phys. B. **15** [1982] 1841/54).
[9] Mohan, S.; Rajaraman, S. (Indian J. Pure Appl. Phys. **20** [1983] 230/2).
[10] Cyvin, S. J.; Cyvin, B. N.; Brunvoll, J. (Spectrosc. Letters **17** [1984] 511/23).

[11] Cyvin, S. J.; Cyvin, B. N. (Spectrosc. Letters **17** [1984] 525/35).
[12] Ibbs, K. G. (Quantum Electron. Electro-Opt. Proc. 5th Natl. Quantum Electron. Conf., Hull, U.K., 1981 [1983], pp. 103/5 from C.A. **99** [1983] No. 149277).
[13] Combescure, C.; Armas, B.; Alnot, M.; Weber, B. (J. Electrochem. Soc. **128** [1981] 358/61).
[14] Combescure, C.; Armas, B.; Alnot, M.; Weber, B. (Proc. Electrochem. Soc. **79** Pt. 3 [1979] 351/9 from C.A. **95** [1981] No. 33584).
[15] Savel'ev, B. A.; Anikin, V. N.; Pozdnyakova, O. P. (Kachestvo Effekt. Primeneniya Tverd. Splavov M. **1984** 107/9 from C.A. **101** [1984] No. 232553).
[16] Bartsch, K.; Wolf, E. (Cryst. Res. Technol. **17** [1982] 405/9 from C.A. **97** [1982] No. 16163).
[17] Bartsch, K.; Wolf, E. (Cryst. Res. Technol. **17** [1982] 327/32 from C.A. **96** [1982] No. 190857).
[18] Armas, B.; Combescure, C.; Bernard, C. (J. Less-Common Metals **95** [1983] 247/57).

[19] Dusseau, J. M.; Robert, J. L.; Armas, B.; Combescure, C. (J. Less-Common Metals **82** [1981] 137/42).
[20] Plaenitz, H.; Treffer, G.; König, H.; Marx, G. (Neue Hütte **27** No. 6 [1982] 228/30 from C.A. **97** [1982] No. 186216).

[21] Bochmann, G.; Spoerl, T.; Ritzel, B.; Wiesner, S.; Wagner, W.; Marx, G. (Neue Hütte **29** No. 1 [1984] 26/8 from C.A. **100** [1984] No. 124867).
[22] Plaenitz, H.; Treffer, G.; König, H.; Marx, G.; Bahr, D. (Ger. [East] 208632 [1984] from C.A. **101** [1984] No. 115091).
[23] Plaenitz, H.; Treffer, G.; König, H.; Marx, G. (Ger. [East] 139602 [1980] from C.A. **95** [1981] No. 224125).
[24] Kemnitz, H.; Morgenstern, B.; Otto, W.; Schönberg, P.; Gebes, B. (Ger. [East] 159647 [1983] from C.A. **99** [1983] No. 108948).
[25] Shorshorov, M. Kh.; Savvateeva, S. M.; Tsirlin, A. M.; Mosin, A. M.; Pronin, Yu. E.; Pletyushkin, A. A.; Ivanova, L. M.; Sultanova, T. N.; Sinitsyn, E. M. (Fiz. Khim. Obrab. Mater. **1981** No. 5, pp. 134/9 from C.A. **96** [1982] No. 20777).
[26] Blazewicz, S.; Chlopek, J. (Koks Smola Gaz **28** [1983] 81/3 from C.A. **102** [1985] No. 150780).
[27] Guo, S. F.; Chen, W. S. (J. Electrochem. Soc. **129** [1982] 1592/6).
[28] Hitachi, Ltd. (Japan. Kokai Tokkyo Koho 84-129422 [1984] from C.A. **102** [1985] No. 16111).
[29] Matsushita Electronics Corp. (Japan. Kokai Tokkyo Koho 84-90925 [1984] from C.A. **101** [1984] No. 181215).
[30] Kawachi, M.; Sudo, S.; Edahiro, T. (Japan. J. Appl. Phys. **20** [1981] 709/12 from C.A. **94** [1981] No. 202045).

[31] Furukawa Electric Co., Ltd.; Nippon Telegraph and Telephone Public Corp. (Japan. Kokai Tokkyo Koho 84-92936 [1984] from C.A. **101** [1984] No. 176234).
[32] Mori, H.; Shimizu, N. (IEEE J. Quantum Electron. **18** [1982] 776/81).
[33] Sevast'yanova, T. N.; Suvorov, A. V. (Vestn. Leningr. Univ. Fiz. Khim. **1980** No. 4, pp. 103/5 from C.A. **94** [1981] No. 95095).
[34] Ahlgren, D. C. (Proc. Electrochem. Soc. **81** Pt. 5 [1981] 832/43 from C.A. **95** [1981] No. 53504).
[35] Jourdain, J. L.; LeBras, G.; Combourieu, J. (J. Phys. Chem. **85** [1981] 655/8).
[36] Jourdain, J. L.; LeBras, G.; Combourieu, J. (Actes 1st Colloq. Intern. Berthelot-Vieille-Mallard-Le Chatelier, Talence, Fr., 1981, Vol. 1, pp. 296/300 from C.A. **98** [1983] No. 128732).
[37] Robinson, L. F.; Morrison, A. W.; Worrall, R. L. (Brit. Appl. 2077715 [1980] from C.A. **96** [1982] No. 183717).
[38] Shore, S. G.; Toft, M. A.; Himpsl, F. L. (U.S. 4338289 [1982] from C.A. **97** [1982] No. 94883).
[39] Leach, J. B.; Toft, M. A.; Himpsl, F. L.; Shore, S. G. (J. Am. Chem. Soc. **103** [1981] 988/9).
[40] Toft, M. A.; Leach, J. B.; Himpsl, F. L.; Shore, S. G. (Inorg. Chem. **21** [1982] 1952/7).

[41] Heppert, J. A.; Kulzick, M. A.; Gaines, D. F. (Inorg. Chem. **23** [1984] 14/8).
[42] Dolan, P. J.; Moody, D. C.; Schaeffer, R. (Inorg. Chem. **20** [1981] 745/8).
[43] Geismar, G.; Westphal, U. (Z. Anorg. Allgem. Chem. **487** [1982] 207/16).
[44] Lentz, D. (J. Fluorine Chem. **24** [1984] 523/30).
[45] Pawelke, G.; Heyder, F.; Buerger, H. (J. Fluorine Chem. **20** [1982] 53/63).
[46] Arnau, A.; Bertran, J.; Silla, E. (Anales Quim. A **80** [1984] 542/5).
[47] Olah, G. A.; Laali, K.; Farooq, O. (Organometallics **3** [1984] 1337/40).

[48] Olah, G. A.; Field, L. D. (Organometallics **1** [1982] 1485/7).
[49] Olah, G. A.; Laali, K.; Farooq, O. (J. Org. Chem. **49** [1984] 4591/4).
[50] Lindeman, L. P.; Wilson, M. K. (J. Chem. Phys. **24** [1956] 242).

[51] Levin, I. W.; Abramovitz, S. (J. Chem. Phys. **43** [1965] 4213/22).
[52] Wentink, T.; Tiensuu, V. H. (J. Chem. Phys. **28** [1958] 826/38).
[53] Konaka, S.; Murata, Y.; Kuchitsu, K.; Morino, Y. (Bull. Chem. Soc. Japan **39** [1966] 1134/46).
[54] Kuchitsu, K.; Konaka, S. (J. Chem. Phys. **45** [1966] 4342/7).
[55] Hedberg, K. (Trans. Am. Crystallogr. Assoc. **2** [1966] 79/89).
[56] Konaka, S.; Ito, T.; Morino, Y. (Bull. Chem. Soc. Japan **39** [1966] 1146/54).
[57] Kabubari, H.; Konaka, S.; Kimura, M. (Bull. Chem. Soc. Japan **47** [1974] 2337/8).
[58] Cook, N. D.; Timms, P. L. (J. Chem. Soc. Dalton Trans **1983** 239/43).
[59] Burns, R. C.; O'Donnell, T. A. (J. Fluorine Chem. **23** [1983] 1/14).
[60] Brown, D.; Berry, J. A.; Holloway, J. H. (J. Chem. Soc. Dalton Trans **1982** 1385/8).

[61] Bacher, W.; Jacob, E. (Ger. Offen. 3009933 [1981] from C.A. **96** [1982] No. 59735).
[62] Burns, R. C.; O'Donnell, T. A.; Randall, C. H. (J. Inorg. Nucl. Chem. **43** [1981] 1231/8).
[63] Daneshvar, N.; Straughan, B. P.; Gardiner, D. J. (J. Inorg. Nucl. Chem. **43** [1981] 1147/9).
[64] Costes, J. P.; Cros, G.; Laurent, J.-P. (Syn. React. Inorg. Metal-Org. Chem. **11** [1981] 383/92).
[65] Nöth, H.; Weber, S. (Angew. Chem. **96** [1984] 998/9; Angew. Chem. Intern. Ed. Engl. **23** [1984] 994).
[66] Nöth, H.; Weber, S. (Z. Naturforsch. **38b** [1983] 1460/5).
[67] Nöth, H.; Rasthofer, B.; Weber, S. (Z. Naturforsch. **39b** [1984] 1058/68).
[68] Anton, K.; Nöth, H.; Pommerening, H. (Chem. Ber. **117** [1984] 2479/94).
[69] Anton, K.; Nöth, H.; Pommerening, H. (Chem. Ber. **117** [1984] 2495/503).
[70] Nöth, H.; Staudigl, R.; Taeger, T. (Chem. Ber. **114** [1981] 1157/75).

[71] Nöth, H.; Staudigl, R. (Chem. Ber. **115** [1982] 3011/24).
[72] Meller, A.; Hirninger, F.-J.; Noltemeyer, M.; Maringgele, W. (Chem. Ber. **114** [1981] 2519/35).
[73] Maringgele, W. (J. Organometal. Chem. **222** [1981] 17/32).
[74] Ponomarev, S. V.; Nikolaeva, S. N.; Molchanova, G. N.; Kostyuk, A. S.; Grishin, Yu. K. (Zh. Obshch. Khim. **54** [1984] 1817/21; J. Gen. Chem. [USSR] **54** [1984] 1620/3).
[75] Pachaly, B.; West, R. (Angew. Chem. **96** [1984] 444/5; Angew. Chem. Intern. Ed. Engl. **23** [1984] 454).
[76] Auner, A.; Grobe, J. (Z. Anorg. Allgem. Chem. **500** [1983] 132/60).
[77] Jutzi, P.; Steiner, W.; Stroppel, K. (Chem. Ber. **113** [1980] 3357/65).
[78] Jutzi, P.; Stroppel, K. (Chem. Ber. **113** [1980] 3366/8).
[79] Brown, H. C.; Bhat, N. G.; Basavaiah, D. (Synthesis **1983** 885/6).
[80] Brown, H. C.; Basavaiah, D.; Bhat, N. G. (Organometallics **2** [1983] 1309/11).

[81] Satoh, Y.; Serizawa, H.; Hara, S.; Suzuki, A. (Syn. Commun. **14** [1984] 313/9).
[82] Zaichikova, L. S.; Shestakova, T. G.; Zyk, N. V.; Borisenko, A. A.; Kirpichenok, N. A.; Zefirov, N. S. (Zh. Org. Khim. **17** [1981] 1879/86; J. Org. Chem. [USSR] **17** [1981] 1679/84).
[83] Sugiyama, T. (Bull. Chem. Soc. Japan **55** [1982] 1504/8).
[84] Hoberg, H.; Froehlich, C. (Synthesis **1981** 830/1).
[85] Green, M.; Porter, S. J.; Stone, F. G. A. (J. Chem. Soc. Dalton Trans. **1983** 513/7).
[86] Nguyen, Q. D.; Fevrier, H.; Jouan, M.; Fischer, E. O.; Roell, W. (J. Organometal. Chem. **275** [1984] 191/207).

[87] Kress, J. R. M.; Osborn, J. A.; Wesolek, M. G. (Fr. Demande 2499083 [1982] from C.A. **99** [1983] No. 5176).
[88] Richmond, T. G.; Shriver, D. F. (Organometallics **2** [1983] 1061/2).
[89] Richmond, T. G.; Shriver, D. F. (Organometallics **3** [1984] 305/14).
[90] Lukehart, C. M.; Raja, M. (Inorg. Chem. **21** [1982] 2100/1).

[91] Kafitz, W.; Weller, F.; Dehnicke, K. (Z. Anorg. Allgem. Chem. **490** [1982] 175/81).
[92] Weber, R.; Dehnicke, K.; Schweda, E.; Straehle, J. (Z. Anorg. Allgem. Chem. **490** [1982] 159/70).
[93] Zheng, L.; Baker, R. T.; Knobler, C. B.; Walker, J. A.; Hawthorne, M. F. (Inorg. Chem. **22** [1983] 3350/5).
[94] Kulkarni, S. U.; Patil, V. D. (Heterocycles **18** [1982] (Spec. Issue) 163/7).
[95] Niwa, H.; Hida, T.; Yamada, K. (Tetrahedron Letters **22** [1981] 4239/40).
[96] Ochiai, M.; Nishide, K.; Node, M.; Fujita, E. (Chem. Letters **1981** 283/4).
[97] Bonner, T. G.; Lewis, D.; Rutter, K. (J. Chem. Soc. Perkin Trans. I **1981** 1807/10).
[98] Wakselman, M.; Zrihen, M. (Chem. Letters **1982** 333/6).
[99] Gallos, J.; Varvoglis, A. (J. Chem. Res. S **1982** 150/1; J. Chem. Res. M **1982** 1649/60).
[100] Malfer, D. J.; Loncrini, D. F. (PCT Intern. Appl. 81-00409 [1981] from C.A. **95** [1981] No. 150979).

[101] Enoki, A.; Yaku, F.; Koshijima, T. (Holzforschung **37** [1983] 135/41 from C.A. **99** [1983] No. 55235).
[102] Carson, D.; Crombie, L.; Kilbee, G. W.; Moffatt, F.; Whiting, D. A. (J. Chem. Soc. Perkin Trans. I **1982** 779/88).
[103] Genieser, H. G.; Gabel, D.; Jastorff, B. (J. Chromatog. **215** [1981] 235/42).
[104] Woolsey, N. F.; Baltisberger, R. J. (DOE-ET-13380-T1 [1981] 1/68 from C.A. **96** [1982] No. 71567).
[105] Drevin, I.; Johannson, B. L. (J. Chromatog. **295** [1984] 210/4).
[106] Hines, H. B.; Burlingame, A. L. (MBL Lect. Biol. **3** [1984] 391/410 from C.A. **101** [1984] No. 68810).
[107] Guindon, Y.; Atkinson, J. G.; Morton, H. E. (J. Org. Chem. **49** [1984] 4538/40).
[108] Hartke, K.; Akguen, E. (Liebigs Ann. Chem. **1981** 47/51).
[109] Olah, G. A.; Narang, S. C.; Field, L. D.; Karpeles, R. (J. Org. Chem. **46** [1981] 2408/10).
[110] Brooks, R. A.; Long, M. A.; Garnett, J. L. (J. Labelled Compounds Radiopharm. **19** [1982] 659/67 from C.A. **97** [1982] No. 145105).

[111] Al-Rawi, J. M. A.; Elvidge, J. A.; Jones, J. R.; Mane, R. B.; Saieed, M. (J. Chem. Res. S **1980** 298/9).
[112] Choi, H. W.; Muetterties, E. L. (Inorg. Chem. **20** [1981] 2664/7).
[113] Herchen, S. R.; Widiger, G. N. (U.S. 4386150 [1983] from C.A. **99** [1983] No. 96790).
[114] Tsukahara, I.; Yamamoto, T. (Furukawa Denko Jiho No. 68 [1980] 137/41 from C.A. **94** [1981] No. 76102).
[115] Malhotra, K. C.; Sud, R. G.; Mahajan, R. K.; Chaudry, S. C.; Mehrotra, G. (J. Indian Chem. Soc. **58** [1981] 143/8).
[116] Ganyushin, L. A.; Romm, I. P.; Gur'yanova, E. N.; Shifrina, R. R. (Zh. Obshch. Khim. **50** [1980] 2164/71; J. Gen. Chem. [USSR] **50** [1980] 1739/45).
[117] Hunt, R. L.; Ault, B. S. (Spectrosc. Intern. J. **1** [1982] 31/44).
[118] Tandon, S. K. (Indian J. Chem. A **23** [1984] 65/6).
[119] Tsvetkov, V. G. (Termodin. Org. Soedin. **1983** 53/7).
[120] Leinwand, D. A.; Lefkowitz, S. M.; Brenner, H. C. (J. Photochem. **26** [1984] 141/7 from C.A. **101** [1984] No. 160527).

[121] Martin, D. R.; Mondal, J. U.; Williams, R. D.; Iwamoto, J. B.; Massey, N. C.; Nuss, D. M.; Scott, P. L. (Inorg. Chim. Acta. **70** [1983] 47/51).
[122] Vezzosi, I. M.; Zanoli, A. F.; Peyronel, G. (Spectrochim. Acta A **37** 593/5).
[123] Kharabaev, N. N.; Kogan, V. A.; Osipov, O. A. (Zh. Obshch. Khim. **52** [1982] 2617/20; J. Gen. Chem. [USSR] **52** [1982] 2314/6).
[124] Sebestyen, A.; Megyeri, L.; Vizi, B. (J. Mol. Struct. **89** [1982] 259/68).
[125] Terao, H.; Fukura, M.; Okuda, T.; Negita, H. (Bull. Chem. Soc. Japan **56** [1983] 1728/31).
[126] Chiba, T. (J. Phys. Soc. Japan **13** [1958] 860/8).
[127] Drake, J. E.; Hencher, J. L.; Kashrou, L. N. (Can. J. Chem. **59** [1981] 2898/908).
[128] Iijima, K.; Koshimizu, E.; Shibata, S. (Bull. Chem. Soc. **54** [1981] 2255/9).
[129] Diemert, K.; Kuchen, W.; Kutter, J. (Phosphorus Sulfur **15** [1983] 155/64).
[130] Toshiba Corp. (Japan. Kokai Tokkyo Koho 83-119655 [1983]; C.A. **99** [1983] No. 185943).

[131] Drake, J. E.; Kashrou, L. N.; Majid, A. (Can. J. Chem. **59** [1981] 2417/28).
[132] Harris, R. K.; Root, A.; Dillon, K. B. (Spectrochim. Acta A **39** [1983] 309/10).
[133] Mohan, S.; Mukunthan, A. (Indian J. Pure Appl. Phys. **20** [1982] 315/7).
[134] Sheikh, S. U.; Ghafoor, A. (Pakistan J. Sci. Ind. Res. **23** [1980] 23/4).
[135] Fischer, E. O.; Chen, J.; Schubert, U. (Z. Naturforsch. **37b** [1982] 1284/8).
[136] Fischer, E. O.; Chen, J.; Scherzer, K. (J. Organometal. Chem. **253** [1983] 231/41).
[137] Nöth, H.; Pommerening, H. (Chem. Ber. **114** [1981] 398/9).
[138] Preetz, W.; Fritze, J. (Z. Naturforsch. **39b** [1984] 1472/7).
[139] Markwell, A. J.; Massey, A. G.; Portal, P. J. (Polyhedron **1** [1982] 134/5).
[140] Haluska, L. A. (U.S. 4482689 [1984] from C.A. **102** [1985] No. 114168).

[141] Fieldhouse, J. W.; Graves, D. F. (U.S. 4226840 [1980] from C.A. **94** [1981] No. 16385).
[142] Fieldhouse, J. W.; Graves, D. F. (ACS Symp. Ser. No. 171 [1981] 315/20 from C.A. **96** [1982] No. 52730).
[143] Kharabaev, N. N.; Breus, S. F.; Kogan, Y. A.; Osipov, O. A. (Koord. Khim. **9** [1983] 322/5; Soviet J. Coord. Chem. **9** [1983] 193/6; C.A. **98** [1983] No. 204614).
[144] Hunt, R. L.; Ault, B. S. (Spectrosc. Intern. J. **1** [1982] 45/61).

7.2 With Hydrogen or Organic Substituents

$HBBr_2$. For the preparation of $HBBr_2$, BBr_3 (at 1 to 3 Torr pressure) is slowly passed through a 12-mm outer diameter Pyrex tube loosely packed for 15 cm with crystalline $Na[BH_4]$; at 250°C, $HBBr_2$ is formed quantitatively. There is no evidence for the formation of B_2H_5Br or H_2BBr under these conditions. If the flow rate is too slow, B_2H_6 is obtained; and if it is too fast, unchanged BBr_3 passes through [1, 2].

The following thermochemical data are predicted from the analysis of published data and assuming a single-bond enthalpy term E(B–H) = 371 ± 12 kJ/mol: standard enthalpy of formation $\Delta H_f^\ominus = -125$ kJ/mol; standard enthalpy of atomization $\Delta H_a^\ominus = 1126.8$ kJ/mol [3].

$(CH_3)_2S{-}BHBr_2$ is widely used in the hydroboration of alkenes and alkynes (in the presence of BBr_3) to yield alkyldibromoboranes and *cis*-alkenyldibromoboranes (in the form of their dimethylsulfide adducts) [4 to 12]. The hydroboration proceeds by a prior dissociation of the reagent, hydroboration by $HBBr_2$, and subsequent adduct formation of the product with dimethylsulfide [5]. For the reduction of $RBBr_2 \cdot S(CH_3)_2$ by $Li[AlH_4]$ to give $RBHBr \cdot S(CH_3)_2$, see [13].

H_2BBr. The standard enthalpy of formation $\Delta H_f^\ominus = -35.9$ kJ/mol and the standard enthalpy of atomization $\Delta H_a^\ominus = 1143.8$ kJ/mol (for the gaseous compound) have been calculated from published thermochemical data. The single bond enthalpy term E(B–H) is assumed to be 371±12 kJ/mol [3].

$[N(n\text{-}C_4H_9)_4][HBBr_3]$ is prepared as follows: 0.59 g (2.3 mmol) of $[N(C_4H_9)_4][BH_4]$ contained in a 30 mL reaction flask are covered with 2.35 mmol of BBr_3 and 5 mL of CH_2Cl_2 at −196°C. The reaction mixture is allowed to reach room temperature within 2 h (under stirring). The flask is then cooled to −78°C and volatiles are fractionated through U-traps at −140 and −196°C (where B_2H_6 is collected). The creamy colorless $[N(n\text{-}C_4H_9)_4][HBBr_3]$ is freed from the remaining solvent in vacuum at room temperature. The yield is 85%. In the ^{11}B NMR spectrum, $\delta = -13.0$ ppm ($J(^{11}B\text{–}^1H) = 176$ Hz); $\nu(BH) = 2520\ cm^{-1}$ [68].

$[(CH_3)_3PC_6H_5][HBBr_3]$ is prepared in analogous fashion [68].

7.3 (Organyl)bromoboranes

7.3.1 (Organyl)dibromoboranes, $RBBr_2$

CH_3BBr_2. The synthesis of $CH_3^{10}BBr_2$ starting from BBr_3 and $K[^{10}BF_4]$ to give $^{10}BBr_3$ (in 76% yield) and reaction of the latter with $(CH_3)_4Sn$ to yield $CH_3^{10}BBr_2$ (92%) is described in detail [14]. Dibromo(methyl)borane is also obtained according to: $(CH_3)_3SnBr + BBr_3 \rightarrow (CH_3)_2SnBr_2 + CH_3BBr_2$ (in liquid BBr_3). When $(CH_3)_2SnBr_2$ is reacted with BBr_3 at 90°C, some CH_3BBr_2 is also formed [15]. The following thermochemical data for CH_3BBr_2 are predicted (by analysis of published data): Standard enthalpy of formation $\Delta H_f^\ominus = -199.4$ kJ/mol; standard enthalpy of atomization $\Delta H_a^\ominus = 2353.9$ kJ/mol (for the gaseous state). E(B–C) = 350±10 kJ/mol has been used for the boron-carbon single bond enthalpy term [3]. With 1,2,3-trimethyl-1,3,2-diazaborolidine, CH_3BBr_2 gives a 1:1 adduct by coordination to one of the nitrogen atoms [16]. The reaction of dibromo(methyl)borane with the graphite intercalate C_8K is claimed to produce methylborylene, CH_3B, which can be trapped by $(t\text{-}C_4H_9)C{\equiv}C(t\text{-}C_4H_9)$ to give borirene and diboretene derivatives [17, 18], or with di-*n*-alkylacetylenes to give 1,4-diboracyclohexa-2,5-diene derivatives [18]; however, compare [69]. With cyclohexene, the same reagent system leads to the formation of a heterocycle derived from a borafluorene structure [19]. The reaction between CH_3BBr_2 and $Li_2[\{\text{–}(CH)_4\text{–}\}B\text{–}N(i\text{-}C_3H_7)_2]$ gives a $C_4B_2H_6$ *nido*-tetracarbahexaborane derivative [20]. $CH_3B[Si(C_6H_5)_3]_2$ is prepared by reacting CH_3BBr_2 with $(C_6H_5)_3SiLi$ [21]. To the amino-iminoborane (tmp)B=N(t-C_4H_9)(tmp = 2,2,6,6-tetramethylpiperidinyl), CH_3BBr_2 is added to give a four-membered B_2N_2 ring containing both sp^2- and sp^3-hybridized B and N atoms [22]. The reaction of CH_3BBr_2 with $LiCH{=}CHCH_2\text{–}N[Si(CH_3)_3]Li$ gives 2-methyl-1-trimethylsilanyl-Δ^3-1,2-azaboroline [23], and with $[(CH_3)_3Si]NLi\text{–}CH_2\text{–}CH_2NLi[Si(CH_3)_3]$, 2-methyl-1,3-bis(trimethylsilanyl)-Δ^4-1,3,2-diazaboroline is obtained; 1,3-di-*t*-butyl-2-methyl-Δ^4-1,3,2-diazaboroline is formed in the reaction of CH_3BBr_2 with 1,4-di-*t*-butyl-1,4-diaza-1,3-butadiene [24]. CH_3BBr_2 and $R[(CH_3)_3Si]NCH_2C{\equiv}CH$ react to give $R(CH_3BrB)CH_2C{\equiv}CH$ ($R = C_6H_5CH_2$); with $RHNCH_2\text{–}C{\equiv}CH$, the alkyne bond inserts into a boron-bromine bond and the product forms a 5-membered ring by intramolecular coordination; with $RN(CH_2\text{–}C{\equiv}CH)_2$, a bicyclic species is formed accordingly by insertion into both B–Br bonds [25]. While Si–N bond cleavage by (alkyl)dibromoboranes is a frequently used synthetic tool (see, for instance [25]), the Si–N-bond in 3,5-dibromo-4-trimethylsilanyl-1,2-dithia-4-aza-3,5-diborolidine is unaffected by treatment with CH_3BBr_2 [26]. Reactions of 1,1-dibutyl-stannacyclohexa-2,5-dienes with CH_3BBr_2 lead to corresponding methylboracyclohexa-2,5-dienes [27].

$C_2H_5BBr_2$ is prepared by the BBr_3-promoted reaction of $(CH_3)_2S\text{–}BHBr_2$ with C_2H_4 in pentane [4, 13].

$[(CH_3)_3Si][Br_2B]C{=}C{=}O$, obtained from BBr_3 and $(CH_3)_3SiC{\equiv}COC_2H_5$ at −50°C (in 67% yield), has a boiling point of 5°C at 1 Torr. In the NMR spectrum, the proton signal appears at

$\delta = 0.20$ ppm, the ^{11}B signal at $\delta = 47.3$ ppm, and ^{13}C signals are observed at $\delta = 0.1$ ppm ($SiCH_3$), 37.3 ppm (>C=), and 168.2 ppm (=C=O). In the IR spectrum, the (C=C=O) group frequency is at 2100 cm^{-1} [28]. Similar data hold true for $[(C_2H_5)_3Si][Br_2B]C{=}C{=}O$ [28].

$C_6H_5BBr_2$. The bond order-dependent enthalpy term for the B–Br bond has been calculated to be E(B–Br) = 384.9 kJ/mol, and the standard enthalpy of formation for the gaseous compound is given with $\Delta H_f^\ominus = -129.3$ kJ/mol [3]. For a comparison of dipole moments with other phenyl and trimethylphenyl derivatives, see [35].

Additional (organyl)dibromoboranes are surveyed in Table 7/5.

Table 7/5
Survey of Additional (Organyl)dibromoboranes, $RBBr_2$.

R	comments	Ref.
$BrC(CH_3){=}CH$	preparation from $CH_3C{\equiv}CH$ and BBr_3; reaction with C_6H_5NCO	[29]
n-C_3H_7	(adduct with $(CH_3)_2S$) preparation from $(C_2H_5)_3B$ + 2 BBr_3 in the presence of $(CH_3)_2S{-}BH_3$	[4]
n-C_4H_9	calculated: $\Delta H_f^\ominus = -255.3$ kJ/mol; $\Delta H_a^\ominus = 5867.9$ kJ/mol	[3]
cyclo-C_5H_9	preparation from (*cyclo*-$C_5H_9)_3B$ + 2 BBr_3 in the presence of $(CH_3)_2S{-}BH_3$	[4]
	reaction with alkynes; hydrolysis, alcoholysis	[6, 30]
n-C_5H_{11}	(adduct with $(CH_3)_2S$) controlled hydridation, sequential hydroboration	[31]
n-$C_4H_9CBr{=}CH$	(Z) preparation from n-$C_4H_9C{\equiv}CH$ + BBr_3; reaction with C_6H_5NCO	[29]
t-$C_4H_9CBr{=}CH$	(Z) preparation from t-$C_4H_9C{\equiv}CH$ + BBr_3; reaction with C_6H_5NCO	[29]
n-$C_4H_9CH{=}CH$	(E) adduct with $(CH_3)_2S$; preparation from (n-$C_4H_9)C{\equiv}CH$ + $(CH_3)_2S{-}BHBr_2$; hydrolysis, alcoholysis	[6]
$H_3C{-}C(CH_3)_2{-}CH{=}CH$	(E) adduct with $(CH_3)_2S$; preparation from (t-$C_4H_9)C{\equiv}CH$ + $(CH_3)_2S{-}BHBr_2$; hydrolysis, alcoholysis	[6]
cyclo-C_6H_{11}	preparation from (*cyclo*-$C_6H_{11})_3B$ + BBr_3 in the presence of $(CH_3)_2S{-}BH_3$	[4]
	reaction with alkynes; hydrolysis, alcoholysis	[6, 30]
n-C_6H_{13}	preparation from (n-$C_6H_{13})_3B$ + 2 BBr_3 in the presence of $(CH_3)_2S{-}BH_3$	[4]
	reaction with alkynes	[32]
	(adduct with $(CH_3)_2S$), preparation from 1-hexene + $(CH_3)_2S$-BHBr in the presence of BBr_3; reaction with 1-octyne; hydrolysis, alcoholysis; controlled hydridation, sequential hydroboration	[5, 6, 30, 31]
n-$C_3H_7CH(CH_3)CH_2$	preparation from R_3B + 2 BBr_3 in the presence of $(CH_3)_2S{-}BH_3$	[4]
	(adduct with $(CH_3)_2S$), preparation by hydroboration of 2-methyl-1-pentene; reaction with 1-pentyne; controlled hydridation, subsequent hydroboration	[30, 33]

References for 7.2 and 7.3 on pp. 89/91

Table 7/5 (continued)

R	comments	Ref.
$(C_2H_5)(n\text{-}C_3H_7)CH$	preparation from $R_3B + 2BBr_3$ in the presence of $(CH_3)_2S\text{-}BH_3$	[4]
	controlled hydridation, subsequent hydroboration, thermal isomerization	[9, 30]
	hydrolysis, alcoholysis	[6]
$C_6H_5CH_2$	bromoboration of alkynes	[34]
$C_6H_5CBr{=}CH$	(Z) preparation from $C_6H_5C{\equiv}CH + BBr_3$; reaction with C_6H_5NCO	[29]
$n\text{-}C_6H_{13}CBr{=}CH$	(Z) preparation from $n\text{-}C_6H_{13}C{\equiv}CH + BBr_3$; reaction with C_6H_5NCO	[29]
$n\text{-}C_6H_{13}CH{=}CH$	(adduct with $(CH_3)_2S$) preparation from 1-octyne and $(CH_3)_2S\text{-}BHBr_2$; hydrolysis, alcoholysis	[6]
$n\text{-}C_8H_{17}$	(adduct with $(CH_3)_2S$) preparation from 1-octene + $(CH_3)_2S\text{-}BHBr_2$; reaction with 1-pentyne	[30]
$2,4,6\text{-}(CH_3)_3C_6H_2$	reaction with $(C_6H_5)_3SiLi$; dipole moment	[21, 35]
$n\text{-}C_{13}H_{27}$	(adduct with $(CH_3)_2S$) preparation from 1-tridecene + $(CH_3)_2S\text{-}BHBr_2$; reaction with 1-decyne	[30]
$t\text{-}C_4H_9CH{=}(C_6H_5CH_2)C$	preparation from $C_6H_5CH_2BBr_2 + HC{\equiv}C(t\text{-}C_4H_9)$; mass spectrum, NMR ($^1H$, ^{11}B)	[34]
$(CH_3)_2SiBr(CH_2)_3$	preparation from silacyclobutane + BBr_3; mass spectrum, NMR (^1H), IR	[36]
$(C_2H_3)_2SiBr(CH_2)_3$	preparation from silacyclobutane + BBr_3; NMR (^1H)	[36]
$CH_3(C_6H_5)SiBr(CH_2)_3$	preparation from silacyclobutane + BBr_3; NMR (^1H)	[36]
$(C_6H_5)SiBr(CH_2)_3$	preparation from silacyclobutane + BBr_3; NMR (^1H)	[36]
$(CH_3)_2SiBrCH_2Si\text{-}(CH_3)_2CH_2$	preparation from silacyclobutane + BBr_3; mass spectrum, NMR (^1H), IR	[36]

7.3.2 (Organyl)(bromo)hydroboranes, RBHBr

Derivatives **RBHBr** in the form of the dimethylsulfide adducts are prepared from the corresponding (organyl)dibromoboranes by controlled hydridation. This is achieved by treating $(CH_3)_2S\text{-}B(R)Br_2$ with an equimolar amount of $Li[AlH_4]$ in diethyl ether [31]. The dimethylsulfide adducts of the following (organyl)(bromo)hydroboranes have been described: C_2H_5BHBr, $n\text{-}C_3H_7BHBr$, *cyclo*-C_5H_9BHBr, $n\text{-}C_5H_{11}BHBr$, $n\text{-}C_6H_{13}BHBr$, $n\text{-}C_3H_7CHCH_3CH_2\text{-}BHBr$, $(C_2H_5)(n\text{-}C_3H_7)CHBHBr$, $(CH_3)_2CHC(CH_3)_2BHBr$, $n\text{-}C_8H_{17}BHBr$ [13, 31].

The ^{11}B NMR spectrum of $n\text{-}C_6H_{13}BHBr\cdot S(CH_3)_2$ exhibits a doublet near $\delta = 0.7$ ppm ($J_{BH} = 136$ Hz), and in the IR spectrum $\nu(BH)$ appears at 2450 cm^{-1}. Since there are no absorption bands near 1550 cm^{-1}, this excludes the formation of $(n\text{-}C_6H_{13}BH_2)_2$. The compounds react with alkenes or alkynes to give mixed organylboranes [13, 31, 33].

7.3.3 (Diorganyl)bromoboranes, R_2BBr

$(CH_3)_2BBr$ is formed in the reaction of trimethylborane with 3,5-dibromo-1,2,4,3,5-trithiadiborolane [37].

The standard enthalpy of formation of $(CH_3)_2BBr$ has been obtained by analysis of published thermochemical data with $\Delta H_f^\ominus = -184.7$ kJ/mol (for the gaseous state), while the standard enthalpy of atomization is given as $\Delta H_a^\ominus = 3598.0$ kJ/mol [3].

$(CH_3)_2BBr$ reacts with 1,2,3-trimethyl-1,3,2-diazaborolidine to form a 1:1 adduct by coordination to one of the nitrogen atoms [16]. Bromo(dimethyl)borane is added across the B–N bond of (tmp)B=N(t-C_4H_9) (tmp = 2,2,6,6-tetramethylpiperidinyl) to give the diazaboretidinium salt [tmp(–BCH_3–)$_2$N(t-C_4H_9)]Br [22]. The reaction of $(CH_3)_2BBr$ with $[(CH_3)_3Si]_3SiLi$ affords $(CH_3)_2B$–Si$[Si(CH_3)_3]_3$ [38], and with $(CH_3)_2N$–NLi–$B(CH_3)_2$, 1,1-bis(dimethylboryl)-2,2-dimethylhydrazine is obtained [39].

The reaction of urea and thiourea derivatives, $(RHN)_2CX$ (X = O, S), with $(CH_3)_2BBr$ leads to a variety of cyclic ureidoboranes [40]. Such cyclic ureidoboranes have also been prepared from unsymmetrical ureas such as $(C_2H_5)_2NCONHCH_3$ and silylated ureas, i.e., $[(CH_3)_3SiNR$-$C(O)]_2NCH_3$ [41]. A number of cyclic and bicyclic compounds (e.g., 2,6-dioxonia-4,8-diaza-3,7-diboratabicyclo[3.3.0]octanes) result from the reaction of $(CH_3)_2BBr$ with symmetrically and unsymmetrically substituted ethanediamides such as RNH–CO–CO–NHR [42]. With $(CH_3)_3SiSLi$, $(CH_3)_2BBr$ reacts to form $(CH_3)_3SiSB(CH_3)_2$ [43], and reaction with $(CH_3)_3SiNR$–$CH_2C{\equiv}CH$ gives $(CH_3)_2B$–NR–$CH_2C{\equiv}CH$ [25]. Treatment of the carbamoyl complex $[(CO)_4FeC(O)N(CH_3)_2][C\{N(CH_3)_2\}_3]$ with bromo(dimethyl)borane yields the boroxycarbene complex $(CO)_4Fe[C\{OB(CH_3)_2\}N(CH_3)_2]$ [47].

$(CH_3)_2BBr$ is used for the cleavage of carbon-oxygen bonds in ethers [44], acetals, and ketals [45, 46]. Cyclosilabutanes are also cleaved by $(CH_3)_2BBr$ [36].

$(C_2H_5)_2BBr$ has been used in the haloboronation of alkynes (reversible) to give (Z)-alkenes. At elevated temperatures, the reaction produces 1,1-organoboronated alkenes [34]. The hydridation in the presence of 2-butyne yields (2Z)-3-methyl-2-pentene [48]. The adduct with dimethylsulfide is prepared from $B(C_2H_5)_3$ and BBr_3 (catalyzed by $(CH_3)_2S$–BH_3) [4]. The isocyanoborane complex $(CO)_5CrCNB(C_2H_5)_2$ is formed in the reaction of $(C_2H_5)_2BBr$ with $K[(CO)_5CrCN]$ [54].

$(C_6H_5)_2BBr$. The bond order-dependent enthalpy term for the B–Br bond has been calculated to be E(B–Br) = 394.6 kJ/mol, and the standard enthalpy of formation is $\Delta H_f^\ominus = -9.4$ kJ/mol (for the gaseous compound) [3]. Bromo(diphenyl)borane forms a 1:1 adduct with 2-chloro-1,3-dimethyl-1,3,2-diazaborolidine by coordination to one of the nitrogen atoms [16]. With alkynes, a regiospecific addition across the C≡C triple bond occurs which proceeds by (reversible) haloboronation to give Z-alkenes. However, at elevated temperatures the irreversible 1,1-organoboronation predominates [34].

$(C_6H_5)_2BBr$ and $(CH_3)_3SiN_3$ interact to give $(C_6H_5)_2BN_3$ [49], with $(CH_3)_2N$–NLi–$B(CH_3)_2$ the product is $(CH_3)_2N$–$N[B(CH_3)_2][B(C_6H_5)_2]$ [39], and with $[C(CH_3)_3Si]_3CLi$ the organylborane $[(CH_3)_3Si]_3CB(C_6H_5)_2$ is obtained [50]. The product of the interaction with pentadienyllithium reacts with carbonyl compounds to give 3-(1-hydroxyalkyl)penta-1,4-dienes in a highly regioselective reaction [51]. Thallium diacylcyclopentadienides can be cyclized with $(C_6H_5)_2$ BBr to give corresponding dioxaborepines [52], and boron-bridged Fe^{II} dioxime derivatives are prepared by replacement of the O–H–O bridging hydrogen atom using $(C_6H_5)_2BBr$ [53].

Bromo(diphenyl)borane is used for the cleavage of acetals and ketals to give the corresponding aldehydes and ketones [45, 46], and the carbamoyl complex $[(CO)_4FeC(O)N(CH_3)_2]$-$[C\{N(CH_3)_2\}_3]$ is converted to the boroxycarbene complex $(CO)_4FeC[OB(C_6H_5)_2]N(CH_3)_2$ [47].

References for 7.2 and 7.3 on pp. 89/91

A survey of additional (diorganyl)bromoboranes is given in Table 7/6.

Table 7/6
Survey of Additional (Diorganyl)bromoboranes, R_2BBr.

compound	comments	Ref.
$(n\text{-}C_3H_7)_2BBr$	reaction with acetylacetone	[55]
$(n\text{-}C_4H_9)_2BBr$	standard enthalpy of formation $\Delta H_f^{\ominus}(gas) = -305.2$ kJ/mol; bond enthalpy term E(BBr) = 410 kJ/mol	[3]
$(t\text{-}C_4H_9)_2BBr$	preparation from $B(OCH_3)_3$ and $t\text{-}C_4H_9Li$ and treatment with BBr_3; reaction with Na/K	[56]
$(cyclo\text{-}C_5H_9)_2BBr$	preparation from $(cyclo\text{-}C_5H_9)_3B + BBr_3$ catalyzed by $(CH_3)_2S\text{–}BH_3$	[4]
	reduction with NaH in diglyme; reduction with $Li[AlH_4]$	[48, 57, 58]
$(n\text{-}C_5H_{11})_2BBr$	reduction with NaH in diglyme; reduction with alkenes	[57]
$(cyclo\text{-}C_6H_{11})_2BBr$	preparation from $(cyclo\text{-}C_6H_{11})_3B + BBr_3$ catalyzed by $(CH_3)_2S\text{–}BH_3$	[4]
$(n\text{-}C_6H_{13})_2BBr$	preparation from $(n\text{-}C_6H_{13})_3B + BBr_3$ catalyzed by $(CH_3)_2S\text{–}BH_3$	[4]
	reduction by NaH or KH in diglyme; reduction with alkenes	[57]
$[n\text{-}C_3H_7CH(CH_3)\text{–}CH_2]_2\text{-}BBr$	preparation from $(2\text{-methyl-1-pentyl})_3B + BBr_3$ (in the presence of $(CH_3)_2S\text{–}BH_3$)	[4]
$(n\text{-}C_5H_{11})(n\text{-}C_6H_{13})BBr$	preparation by stepwise hydridation of $n\text{-}C_6H_{13}BBr_2\text{–}S(CH_3)_2$ and reaction with 1-pentene	[31]
$(n\text{-}C_6H_{13})(n\text{-}C_{12}H_{25})BBr$	preparation by stepwise hydridation of $n\text{-}C_6H_{13}BBr_2\text{–}S(CH_3)_2$ and reaction with 1-dodecene	[31]
$C_2H_5(t\text{-}C_4H_9CH{=}CC_2H_5)\text{-}BBr$	(Z + E) preparation from $(C_2H_5)_2BBr + HC{\equiv}C(t\text{-}C_4H_9)$; mass spectrum, NMR (^{1}H, ^{11}B)	[34]
$C_6H_5[RCH{=}(C_6H_5)C]BBr$	$R = n\text{-}C_4H_9$, $t\text{-}C_4H_9$; (Z + E) preparation from $(C_6H_5)_2BBr$ and $HC{\equiv}CR$; mass spectrum, NMR (^{1}H, ^{11}B)	[34]
$C_6H_5[RCX{=}(Y)C]BBr$	$R = n\text{-}C_4H_9$, $X = (CH_3)_3Si$, $Y = C_6H_5$; $R = t\text{-}C_4H_9$, $X = C_6H_5$, Y = H; preparation by organoboronation of alkynes; mass spectrum, NMR (^{1}H, ^{11}B)	[34]
$C_6H_5CH_2(t\text{-}C_4H_9CBr{=}CH)\text{-}BBr$	(Z + E) preparation from $C_6H_5BBr_2 + HC{\equiv}C(t\text{-}C_4H_9)$; mass spectrum, NMR (^{1}H, ^{11}B)	[34]
$CH_3[Br(CH_3)_2Si\text{–}CH_2\text{–}Si\text{-}(CH_3)Br\text{–}CH_2]BBr$	preparation by cleavage of silacyclobutane derivatives with BBr_3; NMR (^{1}H)	[36]
cyclic species		
$C_8H_{14}BBr$	(9-bromo-9-borabicyclo[3.3.1]nonane (9-Br-9-BBN)) formation from dibromotrithiadiborolane + 9-H-9-BBN	[59]
	reduction by $9\text{-LiH}_2\text{-9-BBN}$ to 9-H-9-BBN	[60]
	^{13}C NMR study; reaction with $[(CH_3)_3Si]_3SiLi$	[38, 61]
	haloboronation of 1-alkynes	[62, 63]

Table 7/6 (continued)

compound	comments	Ref.
$C_9H_{16}BBr$	(3-bromo-7-methyl-3-borabicyclo[3.3.1]nonane) reaction with i-C_3H_7MgBr	[64]
	NMR (^1H, ^{13}C; also of pyridine adduct)	[65]
$C_9H_{15}BBr_2$	(3-bromo-7-bromomethyl-3-borabicyclo[3.3.1]nonane) reaction with i-C_3H_7MgBr	[65]
$C_{12}H_{22}BBr$	(3-bromo-4,4,8-trimethyl-3-borabicyclo[4.3.1]decane) preparation; NMR (^1H)	[65]
$C_{12}H_{21}BBr_2$	(3-bromo-8-bromomethyl-4,4-dimethyl-3-borabicyclo-[4.3.1]decane) preparation; NMR (^1H)	[65]
$C_9H_{14}BBr$	(2-bromo-2-boraadamantane) preparation by bromination of the 2-isopropyl derivative	[66]
$S[-C(C_6H_5)=CH-]_2BBr$	formation from $C_6H_5C{\equiv}CH$ and 3,5-dibromo-1,2,4-trithia-3,5-diborolane; mass spectrum, NMR (^1H, ^{11}B)	[67]

References for 7.2 and 7.3:

[1] Frost, D. C.; Kirby, C.; McDowell, C. A.; Westwood, N. P. C. (J. Am. Chem. Soc. **103** [1981] 4428/32).

[2] Westwood, N. P. C. (NATO ASI Ser. B **90** [1983] 275/8 from C.A. **99** [1983] No. 45662).

[3] Holbrook, J. B.; Smith, B. C.; Housecroft, C. E.; Wade, K. (Polyhedron **1** [1982] 701/6).

[4] Brown, H. C.; Basavaiah, D.; Bhat, N. G. (Organometallics **2** [1983] 1309/11).

[5] Brown, H. C.; Chandrasekaran, J.; (Organometallics **2** [1983] 1261/3).

[6] Brown, H. C.; Bhat, N. G.; Somayaji, V. (Organometallics **2** [1983] 1311/6).

[7] Brown, H. C.; Cha, J. S.; Nazer, B.; Kim, S.-C.; Krishnamurthy, S. (J. Org. Chem. **49** [1984] 885/92).

[8] Brown, H. C.; Imai, T. (Organometallics **3** [1984] 1392/5).

[9] Brown, H. C.; Racherla, U. S. (J. Org. Chem. **48** [1983] 1389/91).

[10] Brown, H. C.; Basavaiah, D. (Synthesis **1983** 283/4).

[11] Brown, H. C.; Chandrasekaran, J. (J. Org. Chem. **48** [1983] 644/8).

[12] Brown, H. C.; Imai, T. (J. Am. Chem. Soc. **105** [1983] 6285/9).

[13] Brown, H. C.; Basavaiah, D.; Bhat, N. G. (Organometallics **2** [1983] 1468/70).

[14] Nöth, H.; Staudigl, R. (Inorg. Syn. **22** [1983] 218/26).

[15] Daneshvar, N.; Straughan, B. P.; Gardiner, D. J. (J. Inorg. Nucl. Chem. **43** [1981] 1147/9).

[16] Anton, K.; Nöth, H.; Pommerening, H. (Chem. Ber. **117** [1984] 2479/94).

[17] Van der Kerk, S. M.; Budzelaar, P. H. M.; van der Kerk-van Hoof, A.; van der Kerk, G. J. M.; Ragné Schleyer, P. v. (Angew. Chem. **95** [1983] 61; Angew. Chem. Intern. Ed. Engl. **22** [1983] 48).

[18] Van der Kerk, S. M.; Budzelaar, P. H. M.; van Eekeren, L. M.; van der Kerk, G. J. M. (Polyhedron **3** [1984] 271/80).

[19] Van der Kerk, S. M.; Roos-Venekamp, J. C.; van Beijnen, A. J. M.; van der Kerk, G. J. M. (Polyhedron **2** [1983] 1337/43).

[20] Herberich, G. E.; Ohst, H.; Mayer, H. (Angew. Chem. **96** [1984] 975/6; Angew. Chem. Intern. Ed. Engl. **23** [1984] 969).

[21] Pachaly, B.; West, R. (Angew. Chem. **96** [1984] 444/5; Angew. Chem. Intern. Ed. Engl. **23** [1984] 454).
[22] Nöth, H.; Weber, S. (Angew. Chem. **96** [1984] 998/9; Angew. Chem. Intern. Ed. Engl. **23** [1984] 994).
[23] Schulze, J.; Boese, R.; Schmid, G. (Chem. Ber. **114** [1981] 1297/305).
[24] Schulze, J.; Schmid, G. (Chem. Ber. **114** [1981] 495/504).
[25] Meller, A.; Hirninger, F. J.; Noltemeyer, M.; Maringgele, W. (Chem. Ber. **114** [1981] 2519/35).
[26] Meller, A.; Habben, C. (Monatsh. Chem. **113** [1982] 139/53).
[27] Ashe III, A. J.; Abu-Orabi, S. T.; Eisenstein, O.; Sandford, H. F. (J. Org. Chem. **48** [1983] 901/3).
[28] Ponomarev, S. V.; Nikolaeva, S. N.; Molchanova, G. N.; Kostyuk, A. S.; Grishin, Yu. K. (Zh. Obshch. Khim. **54** [1984] 1817/21; J. Gen. Chem. [USSR] **54** [1984] 1620/3).
[29] Satoh, Y.; Serizawa, H.; Hara, S.; Suzuki, A. (Syn. Commun. **14** [1984] 313/9).
[30] Brown, H. C.; Basavaiah, D. (J. Org. Chem. **47** [1982] 3806/8).

[31] Kulkarni, S. U.; Basavaiah, D.; Zaidlewicz, M.; Brown, H. C. (Organometallics **1** [1982] 212/4).
[32] Hara, S.; Kato, T.; Suzuki, A. (Synthesis **1983** 1005/6).
[33] Brown, H. C.; Basavaiah, D.; Kulkarni, S. U. (J. Org. Chem. **47** [1982] 3808/10).
[34] Binnewirtz, R. J.; Klingenberger, H.; Welte, R.; Paetzold, P. (Chem. Ber. **116** [1983] 1271/84).
[35] Exner, O.; Folli, U.; Marcacciolli, S.; Viarelli, P. (J. Chem. Soc. Perkin Trans. II **1983** 757/60).
[36] Auner, N.; Grobe, J. (Z. Anorg. Allgem. Chem. **500** [1983] 132/60).
[37] Nöth, H.; Staudigl, R.; Taeger, T. (Chem. Ber. **114** [1981] 1157/75).
[38] Biffar, W.; Nöth, H. (Z. Naturforsch. **36b** [1981] 1509/15).
[39] Fußstetter, H.; Nöth, H. (Liebigs Ann. Chem. **1981** 633/41).
[40] Maringgele, W. (J. Organometal. Chem. **222** [1981] 17/32).

[41] Maringgele, W. (Chem. Ber. **115** [1982] 3271/89).
[42] Maringgele, W.; Sheldrick, G. M.; Meller, A.; Noltemeyer, M. (Chem. Ber. **117** [1984] 2112/31).
[43] Hennemuth, K.; Meller, A.; Wojnowska, M. (Z. Anorg. Allgem. Chem. **489** [1982] 47/52).
[44] Guindon, Y.; Yoakim, C.; Morton, H. E. (Tetrahedron Letters **24** [1983] 2969/72).
[45] Guindon, Y.; Morton, H. E.; Yoakim, C. (Tetrahedron Letters **24** [1983] 3969/72).
[46] Guindon, Y.; Yoakim, C.; Morton, H. E. (J. Org. Chem. **49** [1984] 3912/20).
[47] Petz, W. (Z. Naturforsch. **36b** [1981] 335/8).
[48] Brown, H. C.; Basavaiah, D.; Kulkarni, S. U. (J. Org. Chem. **47** [1982] 171/3).
[49] Paetzold, P.; Truppat, R. (Chem. Ber. **116** [1983] 1531/9).
[50] Eaborn, C.; El-Kheli, M. N. A.; Hitchcock, P. B.; Smith, J. D. (J. Organometal. Chem. **272** [1984] 1/9).

[51] Hutchings, M. G.; Paget, W. E.; Smith, K. (J. Chem. Res. S **1983** 31 from C.A. **99** [1983] No. 21896).
[52] Hartke, K.; Kohl, A.; Kaemchen, T. (Chem. Ber. **116** [1983] 2653/7).
[53] Verhage, M.; Hoogwater, D. A.; van Bekkum, H.; Reedijk, J. (Rec. J. Roy. Neth. Chem. Soc. **101** [1982] 351/7).
[54] Hoefler, M.; Loewenich, H. (J. Organometal. Chem. **226** [1982] 229/37).
[55] Costes, J.-P.; Cros, G.; Laurent, J.-P. (Syn. React. Inorg. Metal-Org. Chem. **11** [1981] 383/92).
[56] Klusik, H.; Berndt, A. (J. Organometal. Chem. **232** [1982] C21/3).

[57] Maddocks, P. J.; Pelter, A.; Rowe, K.; Smith, K.; Subramanyam, Ch. (J. Chem. Soc. Perkin Trans. I **1981** 653/6).
[58] Brown, H. C.; Basavaiah, D. (J. Org. Chem. **47** [1982] 754/6).
[59] Nöth, H.; Staudigl, R. (Z. Anorg. Allgem. Chem. **481** [1981] 41/50).
[60] Brown, H. C.; Kulkarni, S. U. (J. Organometal. Chem. **218** [1981] 299/307).

[61] Blue, C. D.; Nelson, D. J. (J. Org. Chem. **48** [1983] 4538/42).
[62] Hara, S.; Dojo, H.; Takinami, S.; Suzuki, A. (Tetrahedron Letters **24** [1983] 731/4).
[63] Hara, S.; Satoh, Y.; Ishiguro, H.; Suzuki, A. (Tetrahedron Letters **24** [1983] 735/8).
[64] Vasilyev, L. S.; Veselovskii, V. V.; Struchkova, M. I.; Mikhailov, B. M. (J. Organometal. Chem. **226** [1982] 115/28).
[65] Gurskii, M. E.; Shashkov, A. S.; Mikhailov, B. M. (Izv. Akad. Nauk SSSR Ser. Khim. **1981** 341/52; Bull. Acad. Sci. USSR Div. Chem. Sci. **30** [1981] 264/73).
[66] Shchegoleva, T. A.; Shashkova, E. M.; Mikhailov, B. M. (Izv. Akad. Nauk SSSR Ser. Khim. **1980** 2426/7).
[67] Habben, C.; Maringgele, W.; Meller, A. (Z. Naturforsch. **37b** [1982] 43/53).
[68] Toft, M. A.; Leach, J. B.; Himpsl, F. L.; Shore, S. G. (Inorg. Chem. **21** [1982] 1952/7).
[69] Schlögl, R.; Wrackmeyer, B. (Polyhedron **4** [1985] 885/92).

7.4 With Oxygen

BrBO. The enthalpy of formation in the gaseous state as well as the entropy and heat capacity of gaseous BrBO at 298 K have been evaluated by using a combination of tungsten transport measurements and theoretical estimations to be $\Delta H_{298} = -251$ kJ/mol; $S_{298} = 249\ J \cdot mol^{-1} \cdot K^{-1}$; $C_{p,298} = 48\ J \cdot mol^{-1} \cdot K^{-1}$ [30].

Compound **1** is prepared from 7.2 g (0.037 mol) of (2,4-pentanedionato)dipropylborane (dissolved in 50 mL of hexane) by dropwise addition of 9.2 g (0.037 mol) of BBr_3 to the stirred solution at −20°C. After 3 h at ambient temperature, the reaction mixture is filtered and the solid material is dissolved in $CHCl_3$, precipitated with hexane, and dried under vacuum. The yield is 50%. NMR data: $\delta^1H = 2.57$ and 6.23 ppm; $\delta^{11}B = 3.9$ ppm [1].

CH3 O Br B Br O CH3

1

7.5 With Nitrogen

$(CH_3)_2N–BBr_2$. $\delta^{11}B = 25.7$ ppm, $\delta^{13}C = 42.00$ ppm [2].

$(C_2H_5)_2N–^{10}BBr_2$ is formed in the reaction between $(–S–CH_2–CH_2–S–)BN(C_2H_5)_2$ and $^{10}BBr_3$; for the reaction mechanism, see [3]. NMR data: $\delta^{11}B = 26.7$ ppm, $\delta^{13}C = 47.2$ and 15.8 ppm [2]. The reaction with tmpLi yields $[tmp][(C_2H_5)_2N]BBr$ (tmp = 2,2,6,6-tetramethylpiperidinyl) [4, 5].

$(i\text{-}C_3H_7)_2N–BBr_2$ is formed in the decomposition (at ambient temperature) of the adduct of 2-diisopropylamino-1,3,2-dithiadiborolane with BBr_3 [3].

References for 7.4 to 7.7 on pp. 96/7

$[Cl_2P(O)N(CH_3)BBr_2]_2$ is prepared by the dropwise addition of 15.1 g (0.061 mol) of BBr_3 to 9.0 g (0.061 mol) of $Cl_2P(O)NHCH_3$ in 8 mL of CH_2Cl_2 under stirring; HBr is evolved vigorously. After 4 h of reflux, the solvent is evaporated and the solid residue washed with a small amount of CH_2Cl_2 and recrystallized from this same solvent. The yield is 86% and the melting point is 116 to 118°C. In the mass spectrum, the molecular ion is detected at M/z 317 (for the monomeric form in the gas phase). NMR data: $\delta^1H = 2.62$ ppm (doublet, $^3J(PNCH_3) = 24$ Hz); $\delta^{11}B = 6.8$ ppm; $\delta^{13}C = 34.1$ ppm ($J(^{31}P^{13}C) = 7$ Hz); $\delta^{14}N = -298$ ppm (relative to NO_3^-); $\delta^{17}O = 123$ ppm (relative to ClO_4^-); $\delta^{31}P = 34.2$ ppm (relative to H_3PO_4). The dimer (in the solid state) forms an eight-membered $(-B-N-P-O-)_2$ ring system [6].

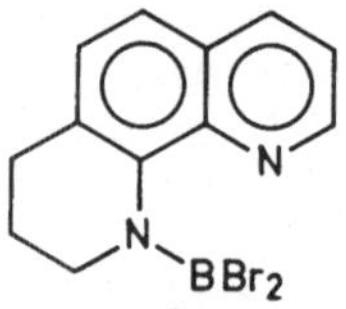

N
N
BBr_2
2

2 is prepared from the corresponding N-trimethyl-silylated compound and BBr_3; for NMR (^1H, ^{13}C) data, see [26].

$[(CH_3)_2N]_2BBr$ is formed on treatment of $(CH_3)_3Sn-B[N(CH_3)_2]_2$ with Br_2 [25]. The vibrational spectrum has been assigned using group frequency correlation with related compounds [7]. (Since no isotopically labelled molecules have been included in the study, there are serious doubts about these assignments since coupling effects between inner modes of the alkyl groups and the skeleton vibrations must play an important role in the spectrum; see "Borverbindungen" 5, 1975, pp. 57 to 168.)

$[(CH_3)_2N]_2BOCO(t\text{-}C_4H_9)$ is obtained from reaction of $[(CH_3)_2N]_2BBr$ with the lead salt of pivalic acid. For analogous reactions with other salts of carboxylic acids, see [8]. Reaction of $[(CH_3)_2N]_2BBr$ with tmpLi (tmp = 2,2,6,6-tetramethylpiperidinyl) yields $[tmp][(CH_3)_2N]BBr$ [5], and the reaction with $t\text{-}C_4H_9N{=}C(R)Li$ gives the corresponding bis(dimethylamino)carbiminoborane [9]. C_2H_5Br is eliminated in the reaction of $[(CH_3)_2N]_2BBr$ with $(CH_3)_3SiC{\equiv}COC_2H_5$ to yield $[(CH_3)_3Si][\{(CH_3)_2N\}_2B]C{=}C{=}O$ [10]; however, if $(CH_3)_3GeC{\equiv}COC_2H_5$ is the reactant, the product is $[(CH_3)_2N]_2B{-}C{\equiv}COC_2H_5$ [10].

$[(C_2H_5)_2N]_2BBr$. With respect to the assignment of the vibrational spectrum by [7], see the remarks made for $[(CH_3)_2N]_2BBr$. Reaction of $[(C_2H_5)_2N]_2BBr$ with metal (K^+, Ag^+, Pb^{2+}) salts of carboxylic acids yields species of the type $[(C_2H_5)_2N]_2BOCOR$ (R = CH_3, CF_3, $t\text{-}C_4H_9$) [8]. Bromobis(diethylamino)borane has been used in the synthesis of monomeric carbiminoboranes [9].

$(-NCH_3-CH_2-CH_2-NCH_3-)BBr$ forms 1:1 adducts with BBr_3 [11] or $AlBr_3$ [12]. The Lewis acid is coordinated to one of the nitrogen atoms. NMR data (^1H, ^{11}B, ^{13}C, ^{14}N) of the adducts are available. For methathesis reactions, see [11, 12].

$(-BBr-NCH_3-)_3$ is formed as a by-product in the reaction of $(CH_3HN)_2CS$ with BBr_3 [13]. NMR data: $\delta^{11}B = 32.0$ ppm, $\delta^{13}C = 40.4$ ppm [2].

$(tmp)(R_2N)BBr$ (tmp = 2,2,6,6-tetramethylpiperidinyl; R = CH_3, C_2H_5) species are prepared from equimolar amounts of tmpLi and R_2NBBr_2 in hexane at 0°C (about 65% yield). The boiling points are 69°C/0.1 Torr (R = CH_3) and 80°C/0.1 Torr (R = C_2H_5); NMR data (^1H, ^{11}B, ^{13}C) are tabulated [5], $\delta^{11}B \approx 30.0$ ppm [4, 9]. The reaction with $AlBr_3$ produces the corresponding diamidoboron cations $[(tmp)B{=}NR_2][AlBr_4]$ [4, 5, 14, 15].

$[(CH_3)_2N][CH_3]BBr$ reacts with t-$C_4H_9N{=}C(n$-$C_4H_9)Li$ to give t-$C_4H_9N{=}C(n$-$C_4H_9)B[CH_3]$-$[N(CH_3)_2]$ [9].

$[(CH_3)_2N][C_6H_5]BBr$ reacts with C_6H_5NCO by insertion into the B–N bond [16, 17].

(tmp)(R)BBr (tmp = 2,2,6,6-tetramethylpiperidinyl; R = CH_3, C_6H_5) species are prepared from equimolar quantities of tmpLi and $RBCl_2$ in hexane at 0°C and subsequent reaction at room temperature (12 h, R = CH_3) or under reflux (2 h, R = C_6H_5) in 40% and 55% yield, respectively. $(tmp)(CH_3)BBr$ has a boiling point of 95°C/6 Torr; $(tmp)(C_6H_5)BBr$ has a melting point of 113 to 115°C and can be purified by sublimation at 100 to 110°C/0.01 Torr. NMR data (^1H, ^{11}B, ^{13}C) are tabulated; $\delta^{11}B = 41.7$ (R = CH_3), 40.4 ppm (R = C_6H_5) [5]. Reaction with $AlBr_3$ yields the heteroallene-type boron cations $[tmp{=}B{=}R]^+$ [5, 14].

Compounds of type **3**, **4**, and **5** have been prepared from N-organyl-N-trimethylsilyl-propargylamines or propargylamines and $XBBr_2$; mass spectral and NMR (^1H, ^{11}B) data are reported [18].

$HC{\equiv}C{-}CH_2N(R)B(CH_3)Br$

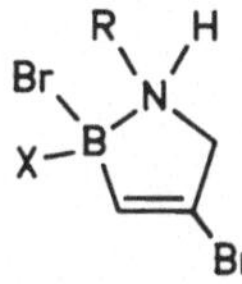

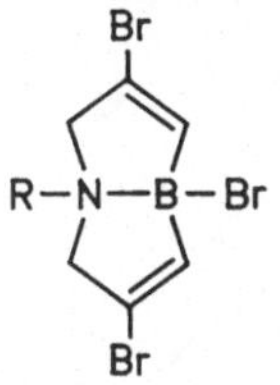

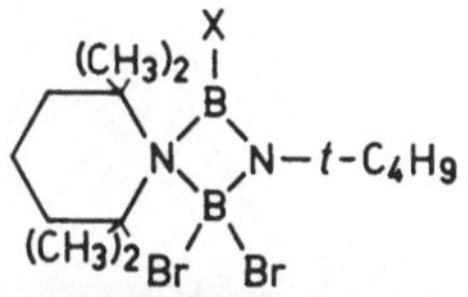

3a: R = CH_3
3b: R = C_2H_5
3c: R = $C_6H_5CH_2$

4	X	R
a	CH_3	n-C_3H_7
b	Br	i-C_3H_7
c	CH_3	i-C_3H_7
d	Br	$C_6H_5CH_2$

5	R
a	CH_3
b	C_2H_5
c	n-C_3H_7
d	i-C_3H_7
e	n-C_4H_9

6a: X = Br
6b: X = CH_3

Compounds of type **6** result from the addition of BBr_3 (CH_3BBr_2) to the amino-iminoborane $(tmp)B{=}N(t$-$C_4H_9)$ [19, 20] (tmp = tetramethylpiperidinyl). Thus, **6a** is obtained as follows: 0.23 mL (0.0025 mol) of BBr_3 in pentane are added dropwise at −30°C to 4.6 mL of a 0.543 molar solution of the amino-iminoborane in pentane (0.0025 mol). After warming to ambient temperature a colorless precipitate is filtered off and 0.49 g of **6a** crystallize from the filtrate at −20°C. The melting point is 121°C. In the mass spectrum the isotopic pattern of the molecular ion is observed at M/z 472. NMR (in ppm): $\delta^{11}B = 25.9$ and −0.6; $\delta^1H = 1.82$ (singlet, $C(CH_3)_3$), 1.63 (singlet, 6H = H(2,3,4)), 1.58 (singlet, 12H = 4 × CH_3); $\delta^{13}C = 64.1$ (C(1,5)), 54.3 ($C(CH_3)_3$), 40.1 (C(1,5)), 32.6, (2 × CH_3), 30.4 (3 × CH_3), 27.0 (2 × CH_3), 15.5 (C(3)); $\delta^{14}N = -212$ and −299 ppm [19].

The crystal and molecular structure of **6a** has been determined by X-ray diffraction. The crystals are orthorhombic, space group Pnma-D_{2h}^{16} (No. 62), Z = 4; a = 13.6618, b = 12.6076, and c = 10.8544 Å; $\alpha = \beta = \gamma = 90°$; V = 1869.6 Å^3; $D_{calc} = 1.68$ g/cm^3. The structure is depicted in **Fig. 7-1** and characteristic bond lengths and bond angles are listed in Table 7/7 [19]. The reaction of compounds **6** with $AlBr_3$ gives cationic species (see compound **9**, p. 95) [20].

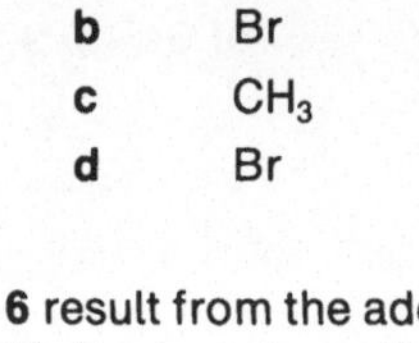

7 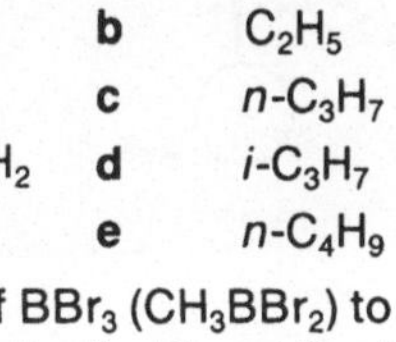 **8**

References for 7.4 to 7.7 on pp. 96/7

The species **$[(CH_3)_3Si][CH_3(CH_2)_2CH{=}]C{-}N(\textit{n}{-}C_4H_9){-}B[N(CH_3)_2]Br$** (mass spectrum, IR) [9] and the cyclic thioureidoborane **7** and **8** (see p. 93) [13] have been described.

$[N(\textit{n}{-}C_4H_9)_4][BBr_4]$ undergoes ligand exchange with $[N(\textit{n}{-}C_4H_9)_4][BH_4]$ in molar ratios of 3:1, 1:1, and 1:3 (in benzene) to yield the anions $[H_3BBr]^-$, $[H_2BBr_2]^-$, and $[HBBr_3]^-$. **$[N(\textit{n}{-}C_4H_9)_4][H_3BBr]$** is unstable and dissociates at 20°C in vacuum to give $[N(\textit{n}{-}C_4H_9)_4]Br$. **$[N(\textit{n}{-}C_4H_9)_4][H_2BBr_2]$** and **$[N(\textit{n}{-}C_4H_9)_4][HBBr_3]$** are stable at ambient temperature but decompose vigorously at 140 and 160°C. IR spectra of the species are tabulated [21].

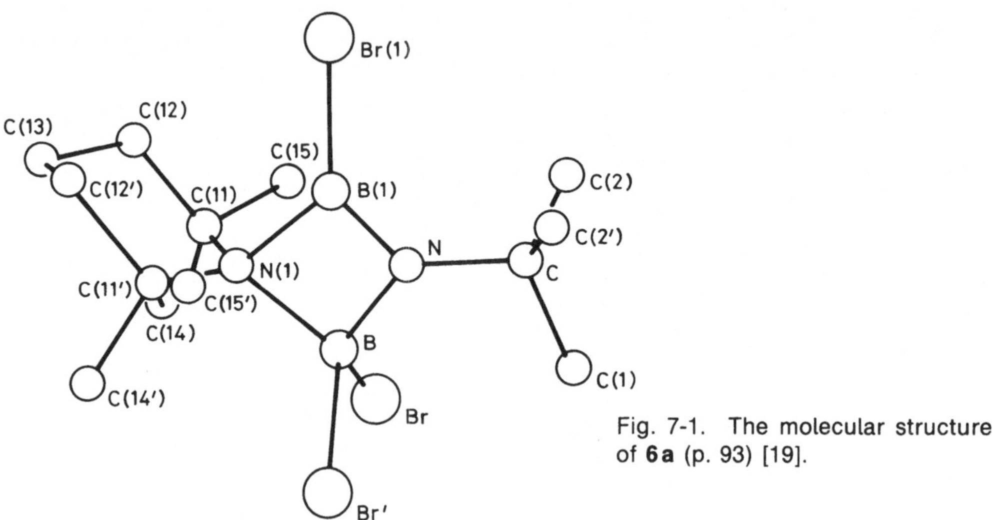

Fig. 7-1. The molecular structure of **6a** (p. 93) [19].

Table 7/7
Selected Bond Lengths (in Å) and Bond Angles (in degrees) for **6a** (p. 93) [19].

distances		angles		angles	
B–Br	2.040(7)	B–N–B(1)	95.3(8)	Br–B–Br'	101.7(5)
B(1)–Br(1)	1.915(12)	B–N–C	132.5(8)	N–B–N(1)	87.0(7)
B–N	1.474(14)	B(1)–N–C	132.2(9)	N(1)–B(1)–Br(1)	127.4(7)
B–N(1)	1.719(13)	B–N(1)–B(1)	78.1(7)	N–B(1)–Br(1)	133.0(8)
B(1)–N	1.340(14)	B–N(1)–C(11)	118.0(4)	N–B(1)–N(1)	99.6(8)
B(1)–N(1)	1.564(13)	B(1)–N(1)–C(11)	112.2(4)	N(1)–C(11)–C(12)	111.1(6)
N–C	1.512(13)	C(11)–N(1)–C(11')	113.3(7)	N(1)–C(11)–C(14)	109.7(6)
N(1)–C(11)	1.591(8)	N–C–C(1)	111.1(9)	N(1)–C(11)–C(15)	109.2(6)
C–C(1)	1.542(7)	N–C–C(2)	107.7(6)	C(14)–C(11)–C(15)	110.3(7)
C–C(2)	1.537(12)	C(1)–C–C(2)	110.2(6)	C(12)–C(11)–C(15)	106.3(7)
C(11)–C(12)	1.537(12)	C(2)–C–C(2')	109.8(10)	C(12)–C(11)–C(14)	110.2(7)
C(11)–C(14)	1.549(12)	C(11)–C(12)–C(13)	115.5(9)		
C(11)–C(15)	1.526(12)	Br–B–N	116.9(4)		
C(12)–C(13)	1.507(13)	Br–B–N(1)	117.7(4)		

Cationic species such as [(tmp)B=tmp][BBr_4] (tmp = 2,2,6,6-tetramethylpiperidinyl) and [(tmp)B=N(C_2H_5)$_2$][BBr_4] [4, 15] have also been reported. Data for the corresponding cations (containing two-coordinate boron) are presented in "Boron Compounds" 3rd Suppl. Vol. 3, 1988, Section 4.2.11.1, pp. 204/6, $[BBr_4]^-$ is discussed in Section 7.1.3 (p. 78).

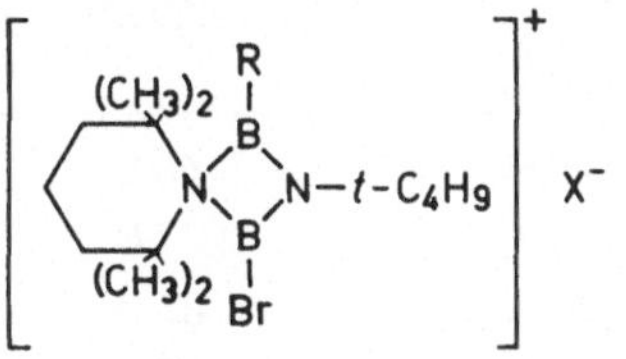

9	R	X
a	Br	[$AlBr_4$]
b	CH_3	[$AlBr_4$]

Cations with three-coordinate boron of type **9** have been obtained by reacting compound **6a** and **6b** (see p. 93), respectively, with $AlBr_3$ [20]. The following data are given:

9a: decomposition point 74 to 76°C. NMR data (in ppm): $\delta^1H = 2.02$ (H(2,3,4)), 1.70 (4 × CH_3), 1.58 (3 × CH_3); $\delta^{11}B = 36.8$; $\delta^{27}Al = 79.7$ [20].

9b: decomposition point 117 to 118°C. NMR data (in ppm): $\delta^1H = 1.92$ (H(2,3,4)), 1.66 + 1.52 (4 × CH_3), 1.51 (3 × CH_3), 1.44 (BCH_3); $\delta^{11}B = 47.8 + 36.5$; $\delta^{27}Al = 79.9$ [20].

7.6 With Nitrogen and Oxygen

The reaction of (tmp)BrB–O–C_6H_2-2,4,6-(t-C_4H_9)$_3$ with $AlBr_3$ to yield [(tmp)B=O–C_6H_2-(t-C_4H_9)$_3$][$AlBr_4$] (tmp = 2,2,6,6-tetramethylpiperidinyl) has been reported in a review [15]; no details are available.

7.7 With Halogen

The force fields of **$BBrF_2$**, **$BBrCl_2$**, **BBr_2F**, and **BBr_2Cl** have been computed using standardized structural parameters; in the case of $BBrF_2$ two conflicting assignments have been employed with almost equally acceptable results (see Table 7/8, p. 96). For a discussion, see [27]. BBr_2F is formed in the reaction of (tmp)$_2$BF with BBr_3 (tmp = 2,2,6,6-tetramethylpiperidinyl) [4].

(CH_3)BBrCl is formed when (CH_3)$_3$SnCl is reacted with liquid BBr_3 [24].

ClBrB–NCH_3–CH_2–CH_2–NCH_3–BBrCl and **ClBrB–NCH_3–CH_2–CH_2–NCH_3–BBr_2** are identified by NMR spectroscopy in solutions of adducts of 2-chloro-1,3-dimethyl-1,3,2-diazaborolidines with BBr_3 [11].

(C_6H_5)(CH_3)$_2$N–BF_2Br and **(C_6H_5)(CH_3)$_2$N–$BFBr_2$** have been prepared and studied by NMR (^{11}B, ^{13}C, ^{19}F); see Tables 6/34 and 6/35 in Chapter 6.5, p. 67; ^{13}C complexation shifts are tabulated [22]. It has been shown that the ^{19}F chemical shifts have a marked dependence on the steric effect of amine substituents (as estimated from the "cone angle" of the amines). Base strengths of the amines primarily influence the halogen distribution rate [22].

References for 7.4 to 7.7 on pp. 96/7

Table 7/8
Calculated Valence Force Constants (in $N \cdot m^{-1}$) for $BBrF_2$, $BBrCl_2$, BBr_2F, and BBr_2Cl [27] (for $BBrF_2$, set I is based on the assignments of [28] and set II on those of [29]).

force constant	$BBrF_2$ set I	$BBrF_2$ set II	$BBrCl_2$	BBr_2F	BBr_2Cl
f_r	732.0	707.5	449.6	362.4	308.1
f_R	383.7	396.4	247.8	664.6	600.5
f_α	79.4	85.2	87.0	84.0	87.1
f_β	64.5	62.4	56.3	51.0	56.0
f_Θ	42.6	42.6	39.2	42.3	37.5
f_{rr}	28.3	52.7	91.2	41.0	−29.1
f_{rR}	45.2	45.3	50.4	23.5	148.4
$f_{r\alpha}$	14.1	14.1	8.1	27.5	6.5
$f_{r\beta}$	−49.4	−31.9	−28.4	−36.5	−37.1
$f'_{r\beta}$	35.2	17.9	20.3	8.9	30.5
$f_{R\alpha}$	−19.3	−25.6	−28.6	−17.5	−38.8
$f_{R\beta}$	9.6	12.8	14.3	8.7	19.4
$f_{\alpha\beta}$	−39.7	−42.6	−43.5	−42.0	−43.6
$f_{\beta\beta}$	−24.8	−19.8	−12.8	−9.0	−12.4

Tertiary-amine (= D) adducts D–$BFBr_2$ are unreactive toward bromide ion displacement to form cationic species; however, on treatment of D–BF_2Br with tertiary amines of low steric requirements, the bis(*t*-amine)difluoroboron cations $[D_2BF_2]^+$ or $[DD'BF_2]^+$ are formed; quinuclidine, $HC(-CH_2-CH_2-)_3N$, is particularly favorable in this reaction [23]. Complexes have been prepared with the following amines $(CH_3)_3N$, $(C_2H_5)(CH_3)_2N$, C_5H_5N, and $HC(-CH_2-CH_2-)_3NH$. In the case of $[DD'BF_2]^+$, one amine is always quinuclidine while D is one of the cited amines and also $(CH_3)(C_2H_5)_2N$ and $(C_2H_5)_3N$ (with which complex cations $[D_2BF_2]^+$ could not be obtained). ^{11}B and ^{19}F NMR data of the various species are reported. As an example, $[\{(CH_3)_3N\}_2BF_2]Br$ exhibits $\delta^{11}B = 1.4$ ppm and $\delta^{19}F = -165.4$ ppm (relative to $CFCl_3$); $J(^{11}B-^{19}F) = 39.5$ Hz. The time to complete the reaction $D-BF_2Br + D \rightarrow [D_2BF_2]^+Br$ is 36 h at 25°C for $D = (CH_3)_3N$ [23].

References for 7.4 to 7.7:

[1] Costes, J. P.; Cros, G.; Laurent, J. P. (Syn. React. Inorg. Metal-Org. Chem. **11** [1982] 383/92).
[2] Nöth, H.; Wrackmeyer, B. (Chem. Ber. **114** [1981] 1150/6).
[3] Nöth, H.; Staudigl, R. (Chem. Ber. **115** [1982] 3011/24).
[4] Nöth, H.; Rasthofer, B.; Weber, S. (Z. Naturforsch. **39b** [1984] 1058/68).
[5] Nöth, H.; Staudigl, R.; Wagner, H.-U. (Inorg. Chem. **21** [1982] 706/16).
[6] Nöth, H.; Storch, W. (Chem. Ber. **117** [1984] 2140/56).
[7] Davidson, G.; Phillips, S. (J. Chem. Soc. Dalton Trans. **1981** 306/10).
[8] Bessler, E.; Pires da Silva, E. (Z. Anorg. Allgem. Chem. **480** [1981] 117/21).
[9] Sicker, U.; Meller, A.; Maringgele, W. (J. Organometal. Chem. **231** [1982] 191/203).
[10] Ponomarev, S. U.; Nikolaeva, S. N.; Molchanova, G. N.; Kostyuk, A. S.; Grishin, Yu. K. (Zh. Obshch. Khim. **54** [1984] 1817/21; J. Gen. Chem. [USSR] **54** [1984] 1620/3).

[11] Anton, K.; Nöth, H.; Pommerening, H. (Chem. Ber. **117** [1984] 2479/94).
[12] Anton, K.; Euringer, C.; Nöth, H. (Chem. Ber. **117** [1984] 1222/34).
[13] Maringgele, W. (J. Organometal. Chem. **222** [1981] 17/32).
[14] Nöth, H.; Staudigl, R. (Angew. Chem. **93** [1981] 830/31; Angew. Chem. Intern. Ed. Engl. **20** [1981] 794/5).
[15] Nöth, H.; Weber, S.; Rasthofer, B.; Narula, C.; Konstantinov, A. (Pure Appl. Chem. **55** [1983] 1453/61).
[16] Cragg, R. H.; Miller, T. J. (J. Organometal. Chem. **255** [1983] 143/52).
[17] Cragg, R. H.; Miller, T. J. (J. Organometal. Chem. **260** [1984] 1/5).
[18] Meller, A.; Hirninger, F. J.; Noltemeyer, M.; Maringgele, W. (Chem. Ber. **114** [1981] 2519/35).
[19] Nöth, H.; Weber, S. (Z. Naturforsch. **38b** [1983] 1460/5).
[20] Nöth, H.; Weber, S. (Angew. Chem. **96** [1984] 998; Angew. Chem. Intern. Ed. Engl. **23** [1984] 994).

[21] Gavrilova, L. A.; Titov, L. V.; Rosolovskii, V. Ya. (Zh. Neorgan. Khim. **26** [1981] 1769/74; Russ. J. Inorg. Chem. **26** [1981] 955/8).
[22] Fox, A.; Hartmann, J. S.; Humphries, R. E. (J. Chem. Soc. Dalton Trans. **1982** 1275/83).
[23] Farquharson, M. J.; Hartmann, J. S. (J. Chem. Soc. Chem. Commun. **1984** 256/7).
[24] Daneshvar, N.; Straughan, B. P.; Gardiner, D. J. (J. Inorg. Nucl. Chem. **43** [1981] 1147/9).
[25] Nöth, H.; Schwerthöffer, R. (Chem. Ber. **114** [1981] 3056/62).
[26] Klebe, G.; Hensen, K.; v. Jouanne, J. (J. Organometal. Chem. **258** [1983] 137/46).
[27] Aron, J.; Ford, T. A. (S. African J. Chem. **35** [1982] 129/38).
[28] Lindeman, L. P.; Wilson, M. K. (J. Chem. Phys. **24** [1956] 242/9).
[29] Wolfe, D. F.; Humphrey, G. L. (J. Mol. Struct. **3** [1969] 293/303).
[30] Dittmer, G.; Niemann, U. (Philips J. Res. **37** [1982] 1/30).

8 The System Boron-Iodine

Anton Meller
Institut für Anorganische Chemie, Universität Göttingen
Göttingen, Federal Republic of Germany

8.1 Binary Species

8.1.1 BI and BI_2

When BI_3 vapor mixed with argon or helium is passed through a microwave discharge, four new bands (at 349.8, 350.8, 3651.2, and 352.5 nm) degrading to the red have been observed and assigned to **BI**. It is assumed that the formation of B_2I_4 in this process occurs via the donation of the lone pair of BI into the empty 2p orbital of BI_3 [1].

Thermodynamic data for gaseous BI and $\mathbf{BI_2}$ have been obtained by a combination of tungsten transport measurements and theoretical estimations. The following values are given [2]: BI: $\Delta H_{298} = 321$ kJ/mol; $S_{298} = 233$ $J \cdot mol^{-1} \cdot K^{-1}$; $C_{p,298} = 34$ $J \cdot mol^{-1} \cdot K^{-1}$. BI_2: $\Delta H_{298} = 208$ kJ/mol; $S_{298} = 309$ $J \cdot mol^{-1} \cdot K^{-1}$; $C_{p,298} = 51$ $J \cdot mol^{-1} \cdot K^{-1}$.

8.1.2 Triiodoborane, BI_3, and Its Adducts with Lewis Bases

Structural and Physical Data

The core binding energies for $\mathbf{BI_3}$ have been redetermined with probable uncertainties of ± 0.05 eV. Thus, for B 1s $E_B = 197.92$ eV with a full width at half maximum (fwhm) = 1.21 eV; for I $3d_{5/2}$ $E_B = 626.82$ eV with fwhm = 1.32 eV. These data were used in conjunction with published valence ionization potentials to establish the I–B π-bonding (by evaluating the difference between the $1a_2''$ ionization potential and the localized orbital ionization potential for the halogen valence p-orbital). The $1a_2''$ MO is composed of the pπ-orbitals of all four atoms and the difference is a measure of the stabilization of the halogen pπ-orbital by π-bonding to the boron atom. This difference is 1.69 eV for BI_3 (1.99 for BBr_3, 2.08 for BCl_3, and 2.57 for BF_3) [3].

The force field of BI_3 has been calculated once again with the aid of the parametric representation method. Resultant force constants, mean amplitudes of vibration, shrinkage constants, Coriolis coupling constants, and rotational distortion constants are tabulated [4]. Final E′ force constants for trihaloboranes have been calculated. For their values in terms of valence-central and Keating coordinates, see Table 7/2, p. 72 [5]. The force field given in Table 7/2 has been employed to calculate the mean amplitudes of vibration for the trihaloboranes. Data are listed in Table 7/3, p. 73 [6].

The molecular structure of $\mathbf{(CH_3)_3N{-}BI_3}$ has been determined from gas phase electron diffraction data [7]. The bond lengths (in Å) are r(BI) = 2.245 ± 0.004, r(NB) = 1.663 ± 0.0013, r(CN) = 1.497 ± 0.005, r(CH) = 1.103 ± 0.010. Bond angles (in degrees) are ∢(IBI) = 108.6 ± 0.4; ∢(CNC) = 106.0 ± 0.8. The rotational barrier about the B–N bond is estimated to be 3.5 kcal/mol [7].

Mean amplitudes of vibration of $\mathbf{H_3P{-}BI_3}$ (and 11 other phosphane-boranes) have been computed on the basis of structural and spectroscopic data from the literature. For a tabulation of the results, see [8]. Infrared and Raman spectra of $\mathbf{(CH_3)_2HP{-}BI_3}$, $\mathbf{(CH_3)_2DP{-}BI_3}$, and $\mathbf{(CD_3)_2HP{-}BI_3}$ were recorded in solution and in solid state at room temperature. A modified Urey-Bradley force field was applied in supporting the assignments made (for C_s symmetry). Selected observed fundamentals (calculated frequencies in parentheses); ν(BP) at

756(745) cm^{-1}, $\nu_{as}(BI_3)$ at 570(557), $\nu_s(BI_3)$ at 207(200), $\delta_{as}(BI)_3$ at 130(132), $\delta_s(BI_3)$ at 112(120), and $\varrho(BI_3)$ at 104(102) cm^{-1}. For a comprehensive tabulation of the results, see [9].

$(CH_3)_3As\text{–}BI_3$ and **$(C_6H_5)_3As\text{–}BI_3$** have been prepared from their individual components and characterized by NMR (^1H, ^{13}C), IR, and Raman spectroscopy. Assignments are supported by a modified Urey-Bradley force field calculation. A comprehensive tabulation of data is presented [10]. Selected data for $(CH_3)_3As\text{–}BI_3$: $\delta^1H = 1.68$, $\delta^{13}C = 5.33$ ppm; vibrational data (calculated values in parentheses): ν(BAs) at 656(651) cm^{-1}, $\nu_{as}(BI_3)$ at 542(546), $\nu_s(BI_3)$ at 170(172), $\delta_{as}(BI_3)$ at 117(108), $\delta_s(BI_3)$ at 95(95), $\varrho(BI_3)$ at 80(80) cm^{-1} [10]. $(C_6H_5)_3As\text{–}BI_3$: ν_{as}(BAs) near 630 cm^{-1} and $\nu_{as}(BI_3)$ at 550 cm^{-1} [10].

Chemical Behavior

BI_3 forms a 1:1 complex with B_6H_{10}, which exchanges halogen to give B_6H_8I. Reaction of B_6H_{10} with an excess of BI_3 gives $B_{13}H_{19}$ in 40% yield [11]. BI_3 reacts with $(\text{–}NCH_3\text{–}CH_2\text{–}CH_2\text{–}CH_3N\text{–})BCH_3$ to give a 1:1 complex; one of the nitrogen atoms acts as the Lewis base [12]. If the diborane(4) derivative $[(\text{–}NCH_3\text{–}CH_2\text{–}CH_2\text{–}CH_3N\text{–})B]_2$ is treated with two molar equivalents of BI_3, BI_3 is coordinated to one of each of the diazaborolidine rings [13]. Derivatives of sulfonic acid can be reduced with iodide in the presence of BI_3 to yield disulfanes; a reaction mechanism has been proposed [14].

(η^5-Pentamethylcyclopentadienyl)tin trifluoromethanesulfonate, $[(CH_3)_5C_5Sn]^+$ (pentagonal pyramid), reacts with BI_3 in CH_2Cl_2 to eliminate SnI_2 and the corresponding BI-substituted cluster **1** is obtained [15].

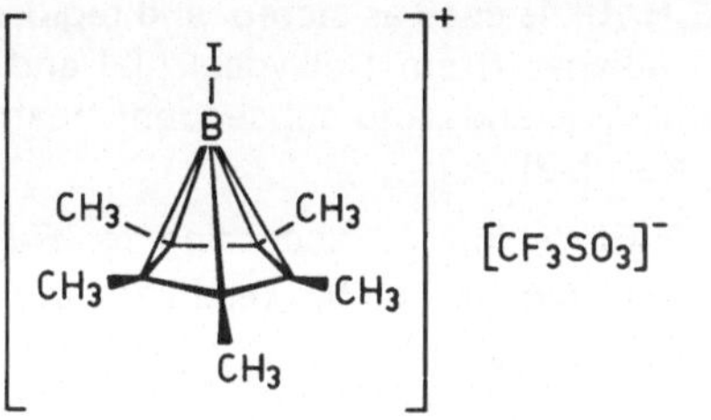

$[CF_3SO_3]^-$

1 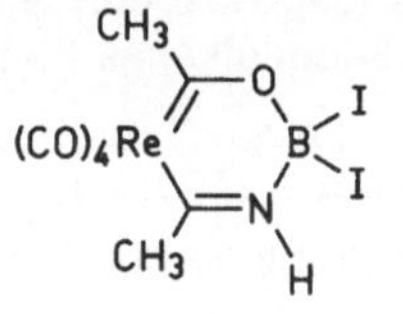 **2**

The reaction of $(CO)_5Cr\text{–}Ge(S\,mes)_2$ with BI_3 gives the diiodogermylene complex $(CO)_5Cr\text{–}GeI_2$ and $I_nB(S\,mes)_{3-n}$ (n = 0, 1; mes = C_6H_2-2,4,6-$(CH_3)_3$) [16]. Compound **2** is obtained when *cis*-$(CO)_4Re(CH_3CO)(CH_3C{=}NH_2)^+$ is reacted with NaH and subsequently treated with BI_3 [17]. Halogen exchange has been observed between RuF_6, OsF_6, or IrF_6 [18] and BI_3.

Facile halogen exchange takes place upon reaction of transition metal perfluoroalkyl carbonyl complexes with BI_3 according to $(CO)_5MNCF_3 + BI_3 \rightarrow (CO)_5MnCI_3$ [20, 21]. The reaction with $[\eta^5\text{-}C_5(CH_3)_5]_2UCl_2$ gives $[\eta^5\text{-}C_5(CH_3)_5]UI_2$ [22]. When *cis*-3,4-dichlorocyclobutene is reacted with BI_3, *trans*-3,4-diiodocyclobutene is obtained [23].

BI_3 is used in order to remove adhesive deposits in equipment for handling UF_6 [19]. A solid state electrolyte with an interdiffusion mixture of BI_3, LiI, and Al_2O_3 is used in lithium batteries [24].

References for 8 on p. 102

8.1.3 Additional Binary Species

For species containing boron-boron bonds see also "Boron Compounds" 3rd Suppl. Vol. 1, 1987, p. 66. Only some selected data are given here. B_2Cl_4 and BI_3 react almost quantitatively to give B_2I_4; mass spectral, NMR ($\delta^{11}B = 67.5$ ppm), and vibrational data are reported [25]. Thermal decomposition of B_2I_4 (100 to 400°C) yields B_8I_8 and B_9I_9 [26]. B_6I_6 is obtained by treatment of $Na_2[B_6H_6]$ with I_2 in aqueous alkaline solution; NMR, IR, and Raman spectra are given [27]. B_9I_9 has been prepared by reaction of $[B_9I_9]^-$ with thallium(III) trifluoroacetate; mass spectral, IR, and UV data are available [28].

8.2 With Hydrogen

$(CH_3)_3N–BH_2I$ reacts with 2-aminoethylpyridine with formation of a boronium salt [30]. The hyperlipidemic activity of trimethylamine-iodoborane and that of $[Na(H_3N–BH_2CN)_6]I$ have been tested [29].

8.3 (Organyl)iodoboranes

$(n\text{-}C_3H_7)(C_2H_5)CBI_2 \cdot S(CH_3)_2$ was synthesized by direct hydroboration of *cis*-3-hexene with $(CH_3)_2S–BHI_2$ (in CH_2Cl_2, 40°C, 5 h). For data on the thermal isomerization (at 150°C in *o*-dichlorobenzene), see [31].

9-Iodo-9-borabicyclo[3.3.1]nonane, $C_8H_{14}BI$, is used as stereo- and regioselective haloborating agent in the synthesis of 2-iodo-1-alkenes (from 1-alkynes) [32] and of (Z)-1-alkynyl-2-iodo-1-alkenes [33]. Reaction with terminal allenes and subsequent treatment with acetic acid also yields 2-iodoalkenes [34]; see also [42].

A series of (diorganyl)iodoboranes, R_2BI, is formed according to: $R_3B + H_2C{=}CHMgBr \rightarrow [R_2BCH{=}CH_2]MgBr \xrightarrow[+I_2]{-78°C} RCH{=}CH_2 + R_2BI + MgBrI$. For the relative migratory aptitudes of the alkyl groups in this iodination reaction, see [35].

8.4 With Oxygen

IBO. The enthalpy of formation as well as the entropy and heat capacity of gaseous IBO have been made available by using a combination of tungsten transport measurements and theoretical estimations. $\Delta H^{\circ}_{298} = -160$ kJ/mol; $S_{298} = 265\ J \cdot mol^{-1} \cdot K^{-1}$; $C_{p,298} = 49\ J \cdot mol^{-1} \cdot K^{-1}$ [2].

8.5 With Nitrogen

$I_2BN(CH_3)_2$. The following NMR data are reported: $\delta^{11}B = 4.9$ ppm, $\delta^{13}C = 46.10$ ppm [36].

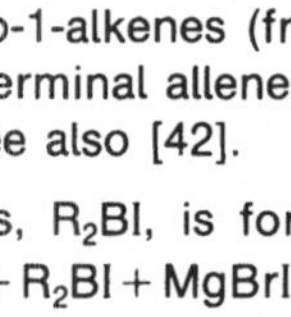

N
N
BI_2

3

3 can be prepared from the corresponding N-trimethylsilylated compound and BI_3; for NMR (^1H, ^{13}C) data, see [43].

$IB[N(CH_3)_2]_2$ is formed in the reaction between $[(CH_3)_2N]_2B–Sn(CH_3)_3$ and I_2 in ether at ambient temperature. The boiling point is 55°C/10 Torr (yield 89%); $\delta^{11}B = 25.5$ ppm [37].

$(-BI-NCH_3-)_3$. $\delta^{11}B = 28.5$ ppm, $\delta^{13}C = 49.9$ ppm [36].

$IB[tmp][N(C_2H_5)_2]$ (tmp = 2,2,6,6-tetramethylpiperidinyl) has been prepared from equimolar quantities of tmpLi and $I_2BN(C_2H_5)_2$ in hexane (ambient temperature, stirring) and subsequent refluxing (20 h). The boiling point is 104°C/0.03 Torr, the yield 22%. NMR data (in ppm): $\delta^1H = 3.53$ (quartet, CH_2), 1.47 (multiplet, C(2,3,4)), 1.35 ($4 \times CH_3$), 1.20 ($2 \times CH_3$); $\delta^{11}B = 26.0$ ppm. Upon treatment of the compound with BI_3 (equimolar amount in CH_2Cl_2, 65°C), diethylamido-2,2,6,6-tetramethylpiperidinoboron(1+) tetraiodoborate, $[tmp{=}B{=}N(C_2H_5)_2][BI_4]$, is formed. NMR data for this salt: $\delta^1H = 3.50$ (quartet, CH_2), 1.67 (multiplet, H(2,3,4)), 1.53 ($4 \times CH_3$), 1.45 ppm ($2 \times CH_3$); $\delta^{11}B = 37.7$ and -127.1 ppm [38].

8.6 With Halogens

Halogen redistribution equilibria are very unfavorable for the formation of tertiary amine (= D) adducts $D–BF_2I$ and $D–BFI_2$, since the BI_3 and BF_3 species are much favored (as shown by a multinuclear NMR study) [39]. In case of $D = (C_6H_5CH_2)(CH_3)(C_2H_5)N$, however, it appears that the salt $[D_2BF_2]I$ is formed in the equilibrium [39, 40].

The force fields of **BF_2I, BCl_2I, BBr_2I, BFI_2, $BClI_2$**, and **$BBrI_2$** have been computed using standarized structural parameters and the valence force constants obtained have been interpreted in terms of effective group electronegativities and intramolecular coupling of the stretching modes (see Table 8/1). For a discussion in relation to previously reported data, see [41].

Table 8/1

Calculated Valence Force Constants (in $N \cdot m^{-1}$) of BF_2I, BCl_2I, BBr_2I, BFI_2, $BClI_2$, and $BBrI_2$ [41].

force constant	BF_2I	BCl_2I	BBr_2I	BFI_2	$BClI_2$	$BBrI_2$
f_r	719.9	379.9	366.4	296.8	251.5	229.7
f_R	407.2	235.1	178.5	727.5	430.7	426.3
f_α	49.9	70.2	70.9	70.3	72.1	75.8
f_β	57.8	64.2	54.7	43.4	51.4	52.2
f_Θ	38.8	36.1	33.4	31.9	32.2	31.0
f_{rr}	149.1	112.1	72.4	32.1	5.3	−9.5
f_{rR}	98.6	50.2	30.3	72.2	67.5	67.5
$f_{r\alpha}$	8.6	6.2	8.1	18.6	13.8	12.2
$f_{r\beta}$	−3.8	−4.3	−28.3	−30.2	−31.7	−29.6
$f'_{r\beta}$	−4.9	−1.9	20.1	11.7	17.9	17.4
$f_{R\alpha}$	−25.1	−23.8	−19.3	−12.7	−17.8	−30.5
$f_{R\beta}$	12.6	11.9	9.6	6.3	8.9	15.3
$f_{\alpha\beta}$	−25.0	−35.1	−35.5	−35.2	−36.1	−37.9
$f_{\beta\beta}$	−32.8	−29.1	−19.2	−8.2	−15.3	−14.2

References for 8 on p. 102

References for 8:

[1] Briggs, A. G.; Massey, A. G.; Reason, M. S.; Portal, P. J. (Polyhedron **3** [1984] 369/71).
[2] Dittmer, G.; Niemann, U. (Philips J. Res. **37** [1982] 1/30).
[3] Beach, D. B.; Jolly, W. L. (J. Phys. Chem. **88** [1984] 4647/9).
[4] Mohan, S.; Rajaraman, S. (Indian J. Pure Appl. Phys. **20** [1982] 230/2).
[5] Cyvin, S. J.; Cyvin, B. N.; Brunvoll, J. (Spectrosc. Letters **17** [1984] 511/23).
[6] Cyvin, S. J.; Cyvin, B. N. (Spectrosc. Letters **17** [1984] 225/35).
[7] Iijima, K.; Shibata, Sh. (Bull. Chem. Soc. Japan **56** [1983] 1891/5).
[8] Sebesty'en, A.; Megyeri, L.; Vizi, B. (J. Mol. Struct. **89** [1982] 259/68 [THEOCHEM **6**]).
[9] Drake, J. E.; Hencher, J. L.; Khasrou, L. N. (Can. J. Chem. **59** [1981] 2898/908).
[10] Drake, J. E.; Khasrou, L. N.; Majid, A. (Can. J. Chem. **59** [1981] 2417/28).

[11] Dolan, P. J.; Moody, D. C.; Schaeffer, R. (Inorg. Chem. **20** [1981] 745/8).
[12] Anton, K.; Nöth, H.; Pommerening, H. (Chem. Ber. **117** [1984] 2479/94).
[13] Anton, K.; Nöth, H.; Pommerening, H. (Chem. Ber. **117** [1984] 2495/503).
[14] Olah, G. A.; Narang, S. C.; Field, L. D.; Karpeles, R. (J. Org. Chem. **46** [1981] 2408/10).
[15] Kohl, F.; Jutzi, P. (Angew. Chem. **95** [1983] 55; Angew. Chem. Intern. Ed. Engl. **22** [1983] 56).
[16] Jutzi, P.; Steiner, W.; Stroppel, K. (Chem. Ber. **113** [1980] 3357/65).
[17] Lukehart, C. M.; Raja, M. (Inorg. Chem. **21** [1982] 2100/1).
[18] Burns, R. C.; O'Donnell, T. A. (J. Fluorine Chem. **23** [1983] 1/14).
[19] Bacher, W.; Jakob, E. (Ger. Offen. 3009933 [1981] from C.A. **96** [1982] No. 59735).
[20] Richmond, T. G.; Shriver, D. F. (Organometallics **3** [1984] 305/14).

[21] Richmond, T. G.; Shriver, D. F. (Organometallics **2** [1983] No. 70792).
[22] Finke, R. G.; Hirose, Y.; Gaughan, G. (J. Chem. Soc. Chem. Commun. **1981** 232/4).
[23] Hoberg, H.; Froehlich, Ch. (Synthesis **1981** 830/1).
[24] Joshi, A. V.; Jatkar, A. D. (U.S. 4298664 [1981] from C.A. **96** [1982] No. 71941).
[25] Haubold, W.; Jacob, P. (Z. Anorg. Allgem. Chem. **507** [1983] 231/4).
[26] Massey, A. G.; Portal, P. J. (Polyhedron **1** [1982] 319).
[27] Preetz, W.; Fritze, J. (Z. Naturforsch. **39b** [1984] 1472/7).
[28] Wong, E. H. (Inorg. Chem. **20** [1981] 1300/2).
[29] Hal, I. H.; Das, M. K.; Harchelroad, F., Jr.; Wisian-Neilson, P.; McPhail, A. T.; Spielvogel, B. F. (J. Pharm. Sci. **70** [1981] 339/41).
[30] Niedenzu, K.; Read, R. B. (Z. Anorg. Allgem. Chem. **473** [1981] 139/52).

[31] Brown, H. C.; Racherla, U. S. (J. Org. Chem. **48** [1983] 1389/91).
[32] Hara, S.; Dojo, H.; Takinami, S.; Suzuki, A. (Tetrahedron Letters **24** [1983] 731/4).
[33] Hara, S.; Satoh, Y.; Ishiguro, H.; Suzuki, A. (Tetrahedron Letters **24** [1983] 735/8).
[34] Hara, S.; Takinami, S.; Hyuga, S.; Suzuki, A. (Chem. Letters **1984** 345/8).
[35] Slayden, S. W. (J. Org. Chem. **47** [1982] 2753/7).
[36] Nöth, H.; Wrackmeyer, B. (Chem. Ber. **114** [1981] 1150/6).
[37] Nöth, H.; Schwerthöffer, R. (Chem. Ber. **114** [1981] 3056/62).
[38] Nöth, H.; Rasthofer, B.; Weber, S. (Z. Naturforsch. **39b** [1984] 1058/68).
[39] Fox, A.; Hartmann, J. S.; Humphries, R. E. (J. Chem. Soc. Dalton Trans. **1982** 1275/83).
[40] Farquharson, M. J.; Hartman, J. S. (J. Chem. Soc. Chem. Commun. **1984** 256/7).

[41] Aron, J.; Ford, T. A. (S. African J. Chem. **35** [1982] 129/38).
[42] Hara, S.; Kato, T.; Suzuki, A. (Synthesis **1983** 1005/6).
[43] Klebe, G.; Hensen, K.; v. Jouanne, J. (J. Organometal. Chem. **258** [1983] 137/46).

9 The System Boron-Sulfur

Gert Heller
Institut für Anorganische und Analytische Chemie, Freie Universität Berlin
Berlin, Federal Republic of Germany

9.1 General Remarks

This presentation continues the earlier discussion of the system boron-sulfur in "Boron Compounds" 2nd Suppl. Vol. 2, 1982, pp. 155/216, and is arranged in similar fashion but organic derivatives are incorporated with the basic species; for a survey of the treatment of boron-sulfur compounds within the Gmelin Handbook, see l.c., p. 155.

9.2 Binary Species

9.2.1 Neutral Compounds

BS. Ab initio SCF calculations have been performed in order to gain an understanding of the unusual behavior of some γ and p doubling parameters of the BS molecule. The CI wave functions of the lowest states at their equilibrium separations are given in Table 9/1. When a configuration yields more than one state of a given symmetry, the coefficient of the CI wave function appearing in the table is a global one, calculated as the square root of the sum of the squares of all coefficients related to this configuration. It should be emphasized that the coefficients in the wave functions change very fast on going towards shorter separation (except for $X^2\Sigma^+$ and $D^2\Delta_i$). The spectroscopic constants of BS as given for six states are listed in Table 9/2 [1].

Table 9/1
CI Wave Functions of the Lowest States of BS [1].

state	$\|6\sigma^2 2\pi^4 7\sigma\|$	$\|6\sigma 2\pi^4 7\sigma^2\|$	$\|6\sigma^2 2\pi^3 7\sigma 3\pi\|$	$\|6\sigma^2 2\pi^3 7\sigma^2\|$	$\|6\sigma^2 2\pi^4 3\pi\|$
$X^2\Sigma^+$	0.98	—	0.17	—	—
$A^2\Pi_i$	—	—	—	0.90	0.24
$B^2\Sigma^+$	0.14	0.47	0.80	—	—
$C^2\Pi_r$	—	—	—	0.23	0.90
$D^2\Delta_i$	—	—	0.96	—	—
$E^2\Sigma^+$	0.10	0.66	0.74	—	—

Table 9/2
Spectroscopic Constants of BS [1].

state	T_e in cm^{-1}		B_e in cm^{-1}		ω_e in cm^{-1}		$\omega_e\chi_e$ in cm^{-1}	
	calc	exp	calc	exp	calc	exp	calc	exp
$X^2\Sigma^+$	0	0	0.787	0.794	1180.33	1180.17	6.6	6.29
$A^2\Pi_i$	30930.26	15828.6	0.587	0.622	751.72	754.4	4.4	4.78
$B^2\Sigma^+$	49027.9	36017.5	0.580	0.630	670.2	760	6.41	—
$C^2\Pi_r$	41400.8	38839.6	0.697	0.702	874.11	892.6	8.60	6.74
$D^2\Delta_i$	59839.3	47649	0.560	0.601	607.4	676	0.45	—
$E^2\Sigma^+$	65894.41	47929.3	0.612	0.671	674	801	6.39	10±5

References for 9.2 on p. 109

The deperturbed vibrational and rotational spectroscopic constants of the $A^2\Pi$ and $X^2\Sigma^+$ states of BS as obtained by spectral studies between 4400 and 6150 Å are listed in Table 9/3. The interaction parameters are defined by (SO = spin-orbit; RE = rotation-electronic): $\mathscr{H}^{SO}_{el} = \alpha|\langle v'_\pi|v''_\Sigma)$ where $\alpha = 1/2 \langle \pi, v'/AL + |\Sigma, v''\rangle$; $\mathscr{H}^{RE}_{el} = \beta|\langle v'_\pi|B|v''_\Sigma)$ where $\beta = \langle \pi, v'|B(R)L + |\Sigma, v'')$.

Table 9/3

Constants of Vibration and Rotation of the States $X^2\Sigma^+$ and $A^2\Pi$ of BS (in cm^{-1}; [a] means reference of energies $X^2\Sigma^+$, v = 0, N = 0; N is the number of bands; [b] value of p for v' = 0; [c] $A_{v'} = -328.10 + 0.025\ (v + 1/2) + 0.014\ (v + 1/2)^2)$ [2].

constants	$X^2\Sigma^+$ ($0 \le v'' \le 25$)	$A^2\Pi$ ($0 \le v' \le 12$)
T_e [a]	0	15451.045
ω_e	1179.91	754.239
$\omega_e x_e$	6.250	4.859
$\omega_e y_e$	−0.0083	0.019
B_e	0.79478	0.62295
α_e	0.00578	0.00608
γ_e	−0.000014	0.00002
r_e in Å	1.6092	1.8177
D_e	1.40×10^{-6}	1.60×10^{-6}
γ, p	0.0136	0.0176 [b]
A_e [c]		−328.10
	$\mathscr{H}^{SO}_{el} = -180$	$\mathscr{H}^{RE}_{el} = 0.236$

The perturbations in the level of the $A^2\Pi$ state observed in the $A^2\Pi_i$–$X^2\Sigma^+$ system are the result of interactions with high levels of the $X^2\Sigma^+$ state. The electronic factors of the spin-orbit and rotation-electronic perturbation matrix elements for the A ↔ X interaction are elevated. Table 9/4 shows the spectroscopic constants derivated from two $A^2\Pi$ niveaus; calculated deperturbed spin-orbit constants of the $A^2\Pi$ state are listed in Table 9/5 [2].

Table 9/4

Spectroscopic Constants (in cm^{-1}) of Two Levels of BS (values in parentheses are the deviations of each constant) [2].

	level $A^2\Pi$, v' = ~3 $X^2\Sigma^+$, v'' = 17	level $A^2\Pi$, v' = ~6 $X^2\Sigma^+$, v'' = 19
T_Π	18034.216(29)	20154.210(97)
A	−331.843(35)	−328.940(102)
B_Π	0.601815(17)	0.584212(40)
D_Π	1.6×10^{-6} fixed	1.6×10^{-6} fixed
p	0.0126 fixed	0.0126 fixed
T_Σ	18101.397(383)	19981.187(63)
B_Σ	0.68864(40)	0.67685(42)
D_Σ	1.4×10^{-6} fixed	1.4×10^{-6} fixed
γ	0.0136 fixed	0.0136 fixed
$\mathscr{H}^{SO}_{el}$	−214.48(1.78)	−193.84(97)
$\mathscr{H}^{RE}_{el}$	−0.053(5)	0.236(5)
deviation	0.138	0.125

Table 9/5

Spin-Orbit Constants $A_{v'}$ of the State $A^2\Pi$ of BS (in cm^{-1}; [a] the values $A_{v'}$ have been calculated from the expression given in Table 9/3 under [c]; [b] values from [3]).

v'	$A_{v'}$*	$\Delta T_{v'A}$	$A_{v'}$	$A_{v'}$ calc [a]
0 [b]	−330.91	3.00	−327.91	−328.08
1	−331.27	3.37	−327.90	−328.03
2	−331.89	3.88	−328.01	−327.95
3	−333.19	5.22	−327.97	−327.84

Table 9/5 (continued)

v'	$A_{v'}$*	$\Delta T_{v'A}$	$A_{v'}$	$A_{v'}$ calc [a)]
4	−332.45	4.89	−327.56	−327.70
5	−334.44	6.87	−327.57	−327.54
6	−338.09	10.66	−327.43	−327.35
7	−352.38	25.25	−327.13	−327.15
8	−310.15	−16.85	−327.00	−326.88
9	−322.67	−3.96	−326.63	−326.60
10	−326.87	0.84	−326.03	−326.29
11	—	—	—	−325.96
12	−325.10	—	—	−325.60

The perturbations observed in the 36000 to 46000 cm^{-1} energy range, particularly in the $C^2\Pi$ levels, are discussed; transitions $C^2\Pi_v \rightarrow X^2\Sigma^+$ occur between 2300 and 2900 Å (Table 9/6). For the $F^2\Delta$ state, the following spectroscopic constants had been used: $T_e = 39630$, $\omega_e = 772$, and $B_e = 0.60$ cm^{-1}. A study of the structure of the $B^2\Sigma^+$ state has also been performed on the $B^2\Sigma^+ \rightarrow A^2\Pi_i$ transition at 5000 Å; the spectroscopic constants are shown in Table 9/7, p. 106. The location of the different perturbing states allows a discussion of all the states of the first excited electronic configurations. Table 9/8, p. 106, shows the constants of the electronic states of BS [4].

Table 9/6
Energetic Positions of the $C^2\Pi$ and $F^2\Delta$ Levels of BS [4].

	level $C^2\Pi$		level $F^2\Delta$	perturbations	
v = 2	$T_{1/2} = 40532.4$ $T_{3/2} = 40646.8$	v = 1	$T_{3/2} = 40766.2$ $T_{5/2} = 40820.5$	$^2\Delta_{3/2} \sim {}^2\Pi_{3/2}$	$J \geq 40.5$
v = 3	$T_{1/2} = 41378.8$ $T_{3/2} = 41485.6$	v = 2	$T_{3/2} = 41525$ $T_{5/2} \sim 41578$	$^2\Delta_{3/2} \sim {}^2\Pi_{3/2}$ $^2\Delta_{3/2} \sim {}^2\Pi_{1/2}$ $^2\Delta_{5/2} \sim {}^2\Pi_{3/2}$	J = 19.5 J = 33.5 J > 37.5 not observed
v = 4	$T_{1/2} = 42218.0$ $T_{3/2} = 42332$	v = 3	$T_{3/2} \sim 42270$ $T_{5/2} \sim 42324$	$^2\Delta_{3/2} \sim {}^2\Pi_{1/2}$ $^2\Delta_{5/2} \sim {}^2\Pi_{3/2}$	not observed J negative
v = 5	$T_{1/2} = 43035$ $T_{3/2} = 43138.5$	v = 4	$T_{3/2} \sim 43010$ $T_{5/2} \sim 43060$		—

The dissociation energies E_D of similar molecules with isostructural atoms on their atomic nuclear charges have been calculated from the hyperbolic dependence of E_D. The published E_D of BS is 138 ± 3 kcal/mol, the calculated value is also 138 kcal/mol; extrapolation (linear according to Birge-Sponer) for the $C^2\Pi$ state gives 156 ± 30 kcal/mol; Mass spectral investigation of the equilibrium CS + B → BS + C gives the value 140 ± 6 kcal/mol, and the equilibrium GeS + B → BS + Ge results in the value 145 ± 6 kcal/mol [6].

For BS charge-stripping reactions of the type $m^+ + N \rightarrow m^{2+} + N + e^-$ have been studied [7]. Weighted column densities of BS have been calculated in six model atmospheres of red-giant stars with solar composition [8].

References for 9.2 on p. 109

Table 9/7

Spectroscopic Constants of the State $B^2\Sigma^+$ of BS (in cm^{-1}; a) reference of energies $X^2\Sigma^+$, $v=0$, $N=0$; b) values in parentheses for the niveau $v=0$ are from [3, 4]).

v	T a)	B	$D \cdot 10^6$	γ
0	36016.64	0.63160 (0.6311) b)	1.66 (1.53)	−0.0878 (−0.0901)
1	36778.91	0.62657	2.05	−0.0499
2	37534.06	0.62050	1.69	−0.0373
3	38281.53	0.61442	1.68	−0.0361
4	39020.5	0.6076	1.68	−0.035

$T_e = 35633.065$
$\omega_e = 768.708$
$\omega_e\chi_e = 3.048$
$\omega_e\gamma_e = -0.109$
$B_e = 0.63422$
$\alpha_e = 0.00495$
$\gamma_e = -0.00020$
$r_e = 1.8014$ Å

Table 9/8

Constants of the Electronic States for BS (in cm^{-1}; a) reference of energies $X^2\Sigma^+$, $v=0$, $N=0$; b) constants of Bell and McLean [5]; c) values are for $v=0$ [4]).

state	T_e a)	ω_e	$\omega_e\chi_e$	$\omega_e\gamma_e$	B_e	α	r_e in Å	remarks
$E^2\Sigma^+$ b)	47929.3 c)	781			0.671	0.008	1.752	
$D^2\Delta_i$ b)	47649.2 c)	676			0.6019 c)		1.849 c)	$A_0 = -176$
$G^2\Sigma^+$	40647 c)				0.574 c)		1.880 c)	
$F^2\Delta_r$	39630	772			0.60		1.825	$A_0 = 75.1$ $A_1 = 54.3$
$C^2\Pi_r$	38781.64 c)	>887.6	>9.1		0.70374 c)		1.7101 c)	$A_0 = 116.57$ $p = 0.0115$
$g^4\Sigma^-$	~36250	~801	~3.5					
$f^4\Delta$	33750 c)							
$e^4\Sigma^+$	31250 c)							
$B^2\Sigma^+$	35633.06	768.71	3.048	−0.109	0.63422	0.00495	1.8014	$\gamma = -0.0878$ c)
$A^2\Pi_i$	15451.05	754.24	4.859	0.019	0.62295	0.00608	1.8177	$A = -328.10$ $p = 0.0176$ c)
$X^2\Sigma^+$	0	1179.91	6.250	−0.0083	0.79478	0.00578	1.6092	$\gamma = 0.0136$

BS_2 exists in the solid state in the form of **B_8S_{16}** and in the polymeric form **$(BS_2)_n$**. While the previously described B_8S_{16} (see p. 107) has large π-bonding ratios in the B–S bonds of the porphin-like structure, $(BS_2)_n$ consists of trithiadiborolane rings, which are connected by S atoms in a chain-type arrangement (**Fig. 9-1**). Both compounds are prepared by the reaction of 3,5-dibromo-1,2,4-trithia-3,5-diborolane with $S{=}C(SH)_2$ between 120 and 250°C [27 to 29].

Fig. 9-1. The structure of $(BS_2)_n$ (distances in Å, angles in degrees). The distance S(2)...S(3) is 3.167 Å) [29].

For gaseous $^{11}BS_2$ with $D_{\infty h}$ symmetry, the following absorption bands of $^2\Sigma_u^+$ symmetry have been observed in a Ne medium: a symmetrical stretching mode at 510 cm^{-1} with E_g^+ vibrational symmetry, a bending mode at ca. 120 cm^{-1} (estimated from isotopic shift in origin of A–X transition) with Π_u vibrational symmetry (in the UV), and a strong asymmetric stretching mode at 1015 cm^{-1} (in the IR) with Σ_u^+ vibrational symmetry [30].

B_2S_3 crystals have been prepared by controlled thermal decomposition of $Ag_3B_5S_9$ at 750°C [29]. B_2S_3 glass has been prepared from B_2S_3 powder with special purity requirements in a quartz tube and then outgassing under vacuum at ca. 200°C overnight, placing the tube into a vertical muffle furnace under argon flow and heating to 650°C for about five hours. Subsequently the probe was air-quenched, and the light (not the dark) amber glass rod was broken out of the tube in one piece. In a second preparative method, stoichiometric amounts of finely powdered elemental boron and sulfur were well mixed in a slurry with acetone, and then outgassed under vacuum at 100°C overnight. The material was placed in quartz ampoules which were sealed under vacuum and placed in a vertical muffle furnace with a temperature gradient of ca. 10°C/cm in order to prevent evaporation of the melt. After five days at 800°C the tubes were opened and the amber glass was removed from unreacted sulfur and boron. Differential-scanning calorimetry shows glass transition temperatures between 249 and 270°C for different vitreous samples. By comparison with $(HBS_2)_3$, the room temperature Raman spectra (in vacuum-sealed quartz ampoules) of all vitreous samples showed broad strong luminescence bands in the green to red, but the material made from the elements was much less luminescent; at 457.9 nm the luminescence interfered least with the Raman spectra. A broad line at 325 cm^{-1} is due to the symmetric stretch of sulfur atoms bridging between rings, the sharp dominant line at 445 cm^{-1} is associated with the symmetric stretching motion of the sulfur atoms in the trimer thioboric acid (boroxin-like) rings, and the sharp line at 512 cm^{-1} is assigned to the symmetric stretch motion of sulfur in dimeric thioboric acid rings. The conclusion was reached that the vitreous B_2S_3 network can incorporate a significant number of HS groups, and that the Raman spectra are consistent with the existence of planar trimeric thioboric acid (boroxin-like) rings, dimeric thioboric acid rings (formed by edge-sharing triangles), and sulfur atoms which bridge between the rings [9].

Vitreous species $B_2S_3 \cdot \leqq 0.5\,Li_2S$ or $B_2S_3 \cdot \leqq 0.4\,Na_2S$, respectively, were obtained when mixtures of powdered B_2S_3 and Li_2S or Na_2S were melted between 700 and 900°C in graphite crucibles sealed in evacuated Si ampoules. The electric conductivity of the Na sulfide glass is lower than that of the Li glass [12]. A glass of greater electric conductivity is supposedly formed by mixing B_2S_3 and Li_2S at 650°C [13]. Other B_2S_3–Na_2S compounds have been prepared by a special new heating method; in comparison with oxide glasses of the same composition replacement of O by S improves the conductivity; batteries or fuel cells have been prepared from such systems [14, 15]. Pressed pellets B_2S_3–Li_2S–LiI have a good ionic conductivity; in this system, fast Li^+ ion transport is possible, so that new secondary all-solid batteries have been prepared [16 to 21].

Reaction of B_2S_3 in refluxing $CHCl_3$ with selenopyranones yields selenopyranthiones [10]. Reaction in a Ni crucible at 300°C with ReF_7 and ReF_6 yields ReF_5S and ReF_4S, respectively [11].

B_2S_4. The muonic Coulomb capture ratio in B_2S_4 has been calculated to be $A(Z, Z') = 0.45 \pm 0.12$ per atom [22].

B_8S_{16}. Simple and extended Hückel MO calculations have been performed in order to compare the valence electronic structures of B_8S_{16} and porphin [23] and to discuss the bonding in a heretofore unknown Cu^{2+} complex. **Fig. 9-2**, p. 108, shows the energy levels of B_8S_{16} around the HOMO-LUMO gap and compares the Hückel levels with those of the extended Hückel method without and with d atomic orbitals on the sulfur. The Hückel levels were located

References for 9.2 on p. 109

on the eV scale by matching the HOMO and LUMO with those of the extended Hückel results without d atomic orbitals [23, 24].

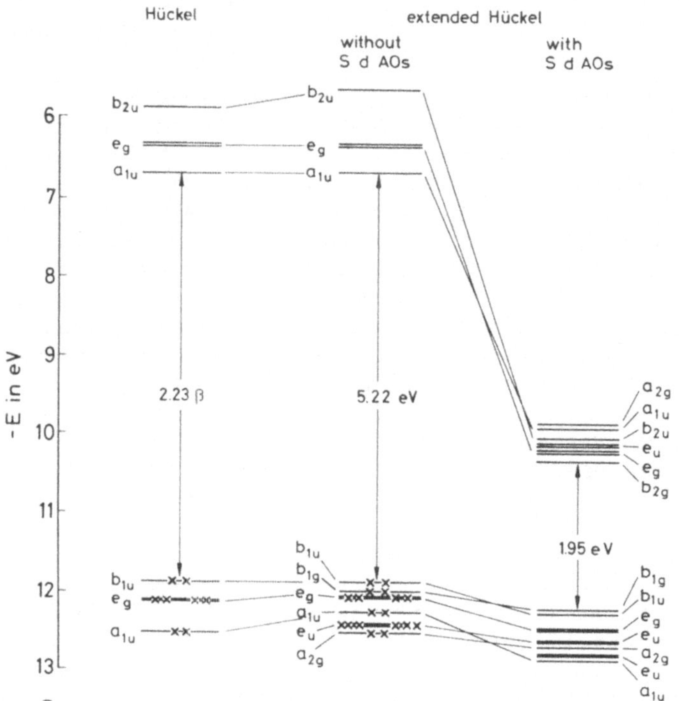

Fig. 9-2. Energy levels of B_8S_{16} around the HOMO-LUMO gap [24].

9.2.2 Ionic Species

Thioborates may exist as M^IBS_2, $M^{II}[BS_2]_2$, $M^I_4B_2S_5$, $M^{II}_2B_2S_5$, $M^I_3BS_3$, $M^I_8B_2S_7$, $M^I_5BS_4$, $M^I_4B_6S_{11}$, $M^I_3B_3S_6$, $M^I_3B_5S_9$ species, the "perthioborates" $M^I_2B_2S_5$ or M^IBS_3, and boron-rich phases such as CuBS or AgBS [29].

Mg, Ca, and Ba salts of thioboric acid containing the anion $\mathbf{[BS_2]^-}$ have been prepared from a mixture of the alkaline earth carbonate, H_3BO_3, and H_2S; they form phosphors with Eu_2O_3 and Ga_2O_3 [25].

The ion $\mathbf{[B_{10}S_{18}]^{6-}}$ (with tetrahedral coordination of boron) has been found in $Ag_6[B_{10}S_{18}]$. The latter was prepared from stoichiometric amounts of Ag_2S, B, and S at 700°C with successive annealing between 580 and 460°C. The orange-yellow compound crystallizes in the monoclinic space group C2/c–C^6_{2h} (No. 15) with a = 2166.3(8), b = 2163.9(8), c = 1657.2(5) pm; β = 129.40(4)°; Z = 8, D_{calc} = 2.948 g/cm³. The anionic part of the structure contains $B_{10}S_{20}$ supertetrahedra consisting of 10 parallel corner-sharing BS_4 tetrahedra; the $B_{10}S_{20}$ groups are linked through corners to form a layer-like arrangement of $(B_{10}S_{16}S_{4/2})^{6n-}_n = (B_{10}S_{18})^{6n-}_n$ polyanions. The mean B–S bond length is 191.5 pm. The Ag^+ ions show a disordered arrangement, making $Ag_6B_{10}S_{18}$ an Ag^+ ionic conductor. The IR spectrum shows B–S stretching vibrations at 610, 640, 685, 735, and 760 cm^{-1} [26, 29].

References for 9.2:

[1] Sennesal, J. M.; Robbe, J. M.; Schamps, J. (Chem. Phys. **55** [1981] 49/56).
[2] Jenouvrier, A.; Pascat, B. (Can. J. Phys. **59** [1981] 1851/61).
[3] McDonald, J. K.; Innes, K. K. (J. Mol. Spectrosc. **29** [1969] 251/72).
[4] Jenouvrier, A.; Pascat, B. (Can. J. Phys. **59** [1981] 1862/78).
[5] Bell, S.; McLean, M. L. (J. Mol. Spectrosc. **63** [1976] 521/6).
[6] Cherkesov, A. I. (Zh. Neorgan. Khim. **26** [1981] 3181/5; Russ. J. Inorg. Chem. **26** [1981] 1703/5).
[7] Porter, C. J.; Proctor, C. J.; Ast, T.; Beyron, J. H. (Croat. Chem. Acta **54** [1981] 407/19).
[8] Johnson, H. R.; Sauval, A. J. (Astron. Astrophys. Suppl. Ser. **49** [1982] 77/87 from C.A. **97** [1982] No. 136174).
[9] Geissberger, A. E.; Galeener, F. L. (Struct. Non-Cryst. Mater. Proc. 2nd Intern. Conf., Cambridge, Engl., 1982 [1983], pp. 381/91 from C.A. **99** [1983] No. 148817).
[10] Es-Sedikki, S.; Le Coustomer, G.; Mollier, Y.; Devaud, M. (Tetrahedron Letters **22** [1981] 2771/2).

[11] Holloway, J. H.; Puddick, D. C.; Staunton, G. M.; Brown, D. (Inorg. Chim. Acta **64** [1982] L209/L210).
[12] Levasseur, A.; Olazcuaga, R.; Kbala, M.; Zahir, M.; Hagenmuller, P. (Compt. Rend. [2] **293** [1981] 563/5).
[13] European Atomic Energy Community (Neth. Appl. 80-03969 [1980/82] 1/4 from C.A. **96** [1982] No. 222190).
[14] Susman, S. S.; Boehm, L.; Volin, K. J.; Delbecq, C. J. (Solid State Ionics **5** [1981] 667/9).
[15] Susman, S. S.; Boehm, L.; Volin, K. J.; Delbecq, C. J. (U.S. Appl. 375525 [1982/83] 1/16 from C.A. **100** [1984] No. 124156).
[16] Burckhardt, W.; Makyta, M.; Levasseur, A.; Hagenmuller, P. (Mater. Res. Bull. **19** [1984] 1083/9).
[17] Makyta, M.; Levasseur, A.; Hagenmuller, P. (Mater. Res. Bull. **19** [1984] 1361/6).
[18] Menetrier, M.; Levasseur, A.; Hagenmuller, P. (J. Electrochem. Soc. **131** [1984] 1971/3).
[19] Menetrier, M.; Levasseur, A.; Delmas, C.; Audebert, J. F.; Hagenmuller, P. (Solid State Ionics **14** [1984] 257/61).
[20] Levasseur, A.; Menetrier, M.; Hagenmuller, P. (EUR-8660 [1984] 45/52 from C.A. **100** [1984] No. 212932).

[21] Levasseur, A.; Delmas, C. (EUR-9405 [1984] 1/67 from C.A. **102** [1985] No. 116564).
[22] Hartmann, F. J.; Bergmann, R.; Daniel, H.; von Egidy, T.; Fottner, G.; Naumann, R. A.; Reidy, J. J.; Wilhelm, W. (Z. Physik A **308** [1982] 103/5).
[23] Gimarc, B. M.; Trinajstič, N. (Inorg. Chem. **21** [1982] 21/5).
[24] Gimarc, B. M.; Zhu, J. K. (Inorg. Chem. **22** [1983] 479/84).
[25] Showa Denko K. K. (Japan. Kokai Tokkyo Koho 81-145199 [1980/81] 1/4 from C.A. **96** [1982] No. 133747).
[26] Krebs, B.; Diercks, H. (Z. Anorg. Allgem. Chem. **518** [1984] 101/14).
[27] Krebs, B.; Hürter, H.-U. (Angew. Chem. **92** [1980] 479/80).
[28] Krebs, B.; Hürter, H.-U. (Acta Crystallogr. A **37** [1981] C163).
[29] Krebs, B. (Angew. Chem. **95** [1983] 113/34).
[30] Jacox, M. E. (J. Phys. Ref. Data **13** [1984] 945/1086).

9.3 With Hydrogen or Organic Substituents

9.3.1 HBS, DBS, and CH_3BS

The two IR stretching fundamentals of both gaseous **HBS** and **DBS** with $C_{\infty v}$ symmetry (prepared from boron and H_2S or D_2S, respectively) have been observed in Ar at 1200 K using a continuous flow system at a pressure of ca. 1 Torr. For $H^{11}BS$, the stretching modes are $\nu_1 = 2735.80$ and $\nu_3 = 1172.35\ cm^{-1}$; and for $D^{11}BS$, $\nu_1 = 2077.71$ and $\nu_3 = 1119.98\ cm^{-1}$; ν_3 of the latter is in strong Fermi resonance with the overtone of the bend $2\nu_2$ at $1098.60\ cm^{-1}$. The bending fundamental ν_2 has not been observed and must be very weak. The analysis gave the equilibrium structure with r(B–H) = 116.98 pm, r(B–S) = 159.78 pm; the harmonic force field constants are f_{rr}(B–H) = 4.38(2) aJ·Å^{-2} and f_{RR}(B–S) = 7.358(34) aJ·Å^{-2}, and the anharmonic force field constants are f_{rrr}(B–H) = –21.66(45) aJ·Å^{-3} and f_{RRR} = –35.57(23) aJ·Å^{-3} [1, 56]. For HBS, the bond lengths r(B–H) = 117.0 pm and r(B–S) = 162.5 pm have been calculated by ab initio SCF calculations on double zeta basis (DZP); the experimental data are 116.9 and 159.9 pm, respectively. The ionization potentials are 10.79 and 14.06 eV. Rotational constants, net charges, and overlap populations have been tabulated [2]. The vertical ionization potentials for HBS have been calculated by applying Rayleigh-Schrödinger perturbation corrections $(\Delta E)^{GA}$ to Koopman's theorem; the results are for 2π: 10.74, for 7σ: 12.92, and for 6σ: 15.44 eV [3].

The following bond lengths have been calculated for $\mathbf{H_3CBS}$ (experimental data in parentheses): r(B–C) 153.6(153.5); r(B–S) 163.5(160.2); r(C–H) 109.5(110.9) pm; ∢(H–C–B) = 110.8 (110.3)° [2]. The molecular polarizability, the diamagnetic susceptibility, and the electron-impact ionization cross-section of CH_3BS have been calculated using a semiempirical quantum-mechanical method [4]. SCF calculations using a double-zeta plus polarization functions basis (DZP) suggest a B–S triple bond due to participation of a sulfur lone pair. The ionization potentials are 10.30 and 13.54 eV [70].

9.3.2 Tris(organylthio)boranes, $B(SR)_3$

$\mathbf{B(SCH_3)_3}$ reacts with 1,2,4-trithia-3,5-diborolanes $(XB)_2S_3$ such as $Br_2B_2S_3$ or $(CH_3)_2B_2S_3$ at 303 K under substituent exchange. The exchange occurs via two processes: an exo exchange provides for mixed species such as $CH_3S(X)B_2S_3$ without site exchange of boron atoms; and an endo process occurs via SCH_3 or X exchange. The exo reaction predominates if X is more basic than the annular S atoms, whereas the endo reaction predominates if X does not contain free electron pairs. In addition, steric factors can influence the ratio of endo to exo process. These studies were based on isotopic labelling of boron [5].

$B(SCH_3)_3$ reacts with $(CH_3)_3Si–N[Sn(CH_3)_3]$ at 110°C to yield an annealed B_6N_7 system with three $B–SCH_3$ and six $N–Si(CH_3)_3$ groups; δ^1H = 0.29 and 2.26 ppm; $\delta^{11}B$ = 39.4 ppm; $\delta^{13}C$ = 4.61 and 13.31 ppm; X-ray analysis gave point group C_3 [6]. At elevated temperatures and in inert solvents $B(SCH_3)_3$ and $\mathbf{B(SC_2H_5)_3}$ react with pyrazole, to form B-organylthiopyrazaboles according to the equation $2B(SR)_3 + 2Hpz \rightarrow 2RSH + (RS)_2B(\mu\text{-pz})_2B(SR)_2$ (Hpz = pyrazole) [7].

9.3.3 Organylbis(organylthio)boranes, $RB(SR')_2$

NMR data for $\mathbf{CH_3B(SCH_3)_2}$: $\delta^{11}B$ = 66.3 ppm; for $\mathbf{C_6H_5B(SCH_3)_2}$: $\delta^{11}B$ = 65.0 ppm; $\delta^{13}C$ (C1) = 138, and (C4) = 129 ppm (in $CDCl_3$). For $\mathbf{C_6H_5B(SC_2H_5)_2}$: $\delta^{13}C$(C1) = 139.2, (C4) = 129.1 ppm [8].

$C_6H_5B(SC_2H_5)_2$ reacts at elevated temperatures in an inert solvent with Hpz = pyrazole to yield the pyrazabole $(C_6H_5)(C_2H_5S)B(\mu\text{-pz})_2B(C_6H_5)(SC_2H_5)$ [7]; reactions with amines, see [9]. The dipole moments (in benzene at 25°C) for $C_6H_5B(SCH_3)_2$ and 2,4,6-$(CH_3)_3C_6H_2B(SCH_3)_2$ are 4.2 and 5.5×10^{-30} cm, respectively, if a 5% correction of the R_D value for the atomic polarization is applied. The values are 3.8 and 5.1×10^{-30}, respectively, if a 15% correction is applied [10].

9.3.4 (Organylthio)diorganylboranes, R_2BSR'

The mass spectra of (organylthio)diorganylboranes, R_2BSR' (R = 2,4,6-$(CH_3)_3C_6H_2$, R′ = CH_3, C_2H_5, C_6H_5, C_6H_4-4-CH_3), have been studied; the only significant B-containing fragment that was observed was the ion R_2B^+ [11].

$(n\text{-}C_3H_7)_2BS(n\text{-}C_4H_9)$ reacts with propargylalcohol to yield propargyldipropylborinate [12].

$(n\text{-}C_4H_9)_2BS(n\text{-}C_4H_9)$ has been prepared in 15% yield by heating of $(n\text{-}C_4H_9)_3B$ and $(n\text{-}C_4H_9)SCH_2{=}C(CH_3)_2$ for 15 h to 150°C [13]. The compound reacts with 2,2′-iminodipyridine to yield the corresponding (amino)dibutylborane [14].

9.3.5 Metathioboric Acid, $(\text{–BSH–S–})_3$

Polycrystalline thioboric acid, $(\text{–BSH–S–})_3$, has been obtained from powdered commercial B_2S_3 by vacuum sublimation from a quartz tube on a hot plate set at 120°C over 24 h; clear needle-like crystals appear on the tube wall at the edge of the hot plate surface. The Raman spectrum of the material shows a strong line at 438 cm^{-1}, assigned to the symmetric stretch motion of the sulfur atoms in B_3S_3 rings; and a characteristic line at 2553 cm^{-1} which is assigned to H–S stretching [15].

9.3.6 Boron-Sulfur Heterocycles

1,2,4-Trithia-3,5-diborolane, **1** with R = H, has been prepared by the reduction of $(C_2H_5SB)_2S_3$ or $[(CH_3)_2NSB]_2S_3$ with thf-BH_3. The reaction was monitored by ^{11}B NMR spectroscopy ($\delta^{11}B$ = 60.9 to 61.3 ppm) in order to find the optimal conditions for the synthesis of **1** with R = H. The compound is volatile and polymerizes readily according to the equation $2H_2B_2S_3 \rightarrow H_4B_4S_6$ to yield bis-μ-[(dithioperthio)thioboronato(2-)-S,SS:SS]-dihydrodiboron, **2**. Therefore, pure monomeric $H_2B_2S_3$ does not exist as a condensed phase; however, it complexes with $N(CH_3)_3$, see Section 9.5.1, p. 123 [16].

1 **2**

The following 3,5-diorganyl-1,2,4-trithia-3,5-diborolanes of type **1** have been studied.

1a: R = CH_3. The dimethyl compound $S_3^{10}B_2(CH_3)_2$ has been obtained by the reaction of $K[^{10}BF_4]$ with $AlBr_3$ to give 76% $^{10}BBr_3$, which was reacted with $Sn(CH_3)_4$ to yield 92% $CH_3^{10}BBr_2$. The latter was reacted with H_2S_x to give a 55% yield of $S_3^{10}B_2(CH_3)_2$ [17]. The compound $S_3^{11}B_2(CH_3)_2$ shows $\delta^{11}B$ = 70.6 ppm [8, 18, 19]. Exchange of the annular disulfide

References for 9.3 on pp. 117/9

group occurs with hydrazines. Intermolecular exchange of different R groups in 1,2,4-trithia-3,5-diborolanes of the types $(XB)_2S_3$ and $(YB)_2S_3$ has been studied employing $^{10}BBr_3$. The exchange proceeds by two mechanisms: an exo exchange provides for mixed species $(XYB_2)S_3$ without site exchange of boron atoms; and an endo process via BX and BY exchange. In general, the exo process is the faster one but in the case of $(CH_3B)_2S_3$ as one of the reactants, only the endo process was observed [5, 18, 19]. The reaction with $^{10}BBr_3$ serves as a test for such reactions [20]. CNDO/MO calculations show the following values for the bonding charges in the $(CH_3B)_2S_3$ ring system (numbering of atoms shown below) [71]: S(1) = 0.021, S(2) = 1.907, B(3) = 0.222; S(4) = 1.830; B(5) = 0.405; and C = 0.382 (R = CH_3).

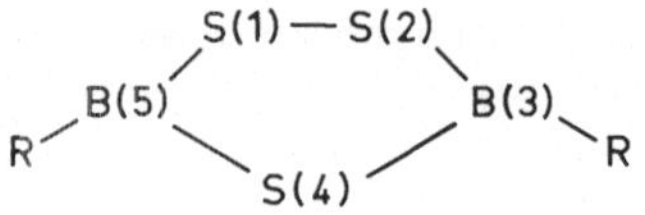

S(1)—S(2)
R–B(5) B(3)–R
S(4)

1a reacts with *p*-fluorophenyl-N-sulfinylamine to yield 5-*p*-fluorophenyl-2,4,6-trimethyl-*cyclo*-5-aza-1,3-dioxa-2,4,6-triborane (crystal structure given) and 3,5-di-*p*-fluoromethyl-2,4,6-trimethyl-*cyclo*-1-oxa-3,5-diaza-2,4,6-triborane (see "Boron Compounds" 3rd Suppl. Vol. 3, 1988, p. 139). Reaction with other sulfinylamines results in the formation of 1,2-dithia-4-aza-3,5-diborolidines and 1,4-dithia-2-aza-3,5-diborolidines [21].

1b: R = C_6H_5. $\delta^{11}B$ = 65.9 ppm [8].

1c: R = SC_2H_5 was obtained from $Br_2B_2S_3$ and $(CH_3)_3SiSC_2H_5$; $\delta^{11}B$ = 64.1 ppm [5].

9.3.7 Heterocycles Containing Annular Carbon

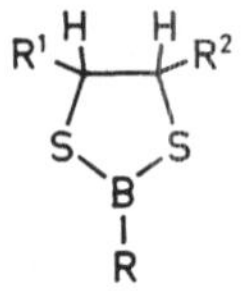

3 (1,3-dithia-2-boracyclopentane ring with R on B, H and R^1 on C4, H and R^2 on C5)

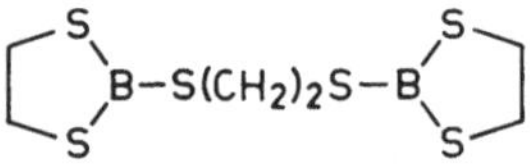

4 $(S_2C_2H_4)B–S(CH_2)_2S–B(S_2C_2H_4)$

Data for the following 1,3-dithia-2-boracyclopentanes of type **3** are available:

3a: R = CH_3, R^1 = R^2 = H has been obtained as a by-product from the reaction of 2-chloro-1,3,2-dithiaborolane with $(CH_3)_3SnLi$; b.p. 26 to 27°C/0.5 Torr; $\delta^{11}B$ = 69.0 ppm [22], 69.1 ppm [22], 69.6 ppm [8]; δ^1H = 0.483 (t, 3H), 2.80 (q, 2H) [22].

3b: R = C_6H_5, R^1 = R^2 = H: $\delta^{11}B$ = 66.2 ppm [8]; IR and Raman spectral data (see Table 9/9) show "local" C_{2v} symmetry [24].

3c: R = *n*-C_4H_9, R^1 = CH_3, R^2 = *n*-C_3H_7 has been prepared from 2-bromo-4-*n*-propyl-1,3,2-dithiaborolane and *n*-C_4H_9Li; b.p. 98°C/0.01 Torr [25].

3d: R = *n*-C_4H_9, R^1 = H, R^2 = C_6H_5; b.p. 118°C/0.01 Torr [25].

3e: R = *n*-C_4H_9, R^1 = R^2 = C_6H_5; b.p. 158°C/0.01 Torr [25].

Compounds of type **3** where R = different pentyl or hexyl groups (R^1 = R^2 = H) react with trichloromethyllithium, $LiCCl_3$, in tetrahydrofuran with formation of an intermediate which is converted to $C_6H_{13}C(S_2C_2H_4)B[O(CH_2)_4Cl]_2$ by the solvent [26].

4 has been obtained as a by-product in the preparation of **3a**; $\delta^{11}B$ = 64.2 ppm [23].

Table 9/9

Vibrational Wavenumbers (in cm^{-1}) for 2-Phenyl-1,3-dithia-2-boracyclopentane (**3b**, liquid; pol = polarized, dp = depolarized, def. = deformation) [24].

IR	Raman	assignment
3075 wm		phenyl CH stretch
3055 wm	3057 s, pol	phenyl CH stretch
3020 wm		phenyl CH stretch
2960 w	2968 m, br	CH_2 stretch (A + B)
2925 m	2928 s, pol	CH_2 stretch (A + B)
2860 vvw	2868 w, pol	1427 + 1438 = 2865
2840 w	2842 w, pol	1490 + 1350 = 2840
	1638 w	1002 + 634 = 1636
1598 m	1598 vs, dp	phenyl ring stretch (A_1 + B_2)
1492 vvw	1490 w, pol	phenyl ring stretch (A_1)
1440 wm, sh 1436 ms	1438 w, dp	phenyl ring stretch (B_2) + CH_2 scissors (A)
1380 w, sh		?2 × 695 = 1390
1368 m		?913 + 454 = 1367
1350 ms		phenyl ring stretch (B_2)
	1334 vw	?2 × 668 = 1336
1310 wm		
1284 m	1284 vw, pol	CH_2 wag (A) + phenyl CH def. (B_2)
1260 ms	1264 wm, ?pol	^{10}B-phenyl stretch (A_1)
1244 ms	1246 wm, ?pol	CH_2 wag (B)
1232 vs	1234 m, pol	^{11}B-phenyl stretch (A_1)
1186 w	1186 wm, pol	phenyl CH def. (A_1)
1159 w	1159 wm, dp	phenyl CH def. (B_2) + CH_2 twist (A)
1140 vw		
1112 vw	1112 vw, dp	CH_2 twist (B)
1090 vvw		
1070 vvw	1070 vw, dp	phenyl CH def. (B_2)
1030 vw	1033 m, pol	phenyl CH def. (A_1)
1002 vw	1002 vs, pol	phenyl "ring breathing" (A_1)
990 vw	990 w, sh	ring stretch (A)
	976 w, sh	phenyl CH def. (A_2)
955 m	952 vw, pol	CH_2 rock (A) + phenyl CH def. (B_1)
935 m, sh		^{10}B–S stretch (B)
913 vs	920 vw	^{11}B–S stretch (B)
906 vs	906 vw	phenyl CH def. (B_2)
842 m	848 w, br, dp	CH_2 rock (B) + phenyl CH def. (A_2)
752 s	752 vw, dp	phenyl CH def. (B_1)
695 s		phenyl ring def. (B_1)
668 w	668 m, pol	ring stretch (A + B)
636 m	640 m, pol	phenyl ring def. (A_1)
620 vw, sh	619 wm, dp	phenyl ring def. (B_2)
582 w		
549 w	548 vw	^{10}B-phenyl def. (B_1)
538 m	536 w, dp	^{11}B-phenyl def. (B_1)
	454 w, dp	ring def. (B)
	434 ms, pol	B–S stretch [A]
	396 vw, dp	?B-phenyl def. (B_2)
	252 m, pol	ring def. (A)
	174 wm, dp	ring def. (B)

References for 9.3 on pp. 117/9

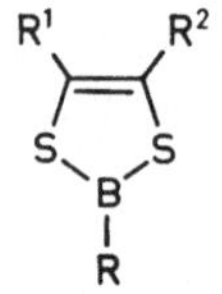

R¹ R²
S B S
R
5

The following 1,3-dithia-2-borol-4-enes of type **5** have been prepared from 2-bromo-1,3,2-dithiaborol-4-enes and Li organyls:

5a: $R = CH_3$, $R^1 = R^2 = H$: $\delta^{11}B = 61.5$ ppm [8].

5b: $R = C_6H_5$, $R^1 = R^2 = H$: $\delta^{11}B = 59.1$ ppm [8].

5c: $R = n\text{-}C_4H_9$, $R^1 = H$, $R^2 = C_6H_5$: b.p. 118°C/0.01 Torr; $\delta^1H = 6.35$ ppm (1H, s); $\delta^{11}B = 59.4$ ppm; IR: $\nu(C{=}C) = 1510\ cm^{-1}$ [25].

5d: $R = n\text{-}C_4H_9$, $R^1 = C_6H_5$, $R^2 = C_6H_5$: b.p. 158°C/0.01 Torr; $\delta^1H = 6.88$ to 7.63 ppm (10 H, br); $\delta^{11}B = 62.4$ ppm; IR: $\nu(C{=}C) = 1600\ cm^{-1}$ (s); mass spectrum, M/z = 310 (base peak 254) [25].

5e: $R = n\text{-}C_4H_9$, $R^1 = CH_3$, $R^2 = n\text{-}C_3H_7$: b.p. 98°C/0.01 Torr; $\delta^1H = 2.02$ ppm (3 H, s); $\delta^{11}B = 62.5$ ppm; IR: $\nu(C{=}C) = 1520\ cm^{-1}$ [25].

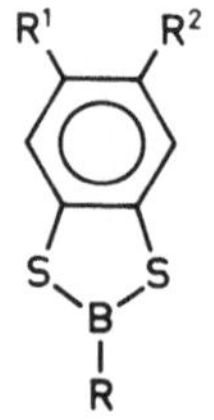

R¹ R²
S B S
R
6

NMR data are given for the following compounds of type **6** [8]:

6a: $R = CH_3$, $R^1 = R^2 = H$: $\delta^{11}B = 62.3$ ppm;

6b: $R = C_6H_5$, $R^1 = R^2 = H$: $\delta^{11}B = 59.8$ ppm;

6c: $R = CH_3$, $R^1 = R^2 = CH_3$: $\delta^{11}B = 62.2$ ppm; $\delta^{13}C$: C(1) = 141.6, C(2) = 126.4, C(3) = 135.1, C(4) = 126.5, C(5) = 125.7, C(6) = 138.3 ppm; C in $(B)CH_3 = 0.2$ ppm; C in $(C)CH_3 = 20.9$ ppm.

6d: $R = C_6H_5$, $R^1 = CH_3$, $R^2 = H$: $\delta^{11}B = 59.8$ ppm; $\delta^{13}C$: C(1) = 141.2, C(2) = 126.7, C(3) = 135.5, C(4) = 126.8, C(5) = 126.1, C(6) = 137.8 ppm; C(B) = 133.2, C(o) = 134.1, C(m) = 128.2, C(p) = 131.5 ppm; C in $CH_3 = 20.9$ ppm.

6e: $R = BS_2C_6H_4$, $R^1 = R^2 = H$: $\delta^{11}B = 58.6$ ppm; $\delta^{13}C$: C(1) = 143.9, C(2) = 126.6, C(3) = 125.7, C(4) = 125.7, C(5) = 126.6, C(6) = 143.9 ppm [8].

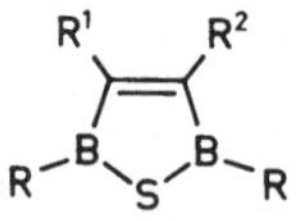

R¹ R²
R B S B R
7

2-Thiadiborol-4-enes (**7**) serve as ligands in triple- and tetradecker transition metal sandwich compounds of complex structures. The electron correlation has been investigated by means of semiempirical INDO calculations [27].

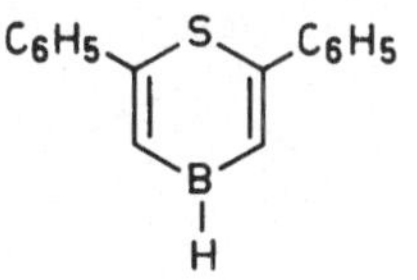

8

The compound **8** has been prepared from $\Delta^{2,5}$-4-brom-2,6-diphenyl-1,4-thiaboracyclohexadiene and *n*-butyllithium; m.p. 45°C, sublimation point 140°C/0.01 Torr; $\delta^1H = 0.93$ to 1.6 (9 H), 7.1 to 7.56 (12 H); $\delta^{11}B = 66.8$ ppm; IR: ν(C═C) = 1590 cm^{-1}; in the mass spectrum, M/z = 304 (base peak 236) [25].

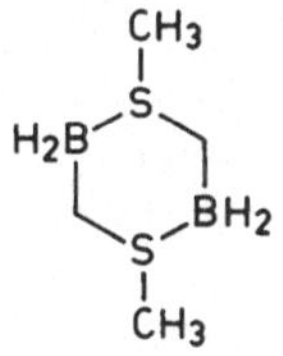

9

The compound **9** has been prepared from $(CH_3)_3N–BH_2–CH_2–SCH_3$ by thermal decomposition at 100°C; m.p. 102 to 104°C; $\delta^1H = 1.6$ to 2.3(BCH_2), 2.16(SCH_3) ppm; $\delta^{11}B = -13.6$ ppm; $\delta^{13}C = 29.6$(BCH_2), and 25.9(SCH_3) ppm (in $CDCl_3$). As revealed by X-ray crystallography, it contains a six-membered ring in chair conformation and the methyl groups are in equatorial positions; it crystallizes rhombohedrally from petroleum ether in space group $P2_1/c\text{-}C^5_{2h}$ (No. 14) with a = 543.4(1), b = 640.4(1), and c = 1260.4(3) pm; β = 100.38(2)°; Z = 2; $D_{calc} = 1.14$ and $D_{meas} = 1.13$ g/cm^3; from 713 observed reflections the data were refined to R = 2.9% [28, 29]. For the molecular structure, see **Fig. 9-3** [28].

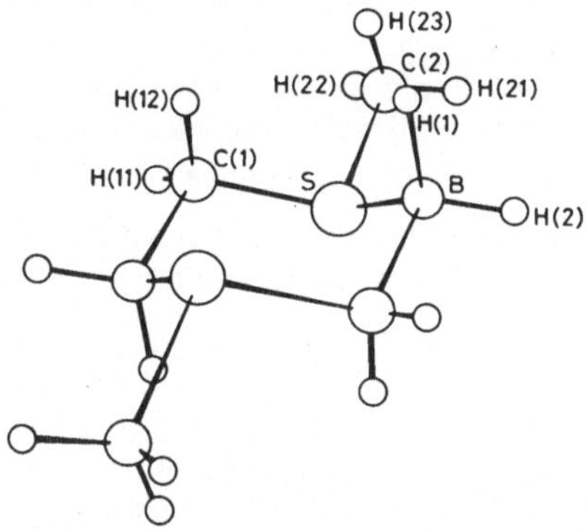

Fig. 9-3. Structure of $[CH_2(SCH_3)(BH)_2]_2$ (compound **9**) [28].

The structure of the cage compound $[CH_2(SBH_2)_2]_2$ (with adamantane skeleton) is shown in **Fig. 9-4**, p. 116. The compound has been obtained from CS_2 and thf-BH_3 after three weeks, or by the action of methanedithiole on thf-BH_3 or thf-BH_2Cl; $\delta^{11}B = -17.0$ ppm (triplet). The crystal structure was determined by X-ray diffraction and was refined to R = 2.7% (2240 independent reflections). The species is monoclinic, space group $P2_1/c\text{-}C^5_{2h}$ (No. 14), with a = 730.7(1), b = 911.5(2), and c = 1462.6(2) pm; β = 90.84(1)°; Z = 4; $D_{calc} = 1.41$ and $D_{meas} = 1.40$ g/cm^3 [30].

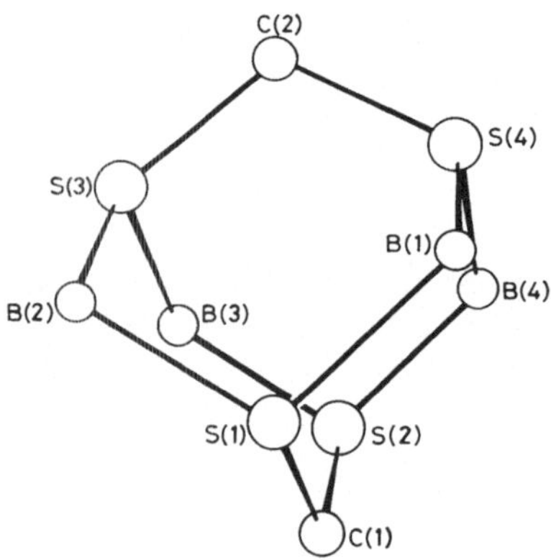

Fig. 9-4. Structure of $[CH_2(SBH_2)_2]_2$ (hydrogen atoms omitted) [30].

9.3.8 Borane Adducts with Sulfur Donor Molecules

The complex **$(CH_3)_2S–BH_3$** reacts with oxygen atoms developing a chemiluminescence which shows the extensive α-band of BO [31]. The complex is a convenient hydroborating agent, e.g., for alkanes [32, 33], alkenes such as $(CH_3)_2C{=}C(CH_3)_2$ [34], α-pinane [35, 36], or α-pinene [37, 38], methylene cyclobutane [39], 1,8-cyclooctadiene [40], and α-aryloxycinnamic acids [41]. It is a reducing agent for α-hydroxy esters in the presence of $Na[BH_4]$ [42]. It reacts with organolithium compounds to give various products [28, 29], and has been utilized for the preparation of $Li[BH_4]$, $Li[H_3CBH_3]$, and $Li[(n\text{-}C_4H_9)BH_3]$ [43 to 45]. Reaction with iodine yields $(CH_3)_2S–BI_3$ [46], with trialkylboranes [37, 38] or borinanes [47] alkylhydroboranes have been obtained, and with heterocycles, e.g., with a bicyclic phosphorane, an adduct with $\delta^{11}B = 49.6$ ppm is formed [48, 49]. It is a reducing agent in the presence of $(CH_3)_2O_2–BF_3$ in tetrahydrofuran [50], e.g., reducing carboxylic acid salts or esters [51, 52], amides or nitriles [53 to 56], 1,2-aminoalcohols [57], or cyanoethylcellulose [58, 59].

$O(CH_2)_2S–BH_3$, thioxane-borane, is formed on interaction of thioxane with B_2H_6. It is an efficient reagent for the hydroboration of alkenes and cycloalkenes [69].

$(CH_3)_2S–BR_3$ (R = H, alkyl, cyclohexyl, or C_6H_5), complex formation has been studied by ^{11}B NMR spectroscopy [60]. **$LiCH_2(CH_3)S–BH_3$** has been prepared from the tetramethylethylenediamine adduct $(CH_3)_2NCH_2CH_2N(CH_3) \cdot LiCH_2(CH_3)S$ and $(CH_3)_3N–BH_3$. The species could not be isolated, but has $\delta^{11}B = -30.5$ ppm; $\delta^{13}C = 20.6(SCH_3)$, $24.1(BCH_2)$ ppm [28, 29].

9.3.9 Ionic Species

Photoelectron spectra for the ion **HBS^+** show the ν_3 B–S stretching mode at 955 ± 40 cm^{-1} [72].

A review of the reactivities and selectivities of the ion **$[BH_2S_3]^-$** (as the sodium salt) in the reduction of organic compounds has been compiled [61].

$Li[B(SC_4H_3)_4]$, lithium tetrakis(2-thienyl)borate, prepared in dimethoxyethane, has an apparent resistivity of $200\,\Omega\cdot cm$ in dioxolane. It is used as an electrolyte for alkali-metal batteries; other efficient electrolytes are $Li[BC_2H_5(SC_4H_3)_3]$ and $Li[B(C_2H_5)_2(SC_4H_3)_2]$ [62]. 2-Thienyl reacts with R_3B (R = *n*-, *i*-, *s*-C_4H_9, C_6H_{11}, C_6H_{13}) to yield $[BR_3(SC_4H_3)]^-$ salts [63].

$Na[BH_4]$ reacts with S_8 in tetrahydrofuran or $Li[BH_4]$ with S_8 in ether via species such as $\mathbf{[H_2B(\text{–}S\text{–}S_x\text{–}S\text{–})]^-}$ which exist in the solution for only a few hours and exhibit two ^{11}B NMR triplets [64]. The reducing species of a $Na[BH_4]$–$HSCH_2CH_2SH$ substrate is considered to have the structure **10** [65].

Na^+ S B S H_2

10

The following data have been calculated for the cationic ligand $\mathbf{[(H_3S)_2B]^+}$ of low electronegativity in the VSEPR (valence shell electron pair repulsion) model: r(S–H) = 134.7 pm, r(S–B) = 187.6 pm; ∢(HSB) = 114.2°, ∢(SBS) = 180°; $\Delta H_f^\circ = -1387.0$ kJ/mol; the force constant for the skeletal bending vibration in the linear cation is $9.1\,N\cdot cm^{-1}$ [66].

Sulfonium tetraphenylborates of the type $[R^1R^2R^3S][B(C_6H_5)_4]$ have been prepared; data are given for: $R^1 = R^2 = CH_3$, $R^3 = 2\text{-}CH_3C_6H_4$ (m.p. 175 to 177°C) [67]; $R^1 = C_6H_5$, $R^2 = CH_3$, $R^3 = 2\text{-}C_5H_{11}$ (m.p. 125 to 127°C, decomposition), ^1H NMR and IR data [68]; $R^1 = C_6H_5$, $R^2 = i\text{-}C_3H_7$, $R^3 = CH_2CH_2CH{=}CH_2$ (m.p. 128 to 131°C), ^1H NMR and IR data [68].

References for 9.3:

[1] Turner, P.; Mills, I. (Mol. Phys. **46** [1982] 161/70).
[2] Ha, T.-K.; Nguyen, M.-T.; Vanquinckenborne, L. G. (J. Mol. Struct. **90** [1982] 99/105).
[3] Frost, D. C.; Kirby, C.; Lau, W. M.; McDowell, C. A.; Westwood, N. P. C. (J. Mol. Struct. **100** [1983] 87/94).
[4] Chopra, J. R.; Pandey, A. N.; Verma, U. P.; Strouch, B. (Acta Phys. Polon. A **65** [1984] 351/4 from C.A. **101** [1984] No. 12472).
[5] Nöth, H.; Staudigl, R.; Taeger, T. (Chem. Ber. **114** [1981] 1157/75).
[6] Gasparin-Ebeling, T.; Nöth, H. (Angew. Chem. **96** [1984] 301).
[7] Hodgkins, T.-G.; Niedenzu, K.; Niedenzu, K. S.; Seelig, S. S. (Inorg. Chem. **20** [1981] 2097/100).
[8] Götze, R.; Nöth, H.; Pommerening, H.; Sedlak, D.; Wrackmeyer, B. (Chem. Ber. **114** [1981] 1884/93).
[9] Cragg, R. H.; Miller, T. J. (J. Organometal. Chem. **241** [1983] 289/300).
[10] Exner, O.; Folli, U.; Marcaccioli, S.; Vivarelli, P. (J. Chem. Soc. Perkin Trans. II [1983] 757/60).

[11] Davidson, F.; Wilson, J. W. (Org. Mass Spectrom. **16** [1981] 467/8).
[12] Mikhailov, B. M.; Baryshnikova, T. N. (J. Organometal. Chem. **219** [1981] 295/300).
[13] Badalyan, K. S.; Bagdasaryan, G. B.; Indzhikyan, M. G. (Arm. Khim. Zh. **37** [1984] 258/61 from C.A. **101** [1984] No. 110981).
[14] Dorokhov, V. A.; Lavrinovich, L. I.; Shashkov, A. S.; Mikhailov, B. M. (Izv. Akad. Nauk SSSR Ser. Khim. **1981** 1371/3; Bull. Acad. Sci. USSR Div. Chem. Sci. **1981** 1097/9).
[15] Geissberger, A. E.; Galeener, F. L. (Struct. Non-Cryst. Mater. Proc. 2nd Intern. Conf., Cambridge, Engl., 1982 [1983], pp. 381/91 from C.A. **99** [1983] No. 148817).
[16] Nöth, H.; Staudigl, R. (Z. Anorg. Allgem. Chem. **481** [1981] 41/50).
[17] Nöth, H.; Staudigl, R. (Inorg. Syn. **22** [1983] 218/26).
[18] Nöth, H.; Staudigl, R.; Brückner, R. (Chem. Ber. **114** [1981] 1871/83).

[19] Nöth, H.; Staudigl, R. (Chem. Ber. **115** [1982] 1555/67).
[20] Nöth, H.; Staudigl, R. (Chem. Ber. **115** [1982] 3011/24).

[21] Meller, A.; Habben, C.; Noltemeyer, M.; Sheldrick, C. M. (Z. Naturforsch. **37b** [1982] 1504/6).
[22] Nöth, H.; Schwerthöffer, R. (Chem. Ber. **114** [1981] 3056/62).
[23] Biffar, W.; Nöth, H.; Schwerthöffer, R. (Liebigs Ann. Chem. **1981** 2067/80).
[24] Davidson, G.; Ewer, K. P. (J. Mol. Struct. **74** [1981] 1981/91).
[25] Habben, C.; Maringgele, W.; Meller, A. (Z. Naturforsch. **37b** [1982] 43/53).
[26] Brown, H. C.; Imai, T. (J. Org. Chem. **49** [1984] 892/8).
[27] Böhm, M. C. (Ber. Bunsenges. Physik. Chem. **85** [1981] 755/68).
[28] Nöth, H.; Sedlak, D. (Chem. Ber. **116** [1983] 1479/86).
[29] Biffar, W.; Nöth, H.; Sedlak, D. (Organometallics **2** [1983] 579/85).
[30] Binder, H.; Diamantikos, W.; Dermentzis, K.; Hausen, H.-D. (Z. Naturforsch. **37b** [1982] 1548/52).

[31] Bauer, S. H.; Jeffers, P. M. (NTIS-16-DFT-3 [1982] 1/23; ARO-18-139-3-CH [1982] 1/23 from C.A. **99** [1983] No. 105310).
[32] Brown, H. C.; Kulkarni, S. U. (J. Organometal. Chem. **218** [1981] 299/307).
[33] Brown, H. C.; Pai, G. G. (J. Organometal. Chem. **250** [1983] 13/22).
[34] Brown, H. C.; Chandrasekharam, J. (J. Am. Chem. Soc. **106** [1984] 1863/5).
[35] Brown, H. C.; Mandal, A., K.; Yoon, N. M.; Singaram, B.; Schwier, J. R. (J. Org. Chem. **47** [1982] 5069/74).
[36] Brown, H. C.; Singaram, B. (J. Org. Chem. **49** [1984] 945/7).
[37] Brown, H. C.; Jadhav, P. K. (J. Am. Chem. Soc. **105** [1983] 2092/3).
[38] Brown, H. C.; Desai, M. C.; Jadhav, P. K. (J. Org. Chem. **47** [1982] 5065/9).
[39] Hill, E. A.; Nylen, P. A.; Fellinger, J. H. (J. Organometal. Chem. **239** [1982] 279/92).
[40] Soderquist, J. A.; Brown, H. C. (J. Org. Chem. **46** [1981] 4599/600).

[41] Ahvonen, T.; Brunow, G.; Kristersson, P. (Acta Chem. Scand. B **37** [1983] 845/9).
[42] Saito, S.; Hasegawa, T.; Inaba, M.; Nishida, R.; Nomizu, S.; Moriwake, T. (Chem. Letters **1984** 1389/92).
[43] Brown, H. C.; Choi, Y. M.; Narasimhan, S. (Inorg. Chem. **21** [1982] 3657/61).
[44] Kim, S.; Lee, S. J.; Kang, H. J. (Syn. Commun. **12** [1982] 723/6).
[45] Kim, S.; Moon, Y. C.; Ahn, K. H. (J. Org. Chem. **47** [1982] 3311/5).
[46] Brown, H. C.; Ravindran, N.; Kulkarni, S. U. (J. Org. Chem. **45** [1980] 384/9).
[47] Brown, H. C.; Pai, G. G. (Heterocycles **17** Spec. Issue [1982] 77/82).
[48] Murillo, A.; Contreras, R.; Klaébé, A.; Wolf, R. (Heterocycles **20** [1983] 1487/9).
[49] Contreras, R.; Houalla, D.; Klaébé, A.; Wolf, R. (Tetrahedron Letters **22** [1981] 3953/4).
[50] Moser, H.; Rihs, G.; Sauter, H. (Z. Naturforsch. **37b** [1982] 451/62).

[51] Yoon, N. M.; Cho, B. T. (Tetrahedron Letters **23** [1982] 2475/8).
[52] Brown, H. C.; Choi, Y. M. (Synthesis **1981** 439/40).
[53] Brown, H. C.; Choi, Y. M. (Synthesis **1981** 441/2).
[54] Brown, H. C.; Narasimhan, S.; Choi, Y. M. (Synthesis **1981** 996/7).
[55] Brown, H. C.; Choi, Y. M.; Narasimhan, S. (Synthesis **1981** 605/6).
[56] Brown, H. C.; Basavaiah, D.; Bhat, N. G. (Organometallics **2** [1983] 1309/11).
[57] Mancilla, T.; Santiesteban, F.; Contreras, R.; Klaébé, A. (Tetrahedron Letters **23** [1982] 1561/4).
[58] Munir, A.; Daly, W. H. (Makromol. Chem. Rapid Commun. **4** [1983] 589/94).

[59] Daly, W. H.; Munir, A. (J. Polym. Sci. Polym. Chem. Ed. **22** [1984] 975/84 from C.A. **100** [1984] No. 211906).
[60] Contreras, R.; Wrackmeyer, B. (Spectrochim. Acta A **38** [1982] 941/51).

[61] Okushi, T. (Kagakuto Kogyo [Osaka] **55** [1981] 8/14 from C.A. **95** [1981] No. 23491).
[62] Klemann, L. P.; Newman, G. H.; Stogryn, E. I.; Whitney, T. A.; Farcasiu, D. (U.S. 4293623 [1980/81] 1/13 from C.A. **96** [1982] No. 38373).
[63] Akimoto, I.; Sano, M.; Suzuki, A. (Bull. Chem. Soc. Japan **54** [1981] 1587/8).
[64] Binder, H.; Diamantikos, W. (Z. Naturforsch. **38b** [1983] 203/7).
[65] Wayne, C.; Entreken, E. E.; Guida, A. R. (J. Org. Chem. **49** [1984] 3024/6).
[66] Cuthbertson, A. F.; Glidewell, C.; Liles, D. C. (J. Mol. Struct. **87** [1982] 273/82).
[67] Caret, R. L.; Vennos, A.; Zapf, M.; Kebel, J. J. (Tetrahedron Letters **22** [1981] 2085/8).
[68] Beak, P.; Sullivan, T. A. (J. Am. Chem. Soc. **104** [1982] 4450/7).
[69] Brown, H. C. (U.S. 4298750 [1980/81] 1/4 from C.A. **96** [1982] No. 52322).
[70] Nguyen, M. T.; Ruelle, P. (J. Chem. Soc. Faraday Trans. II **80** [1984] 1225/34).

[71] Gimarc, B. M.; Zhu, J. K. (J. Inorg. Chem. **22** [1983] 479/84).
[72] Jacox, M. E. (J. Phys. Chem. Ref. Data **13** [1984] 945/1068).

9.4 With Oxygen

9.4.1 Adducts

The compounds $\mathbf{O_3S{-}B_2O_3}$ (decomposition begins at 136°C, maximum 230°C) and $\mathbf{(O_3S)_2{-}B_2O_3}$ (decomposition begins at 148°C, maximum at 242°C) have been synthesized by heating a mixture of liquid SO_3 (in excess) with B_2O_3 in sealed ampoules to 90 to 95°C and 190 to 200°C, respectively; ^1H and ^{11}B NMR and IR spectra show lines or absorption bands indicating distorted BO_4 tetrahedra, BO_3 groups, and a coordinated sulfato group [1, 2].

9.4.2 HOBS and Related Species

Ab initio SCF studies of the B–S bond in HOBS using a double-zeta plus polarization functions (DZP) basis set suggest a triple bond nature. The third bond is formed by donation of a sulfur lone pair. The energies of the localized orbitals are −0.576, −0.566, and −0.550 eV, respectively [15]. The following bond lengths and angles have been calculated for angular H–O–B≡S: r(B≡S) = 164.0, r(B–O) = 131.1, r(O–H) = 95.5 pm; ∢(H–O–B) = 134.3° [3].

$C_6H_5SB(OCH_3)(OH)$ has been used for chiral synthetic studies [4].

9.4.3 3,5-Diorganyloxy-1,2,4-trithia-3,5-diborolanes

S–S
RO–B B–OR
S

11

The thermal stability of compounds of type **11** increases with increasing size of R and from aromatic substituents to aliphatic substituents [5].

References for 9.4 on pp. 121/2

11a: R = CH_3, has been prepared from a solution of the diiodo compound $B_2S_3I_2$ in CS_2 and CH_3O-*t*-C_4H_9 by stirring at 0°C and vaporization; pale yellow solid. Raman data (cm^{-1}; in CS_2 solution): 264 s, 452 w, 490 s, 728 m; δ^1H (in C_6H_6) = 3.96 ppm (s). Thermolytic decomposition products are polymeric [5].

11b: R = C_2H_5, has been prepared from $B_2S_3I_2$ and $C_2H_5OC_2H_5$ in CCl_4 at 10°C and vaporization; pale yellow oil. IR data (in cm^{-1}): 2982 m, 2938 m, 2890 w, 1485 m, 1401 m, 1375 s, 1311 s, 1265 s, 1219 w, 1203 w, 1169 m, 1100 m, 1024 s, 1005 m, 968 s, 925 w, 893 w, 865 w, 804 w, 787 w, 662 w, 481 w; Raman data (in cm^{-1}): 264 m, 287 m, 411 m, 451 m, 505 s, 576 w, 707 w, 871 w; δ^1H (in $CDCl_3$) = 1.30 (3H, t with J = 6.9), 4.09 (2H, q with J = 6.9); $\delta^{11}B$ = 49.2 ppm; thermolytic decomposition products are polymeric [5].

11c: R = C_6H_5, m.p. 81°C, has been prepared from $B_2S_3I_2$ and $CH_3OC_6H_5$ in pentane, vaporization and crystallization from *n*-pentane. IR data (Nujol mull, frequencies in cm^{-1}): 3094 w, 3062 w, 3041 w, 2920 s, 2860 s, 1935 w, 1819 w, 1783 w, 1717 w, 1596 m, 1490 s, 1456 s, 1377 m, 1364 m, 1328 m, 1276 s, 1192 m, 1159 w, 1069 w, 1020 m, 1004 s, 996 s, 925 s, 897 w, 885 w, 876 w, 820 w, 803 m, 754 s, 683 s, 611 w, 579 m, 560 w, 486 w, 459 w; Raman data (in cm^{-1}): 212 m, 360 w, 407 w, 460 m, 519 s, 620 w, 812 w; δ^1H (in CCl_4) = 6.8 to 7.5 ppm; $\delta^{11}B$ (in $CDCl_3$) = 50.7 ppm [5].

11d: R = *i*-C_3H_7: b.p. 78°C/0.1 Torr. IR, Raman, ^1H NMR and thermolysis data are given; $\delta^{11}B$ (in $CDCl_3$) = 48.7 ppm [5].

11e: R = *n*-C_4H_9: ^1H NMR data in benzene solution are available [5].

11f: R = C_6H_3-2,6-$(CH_3)_2$: IR, Raman, ^1H NMR data are given; $\delta^{11}B$ (in $CDCl_3$) = 50.6 ppm [5].

11g: R = $CH_2CH_2OCH_3$: ^1H NMR data (in CCl_4) are given [5].

9.4.4 Heterocycles Containing Annular Carbon

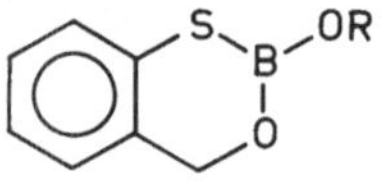

S B OR O

12

Compounds of type **12** have been prepared from B_2O_3, ROH, and 2-$HSC_6H_4CH_2OH$ under reflux in C_6H_6 [6].

12a: R = C_2H_5: b.p. 50 to 52°C/3 Torr; 65% yield; δ^1H = 1.20 (t, 3H), 3.98 (q, 2H), 5.03 (s, 2H), 7 to 7.44 (m, 4H).

12b: R = *i*-C_3H_7: b.p. 55 to 59°C/3 Torr; 60% yield; ^1H NMR data are available.

12c: R = C_6H_5: b.p. 68 to 70°C/3 Torr; 80% yield; ^1H NMR data are available.

CH_3O S B−R O

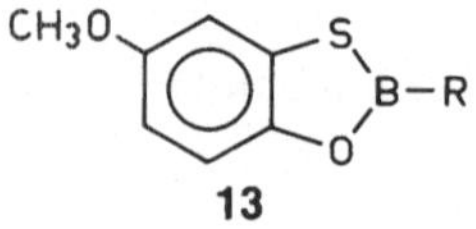

13

Compounds of type **13** have been prepared from 2-chloro-5-methoxy-1,3,2-benzoxathiaborole [7].

13a: R = OH: preparation by reaction with water (1:1) in CH_2Cl_2.

13b: R = OC_6H_5: preparation by reaction with phenol in CH_2Cl_2.

13c: $R = SC_6H_5$: preparation by reaction with thiophenol in CH_2Cl_2.

Several additional species were described [7].

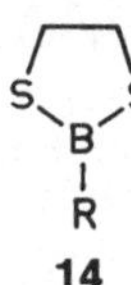

14

The 2-organyloxy-1,3,2-dithiaborolane **14** with $R = i\text{-}C_3H_7$ reacts readily with Schiff bases [8].

9.4.5 Compounds without B–S Bonds

$\mathbf{C_6H_5SCH_2B(OH)_2}$ has been prepared from $C_6H_5SCH_2Li$ and $B(OCH_3)_3$, followed by hydrolysis; it has been used for chiral synthetic studies [4, 5]. Esters thereof, e.g., $\mathbf{C_6H_5SCH_2B(OCH_3)_2}$, have been described and their chemical behavior has been studied [4, 10].

(3-Thiophenyl)dihydroxyborane has been prepared from thienyllithium and alkylborates; the reaction of the species with halothiophenes gave unsymmetrically substituted bithienyls [11]. The cited compound regulates plant growth [12]. Both 2- and 3-thiophenylboronic acid react with bromopyridine [9].

9.4.6 Sulfatoboric Acids and Sulfonatoborates

$\mathbf{H[B(HSO_4)_4]}$, tetrakis(bisulfato)boric acid, prepared from H_3BO_3 and $H_2S_2O_7$ (1:3), increases the kinetics of the sulfonation of aniline by oleum [13].

$\mathbf{(C_2H_5)_2BOSO_2CH_3}$ has been used to prepare (trifluoromethylsulfonato)diorganylboranes, which are used in preparative organic chemistry in order to form enolates for stereoselective aldol condensation [14].

References for 9.4:

[1] Kondrat'ev, S. N.; Mel'nikova, S. I.; Bondar, A. M. (Zh. Neorgan. Khim. **28** [1983] 851/4; Russ. J. Inorg. Chem. **28** [1983/84] 481/3).

[2] Bondar, A. M.; Kondrat'ev, S. N.; Mel'nikova, S. I. (Zh. Neorgan. Khim. **28** [1983] 855/8; Russ. J. Inorg. Chem. **28** [1983/84] 483/7).

[3] Ha, T.-K.; Nguyen, M.-T.; Vanquinckenborne, L. G. (J. Mol. Struct. **90** [1982] 99/105).

[4] Ray, R. (Diss. Washington State Univ. 1981, pp. 1/118 from C.A. **95** [1981] No. 42531).

[5] Schmidt, M.; Sametschek, E. (Z. Naturforsch. **36b** [1981] 1444/50).

[6] Bernard, A. M.; Cocco, M. T.; Plumitallo, A.; Secci, M. (J. Heterocycl. Chem. **19** [1982] 297/8).

[7] Andrä, K.; Straube, A. (Z. Anorg. Allgem. Chem. **490** [1982] 191/6).

[8] Bhal, L.; Singh, R. V.; Tandon, J. P. (Syn. Reactiv. Inorg. Metal-Org. Chem. **14** [1984] 1135/49 from C.A. **102** [1985] No. 185128).

[9] Gronowitz, S.; Lawitz, K. (Chem. Scr. **24** [1984] 5/6).

[10] Matteson, D. S.; Arne, K. H. (Organometallics [Washington] **1** [1982] 280/8).

[11] Gronowitz, S.; Bobosik, V.; Lawitz, K. (Chem. Scr. **22** [1983] 265/6, **23** [1984] 120/2).
[12] Aaberg, B. (Swed. J. Agric. Res. **13** [1983] 101/6 from C.A. **99** [1983] No. 153766).
[13] Khelevin, R. N. (Zh. Org. Khim. **20** [1984] 124/9; J. Org. Chem. [USSR] **20** [1984] 112/6).
[14] Evans, D. A.; Takacs, J.; McGee, L. R.; Ennis, M. D.; Mathre, J. D.; Bartroli, J. (Pure Appl. Chem. **53** [1981] 1109/27).
[15] Nguyen, M.-T.; Ruelle, P. (J. Chem. Soc. Faraday Trans. II **80** [1984] 1225/34).

9.5 With Nitrogen

9.5.1 Compounds Containing B–N and B–S Bonds

The following bond lengths and angles have been calculated for **S≡B–NH$_2$**: r(B≡S) = 164.4, r(B–N) = 137.5, r(N–H) = 100.7 pm; ∢(H–N–B) = 122.6° [1]. The energies of the localized orbitals are −0.581 and −0.488 e.u.; two LUMO's are identical. The ionization potentials are 10.01 and 13.58 eV [39].

A series of (dialkylamino)ethanethiophenylboranes, **C_2H_5S–BC_6H_5–NR_2**, has been prepared and data are given for species with (selection):

R = CH_3: b.p. 68°C/0.1 Torr, preparation from $Pb(SC_2H_5)_2$ and C_6H_5BCl–$N(CH_3)_2$, 70% yield; $\delta^{13}C$ (C atoms of phenyl ring numbered) = 139.2 (C1), 131.2 (C2, C6), 127.8 (C3, C5), 127.8 (C4), 41.3 (CH_3), 23.7 (SCH_2), 17.7 (SCH_3) ppm [2 to 4]; the reactivity with C_6H_5NCO has been compared with that of similar aminoboranes [5].

R = i-C_3H_7: b.p. 88°C/0.1 Torr, preparation analogous to the preceding compound in petroleum ether, 60% yield; $\delta^{13}C$ (C atoms of phenyl ring numbered) = 141.4 (C1), 130.7 (C2, C6), 127.5 (C3, C5), 126.8 (C4), 22.7 (R), 24.0 (SCH_2), 17.4 (SCH_3) ppm [2 to 4].

R = s-C_4H_9: b.p. 110°C/0.1 Torr, 50% yield; $\delta^{13}C$ (C atoms of phenyl ring numbered) = 141.5 (C1), 130.8 (C2), 127.5 (C3), 126.8 (C4), 12.3 (R), 24.1(SCH_2), 17.4 (SCH_3) ppm [2 to 4].

CH_3S–BCH_3–$N(CH_2C_6H_5)(CH_2$–C≡CH): b.p. 115°C/0.01 Torr, has been prepared from $(CH_3S)_2BCH_3$ and $C_6H_5CH_2NH$–CH_2–C≡CH in n-hexane at −78°C by warming to +20°C; $\delta^{11}B$ = 35.7 ppm [6].

$(CH_3S)_2B$–$N(CH_3)_2$. NMR data: δ^1H = 2.23, $\delta^{11}B$ = 43.4 ppm in CCl_4 [7]; $\delta^{11}B$ = 43.2 ppm [8].

Various tris[bis(organylthio)boryl]amines, **$N[B(SR)_2]_3$**, have been prepared. The species with R = CH_3, m.p. 224°C, was obtained from $N(BCl_2)_3$ and $Pb(SCH_3)_2$ (1:3 ratio) in 48% yield. IR and mass spectral data are given; δ^1H (in C_6D_6) = 2.30 ppm, $\delta^{11}B$ = 54.4 ppm, $\delta^{14}N$ = −240 ppm. The NMR data indicate chemically equivalent CH_3S groups in solution, but this equivalence is lost in the solid state: one $(CH_3S)_2B$ group is strongly twisted against the B_3N plane; crystals (see **Fig. 9-5**) have space group C2/c–C^6_{2h} (No. 15); a = 1490.2(18), b = 924.8(11), c = 1417.7(20) pm; β = 122.73(9)°; Z = 4; D_{calc} = 1.33 g/cm^3 [7].

Tris(1,3,2-dithiaborolane-2-yl)amine, m.p. 215 to 218°C, was obtained from $N[Sn(CH_3)_3]_3$ and $BrB(SR)_2$ in 84% yield; δ^1H (in $CDCl_3$) = 2.83, $\delta^{11}B$ = 56.3, $\delta^{13}C$ = 37.47, and $\delta^{14}N$ = −256 ppm; IR, mass spectrometric, and crystallographic data are available. Space group C2/c–C^6_{3h} (No. 15) [7].

Tris(5-methyl-1,3,2-benzodithiaborole-2-yl)amine, m.p. 262°C, was prepared in analogous fashion in 78% yield. IR; mass spectrum; ^1H, ^{11}B, ^{13}C, ^{14}N NMR data are available. The species crystallizes in space group C2/c-C^6_{2h} (No. 15) [7].

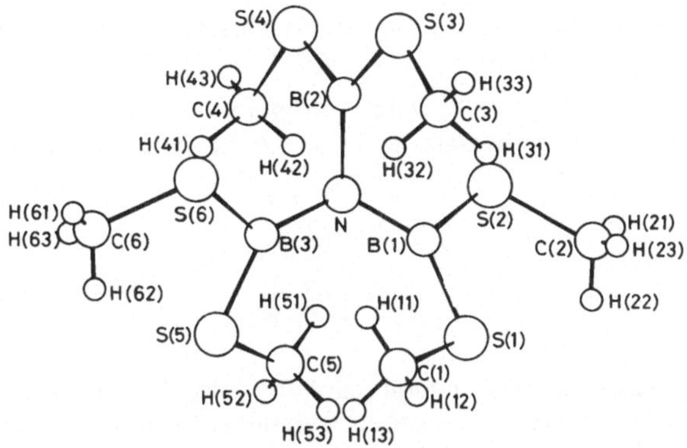

Fig. 9-5. Structure of tris[bis(methylthio)boryl]amine [5].

Tris(1,3,2-dithiaborolene-2-yl)amine, m.p. 102°C, was obtained in 76% yield. IR, mass spectral, and NMR data are available. The compound crystallizes in space group $P\bar{1}$-C_i^1 (No. 2) [7].

In all cases NMR data indicate the existence of chemically equivalent SR groups in solution; however, in the solid state, one of the three $B(SR)_2$ groups is different as is evidenced by a larger B–N distance as compared to the other two B–N bonds [7].

Various B-amino derivatives of **1** (see p. 111) have been described. The species with R = $N(CH_3)_2$, i.e., **$[(CH_3)_2N]_2B_2S_3$**, m.p. 116 to 118°C, was prepared from $^5/_8$ S_8 and $(CH_3)_3Sn$–$B[N(CH_3)_2]_2$ in form of yellow needles in 89% yield; $\delta^{11}B = 43.7$ ppm [10]; m.p. 125 to 130°C, subl. p. 80°C/0.01 Torr, prepared from $B_2Br_2S_3$, $(CH_3)_2NH$ and $(C_2H_5)_3N$; 55% yield; $\delta^{11}B = 46.1$ ppm, $\delta^1H = 2.52$ ppm; mass spectrum M/z = 206 [11]; $\delta^{11}B = 44.7$ ppm, $\delta^{13}C = 41.90$ ppm [8]; preparation from sulfur and $B[N(CH_3)_2]_3$ [12]; m.p. 112 to 114°C, from $B_2Br_2S_3$ and $(CH_3)_3Si$–$N(CH_3)_2$ in CH_2Cl_2, 80% yield; $\delta^{11}B = 44.1$ ppm, $\delta^1H = 2.58$ ppm [13].

Reactions with other ^{10}B-labelled 1,2,4-trithio-3,5-diboranes indicate two different patterns: an exocyclic substituent exchange; and an endocyclic process proceeding via ring opening and substituent exchange [13, 14], in some cases statistically [14]. The compound reacts with thf-BH_3 (or better with 9-borabicyclo-[3.3.1]-nonane) to yield $H_2B_2S_3$ (see Section 9.3.6, p. 111) which polymerizes readily but could be stabilized as the 1:1 and 1:2 adducts with trimethylamine, **$H_2B_2S_3 \cdot N(CH_3)_3$**, decomposes at 70°C (prepared also from $B_2S_3[N(CH_3)_2]_2$ and $(C_3H_7)_2BH$ in thf); IR data are given; NMR data: $\delta^{11}B = 61.3$ ppm, $^1J(BH) = 152 \pm 4$ Hz and $\delta^{11}B = 8.5$ ppm, $^1J(BH) = 130$ Hz. **$H_2B_2S_3 \cdot 2N(CH_3)_3$**: $\delta^{11}B = 8.9$ ppm, $^1J(BH) = 120$ Hz [15].

$R_2B_2S_3$ (**1**) with R = NH(t-C_4H_9): b.p. 125°C/0.01 Torr, from $B_2Br_2S_3$ and t-C_4H_9NHLi, 88% yield; colorless oil; mass spectral and NMR data: M/z = 262; $\delta^1H = 1.33$ ppm; $\delta^{11}B = 44.4$ ppm [11].

$R_2B_2S_3$ (**1**) with R = NCS: m.p. 62°C (decomposition), from $B_2Br_2S_3$ and AgNCS in CS_2, 45% yield; yellow crystals; $\delta^{11}B = 47.0$ ppm [11].

The following dithiaazadiborolidines of type **15** have been studied:

S—S
R–B   B–R
N
R¹

15

References for 9.5 on pp. 132/3

15a: $R = R^1 = CH_3$: Substituent exchange reactions using ^{10}B labelling with BBr_3 follow the endocyclic reaction mechanism [16].

15b: $R = CH_3$, $R^1 = t\text{-}C_4H_9$: b.p. 136°C/7 Torr, preparation from $B_2S_3(CH_3)_2$ and $t\text{-}C_4H_9NSO$; mass spectral data are given; NMR data: $\delta^1H = 0.96$ (s, 6H, B–CH_3), 1.55 (s, 9H, C–CH_3) ppm; $\delta^{11}B = 46.6$ ppm; M/z = 187 [11].

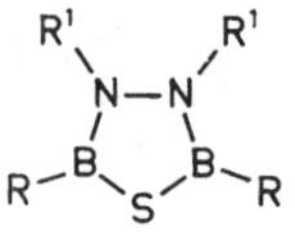

16

16a: $R = R^1 = CH_3$: Substituent exchange reactions using ^{10}B labelling with BBr_3 follow the endocyclic reaction mechanism [16]; reactions with either the dimethyl- or the dibromotrithia-diborolanes show rapid exchange of the ring hydrazino group with the ring disulfide group [17].

16b: $R = N(i\text{-}C_3H_7)_2$, $R^1 = CH_3$: Reaction with $^nBBr_3$ or $^{10}BBr_3$ [17].

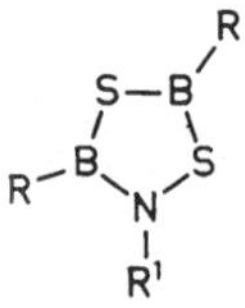

17

17a: $R = CH_3$, $R^1 = C_6H_3\text{-}1,3(CH_3)_2$: b.p. 98°C/0.01 Torr, preparation from $B_2Cl_2S_3$ and $(CH_3)_3SiNSO$ in $1,3\text{-}(CH_3)_2C_6H_3Cl$; $\delta^1H = 0.47$ (B–CH_3), 1.05 (B–CH_3), 2.15 (o-CH_3), 7.02 (C_6H_3) ppm; $\delta^{11}B = 63.7$, 47.1 ppm; M/z = 235 [11].

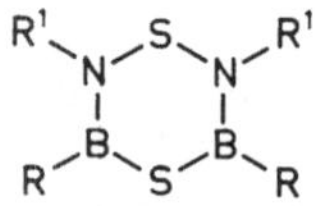

18

18a: $R = CH_3$, $R^1 = t\text{-}C_4H_9$: b.p. 82°C/0.01 Torr, prepared by dropping a solution of N,N'-bis(*t*-butyl)sulfur diimide in CCl_4 into a solution of 2,5-dimethyl-1,3,4-trithia-2,5-diborolidine in CCl_4, heating for 20 h under reflux, cooling and distillation, 70% yield; mass spectrum: M/z = 258; $\delta^1H = 0.94$ (s, 6H), 1.49 (s, 18H) ppm; $\delta^{13}C = 9.60$ (B–CH_3), 32.04 [$C(\underline{C}H_3)_3$], 61.03 [$\underline{C}(CH_3)_3$] ppm; $\delta^{11}B = 50.1$ ppm [18].

9.5.2 Heterocycles Containing Annular Carbon, Boron, and Sulfur

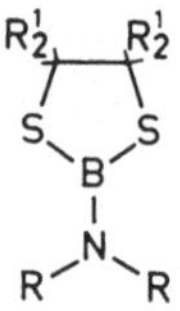

19

Compounds of type **19** have been prepared by conventional methods under inert atmosphere.

19a: $R = CH_3$, $R^1 = H$: $\delta^1H = 2.85$, $\delta^{11}B = 45.9$, $\delta^{13}C = 36.9$, $\delta^{14}N = -304$ ppm (in C_6D_6) [7]; $\delta^{11}B = 46.3$, $\delta^{13}C(S\text{–}C) = 36.90$, $\delta^{13}C(N\text{–}C) = 42.30$ ppm [8]; IR and Raman spectra (see Table 9/10): the internal modes suggest a considerably lower symmetry than C_{2v} [19].

Table 9/10

IR and Raman Spectra (in cm^{-1}) of Compound **19a** (liquid; pol = polarized, dp = depolarized; def. = deformation) [19].

IR	Raman	assignment
2980 w, sh		CH_2 stretch (B)
2960 wm, sh	2965 m, br	CH_2 stretch (A)
2930 s	2927 vs, pol	CH_2 stretch (A + B)
2880 m	2880 w, sh	CH_3 stretch (B_2) or $2 \times 1442 = 2884$
2865 wm	2865 ms, pol	CH_3 stretch (B_2) or $2 \times 1431 = 2862$
2845 wm	2844 m, pol	CH_3 stretch (A_1)
	2814 w, sh	CH_3 stretch (A_2)
2800 m	2796 m, pol	CH_3 stretch (A_1)
2695 m		$1285 + 1410 = 2695$
1515 s	1515 w, br, pol	B–N stretch (A_1)
1460 ms, br	1460 s, dp	CH_3 def. (B_2)
	1442 ms, dp	CH_3 def. (A_2 + ?B_1)
1430 ms	1431 s, dp	CH_2 scissors (A + B) + CH_3 def. (A_1)
1415 ms, sh	1416 ms, dp	CH_3 def. (B_2)
1410 s	1409 s, pol	CH_3 def. (A_1)
1285 ms	1284 m, pol	CH_2 wag (A)
1262 m	1260 vvw	CH_2 wag (B)
1202 s		CH_3 rock (B_1 or B_2)
1168 vw, sh	1167 m, pol	CH_2 twist (A)
1150 m, sh	1150 w, sh	CH_3 rock (A_2)
1140 s	1140 m, pol	CH_3 rock (A_1)
1118 w, sh	1115 m, dp	CH_2 twist (B)
1070 m	1068 vw, dp	CH_3 rock (B_2 or B_1)
1008 m		NC stretch (B_2)
988 wm, sh	989 m, pol	Ring stretch (A)
975 m		$304 + 674 = 978$
950 w, sh	952 w, br, pol	CH_2 rock (A)
940 m	942 vw, pol	NC stretch (A_1)
922 ms		BS stretch (B)
900 m, br	905 w, pol	?
850 ms	848 w, dp	CH_2 rock (B)
815 w, br		
780 w		
760 w	750 w, br, pol	$2 \times 376 = 752$
735 wm		
678 m	680 m, sh, ?dp	Ring stretch (B)
674 wm, sh	674 vs, pol	Ring stretch (A)
645 w, br		
	632 w, pol	$304 + 330 = 634$
550 m		BNC def. (B_1)
515 ms, br	516 vvs, pol	BS ring stretch (A)
500 m	498 w, br	BNC def. (B_2)
	468 ms, sh	$S_2{}^{10}BN$ def. (B)
	462 s, pol	$S_2{}^{11}BN$ def. (B)
455 vvw	452 m, sh, dp	Ring def. (B)
414 wm	412 vs, pol	NC_2 def. (A_1)
	392 vw, pol	?
381 wm	376 vw, ? pol	S_2BN def. (B)
	330 w	?
	304 s, pol	Ring def. (A)
	184 w, br	CH_3 torsion
	176 wm, dp	CH_3 torsion

References for 9.5 on pp. 132/3

19b: $R = C_2H_5$, $R^1 = H$: reactions with $C_2H_4S_2BBr$ [16].

19c: $R = i\text{-}C_3H_7$, $R^1 = H$: reactions with $C_2H_4S_2BBr$ [16].

N,N-Diorganyl-1,3,2-dithiaborol-2-amines of type **20** have been prepared by the reaction of 1,2,4-trithia-3,5-diborolane with alkynes and isocyanates.

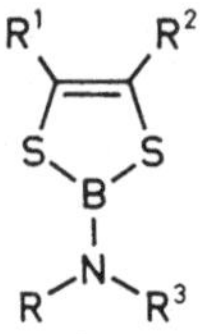

20

20a: $R = R^3 = C_6H_5$, $R^1 = R^2 = H$: m.p. 85°C, b.p. 90°C/0.01 Torr; $\delta^{11}B = 48.0$ ppm; a structural analysis has been performed [20, 21].

20b: $R = R^3 = CH_3$, $R^1 = R^2 = H$ has been prepared from Δ^4-2-bromo-1,3-dithia-2-borolene and $(CH_3)_2NLi$ in *n*-hexane at 0°C [22]; $\delta^1H = 6.43$ ppm; $\delta^{11}B = 44.3$ ppm; $\delta^{13}C(S\text{–}C) = 121.9$ ppm [7]. $\delta^{11}B = 44.1$ ppm; $\delta^{13}C(N\text{–}C) = 42.29$, and $\delta^{13}C(S\text{–}C) = 121.9$ ppm [8].

20c: $R = R^3 = C_2H_5$, $R^1 = H$, $R^2 = C_6H_5$: b.p. 123°C/0.01 Torr; preparation from Δ^4-2-bromo-5-phenyl-1,3-dithia-2-borolene and $(C_2H_5)_2NLi$ in *n*-hexane at 0°C; $\delta^1H = 6.38$ (1H, s) ppm; $\delta^{11}B = 43.1$ ppm; $\nu(C{=}C) = 1540\ cm^{-1}$; M/z = 249 [22].

20d: $R = R^3 = C_2H_5$, $R^1 = CH_3$, $R^2 = n\text{-}C_3H_7$: b.p. 102°C/0.01 Torr; $\delta^1H = 2.0$ (3H, s) ppm, $\delta^{11}B = 43.7$ ppm; $\nu(C{=}C) = 1540\ cm^{-1}$; M/z = 229 [22].

20e: $R = R^1 = R^2 = R^3 = C_2H_5$: b.p. 115°C/0.01 Torr; $\delta^1H = 2.45$ (4H, q) ppm, $\delta^{11}B = 44.8$ ppm; $\nu(C{=}C) = 1500\ cm^{-1}$; M/z = 229 [22].

20f: $R = R^3 = C_2H_5$, $R^1 = R^2 = C_6H_5$: b.p. 163°C/0.01 Torr; $\delta^{11}B = 43.8$ ppm; $\nu(C{=}C) = 1600\ cm^{-1}$; M/z = 325 [22].

20g: $R = R^3 = n\text{-}C_3H_7$, $R^1 = H$, $R^2 = C_6H_5$: b.p. 162°C/0.01 Torr; $\delta^1H = 6.38$ (1H, s) ppm; $\delta^{11}B =$ 51.5 ppm; $\nu(C{=}C) = 1540\ cm^{-1}$; M/z = 277 [22].

20h: $R = t\text{-}C_4H_9$, $R^1 = R^3 = H$, $R^2 = C_6H_5$: b.p. 144°C/0.01 Torr; $\delta^1H = 6.27$ (1H, s) ppm; $\delta^{11}B =$ 45.7 ppm; $\nu(C{=}C) = 1550\ cm^{-1}$; M/z = 249 [22].

20i: $R = t\text{-}C_4H_9$, $R^1 = CH_3$, $R^2 = i\text{-}C_3H_7$, $R^3 = H$: b.p. 84°C/0.01 Torr; $\delta^1H = 1.72$ (3H, s), 2.50 (2H, t) ppm; $\delta^{11}B = 42.3$ ppm; $\nu(C{=}C) = 1570\ cm^{-1}$; M/z = 229 [22].

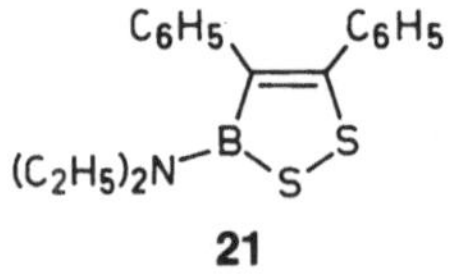

21

3-Diethylamino-4,5-diphenyl-1,2-dithia-3-borole(4), **21**, m.p. 89°C, b.p. 163°C/0.01 Torr, has been prepared by the reaction of 1,2,4-trithia-3,5-diborolane with diphenylalkyne and diethylcarbodiimide; $\delta^{11}B = 43.8$ ppm; structural analysis. 3-(2,6-Dimethylphenyl)-amino-4,5-diethyl-1,2-dithia-3-borole(4), m.p. 71°C, b.p. 90°C/0.01 Torr; $\delta^{11}B = 44.7$ ppm [20, 21]. The compound with a $Si(CH_3)_3$ group instead of the C_2H_5 group reacts with Na under ring contraction [20, 21].

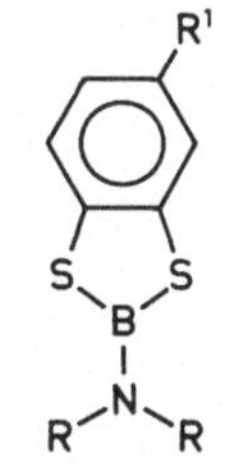

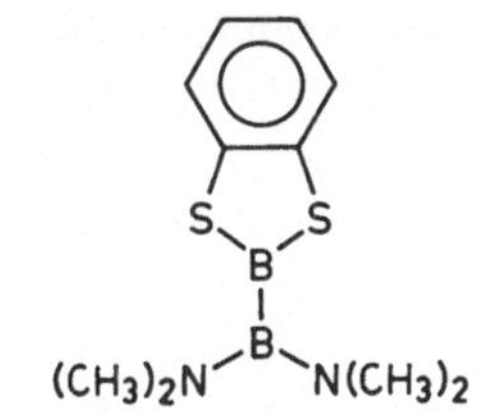

22 23

Compounds of type **22** have been prepared in analogy to those of type **20**.

22a: R = CH_3, R^1 = H: $\delta^{11}B$ = 43.1 ppm; $\delta^{13}C(NCH_3)$ = 42.1 ppm; $\delta^{13}C$(ring) = 139.1 (C1, C6), 124.9 (C2, C5), 125.9 (C3, C4) ppm [9].

22b: R = R^1 = CH_3: δ^1H = 2.15 and 7.35 ppm; $\delta^{11}B$ = 43.0 ppm; $\delta^{13}C(NC)$ = 41.8 ppm; $\delta^{13}C$ (ring) = 139.1 (C1), 126.5 (C2), 134.7 (C 3), 126.0 (C4), 125.6 (C5), 135.7 (C6) ppm [7].

23: $\delta^{11}B$ = 33.6 and 63.4 ppm; $\delta^{13}C(NC)$ = 41.9 ppm, $\delta^{13}C$(phenyl ring) = 143.9 (C1, C6), 124.9 (C2, C5), 125.9 (C3, C4) ppm [9].

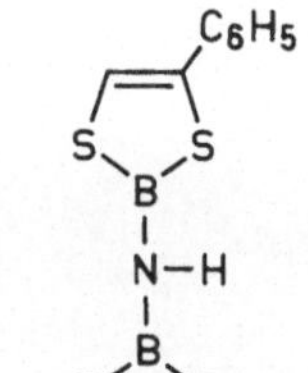

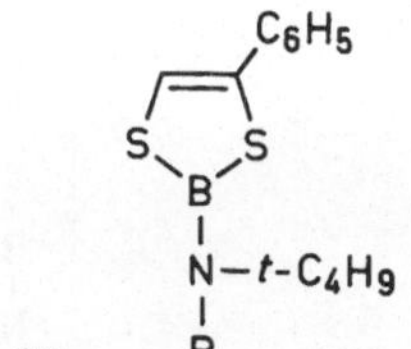

24 25

Bis(5-phenyl-1,3-dithia-2-borolen-yl)-amine, **24**, has been prepared from 2-bromo-5-phenyl-1,3-dithia-2-borolene and 2-*t*-butylamino-5-phenyl-1,3-dithia-2-borolene in CCl_4 at 80°C within 28 h; reddish solid, sublimes at 160°C, m.p. 140 to 142°C; δ^1H(=C–H) = 5.23 (2H, s); $\delta^{11}B$ = 57.6 ppm; ν(C=C) = 1500 cm^{-1} [22].

5-Phenyl-1,3-dithia-2-borenyl-(1,3-dimethyl-1,3-diaza-2-boracyclopentyl)-*t*-butylamine, **25**, has been prepared from compound **20h**, $N(CH_3)_3$, and 2-chlor-1,3-dimethyl-1,3-diaza-2-boracyclopentane; b.p. 165°C/0.01 Torr; $\delta^{11}B$ = 28.6 and 44.1 ppm; ν(C=C) = 1540 cm^{-1}; M/z = 345 [22].

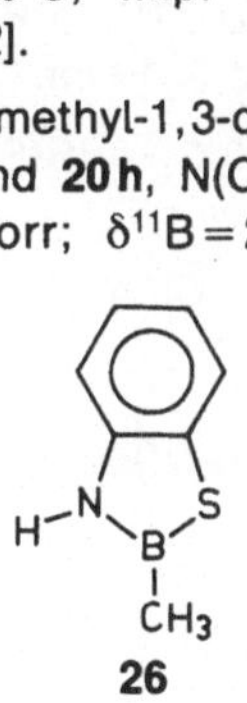

26

2-Methyl-1,3,2-benzothiazaborolidine, **26**, has the following NMR data: $\delta^{11}B$ = 45.1 ppm; $\delta^{13}C$ = 145.7 (C1), 113.1 (C2), 125.8 (C3), 121.0 (C4), 124.7 (C5), 129.3 (C6) [9]. Phosphorescence and ODMR spectroscopic measurements in *n*-hexane at 4.2 K allow conclusions with respect to the distribution and conjugation of π-electrons in the T_1 state of 2,3-dihydro-2-methyl-1,3,2-benzothiazaborole. The triplet state is dominated by the local C_{2v} symmetry

References for 9.5 on pp. 132/3

about the S atom, but reduction of the overall symmetry is to C_s; an occupied p-orbital and an empty d-orbital at the S atom are proposed [23].

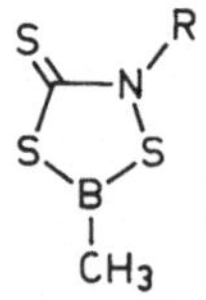

27

2-Organyl-5-methyl-1,4-dithia-2-aza-5-boracyclopentanethiones(3) of type **27** have been prepared from 3,5-dimethyl-1,2,4-trithia-3,5-diborolane and organyl isocyanate in CCl_4 at 80°C in 3 to 4d.

27a: $R = CH_3$: m.p. 36°C, b.p. 138°C/7 Torr, colorless solid; $\delta^1H = 1.02$ (3H, s) ppm; $\delta^{11}B = 53.7$ ppm; M/z = 163.

27b: $R = C_2H_5$: b.p. 142°C/7 Torr, colorless liquid; $\delta^1H = 0.58$ (3H, s) ppm, $\delta^{11}B = 53.0$ ppm; M/z = 177.

27c: $R = i\text{-}C_3H_7$: b.p. 92°C/0.01 Torr, colorless liquid; $\delta^1H = 1.12$ (3H, s) ppm; $\delta^{11}B = 54.6$ ppm; M/z = 191 [22].

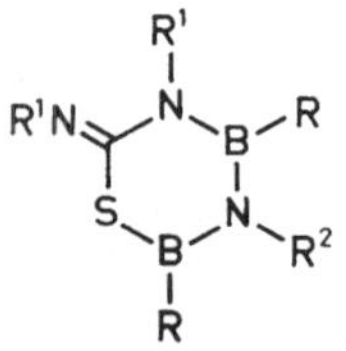

28

2,4-Diorganyl-5-organyl-6-organylimino-1-thia-2,4-dibora-3,5-diazacyclohexanes of type **28** have been prepared from R^1–NH–C(=S)–NHR^1 and RBX_2 (X = Cl, Br) in C_6H_5Cl or CCl_4, respectively [24]. Examples:

28a: $R = CH_3$, $R^1 = C_6H_5$, $R^2 = H$: m.p. 220 to 225°C; $\delta^1H = 0.40$ (3H, s) ppm; $\delta^{11}B(S) = 55.2$ and B(N) = 33.0 ppm; ν(C=N) = 1575 cm^{-1} and ν(N–H) = 3320 cm^{-1} [24].

28b: $R = R^1 = n\text{-}C_4H_9$, $R^2 = H$: m.p. 95 to 96°C; $\delta^1H = 0.62$ to 1.73 (br) ppm; $\delta^{11}B = 8.9$ ppm; ν(C=N) = 1560 cm^{-1} and ν(N–H) = 3320 and 3340 cm^{-1} [24].

28c: $R = CH_3$, $R^1 = R^2 = C_6H_5$: m.p. 210°C; $\delta^{11}B(S) = 55.5$ and B(N) = 34.0 ppm; ν(C=N) = 1610 cm^{-1} [24].

28d: $R = R^1 = R^2 = C_6H_5$: m.p. 75 to 77°C; $\delta^{11}B(S) = 55.2$ and B(N) = 33.0 ppm; ν(C=N) = 1580 cm^{-1} [24].

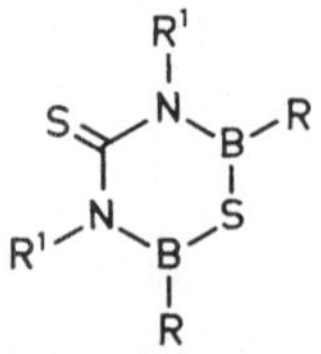

29

3,5-Diorganyl-2,3,5,6-tetrahydro-2,6-diorganyl-4H-1-thia-3,5-diaza-2,6-diborin-4-thiones of type **29** have been prepared from R_2BBr, R^1_2N–C(=S)–NR^1_2, and $n\text{-}C_4H_9Li$ in n-hexane and petroleum ether [25].

29a: $R = CH_3$, $R^1 = C_2H_5$: b.p. 68 to 78°C/0.002Torr; $\delta^1H = 0.63$ (s, 6H), 1.22 (t, 6H), 4.08 (q, 4H) ppm; $\delta^{11}B = 47.3$ ppm; M/z = 214 [25].

29b: $R = CH_3$, $R^1 = n\text{-}C_4H_9$: b.p. 90°C/0.002 Torr; $\delta^1H = 0.62$ (s, 6H), 0.73 to 2.03 and 3.77 to 4.23 (br, 18H); $\delta^{11}B = 53.0$ ppm [25].

29c: $R = CH_3$, $R^1 = cyclo\text{-}C_6H_{11}$: b.p. 125°C/0.002 Torr; $\delta^1H = 0.48$ (s, 6H), 0.63 to 2.28 (br, 22H); $\delta^{11}B = 55.0$ ppm [25].

29d: $R = C_6H_5$, $R^1 = C_2H_5$: b.p. 150°C/0.002 Torr; $\delta^1H = 1.28$ (s, 6H), 4.00 (q, 4H), 7.08 to 7.60 (10H); $\delta^{11}B = 42.5$ ppm [25].

29e: $R = n\text{-}C_4H_9$, $R^1 = CH_3$: b.p. 145°C/0.002 Torr; $\delta^1H = 0.67$ to 1.77 (br, 18H), 3.15 (s, 6H); $\delta^{11}B = 49.0$ ppm [25].

29f: $R = n\text{-}C_4H_9$, $R^1 = C_2H_5$: b.p. 120 to 125°C/0.002 Torr, $\delta^1H = 0.63$ to 2.90 (br, 24H), 4.05 (q, 4H); $\delta^{11}B = 54.7$ ppm [25].

9.5.3 B–N Heterocycles with Exocyclic Sulfur

The following NMR data are given for the borazine **$(\text{-}BSCH_3\text{–}NCH_3\text{-})_3$**: $\delta^{11}B = 37.3$ ppm, $\delta^{13}C(N\text{–}C) = 36.80$, and $\delta^{13}C(S\text{–}C) = 11.68$ ppm [8].

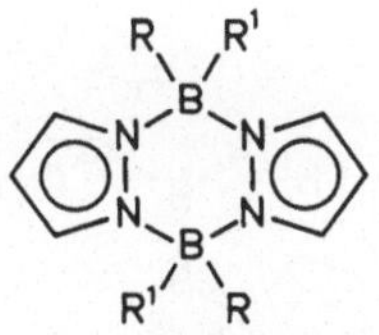

30

4,4,8,8-Tetrakis(alkylthio)pyrazaboles of type **30** have been obtained by the reaction of pyrazole with equimolar quantities of organylthioboranes such as $B(SCH_3)_3$ or $H_5C_6B(SC_2H_5)_2$ and also by reaction of pyrazabole (**30**, $R = R^1 = H$) with $HS(CH_2)_nSH$ (n = 2 or 3) at elevated temperatures in an inert solvent. The boron-bonded R–S group is known to be a weak π-donor and is stable towards nucleophilic attack; therefore, these species exist in pyrazabole structure [26].

30a: $R = R^1 = SCH_3$: m.p. 191 to 192°C; reactions with $^{10}BBr_3$ proceed without exchange of the boron atoms; all four R substituents may be exchanged for Br, and ^{11}B NMR spectra show intermediates; X-ray structural studies: monoclinic; space group C2/c–C^6_{2h} (No. 15); a = 1176.4, b = 848.5, and c = 1703.0 pm, $\beta = 103.74°$; Z = 4 (from 1251 observed reflections to R = 3.4%); NMR data (in $CDCl_3$): $\delta^1H = 8.42$ (d, 2H), 6.68 (t, 1 H), 1.63 (s, 6H) ppm; $\delta^{11}B = 4.6$ ppm; $\delta^{13}C(CH_3) = 10.5$, $\delta^{13}C$ (pyrazole ring) = 138.7 (C3, C5), 108.2 (C4) ppm [26, 27].

30b: $R = R^1 = SC_2H_5$: m.p. 116°C; NMR data (in $CDCl_3$): $\delta^1H = 1.02$ (t, 6H), 2.07 (q, 4H), 6.60 (t, 1H), 8.39 (d, 2H) ppm; $\delta^{11}B = 4.3$ ppm; $\delta^{13}C(C_2H_5) = 16.4$ and 22.1 ppm; $\delta^{13}C$ (pyrazole ring) = 138.9 (C3, C5) and 107.9 (C4) ppm [26].

30c: $R = SC_2H_5$, $R^1 = SC_6H_5$: m.p. 186°C; NMR data (in $CDCl_3$): $\delta^1H = 0.92$ (t, 3H), 1.91 (q, 2H), 6.51 (t, 1H), 7.11 (s, 5H), 8.17 (d, 2H) ppm; $\delta^{11}B = 4.1$ ppm; $\delta^{13}C(C_2H_5) = 16.6$ and 21.4 ppm, $\delta^{13}C$ (pyrazole ring) = 138.6 (C3, C5) and 107.4 (C4) ppm [26].

30d: $R = SCH_3$, $R^1 = CH_3$: m.p. 191 to 192°C; $\delta^1H = 1.63$ (s, 6H), 6.68 (t, 1H), 8.42 (d, 2H) ppm; $\delta^{11}B = 4.6$ ppm; $\delta^{13}C(CH_3) = 10.5$; $\delta^{13}C$ (pyrazole ring) = 138.7 (C3, C5) and 108.2 (C4) ppm [26].

References for 9.5 on pp. 132/3

30e: $RR^1 = SCH_2CH_2S$: m.p. 211 to 214°C; reaction with BBr_3; X-ray data: monoclinic; space group $P2_1/n\text{-}C^5_{2h}$ (No. 14) with a = 1072.8, b = 1060.2, and c = 1340.7 pm, $\beta = 93.43°$; Z = 4 (from 2291 observed reflections to R = 3.4%); NMR data (in $CDCl_3$): $\delta^1H = 3.20$ (s, 4H), 6.53 (t, 1H), 8.48 (d, 2H) ppm; $\delta^{11}B = 8.4$ ppm; $\delta^{13}C(CH_2) = 38.4$, $\delta^{13}C$ (pyrazole ring) = 138.5 (C3, C5) and 106.9 (C4) ppm [26, 27].

30f: $RR^1 = SCH_2CH_2CH_2S$: m.p. 202 to 207°C; $\delta^1H = 1.97$ (p, 2H), 6.55 (t, 1H), 8.45 (d, 2H) ppm; $\delta^{11}B = 4.5$ ppm; $\delta^{13}C(CH_2) = 24.7$, $\delta^{13}C$ (pyrazole ring) = 138.2 (C3, C5), and 107.0 (C4) ppm [26, 27].

9.5.4 Compounds without a B–S Bond

Compounds of the type $\mathbf{(R_2N)_2B\text{–}CH_2\text{–}CH{=}CH\text{–}SCH_3}$ have been prepared from $CH_2{=}CH\text{–}CH_2SH$ by sequential lithiation, boronation with $ClB(NR_2)_2$, and methylation [28]. $\mathbf{(CH_3)_3N\text{–}BH_2\text{–}CH_2\text{–}SCH_3}$ is an intermediate in the reaction of $Li[CH_3S\text{–}CH_2\text{–}BH_3]$ and $[(CH_3)_3NH]Cl$ which could not be isolated, but has $\delta^1H = 1.75$ (t, BCH_2), 2.07 (s, SCH_3), 2.53 (s, NCH_3) ppm; $\delta^{11}B = -3.9$ ppm, $\delta^{13}C = 19.9$ (SCH_3), 23.1 (br, BCH_2), and 52.7 (NCH_3) ppm [29]. $\mathbf{(CH_3)_2B\text{–}NCH_3\text{–}C(S)\text{–}N(C_2H_5)_2}$ has been prepared from $(CH_3)_2BBr$ and $HN(CH_3)\text{–}C(S)\text{–}N(C_2H_5)_2$, m.p. 70°C, in 82% yield [25].

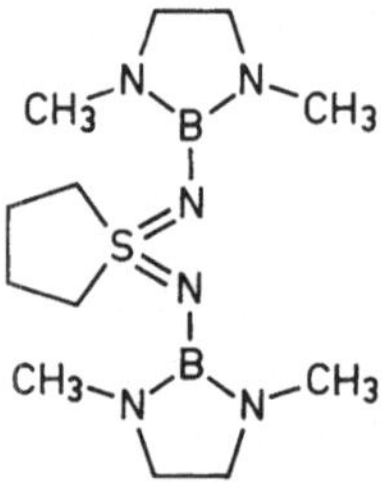

31

N, N¹-Bis(1,3-dimethyl-1,3-diazo-2-borolidin-2-yl)-tetramethylenesulfodiimide, **31**, has been prepared from $ClB(NCH_3\text{–}CH_2)_2$ and $(CH_2\text{–}CH_2)_2S[{=}N\text{–}Si(CH_3)_3]_2$ in 2:1 molar ratio in *n*-hexane; b.p. 72°C; δ^1H (in $CDCl_3$) = 2.45 to 2.66 (m, 4H), 2.63 (s, 12H, NCH_3), 2.95 to 3.13 (s, 12H) ppm; $\delta^{11}B = 24.9$ ppm; IR data are given [30].

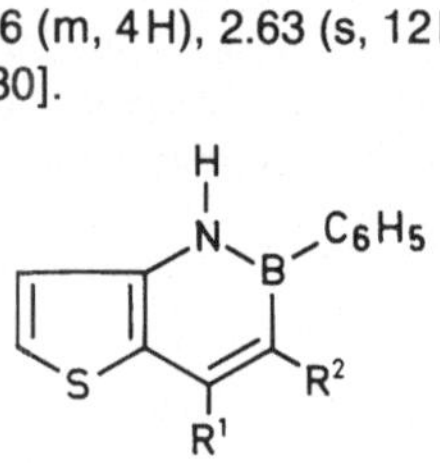

32

Compounds of type **32** have been prepared by reduction of 1-nitro-thieno-2-isopropenyl with $SnCl_2$ and cyclization of the intermediate with $C_6H_5BCl_2$:

32a: $R^1 = CH_3$, $R^2 = H$: m.p. 68.7 to 70.2°C; UV, IR, mass spectrum, ^1H NMR data are given [31].

32b: $R^1 = H$, R^2 = 2-thienyl: m.p. 95 to 97.5°C; UV, IR, mass spectrum, ^1H, ^{13}C NMR spectra are given [31, 32].

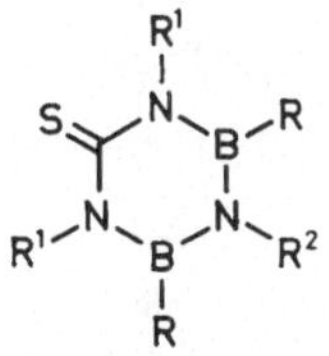

33

1,3,5-Triorganyl-2,6-diorganyl-1,3,5-triaza-2,6-diboracyclohexane-4-thiones of type **33** have been prepared from $R^1NH–C(S)–NHR^2$ and RBX_2 (X = Br, Cl). Selected examples:

33a: $R = CH_3$, $R^1 = R^2 = H$: m.p. 94 to 96°C; $\delta^1H = 0.23$ (s) ppm; $\delta^{11}B = 33.1$ ppm; $\nu(NH) =$ 3420 and 3300 cm^{-1}; M/z = 141 [24].

33b: $R = CH_3$, $R^1 = CH_3$, $R^2 = H$: m.p. 108°C; $\delta^{11}B = 35.2$ ppm; $\nu(NH) = 3290$ cm^{-1} [24].

33c: $R = CH_3$, $R^1 = R^2 = CH_3$: m.p. 103 to 108°C; $\delta^{11}B = 33.4$ ppm [24].

33d: $R = CH_3$, $R^1 = R^2 = C_2H_5$: m.p. 97 to 98°C; $\delta^{11}B = 35.5$ ppm [24]; b.p. 92°C/0.002 Torr; $\delta^1H = 0.56$ (s, 6H), 1.15 (t, 6H), 1.10 (t, 3H), 3.28 (q, 4H), 3.21 (q, 2H); $\delta^{11}B = 35.7$ ppm [25].

33e: $R = C_6H_5$, $R^1 = R^2 = CH_3$: m.p. 118 to 123°C; $\delta^1H = 2.53$ (s, 3H); $\delta^{11}B = 34.3$ ppm; M/z = 307 [24].

33f: $R = C_6H_5$, $R^1 = C_2H_5$, $R^2 = H$: b.p. 121°C/0.002 Torr; $\delta^1H = 1.12$ (t, 6H), 3.25 (q, 4H), 7.10 to 7.83 (br, 10H) ppm; $\delta^{11}B = 35.5$ ppm; $\nu(NH) = 3380$ cm^{-1} [25].

33g: $R = C_6H_5$, $R^1 = R^2 = C_2H_5$: m.p. 95 to 98°C; $\delta^{11}B = 36.4$ ppm [24].

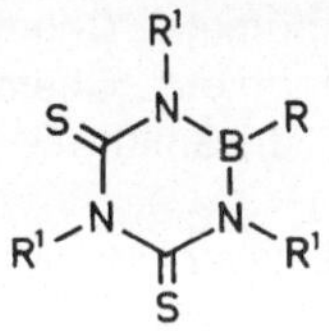

34

1,3,5-Triorganyl-2-organyl-1,3,5-triaza-2-bora-cyclohexane-dithiones(4,6) of type **34** have been prepared from $R^1NH–C(S)–NHR^1$ and RBX_2 [24].

34a: $R = C_6H_5$, $R^1 = CH_3$: m.p. 150 to 154°C; $\delta^{11}B = 33.5$ ppm; M/z = 263 [24].

34b: $R = CH_3$, $R^1 = C_2H_5$: m.p. 132 to 134°C; $\delta^{11}B = 34.7$ ppm [24].

34c: $R = CH_3$, $R^1 = n\text{-}C_4H_9$: m.p. 55 to 57°C; $\delta^{11}B = 33.7$ ppm [24].

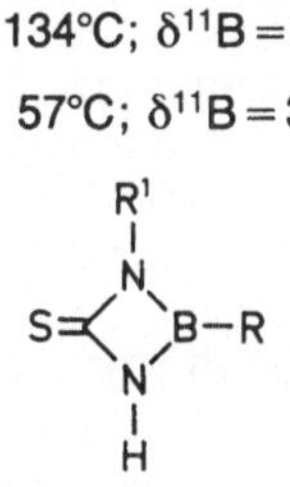

35

1-*t*-Butyl-2-organyl-1,3-diaza-2-bora-cyclobutane-4-thiones of type **35** have been prepared from $t\text{-}C_4H_9NH–C(S)–NH\text{-}t\text{-}C_4H_9$ and RBX_2 [24].

35a: $R = CH_3$, $R^1 = t\text{-}C_4H_9$: m.p. 57 to 58°C; $\delta^{11}B = 33.7$ ppm; $\nu(NH) = 3220$ cm^{-1} [24].

35b: $R = C_6H_5$, $R^1 = t\text{-}C_4H_9$: m.p. 93 to 94°C; $\delta^{11}B = 35.5$ ppm; $\nu(NH) = 3240$ cm^{-1} [24].

35c: $R = n\text{-}C_4H_9$, $R^1 = t\text{-}C_4H_9$: m.p. 98 to 102°C; $\delta^{11}B = 34.5$ ppm; $\nu(NH) = 3240$ cm^{-1} [24].

References for 9.5 on pp. 132/3

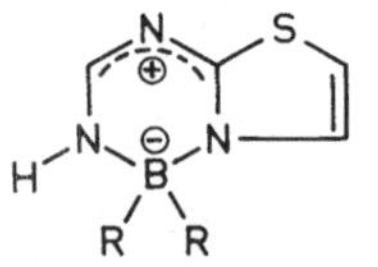

36

Compound **36** ($R = n\text{-}C_4H_9$) has been obtained by reaction of $(n\text{-}C_4H_9)_2BNH_2$ with 2-aminothiazole and $HCONH_2$ in dimethylformamide between 140 and 160°C [33].

In $(CH_3)_2S$ solution, BH_2CN exists in monomeric or dimeric forms as $(CH_3)_2S$ complexes in equilibrium with oligomers [38].

9.5.5 Ionic Species

Electrolytes for an alkali-metal battery contain a pseudohalide-containing salt $M[BR_x(SCN)_{4-x}]$, e.g., $Na[B(C_6H_5)_2(SCN)_2]$ or $K[B(C_6H_5)_3(SCN)]$, in dioxane solution [34].

Various sulfonium tetraphenylborates have been reported [35 to 37] but their discussion is beyond the scope of this presentation.

References for 9.5:

[1] Ha, T.-K.; Nguyen, M.-T.; Vanquinckenborne, L. G. (J. Mol. Struct. **90** [1982] 99/105).
[2] Cragg, R. H.; Miller, T. J. (J. Organometal. Chem. **241** [1983] 289/300).
[3] Cragg, R. H.; Miller, T. J. (J. Organometal. Chem. **243** [1983] 387/92).
[4] Cragg, R. H.; Miller, T. J. (J. Organometal. Chem. **255** [1983] 143/52).
[5] Brown, C.; Cragg, R. H.; Miller, T. J.; Smith, D. O. (J. Organometal. Chem. **244** [1983] 209/15).
[6] Meller, A.; Hirninger, F. J.; Noltemeyer, M.; Maringgele, W. (Chem. Ber. **114** [1981] 2519/35).
[7] Nöth, H.; Staudigl, R.; Storch, W. (Chem. Ber. **114** [1981] 3024/43).
[8] Nöth, H.; Wrackmeyer, B. (Chem. Ber. **114** [1981] 1150/6).
[9] Goetze, R.; Nöth, H.; Pommerening, H.; Sedlak, D.; Wrackmeyer, B. (Chem. Ber. **114** [1981] 1884/93).
[10] Nöth, H.; Schwerthöffer, R. (Chem. Ber. **114** [1981] 3056/62).

[11] Meller, A.; Habben, C. (Monatsh. Chem. **113** [1982] 139/53).
[12] Biffar, W.; Nöth, H.; Schwerthöffer, R. (Liebigs Ann. Chem. **1981** 2067/80).
[13] Nöth, H.; Staudigl, R.; Taeger, T. (Chem. Ber. **114** [1981] 1157/75).
[14] Nöth, H.; Staudigl, R.; Brückner, R. (Chem. Ber. **114** [1981] 1871/83).
[15] Nöth, H.; Staudigl, R. (Z. Anorg. Allgem. Chem. **481** [1981] 41/50).
[16] Nöth, H.; Staudigl, R. (Chem. Ber. **115** [1982] 3011/24).
[17] Nöth, H.; Staudigl, R. (Chem. Ber. **115** [1982] 1555/67).
[18] Habben, C.; Meller, A. (Z. Naturforsch. **39b** [1984] 1022/6).
[19] Davidson, G.; Ewer, K. E. (J. Mol. Struct. **74** [1981] 181/91).
[20] Noltemeyer, M.; Sheldrick, G. M.; Habben, C.; Meller, A. (Z. Naturforsch. **38b** [1983] 1182/91).

[21] Habben, C.; Meller, A. (Chem. Ber. **117** [1984] 2531/7).
[22] Habben, C. Maringgele, W.; Meller, A. (Z. Naturforsch. **37b** [1982] 43/53).
[23] Deeg, F. W.; Bräuchle, C.; Voitländer, J. (Chem. Phys. **64** [1982] 427/36).

[24] Maringgele, W. (J. Organometal. Chem. **222** [1981] 17/32).
[25] Maringgele, W. (Chem. Ber. **115** [1982] 3271/89).
[26] Hodgkins, T. G.; Niedenzu, K.; Niedenzu, K. S.; Seelig, S. S. (Inorg. Chem. **20** [1981] 2097/100).
[27] Niedenzu, K.; Nöth, H. (Chem. Ber. **116** [1983] 1132/53).
[28] Hoffmann, R. W.; Kemper, B. (Tetrahedron Letters **21** [1980] 4883/6).
[29] Nöth, H.; Sedlak, D. (Chem. Ber. **116** [1983] 1479/86).
[30] Haubold, W.; Fehlinger, H. G. (Z. Naturforsch. **36b** [1981] 157/60).

[31] Gronowitz, S.; Ander, I. (Chem. Scr. **15** [1980] 135/44).
[32] Gronowitz, S.; Ander, I.; Zanirato, P. (Chem. Scr. **22** [1983] 55/9).
[33] Drokhov, V. A.; Zolotarev, B.; Mikhailov, B. M. (Izv. Akad. Nauk SSSR Ser. Khim. **1981** 869/74; Bull. Acad. Sci. USSR Div. Chem. Sci. **30** [1981] 649/53).
[34] Klemann, L. P.; Newman, G. H. (U.S. 4279976 [1980/81] 1/11 from C.A. **95** [1981] No. 206819).
[35] Braun, H.; Amann, A.; Barnet, R.; Firl, J. (Z. Naturforsch. **35b** [1980] 1398/405).
[36] Kantlehner, W.; Haug, E.; Hagen, H. (Liebigs Ann. Chem. **1982** 298/305).
[37] Böhme, H.; Ahrens, G. (Liebigs Ann. Chem. **1982** 1030/8).
[38] Gyori, B.; Emri, J.; Feher, I. (J. Organometal. Chem. **255** 1983 17/28).
[39] Nguyen, M.-T.; Ruelle, P. (J. Chem. Soc. Faraday Trans. II **80** [1984] 1225/34).

9.6 Compounds Containing B, S, N, and O

9.6.1 Compounds with a B–S Bond

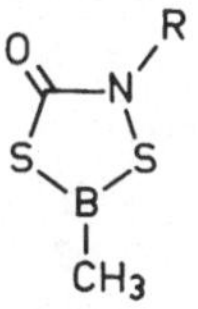

37

2-Alkyl-5-methyl-1,4-dithia-2-aza-5-boracyclopentanones(3), **37**, have been prepared by reaction of 3,5-dimethyl-1,2,4-trithia-3,5-diborolane with isocyanates R–NCO at 80°C in CCl_4 [1]. The species with $R = n\text{-}C_3H_7$ and $R = cyclo\text{-}C_6H_{11}$ are described.

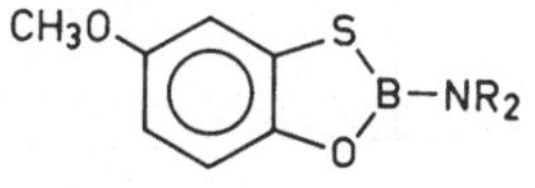

38

2-Dialkylamino-5′-methoxy-1,3,2-benzoxathiaboroles, **38**, have been prepared from the 2-chloro compound and a mixture of NR_3 and $NH(C_2H_5)_2$ in tetrahydrofuran at −60°C, slow warming to 25°C, filtration and stripping of solvent. The compound with $R = C_2H_5$ was obtained in 76% yield, and the one with $R = C_6H_5$ in 69% yield [2].

The compounds (L)=B–OCOCH$_3$, (L)=B–O–B(OCOCH$_3$)$_2$, and (L)=B–O–B=(L) with (L)H$_2$ being a benzothiazoline derivative (with a O,N,S donor system) have been synthesized and characterized [3, 4].

References for 9.6 on p. 135

9.6.2 Compounds without a B–S-Bond

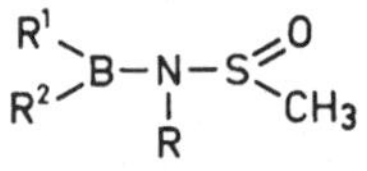

39

Various S-methyl-N-organyl-sulfinamido-diorganylboranes, **39**, have been obtained by reaction of N-organyl-N-trimethylsilanyl-aminoboranes with methylsulfinylchloride in CH_2Cl_2 at −80°C, slow warming to 25°C within 30 h, evaporation of CH_2Cl_2 and $(CH_3)_3SiCl$ under vacuum and drying [5]. Selected examples:

39a: $R = R^1 = R^2 = CH_3$: m.p. 30°C; $\delta^1H = 0.47$ (s, 6H, B–CH_3), 2.56 (s, 3H, S–CH_3), 3.20 (s, 3H, N–CH_3) ppm; $\delta^{11}B = 54.6$ ppm; M/z = 118; ν(S=O) (in CH_2Cl_2) = 1072 cm^{-1} [5].

39b: $R = C_6H_5$, $R^1 = R^2 = CH_3$: yellow oil; $\delta^1H = 0.69$ (s, 6H), 2.30 (s, 3H), 6.9 to 7.6 (br, 5H) ppm; $\delta^{11}B = 53.6$ ppm; M/z = 195; ν(S=O) (capillary film) = 1085 cm^{-1} [5].

39c: $R = CH_3$, $R^1 = R^2 = N(CH_3)_2$: m.p. 95°C (dec); $\delta^1H = 2.73$ (s, 3H), 2.80 (s, 12H), 3.07 (s, 3H, NCH_3) ppm; $\delta^{11}B = 27.8$ ppm; M/z = 191; ν(S=O) = 1060 cm^{-1} [5].

39d: $R = CH_3$, $R^1 = R^2 = N(C_2H_5)_2$: red oil; $\delta^1H = 0.97$ (t, 12H, CCH_3), 2.37 (s, 3H, SCH_3), 2.60 (s, 3H, NCH_3), 2.81 (q, 8H, NCH_2) ppm; $\delta^{11}B = 28.7$ ppm; M/z = 247; ν(S=O) = 1070 cm^{-1} [5].

39e: $R = CH_3$, $R^1 = N(CH_3)_2$, $R^2 = C_6H_5$: yellow oil; $\delta^1H = 2.47$ (s, 3H), 2.69 (s, 6H), 2.79 (s, 3H, NCH_3), 7.31 (br, 5H) ppm; $\delta^{11}B = 33.7$ ppm; M/z = 224; ν(S=O) = 1080 cm^{-1} [5].

39f: $R = CH_3$, $R^1 = N(C_2H_5)_2$, $R^2 = C_6H_5$: yellow oil; $\delta^1H = 1.08$ (t, 6H), 2.43 (s, 3H), 2.77 (s, 3H, NCH_3), 3.04 (q, 4H, NCH_2), and 7.21 (br, 5H) ppm; $\delta^{11}B = 34.5$ ppm; M/z = 252; ν(S=O) = 1080 cm^{-1} [5].

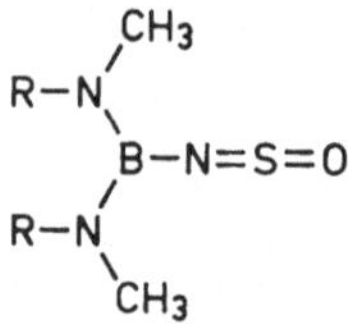

40

The following boryl-sulfinylimides, **40**, have been prepared:

40a: $R = CH_3$: b.p. 78°C/10 Torr; $\delta^1H = 2.46$; $\delta^{11}B = 21.5$ ppm.

40b: $RR = CH_2–CH_2$: b.p. 68°C/11 Torr; $\delta^1H = 2.55$ (NCH_3), 3.20 (CH_2) ppm; $\delta^{11}B = 24.6$ ppm; IR data are given [6].

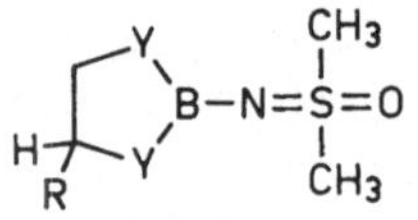

41

The following S,S-dimethyl-N-borolidin-2-yl-sulfoimines, **41**, have been prepared and were studied by NMR spectroscopy [6]:

41a: $Y = NCH_3$, R = H: b.p. 74°C/0.01 Torr; $\delta^1H = 2.63$ (NCH_3), 3.04 (SCH_3), 3.09 (CH_2) ppm; $\delta^{11}B = 23.1$ ppm [6].

41b: Y = O, $R = CH_3$: b.p. 124°C/0.01 Torr; $\delta^1H = 1.31$ (CCH_3), 3.11 (SCH_3) ppm; $\delta^{11}B = 23.4$ ppm [6].

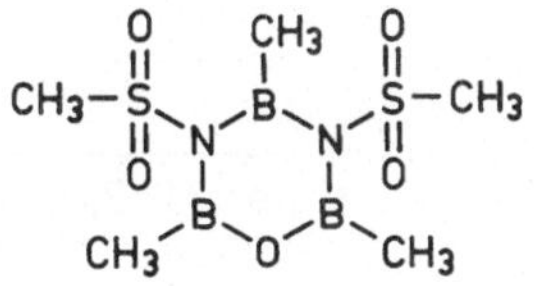

42

Compound **42** has been prepared from $S_3B_2(CH_3)_2$ and CH_3SO_2NSO in CCl_4: b.p. 145°C/ 0.01 Torr, m.p. 182 to 186°C; δ^1H = 0.72 (s, 6H, BCH_3), 1.09 (s, 3H, BCH_3), 3.24 (s, 6H, SO_2CH_3) ppm; $\delta^{11}B$ = 32.9 ppm; M/z = 280 [7]. In the cited reaction, $(\text{-}BCH_3\text{-}NSO_2CH_3\text{-})_2$ was detected in spectroscopic studies but was not isolated [7].

For some 1,2-dihydro-1-hydroxy-2-(organosulfonyl)areno[d]-1,2,3-diazaborines, see [8].

9.6.3 Ionic Species

Various tetraphenylborates of complex sulfur-containing cations have been reported [9 to 11].

References for 9.6:

[1] Habben, C.; Maringgele, W.; Meller, A. (Z. Naturforsch. **37b** [1982] 43/53).
[2] Andrä, K.; Straube, A. (Z. Anorg. Allgem. Chem. **490** [1982] 191/2).
[3] Chaturvedi, K. K.; Singh, R. V.; Tandon, J. P. (Syn. Reactiv. Inorg. Metal-Org. Chem. **13** [1983] 155/7 from C.A. **98** [1983] No. 190585).
[4] Singh, P. K.; Koacher, J. K.; Tandon, J. P. (J. Inorg. Nucl. Chem. **43** [1981] 1755/8).
[5] Meller, A.; Maringgele, W.; Armbrecht, M. (Z. Naturforsch. **36b** [1981] 1411/5).
[6] Haubold, W.; Fehling, H. G. (Z. Naturforsch. **36b** [1981] 157/60).
[7] Meller, A.; Habben, C. (Monatsh. Chem. **113** [1982] 139/53).
[8] Grassberger, M. A.; Turnowsky, F.; Hildebrandt, J. (J. Med. Chem. **27** [1984] 947/54).
[9] Okuma, K.; Higuchi, N.; Ota, H.; Kobayashi, M. (Chem. Letters **1980** 1503/6).
[10] Braun, H.; Amann, A.; Barnert, R.; Firl, J. (Z. Naturforsch. **35b** [1980] 1398/405).

[11] Böhme, H.; Ahrens, G. (Liebigs Ann. Chem. **1982** 1030/8).

9.7 With Halogen

9.7.1 Ternary Systems

The following vertical ionization potentials have been calculated by applying Rayleigh-Schrödinger perturbation corrections $(\Delta E)^{GA}$ to Koopmans' theorem: for **FBS**: 3π = 10.72, 9σ = 13.69, 2π = 17.05, 8σ = 19.34 eV; for **ClBS**: 4π = 10.26, 11σ = 13.10, 3π = 13.15, 10σ = 15.99 eV [1]. Bond length and energy for these species are given in Table 9/11.

Results of ab initio SCF calculations using double zeta (DZ) and double zeta plus polarization function basis sets are also listed in Table 9/11. In addition, net charges and overlap populations have been calculated [2]. The ionization potentials for FBS are 10.97 and 14.83 eV, those for ClBS are 10.74 and 14.42 eV [59].

References for 9.7.1 to 9.7.5 on pp. 140/2

Table 9/11
Geometries of FBS and ClBS.

	4-31G ΔE^{GA} [1]	exper. [1, 2]	SCF(DZ) [2]	STO-4G [3]
F–B≡S				
r(B–F) in pm	130.3	128.4	130.9	130.0
r(B≡S) in pm	159.9	160.6	162.5	163.7
E_T in a.u.	−521.31578	—	−521.64719	−519.0363
Cl–B≡S				
r(B–Cl) in pm	168.1	168.1	172.3	169.2
r(B≡S) in pm	159.9	160.6	162.0	162.7
E_T in a.u.	−880.89840	—	−881.66318	−877.8275

Optimized geometries and energies had also been computed using STO-4G basis functions; the data for the linear X–B≡S compounds (see table 9/11) have been compared with those obtained for the isomeric B–S–X compounds: for B–S–F: r(B–S) = 189.6 and r(S–F) = 160.1 pm; ∢(BSF) = 108.02°; E_T = −518.9251 a.u.; for B–S–Cl: r(B–S) = 188.7 and r(S–Cl) = 198.7 pm; ∢(BSCl) = 97.22°; E_T = −877.7040 a.u. The results suggest that BSX species are potentially stable and overlap populations and ionization potentials have been tabulated [3]. Indeed, the unstable species FBS, $F_2B_2S_2$, and $F_3B_3S_3$ have been obtained by pyrolysis of sulfur fluorides over boron; the identities were confirmed by He(I) photoelectron spectra and in situ quadrupole mass spectroscopy; ab initio MO calculations are reported [4, 5]. PE spectra for $ClBS^+$ show a B≡S stretching mode at 1375 ± 30 cm^{-1} [60].

9.7.2 With Hydrogen or Organic Substituents

Isopropylthiomethyl ethers have been prepared from the corresponding methoxy-ethoxymethyl esters by treatment with $[(CH_3)_2HCS]_2BBr$ [6].

$(CH_3)_2S–BF_3$ reacts with XeF_2 to yield materials containing the cation $[(CH_3)_2SF]^+$ [62].

$(CH_3)HS–BCl_3$. The addition of CH_3SH to $^{10}BCl_3$ or $^{11}BCl_3$ in the gas phase gives rise to new IR absorptions with three peaks at ~1000 cm^{-1} tentatively assigned to the $(CH_3)HS–BCl_3$ complex. The equilibrium constant of the reaction $(CH_3)SH + BCl_3 \rightarrow (CH_3)HS–BCl_3$ is ~1 at room temperature. The IR multiphoton dissociation of this complex has been examined using a CO_2 TEA laser. The dissociation rate of $(CH_3)HS\text{-}^{10}BCl_3$ is ten times larger than that of $(CH_3)HS\text{-}^{11}BCl_3$ (for laser irradiation at 944 cm^{-1}). The dependence of the specific dissociation rate on the absorbed energy suggests two reaction paths with different activation energies, i.e., elimination of Cl atoms or HCl molecules from vibrationally excited complex molecules [7, 8].

$(C_2H_5)_2S–^{10}BCl_3$, $(C_2H_5)_2S–^{11}BCl_3$, $(C_2H_5)HS–^{10}BCl_3$, $(C_2H_5)HS–^{11}BCl_3$, $(n\text{-}C_3H_7)HS–^{10}BCl_3$, and **$(n\text{-}C_3H_7)HS–^{11}BCl_3$** have been prepared and investigated [9].

$CH_3(CH_2Cl)S–BCl_3$ has been prepared from CH_3SCH_2Cl and BCl_3; m.p. 76 to 78°C; IR data are available: δ^1H (in $CDCl_3$) = 3.61 (CH_3), 4.41 (CH_2) ppm; $\delta^{13}C$ = 20.8 (CH_3), 32.6 (CH_2) ppm [10].

$CH_3(CH_2Br)S–BBr_3$: m.p. 75 to 76°C; IR data are available; δ^1H (in CD_3NO_2) = 2.85 (CH_3), 4.73 (CH_2) ppm; $\delta^{13}C$ (in $CDCl_3$) = 21.4 (CH_3), 33.0 (CH_2) ppm [10].

$(CH_3)_2S$–$BHCl_2$ is a convenient reagent for the hydroboration of alkanes [11], alkenes [12], especially 1-methylcyclooctene [13], and 1-alkynes [9]; it has been prepared from $(CH_3)_2S$–BH_3 and $(CH_3)_2S$–BCl_3 (1:2 ratio) at 25°C. It reacts with 2,3-dimethyl-2-butene to give the thexylchloroborane adduct (thexyl = 1,1,2-trimethyl-1-propyl = $(CH_3)_2CH$–$C(CH_3)_2$- = C_6H_{13}), $(CH_3)_2S$–$BHCl(C_6H_{13})$, which (in CH_2Cl_2) is a reagent for the hydroboration of terminal alkenes [14]. For the reaction with trialkylboranes, see [15].

$(CH_3)_2S$–$BHBr_2$ is also a reagent for the hydroboration of alkanes [11], alkenes [12, 16 to 19], 1-methylcyclooctene [13], 1-alkynes [9, 18, 19], and 1-bromoalkynes [20]. It has a lower reactivity towards alkenes than $(CH_3)_2S$–$BHCl(C_6H_{13})$ [21]. It has been used in the stereospecific synthesis of racemic disparlure, the sex pheromone of the female gypsy moth (porthetria dispar L.) [22], and it reacts with AgCN to give isocyanoborates [23].

$(CH_3)_2S$–$BRCl_2$ species have been prepared from $(CH_3)_2S$–$BHCl_2$ and the appropriate olefins, e.g., R = *n*-C_6H_{13} [12]; R = *n*-C_8H_{17}, b.p. 65 to 67°C/2 Torr, 69% yield; R = *trans*-2-methylcyclopentyl, b.p. 45 to 47°C/0.3 Torr, 79% yield [24 to 26].

$(CH_3)_2S$–$BRBr_2$ compounds have been prepared from $(CH_3)_2S$–$BHBr_2$ and alkenes or alkynes [27], e.g., R = cyclopentyl, b.p. 140 to 144°C/2.1 Torr, 93% yield; R = *n*-C_6H_{13}, b.p. 99 to 100°C/1 Torr, 91% yield; δ^1H = 2.45 (CH_3) ppm [12]; R = 3-hexyl, b.p. 73 to 75°C/2.2 Torr, 90% yield; R = 2-methyl-1-pentyl, b.p. 82 to 85°C/1.6 Torr [12, 24]; R = *trans*-2-methylcyclopentyl, b.p. 68 to 69°C/0.5 Torr, 86% yield [12]; R = *n*-C_7H_{15}, R = $C_{10}H_{21}$, R = $C_{11}H_{23}$, R = $C_{14}H_{27}$ [28]. They are used for the hydroboration of 1-alkynes [25, 26] and react with $Li[AlH_4]$ to give $(CH_3)_2S$–BHRBr compounds [26, 29].

$(CH_3)_2S$–BHI_2 has been prepared by the action of I_2 on (commercially available) $(CH_3)_2S$–BH_3 [12]. It reacts with 2-methyl-2-butane under isomerization [30].

$(CH_3)_2S$–BRI_2 has been prepared from $(CH_3)_2S$–BHI_2 and olefins, e.g., R = *n*-octyl, b.p. 125 to 128°C/0.2 Torr, 74% yield; R = *cis*-3-octyl [12, 24].

$(CH_3)_2S$–BH_2Cl reacts with 2,3-dimethyl-2-butane, C_6H_{14}, to yield $(CH_3)_2S$–$BHCl(C_6H_{13})$, which is a stable monohydroborating agent with exceptional regioselectivity [14, 31]; with alkenylcyclopentanes to yield hydroazulenones [32]; and reduces cyclohexyl ketals chemoselectively [33].

$(CH_3)_2S$–BHRCl with R = $(CH_3)_2C$–$CH(CH_3)_2$, i.e., $(CH_3)_2S$–$BHCl(C_6H_{13})$, has been prepared from $(CH_3)_2S$–BH_2Cl or $(CH_3)_2S$–$BHCl_2$ and 2,3-dimethyl-2-butane, C_6H_{14}, [14, 31]. It has a higher Lewis acidity than $(CH_3)_2S$–$BHBr_2$ [21]; is a hydroborating agent for alkynes; and serves also for reactions with –BrC=CBr– compounds. It is used to synthesize dihydrojasmone [34 to 37] and readily reduces carboxylic acids [38].

$(CH_3)_2S$–BH_2Br is a reagent for the hydroboration of 1-alkynes [9] as well as terminal and internal alkynes [24]. It reacts with AgCN to give isocyanoborates [23].

$(CH_3)_2S$–BHRBr compounds have been prepared by hydridation of alkyldibromoboranes [34 to 37]. For example, the species with R = *n*-C_6H_{13} has been prepared by the hydridation of $(CH_3)_2S$–$BRBr_2$; $\delta^{11}B$ = 0.7 ppm; ν(BH) = 2450 cm^{-1} [29]. **$(CH_3)_2S$–BR_2Br** compounds have been prepared from $(CH_3)_2S$–BHRBr compounds by the stepwise hydroboration of two different alkenes. This procedure provides a convenient route to mixed dialkyl haloboranes, e.g., $(CH_3)_2S$–$B(n\text{-}C_5H_{11})(n\text{-}C_6H_{13})Br$ or $(CH_3)_2S$–$B(n\text{-}C_6H_{13})(n\text{-}C_{12}H_{25})Br$ [29]. **$(CH_3)_2S$–BR_3** compounds, e.g., $(CH_3)_2S$–$B(n\text{-}C_5H_{11})(n\text{-}C_6H_{13})(n\text{-}C_{12}H_{25})$, are prepared from three alkenes and are used to prepare mixed trialkylboranes [29].

References for 9.7.1 to 9.7.5 on pp. 140/2

9.7.3 B–S Heterocycles Containing Boron-Bonded Halogen

43

43a with X = Br has been prepared by the reaction of BBr_3 with trithiadiborolanes with $X = SC_2H_5$ [39]. The ^{10}B-labelled compound was obtained employing $^{10}BBr_3$ and the trithiadiborolane with $X = CH_3$ [40]. A ^{10}B-labelling NMR study indicated that intermolecular exchange of Br in **43a** by CH_3, C_2H_5, or $N(CH_3)_2$ proceeds via two mechanisms: an *exo* process giving mixed trithiadiborolanes without exchange of annular boron atoms, and a slower *endo* reaction, in which B exchanged with R = organyl. The reactions are statistically (*endo/exo*) or *endo* only [41]. Reactions of **43a** with H_2S, anhydrous Na_2S, or $SC(SH)_2$ leading to B_8S_{16} [61], or reactions with thiadiazadiborolidines or with triazadiborolidine and reactions with 1,3-diaza-2-borolidines, see [42]; and reactions with RNSO compounds, e.g., $R = C_6F_5$, see [43].

43b with X = I reacts with 2,6-dimethylphenol to yield 3,5-bis(2,6-dimethylphenyloxy)-1,2,4-trithia-3,5-diborolane; with diisopropylether to form 3,5-diisopropyloxy-1,2,4-trithia-3,5-diborolane; with methyl-*t*-butylether to give 3,5-dimethyloxy-1,2,4-trithia-3,5-diborolane; and with methylphenylether to yield 3,5-diphenyloxy-1,2,4-trithia-3,5-diborolane [44].

44

Compound **44** with X = Br forms a well-defined crystalline adduct with 1,2,3-trimethyl-1,3-diaza-2-borolidine (in low yield) when an aged and somewhat impure 3,5-dibromo-1,2,4-trithia-3,5-diborolane and 1,2,3-trimethyl-1,3-diaza-2-borolidine were used [42]; see Section 9.7.7.1, pp. 143/4.

9.7.4 Halogenated B–S Heterocycles Containing Annular Carbon

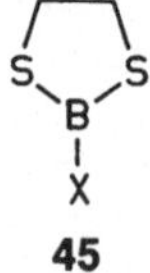

45

45a with X = Cl: IR and Raman spectra, see Table 9/12 [45]. The reaction with $(CH_3)_3SnLi$ leads to 2-methyl-1,3-dithia-2-borolane [47]; the reaction with $N[Sn(CH_3)_3]_3$ to triborylamines (see Section 9.5.1, p. 122) [48]; the compound reacts with $LiSi(CH_3)_3$ to yield $(CH_2S)_2B$–$Si(CH_3)_3$ [49]. Using **45a** at −78°C, β-methoxy ethoxy methylethers are selectively cleaved [46].

45b: X = Br: As for the compound with X = Cl (**45a**), IR and Raman spectra (table in the original paper) show C_2 symmetry of the ring, corresponding to a "twisted-ring" conformation [45].

Table 9/12
Vibrational Wave Numbers (in cm^{-1}) for 2-Chloro-1,3-dithia-2-boracyclopentane (**45a**, liquid; pol = polarized, dp = depolarized, def. = deformation) [45].

IR	Raman	assignment	IR	Raman	assignment
	2982m, sh, dp	CH stretch (B)	920s		^{11}B–S stretch (B)
2970w, br	2973m, pol	CH stretch (A)	840s	840vw, dp	CH_2 rock (B)
2936ms	2936s, pol	CH stretch (A + B)		830vw, pol	
2855vw	2855vw	1427 + 1436 = 2863	662ms	662vs, pol	ring stretches (A+B)
2842w	2840w	1427 × 2 = 2854	471ms	469vvs, pol	B–S stretch (A)
	1437m, sh	CH_2 scissors (A)	458m	456m, dp	ring def. (B)
1429ms	1427ms, dp	CH_2 scissors (B)	430w	426vw	^{10}B–Cl def. (B)
1282ms	1283wm, pol	CH_2 wag (A)	412ms	410wm, dp	^{11}B–Cl def. (B)
1258wm	1258w, dp	CH_2 wag (B)	332wm	330s, pol	ring def. (A)
1165vvw	1165m, ?dp	CH_2 twist (A)		254w	B–Cl def. (B)
1115w	1115m, dp	CH_2 twist (B)		208m, dp	ring def. (B)
1005s		^{10}B–Cl stretch (A)		136w, br, dp	ring def. (A)
987m, sh	988m, pol	ring stretch (A)		136w, br, dp	ring def. (A)
967vs	968w, sh	^{11}B–Cl stretch (A)			
957m, sh					
940s	942wm, ?dp	CH_2 rock (A) + ^{10}B–S stretch (B)			
927m, sh					

46

Δ^4-2-Halo-1,3-dithia-2-borolenes, **46**, have been prepared from 3,5-dihalo-1,2,4-trithia-3,5-diborolanes and alkynes in CCl_4 at 80°C in 3 d. Selected examples are presented below [50].

46a: X = Br, R^1 = H, R^2 = C_6H_5: b.p. 114°C/0.01 mbar; δ^1H = 6.93 (s, 1H), 7.25 to 7.65 (br, 5H); $\delta^{11}B$ = 49.91 ppm; ν(C=C) = 1510 cm^{-1}; M/z = 257 [50].

46b: X = Br, R^1 = H, R^2 = CH_2Br: b.p. 96°C/7 mbar; δ^1H = 6.80 (s, 1H), 4.47 (d, 2H) ppm; $\delta^{11}B$ = 52.0 ppm [50].

46c: X = Br, R^1 = R^2 = C_2H_5: b.p. 58°C/0.01 mbar; yellow liquid; δ^1H = 2.8 (q, 4H), 1.23 (t, 6H) ppm; $\delta^{11}B$ = 50.6 ppm; ν(C=C) = 1510 cm^{-1}; M/z = 237 [50].

46d: X = Br, R^1 = R^2 = C_6H_5: m.p. 85°C, b.p. 165°C/0.01 mbar; δ^1H = 7.18 to 7.67 (br, 10H) ppm; $\delta^{11}B$ = 50.73 ppm; ν(C=C) = 1580 cm^{-1}; M/z = 335 [50].

2-Bromo-1,3-dithia-2-borolene reacts with $N[Sn(CH_3)_3]_3$ to yield tris(1,3-dithia-2-borolen-yl)-amine [48].

References for 9.7.1 to 9.7.5 on pp. 140/2

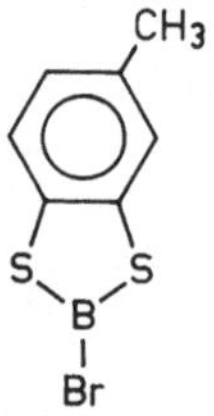

47

Compound **47** reacts with tris(trimethylstannyl)amine to give tris(5-methyl-1,3,2-benzodithiaborol-2-yl) amine [48].

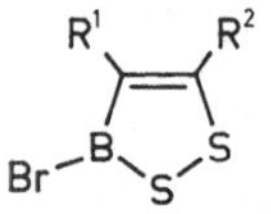

48

Compounds of type **48** ($R^1 = H$, CH_3, C_2H_5, n-C_3H_7; $R^2 = C_2H_5$, n-C_3H_7, n-C_4H_9, C_6H_5) react with $LiN[Si(CH_3)_3]_2$ [52].

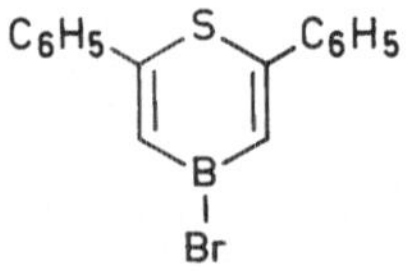

49 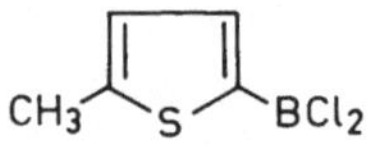 **50**

49 has been prepared by sublimation of the residue after distillation of compound **46a** (see p. 139) [50]. **50** has been used in order to prepare thienoazoborines [51].

9.7.5 Ionic Species

A number of sulfonium haloborates has been studied. They include simple triorganylsulfonium tetrahaloborates such as $[(CH_3)_3S][BF_4]$ [53, 54], $[(C_2H_5)_3S][BF_4]$ [53], $[(C_6H_5)_3S][BF_4]$ [53, 55 to 58], and much more complex species. Their discussion exceeds the scope of this chapter.

References for 9.7.1 to 9.7.5:

[1] Frost, D. C.; Kirby, C.; Lau, W. M.; McDowell, C. A.; Westwood, N. P. C. (J. Mol. Struct. **100** [1983] 87/94).

[2] Ha, T.-K.; Nguyen, M.-T.; Vanquinckenborne, L. G. (J. Mol. Struct. **90** [1982] 99/105).

[3] So, S. P. (J. Mol. Struct. **90** [1982] 29/34).

[4] Cooper, T. A.; Kroto, H. W.; Kirby, C.; Westwood, N. P. C. (J. Chem. Soc. Dalton Trans. **1984** 1047/52).

[5] Westwood, N. P. C. (NATO ASI Ser. B **90** [1983] 275/8 from C.A. **99** [1983] No. 45662).

[6] Corey, E. J.; Hua, D. H.; Seitz, S. P. (Tetrahedron Letters **25** [1984] 3/6).

[7] Ishikawa, Y.; Kurihara, O.; Arai, S.; Nakane, R. (J. Phys. Chem. **85** [1981] 3817/20).

[8] Ishikawa, Y.; Kurihara, O.; Arai, S.; Nakane, R. (Reza Kagaku Kenkyu No. 3 [1981] 69/75 from C.A. **96** [1982] No. 34333).

[9] Brown, H. C.; Basavaiah, D.; Kulkarni, S. U. (J. Organometal. Chem. **225** [1982] 63/9).
[10] Hartke, K.; Akguen, E. (Liebigs Ann. Chem. **1981** 47/51).

[11] Brown, H. C.; Kulkarni, S. U. (J. Organometal. Chem. **218** [1981] 299/307).
[12] Brown, H. C.; Ravindran, N.; Kulkarni, S. U. (J. Org. Chem. **45** [1980] 384/8).
[13] Brown, H. C.; Racherla, U. S. (J. Org. Chem. **48** [1983] 1389/91).
[14] Brown, H. C.; Sikorski, J. A. (Organometallics **1** [1982] 212/4).
[15] Brown, H. C.; Pai, G. G. (J. Organometal. Chem. **250** [1983] 13/22).
[16] Brown, H. C.; Imai, T. (J. Am. Chem. Soc. **105** [1983] 6285/9).
[17] Brown, H. C.; Imai, T. (J. Org. Chem. **49** [1984] 892/8).
[18] Brown, H. C.; Chandrasekharan, J. (J. Org. Chem. **48** [1983] 644/8).
[19] Brown, H. C.; Chandrasekharan, J.; Nelson, D. J. (J. Am. Chem. Soc. **106** [1984] 3768).
[20] Brown, H. C.; Imai, T. (Organometallics **3** [1984] 1392/5).

[21] Sikorski, J. A.; Brown, H. C. (J. Org. Chem. **47** [1982] 872/6).
[22] Brown, H. C.; Basavaiah, D. (Synthesis **1983** 283/4).
[23] Gyori, B.; Emri, J.; Feher, I. (J. Organometal. Chem. **255** [1983] 17/28).
[24] Brown, H. C.; Campbell, J. B. (J. Org. Chem. **45** [1980] 389/92).
[25] Brown, H. C.; Basavaiah, D.; Kulkarni, S. U. (J. Org. Chem. **47** [1982] 3808/10).
[26] Brown, H. C.; Basavaiah, D. (J. Org. Chem. **47** [1982] 3806/8).
[27] Brown, H. C.; Bhat, N. G.; Somayaji, V. (Organometallics **2** [1983] 1311/6).
[28] Brown, H. C.; Basavaiah, D. (J. Org. Chem. **47** [1982] 5407/9).
[29] Kulkarni, S. U.; Basavaiah, D.; Zaidlewicz, M.; Brown, H. C. (Organometallics **1** [1982] 212/4).
[30] Brown, H. C.; Racherla, U. S. (J. Am. Chem. Soc. **105** [1983] 6506).

[31] Brown, H. C.; Sikorski, J. A.; Kulkarni, S. U.; Lee, H. D. (J. Org. Chem. **47** [1982] 863/72).
[32] Stevenson, J. W. S.; Bryson, T. A. (Chem. Letters **1984** 518).
[33] Bordes, R. J.; Bryson, T. A. (Chem. Letters **1984** 9/12).
[34] Brown, H. C.; Basavaiah, D.; Bhat, N. G. (Organometallics **2** [1983] 1309/11; 1468/70).
[35] Brown, H. C.; Bhat, N. G.; Basavaiah, D. (Israel J. Chem. **24** [1984] 72/5).
[36] Brown, H. C.; Bhat, N. G.; Basavaiah, D. (Synthesis **1983** 885/6).
[37] Brown, H. C.; Basavaiah, D.; Racherla, U.S. (Synthesis **1983** 886/8).
[38] Brown, H. C.; Cha, J. S.; Nazer, B.; Yoon, N. M. (J. Am. Chem. Soc. **106** [1984] 8001/2).
[39] Nöth, H.; Staudigl, R.; Taeger, T. (Chem. Ber. **114** [1981] 1157/75).
[40] Nöth, H.; Staudigl, R. (Chem. Ber. **115** [1982] 3011/24).

[41] Nöth, H.; Staudigl, R.; Brückner, R. (Chem. Ber. **114** [1981] 1871/83).
[42] Nöth, H.; Staudigl, R. (Chem. Ber. **115** [1982] 813/7; 1555/67).
[43] Meller, A.; Habben, C. (Monatsh. Chem. **113** [1982] 139/53).
[44] Schmidt, M.; Sametschek, E. (Z. Naturforsch. **36b** [1981] 1444/50).
[45] Davidson, G.; Ewer, K. P. (J. Mol. Struct. **74** [1981] 181/91).
[46] Williams, D. R.; Sakdarat, S. (Tetrahedron Letters **24** [1983] 3965/8).
[47] Nöth, H.; Schwerthöffer, R. (Chem. Ber. **114** [1981] 3056/62).
[48] Nöth, H.; Staudigl, R.; Storch, W. (Chem. Ber. **114** [1981] 3024/43).
[49] Biffar, W.; Nöth, H.; Schwerthöffer, R. (Liebigs Ann. Chem. **1981** 2067/80).
[50] Habben, C.; Maringgele, W.; Meller, A. (Z. Naturforsch. **37b** [1982] 43/53).

[51] Gronowitz, S.; Anders, I. (Chem. Scr. **15** [1980] 135/44).
[52] Habben, C.; Meller, A. (Chem. Ber. **117** [1984] 2531/4).

[53] Ohkubo, K. (Mem. Fac. Eng. Kumamoto Univ. **25** [1980] 1/20 from C.A. **94** [1981] No. 156009).
[54] Seitz, G.; Polier, S. (Arch. Pharm. **315** [1982] 169/74).
[55] Crivello, J. V.; Lam, J. H. (U.S. 4238619 [1977/80] 1/6 from C.A. **94** [1981] No. 140452).
[56] Hitachi Chemical Co., Ltd. (Japan. Kokai Tokkyo Koho 82-30722 [1980/82] 1/6 from C.A. **97** [1982] No. 111396).
[57] Shchepina, N. E.; Sukhov, S. G.; Nefedov, V. D.; Toropova, M. A. (Sint. Osn. Magnii Tsinkorg. Soedin **1980** 113/6 from C.A. **95** [1981] No. 143586).
[58] Shchepina, N. E.; Toropova, M. A.; Nefedov, V. D.; Avrorin, V. V.; Sinotova, E. N.; Zhuravlev, V. E. (U.S.S.R. 860443 [1980/82] from C.A. **97** [1982] No. 127270).
[59] Nguyen, M.-T.; Ruelle, P. (J. Chem. Soc. Faraday Trans. II **80** [1984] 1225/34).
[60] Jacox, M. E. (J. Phys. Chem. Ref. Data **13** [1984] 945/1068).

[61] Krebs, B. (Angew. Chem. **95** [1983] 113/34; Angew. Chem. Intern. Ed. Engl. **22** [1983] 113/34).
[62] Forster, A. M.; Downs, A. J. (J. Chem. Soc. Dalton Trans. **1984** 2827/34).

9.7.6 With Oxygen

The adducts $\mathbf{(CH_3)_2S(O){-}BF_3}$ and $\mathbf{(CD_3)_2S(O){-}BF_3}$ have been prepared; ^{1}H and ^{19}F NMR spectral data reveal that the two CH_3 groups are not equivalent [1].

2-Chloro-5-methoxy-1,3,2-benzooxathiaborole has been prepared from 2-mercapto-4-methoxyphenol and BCl_3 in CH_2Cl_2 at −78°C and subsequent warming to 25°C. The colorless crystals were obtained in 95% yield; $\delta^{11}B = 42.1$ ppm [2].

The application of $\mathbf{B(OSO_2CF_3)_3}$ in aldol reactions has been reviewed [4]. – The species $R_2B(OSO_2CF_3)$ ($R = n\text{-}C_4H_9$) is a reagent for stereocontrolled aldol syntheses [5], or regioselective aldol condensations through a vinyloxyborane [6]. It reacts with acetonitriles [7]; the compound with $R = cyclo\text{-}C_5H_9$ also reacts stereoselectively [8]. The compound with $R = n\text{-}C_3H_7$ is a catalyst for the preparation of macrolides from trimethylsilanyl-ω-trimethylsilanyloxycarboxylates [10].

The adducts $RS(F){=}O{-}BF_3$ ($R = C_6H_5$ or C_6H_4-4-F) have been prepared from RSOF and BF_3; $\delta^{13}C$ NMR spectra reveal O-bridging [3].

The conjugate superacid $\mathbf{[CF_3SO_2OH_2]^+[B(OSO_2CF_3)_4]^-}$ has been prepared from BCl_3 or BBr_3 with excess CF_3SO_2OH; the compound was found to O-protonate ketones [9].

References for 9.7.6:

[1] Jurga, S.; Depireux, J. (J. Magn. Resonance **55** [1983] 12/23).
[2] Andrä, K.; Straube, A. (Z. Anorg. Allgem. Chem. **490** [1982] 191/6).
[3] Ruppert, I. (J. Fluoride Chem. **20** [1982] 79/84).
[4] Mukaiyama, T.; Iwasawa, N. (Kagaku [Kyoto] **37** [1982] 853/5 from C.A. **98** [1983] No. 71421).
[5] Kuwajima, I.; Kato, M.; Mori, A. (Tetrahedron Letters **21** [1980] 4291/4).
[6] Deniff, P.; Macleod, I.; Whitry, D. A. (J. Chem. Soc. Perkin Trans. **1** [1981] 82/7).
[7] Hamana, H.; Sugasawa, T. (Chem. Letters **1982** 1401/4).
[8] Choy, W.; Ma, P.; Masamune, S. (Tetrahedron Letters **22** [1981] 3555/6).
[9] Olah, G. A.; Laali, K.; Farooq, O. (J. Org. Chem. **49** [1984] 4591/4).
[10] Taniguchi, N.; Kinoshita, H.; Inomata, K.; Kotake, H. (Chem. Letters **1984** 1347/8).

9.7.7 With Nitrogen

9.7.7.1 B–S–N Heterocycles

51

3,5-Dihalo-4-organyl-1,2-dithia-4-aza-3,5-diborolanes, **51**, had been prepared from $S_2B_2(CH_3)_2NR$ and BX_3 (X = Br, Cl); the reaction follows an endocyclic mechanism [1].

51a: X = Br, R = CH_3: The reaction mechanism with $^{10}BBr_3$ has been studied [1].

51b: X = Br, R = C_6H_5: b.p. 134°C/0.01 Torr, red oil, prepared from $S_3B_2Br_2$ and N,C-diphenylazomethine, yield 46%; δ^1H = 6.89 to 7.57 ppm (broad); $\delta^{11}B$ = 43.8 ppm; M/z = 337 [13].

51c: X = Br, R = C_6F_5: b.p. 82°C/0.01 Torr, preparation from **51a** and C_6F_5NSO; $\delta^{11}B$ = 57.5 ppm; $\delta^{19}F$ = ≈ −144, ≈ −151, −163.4 ppm (multiplets); M/z = 427 [2].

51d: X = Br, R = C_6H_4-3-CF_3: b.p. 118°C/0.01 Torr, red oil, from $S_3B_2Br_2$ and N-3-trifluoromethylphenyl-C-phenylazomethine, yield 35%; δ^1H = 7.08 to 7.63 ppm (broad); $\delta^{11}B$ = 44.0 ppm; M/z = 405 [16].

51e: X = $N(C_2H_5)_2$, R = C_6F_5: b.p. 126°C/0.01 Torr, from **51c** and $(C_2H_5)_2NH$, yellow oil; δ^1H = 0.95 (t, 12H), 2.92 (q, 8H, CH_2) ppm; $\delta^{11}B$ = 45.9 ppm [2].

52

Compound **52** has been prepared from the methylated thiadiazadiborolidine and BBr_3. The reaction follows the endocyclic reaction mechanism; the ring hydrazino group exchanges rapidly with a ring disulfide group [1]. For the reaction with $^{10}BBr_3$, see [3].

53

Compound **53** has been prepared from the methylated trithiadiborolidine and $F_3CC_6H_4NSO$; it is a colorless oil, b.p. 97°C/0.01 Torr; $\delta^{11}B$ = 48.8, 63.9 ppm; δ^1H = 0.52 (s, 3H), 1.08 (s, 3H), 7.23 to 7.77 (br, 4H) ppm; $\delta^{19}F$ = − 60.8 (s) ppm; M/z = 275 [2].

54

54 is a by-product in the reaction of 2,3,4,5-tetramethyl-1-thia-3,4-diaza-2,5-diborolidine with BBr_3; $\delta^{11}B$ = 43.8 and 1.1 ppm [3].

References for 9.7.7 on pp. 145/6

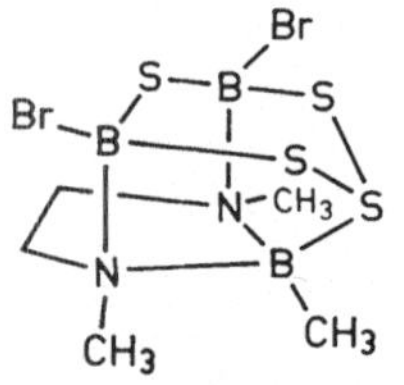

55

55, $S_4B_2Br_2$–$CH_3B(-NCH_3-C_2H_4-NCH_3-)$, has been prepared as a well-defined crystalline (see **Fig. 9-6**) adduct of 4,6-dibromo-1,2,3,5-tetrathia-4,6-diborolane and 1,2,3-trimethyl-1,3-diaza-2-borolidine in low yield using aged and somewhat impure 3,5-dibromo-1,2,4-trithia-3,5-diborolane and 1,2,3-trimethyl-1,3-diaza-2-borolidine, m.p. 112 to 114°C. $\delta^{11}B$ (in CH_2Cl_2) = 15(1B), 8(2B) ppm; the crystal structure (monoclinic space group $P2_1/c$-C^5_{2h} (No. 14)) shows B–S distances corresponding to bond lengths of four-coordinate boron [4].

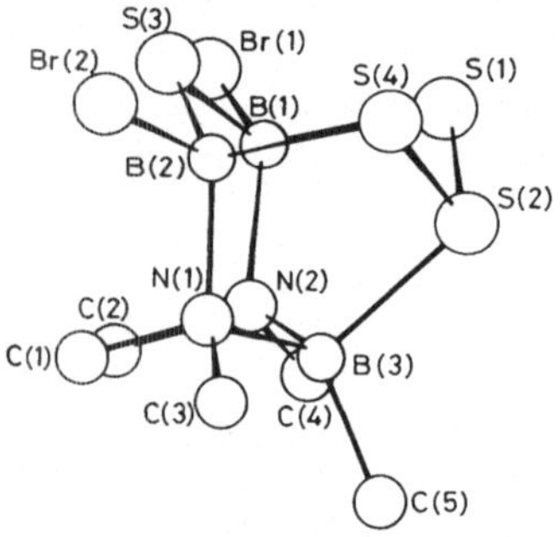

Fig. 9-6. Structure of compound **55** (H atoms omitted) [4].

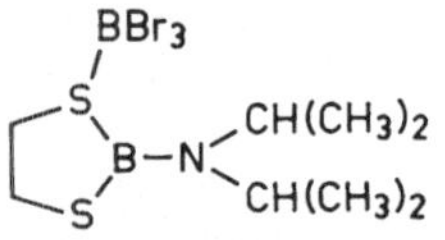

56

Compound **56**, the adduct 2-(di-*i*-propylamino)-1,3-dithia-2-borolanbortribromide, has been prepared from the two components in *n*-pentane at −70°C (stable only to −30°C) in 90% yield. δ^1H = 0.77 (CH_3), 2.64 (CH_2), 3.31 (CH) ppm; $\delta^{11}B$ = 41.6 and −14.5 ppm; $\delta^{13}C$ = 21.8 (CH_3), 38.7 (CH_2), 53.1 (CH) ppm [3].

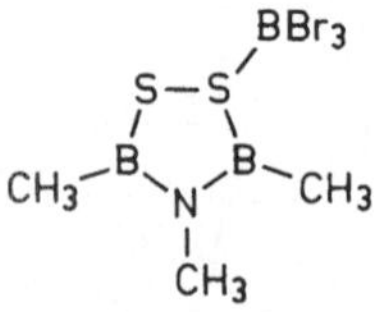

57

Compound **57**, the adduct 3,4,5-trimethyl-1,2-dithia-4-aza-3,5-diborolidinbortribromide, has been prepared from the two components in *n*-pentane (stable only to −20°C) in 77% yield [3].

9.7.7.2 Heterocycles Containing No Annular Sulfur

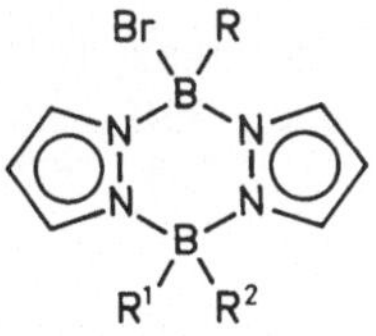

58

The following pyrazaboles of type **58** have been investigated [5]:

58a: $R = R^1 = R^2 = SCH_3$, prepared from $(CH_3S)_2B(\mu\text{-pz})_2B(SCH_3)_2$ (Hpz = pyrazole) and BBr_3 in CH_2Cl_2, m.p. 208 to 210°C, 30% yield; $\delta^{11}B = -2.17$ and -7.16 ppm.

58b: R = Br, $R^1R^2 = SCH_2CH_2S$; $\delta^{11}B = 8.0$ and -6.7 ppm.

In these pyrazaboles, electrophilically induced substitution reactions at the boron atoms using $^{10}BBr_3$ proceed without exchange of the boron atoms; all three substituents R may be exchanged with Br [5].

9.7.7.3 Compounds without a B–S Bond

Various aminosulfonium tetrafluoroborates, $[SRR^1R^2][BF_4]$, have been prepared from $[R^2_2N^+SO][BF_4^-]$ and chiral sulfoxides, $RR^1S{=}O$, the reaction proceeding with retention and/or racemization at S, depending on the structure of the chiral sulfoxides [6 to 13, 16, 17].

$[S_4N_4H]^+[BF_4]^-$ has been prepared from S_4N_4 and $H[BF_4]\text{–}O(C_2H_5)_2$ in CH_2Cl_2; the crystal structure has been determined [14].

$[(CH_3)_3N\text{–}BH_2CH_2S^+(CH_3)_2]I^-$ has been obtained from $(CH_3)_3N\text{–}BH_3$, N,N,N',N'-tetramethylethylenediamine·$LiCH_2SCH_3$, and CH_3I; m.p. 134 to 135°C; δ^1H (solution in $CDCl_3$) = 1.80 (t, BH_2), 2.75 (s, N–CH_3), 2.84 (t, BCH_2), 3.12 (s, SCH_3) ppm; $\delta^{11}B = -7.2$ ppm; $\delta^{13}C = 30.9$ (BCH_2), 28.1 (SCH_3), 53.0 (NCH_3) ppm; for reactions with pyrocatechol and CH_3I, see [15].

References for 9.7.7:

[1] Nöth, H.; Staudigl, R. (Chem. Ber. **115** [1982] 1555/67).
[2] Meller, A.; Habben, C. (Monatsh. Chem. **113** [1982] 139/53).
[3] Nöth, H.; Staudigl, R. (Chem. Ber. **115** [1982] 3011/24).
[4] Nöth, H.; Staudigl, R. (Chem. Ber. **115** [1982] 813/7).
[5] Niedenzu, K.; Nöth, H. (Chem. Ber. **116** [1983] 1132/53).
[6] Schwöbel, A.; Pérez, M. A.; Rössert, M.; Kresze, G. (Liebigs Ann. Chem. **1982** 723/8).
[7] Schwöbel, A.; Kresze, G.; Pérez, M. A. (Tetrahedron Letters **23** [1982] 1243/6).
[8] Pérez, M. A.; Rössert, M.; Kresze, G. (Liebigs Ann. Chem. **1981** 65/9).
[9] Dralowicz, J.; Bujnicki, B.; Mikolajczyk, M. (J. Org. Chem. **46** [1981] 2788/90).
[10] Marel, G.; Le Moing-Orliac, M. A.; Khamsitthideh, S.; Foucoud, A. (Tetrahedron **38** [1982] 527/37).

[11] Braun, H.; Amann, A.; Barnert, R.; Firl, J. (Z. Naturforsch. **35b** [1980] 1398/405); Firl, J.; Braun, H.; Amann, A.; Barnert, R. (Z. Naturforsch. **35b** [1980] 1406/14).
[12] Böhme, H.; Ahrens, G.; Krack, W. (Liebigs Ann. Chem. **1982** 585/94).
[13] Böhme, H.; Ahrens, G. (Liebigs Ann. Chem. **1982** 1030/8).

[14] Cordes, A. W.; Marcellus, C. G.; Noble, M. C.; Oakley, R. T.; Pennington, W. T. (J. Am. Chem. Soc. **105** [1983] 6008/12).
[15] Nöth, H.; Sedlak, D. (Chem. Ber. **116** [1983] 1479/86).
[16] Habben, C.; Meller, A. (Z. Naturforsch. **39b** [1984] 1022/6).
[17] Böhme, H.; Ahrens, G. (Chemiker-Ztg. **106** [1982] 186/7).

10 The System Boron-Selenium

Gert Heller
Institut für Anorganische und Analytische Chemie, Freie Universität Berlin
Berlin, Federal Republic of Germany

For the literature through 1973, see "Borverbindungen" 3, 1975, pp. 75/91; literature from 1974 through 1977, see "Boron Compounds" 1st Suppl. Vol. 3, 1981, pp. 92/101; literature from 1978 through 1980, see "Boron Compounds" 2nd Suppl. Vol. 2, 1982, pp. 217/20 (selenapolyboranes: "Boron Compounds" 2nd Suppl. Vol. 1, 1983, Chapter 2; selenium-containing carboranes: "Boron Compounds" 2nd Suppl. Vol. 2, 1982, Chapter 12).

10.1 Boron Selenides

From the hyperbolic dependence of the dissociation energy E_D of similar molecules with isostructural atoms on their atomic nuclear charges, E_D has been calculated to be 119 kcal/mol for **BSe**; an earlier published value is 109 ± 4 kcal/mol [1].

$(BSe_2)_n$ has been prepared analogously to the preparation of $(BS_2)_2$ (see p. 106). It is isotypical with $(BS_2)_n$ containing annular Se–Se distances of 234.7(4) pm and B–Se distances between 190(3) and 198(3) pm. The Se–Se distance between neighboring B_2Se_3 rings is only 329.5 pm. $(BSe_2)_n$ can be transported by I_2 in a temperature gradient from 300 to 200°C [15, 16].

10.2 Tris(organylseleno)boranes

$B(SeCH_3)_3$ and **$B(SeC_6H_5)_3$** react with α,β-disubstituted epoxides selectively to give β-hydroxyselenides [2].

10.3 Selenonium Tetrafluoroborates

$[(C_6H_5)_3Se][BF_4]$ is a catalyst for the oxidation of hydrocarbons with O_2. UV spectral data provide evidence for interaction between the small d-orbitals of the onium compound and the $(1\pi_g)$ orbital of the O_2; the activated O_2 abstracts hydrogen from the hydrocarbons [3]. A stable solution mixture of $[(C_6H_5)_3Se][BF_4]$ and propylene carbonate has been used as a photohardening catalyst for epoxy resins [4]. The R_f value of $[(C_6H_5)_3Se][BF_4]$ for the thin-layer chromatography on Silufol UV-254 has been determined with a $C_6H_6/C_2H_5OCOC_2H_5/HCOOH$ phase [5]. Tritium-labelled $[(C_6H_5)_2Se(C_6T_5)][BF_4]$ has been prepared by treating C_6T_6 with $(C_6H_5)_2Se$ and KBF_4 between −196 and +25°C for 10 to 15 d [7].

$[(C_6H_4\text{-}4\text{-}OCH_3)_3Se][BF_4]$ has been prepared by treating 2,4,6-tri-4-methoxyphenyl-1-selenocyclohexane-2,6-diselenol with $(C_2H_5)_2O\text{-}BF_3$ in C_6H_6 [6].

$[(C_6H_5Se)(CH_3)_2S][BF_4]$ has been prepared in CH_2Cl_2 at 0°C from C_6H_5SeBr, $(CH_3)_2S$, and $Ag[BF_4]$; m.p. 81 to 83°C; NMR: δ^1H (in CD_3CN) = 2.92 (s, 6H), 7.5 to 8.0 (m, 5H) ppm; IR: ν = 3040 to 680 cm^{-1}. The compound is a highly reactive selenating agent for electron-rich aromatic compounds [8].

$[(C_6H_5CH_2)(CH_3)(C_6H_5)Se][BF_4]$, m.p. 92.5 to 94°C, has been prepared from $C_6H_5CH_2SeC_6H_5$ and $[(CH_3)_3O][BF_4]$ in acetonitrile. Reaction of the compound with $NaNH_2$ in liquid NH_3 gives rearrangement products such as 2-methylbenzylphenyl selenide [9].

References for 10 on pp. 148/9

Additional compounds prepared in a manner similar to that described above are surveyed in Table 10/1.

Table 10/1
Tris(organyl)selenonium Tetrafluoroborates, $[RR^1R^2Se][BF_4]$ [9].

R	R^1	R^2	m.p. in °C	δ^1H in ppm (solution in $CDCl_3$) s, 3H	s, 3H	q, 2H	s, 4H	s, 5H
CH_3	CH_3	$CH_2C_6H_5$	90 to 91.5	—	—	—	—	—
CH_3	CH_3	$CH_2(C_6H_4\text{-}4\text{-}Cl)$	121 to 124	—	—	—	—	—
CH_3	CH_3	$CH_2(C_6H_4\text{-}3\text{-}Cl)$	130 to 132	—	—	—	—	—
CH_3	C_6H_5	$CH_2C_6H_5$	92.5 to 94	—	3.02	4.88	7.17	7.51
CH_3	C_6H_5	$CH_2(C_6H_4\text{-}4\text{-}CH_3)$	101 to 103	2.26	2.98	4.86	6.96	7.50
CH_3	C_6H_5	$CH_2(C_6H_4\text{-}4\text{-}Cl)$	102.5 to 104	—	3.06	4.85	7.06	7.51
CH_3	C_6H_5	$CH_2(C_6H_4\text{-}2\text{-}Cl)$	110 to 112	—	3.19	4.96	7.18	7.48

Rate constants had been determined for the reduction of $[(C_6H_5)(CF_3)CH_3Se][BF_4]$ by stable radicals and anionic radicals [10].

$[(C_6H_5)_2SeY(COR)][BF_4]$ or the corresponding tetraphenylborates (with $Y=Se(C_6H_5)_2$, $S(CH_3)_2$, or $P(C_6H_5)_3$; $R=C_6H_5$ or OC_2H_5) have been prepared and their ^{77}Se NMR chemical shifts have been reported, e.g., for $\{[(C_6H_5)_2Se]_2C(COC_6H_5)\}[B(C_6H_5)_4]$ [11].

Selenopyrylium tetrafluoroborate is a component of an insulating binder for transparent photoconductive reflex exposure [12]. The following NMR data have been reported: $\delta^1H=10.98$ (H2, H6), 8.77 (H3, H5), and 9.03 (H4) ppm; $\delta^{13}C=170.73$ (C2, C6), 137.30 (C3, C5), 149.47 (C4) ppm; $\delta^{77}Se=975.7$ ppm; $^2J(Se,H2)=45.25$, $^3J(Se,H3)=6.16$, $^4J(Se,H4)=-2.36$, $^1J(Se,C2)=-155.4$, $^2J(Se,C3)=-13.3$, and $^3J(Se,C4)=22.9$ Hz [13].

2-Ethylthio-4,5-benzo-1-thia-3-selenolium(1+) tetrafluoroborates of type **1** have been prepared from 4,5-benzo-1-thia-3-selenol-2-thiones and $[(C_2H_5)_3SO][BF_4]$ [14].

1

References for 10:

[1] Cherkesov, A. I. (Zh. Neorgan. Khim. **26** [1981] 3181/5; Russ. J. Inorg. Chem. **26** [1981/82] 1703/5).

[2] Cravador, A.; Krief, A. (Tetrahedron Letters **22** [1981] 2491/4).

[3] Ohkubo, K. (Mem. Fac. Eng. Kumamoto Univ. **25** [1980] 1/20 from C.A. **94** [1981] No. 156009).

[4] Crivello, J. V. (Fr. Demande 2459265 [1979/81] 1/18 from C.A. **95** [1981] No. 82046).

[5] Shchepina, N. E.; Sukhov, S. G.; Nefedov, V. D.; Toropova, M. A. (Sint. Osn. Magnii Tsinkorg. Soedin. **1980** 113/6 from C.A. **95** [1981] No. 143586).

[6] Kharchenko, V. G.; Drevko, B. I. (U.S.S.R. 1051089 [1981/83] from C.A. **100** [1984] No. 120910).

[7] Shchepina, N. E.; Toropova, M. A.; Nefedov, V. D.; Avrorin, V. V.; Sinotova, E. N.; Zhuravlov, V. E. (U.S.S.R. 860443 [1980/82] from C.A. **97** [1982] No. 127270).

[8] Gassman, P. G.; Miura, A.; Miura, T. (J. Org. Chem. **47** [1982] 951/4).
[9] Gassman, P. G.; Miura, T.; Mossman, A. (J. Org. Chem. **47** [1982] 954/9).
[10] Bogillo, V. I.; Levit, A. F.; Kondratenko, N. V.; Maletina, I. I.; Gragerov, I. P. (Teor. Eksperim. Khim. **19** [1983] 508/12; Theor. Exptl. Chem. [USSR] **19** [1983] 472/6 from C.A. **100** [1984] No. 5581).
[11] Kushnarev, D. F.; Kalabin, G. A.; Kyandzhetsian, R. A.; Magdesieva, N. N. (Zh. Org. Khim. **18** [1982] 119/24; J. Org. Chem. [USSR] **18** [1982] 103/7).
[12] Grammatica, S. J. (U.S. 4314012 [1980/82] 1/18 from C.A. **96** [1982] No. 152820).
[13] Sándor, P.; Radics, L. (Org. Magn. Resonance **16** [1981] 148/55).
[14] Poleschner, H.; Böttger, J.; Fanghänel, E. (Synthesis **1984** 667/79).
[15] Krebs, B.; Hürter, H.-U. (Acta Cryst. **A 37** [1981] C163).
[16] Krebs, B. (Angew. Chem. **95** [1983] 113/34).

11 The System Boron-Tellurium

Gert Heller
Institut für Anorganische und Analytische Chemie, Freie Universität Berlin
Berlin, Federal Republic of Germany

For the literature through 1973, see "Borverbindungen" 3, 1975, p. 92; literature from 1974 through 1977, see "Boron Compounds" 1st Suppl. Vol. 3, 1981, pp. 102/4; literature from 1978 through 1980, see "Boron Compounds" 2nd Suppl. Vol. 2, 1982, pp. 221/2.

11.1 Boron Telluride, BTe

From the hyperbolic dependence of the dissociation energy E_D of similar molecules with isostructural atoms on their atomic nuclear charges, E_D has been calculated to be 115 kcal/mol for **BTe**; an earlier published value is 84 ± 5 kcal/mol [1]. However, the actual formation of binary B–Te compounds has not yet been accomplished [18].

11.2 Tris(pentafluorotelluro)borate and Derivatives

$B(OTeF_5)_3$ crystals are hexagonal, space group $P6_3/m$-C_{6h}^2 (No. 176), with a = 921.8(2) and c = 920.7(2) pm; calculated density 3.56 g/cm³ for Z = 2; R = 3.1% for 566 observed reflections. The molecule has D_{3h} symmetry; the environment of boron is strictly planar. Average values for the distances are: B–O 135.8(6), Te–O 187.4(6), and Te–F 181.6(5) pm; the B–O–Te angle is 132.3(4)° [2].

The direct fluorination of $B(OTeF_5)_3$ provides for an improved synthesis of TeF_5OF (see "Tellur" Erg.-Bd. B 2, 1977, p. 60) [3] and leads to the new compound $F_2Te(OTeF_5)_2$ [17]. Reactions of $B(OTeF_5)_3$ with graphite to give an intercalation compound [4], with XeF_4 to yield $Xe(OTeF_5)_4$ [5], with SbF_5 to form $SbF_4(OTeF_5)$ and $SbF_3(OTeF_5)_2$ [6], with MoF_6 to give $MoF_2(OTeF_5)_4$, $MoF_3(OTeF_5)_3$, $MoOF_3(OTeF_5)$, or $MoOF_2(OTeF_5)_2$ [7], with $CsOTeF_5$ to form $Cs[B(OTeF_5)_4]$, and with CH_3CN to yield a 1:1 molar adduct [8] have been described.

B(*cis*-$OTeF_4OCH_3)_3$, m.p. −6°C, has been prepared from *cis*-$HOTeF_4OCH_3$ and BCl_3; NMR data: $\delta^1H = 4.65$ ppm, $\delta^{11}B = -2.17$ (in $CFCl_3$) and −21.3 (in CH_3CN) ppm; in addition, ^{19}F and ^{125}Te NMR as well as IR data are available. **B(*trans*-$OTeF_4OCH_3)_3$** has a m.p. 22°C, $\delta^1H =$ 4.48 ppm, $\delta^{11}B = -4.75$ (in $CFCl_3$) and −22.2 ppm (in CH_3CN) [9].

11.3 Telluronium Tetrafluoroborates

$[(C_6H_5)_3Te][BF_4]$ is a catalyst for the oxidation of hydrocarbons such as cumene or α-pinene with O_2. UV spectral data provide evidence for interaction between the small d-orbitals of the onium compounds and the (1πg) orbital of the O_2; the activated O_2 abstracts hydrogen from the hydrocarbons [10]. The R_f value of $[(C_6H_5)_3Te][BF_4]$ has been determined in a $C_6H_6/C_2H_5OCOOC_2H_5/HCOOH$ phase [11]. Tritium-labelled $[(C_6H_5)_2Te(C_6T_5)][BF_4]$ has been prepared by treating C_6T_6 with $(C_6H_5)_2Te$ and $K[BF_4]$ between −196 and 25°C for 10 to 15 d [12]. Rate constants had been determined for the reduction of $[(C_6H_5)(C_3F_7)(CH_3)Te][BF_4]$ by stable radicals and anionic radicals [13].

The following 1-organyl-3,4-benzocyclopentane-1-telluronium tetrafluoroborates of type **1** have been prepared.

$[BF_4]^-$

1

1a: R = CH_3, m.p. 208°C (dec.); ^1H NMR spectra; mass spectrum; conductivity in DMSO-d_6.

1b: R = C_6H_5, m.p. 184°C (dec.); IR: $\nu(Te\text{-}C_6H_5) = 535\ cm^{-1}$; mass spectrum; conductivity in DMF [14].

Telluropyrylium tetrafluoroborates has been synthesized for dye sensitizers in electron-donating photoconductive films [15].

Unstable salt solutions of species of type **2** have been prepared from 1,3-ditelluracyclopentenes and $[(CH_3)_3O]^+[BF_4]^-$ [16].

$[BF_4]^-$

2

2a: R = H: $\delta^1H = 13.83(H_A)$ and $10.31(H_B)$ ppm; J = 6.9 Hz.

2b: R = CH_3: $\delta^1H = 13.40(H_A)$ and $10.20(H_B)$ ppm; J = 6.6 Hz.

2c: R = i-C_3H_7: $\delta^1H = 13.25(H_A)$ and $10.15(H_B)$ ppm.

2d: R = C_6H_5: $\delta^1H = 13.30(H_A)$ and $10.61(H_B)$ ppm; J = 6.9 Hz [16].

References for 11:

[1] Cherkesov, A. I. (Zh. Neorgan. Khim. **26** [1981] 3181/5; Russ. J. Inorg. Chem. **26** [1981/82] 1703/5).

[2] Sawyer, J. F.; Schrobingen, G. J. (Acta Cryst. B **38** [1982] 1561/3).

[3] Schack, C. J.; Christe, K. O. (Inorg. Chem. **23** [1984] 2922).

[4] Stumpp, E.; Kebschull, R. (Extr. Abstr. Program. 15th Bienn. Conf. Carbon, Clausthal-Zellerfeld, FRG, 1981, pp. 389/90 from C.A. **95** [1981] No. 214402).

[5] Jacob, E.; Lentz, D.; Seppelt, K.; Simon, A. (Z. Anorg. Allgem. Chem. **472** [1981] 7/25).

[6] Leitzke, O.; Sladky, F. (Z. Naturforsch. **36b** [1981] 268/9).

[7] Schröder, K.; Sladky, F. (Z. Anorg. Allgem. Chem. **477** [1981] 95/100).

[8] Knopshofer, H.; Leitzke, O.; Peringer, P.; Sladky, F. (Chem. Ber. **114** [1981] 2644/8).

[9] Toetsch, W.; Aichinger, H.; Sladky, F. (Z. Naturforsch. **38b** [1983] 332/4).

[10] Ohkubo, K. (Mem. Fac. Eng. Kumamoto Univ. **25** [1980] 1/20 from C.A. **94** [1981] No. 156009).

[11] Shchepina, N. E.; Sukhov, S. G.; Nefedov, V. D.; Toropova, M. A. (Sint. Osn. Magnii Tsinkorg. Soedin. **1980** 113/6 from C.A. **95** [1981] No. 143586).

[12] Shchepina, N. E.; Toropova, M. A.; Nefedov, V. D.; Avrorin, V. V.; Sinotova, E. N.; Zhuravlov, V. E. (U.S.S.R. 860443 [1980/82] from C.A. **97** [1982] No. 127270).

[13] Bogillo, V. I.; Levit, A. F.; Kondratenko, N. V.; Maletina, I. I.; Gragerov, I. P. (Teor. Eksperim. Khim. **19** [1983] 508/12; Theor. Exptl. Chem. [USSR] **19** [1983] 472/6 from C.A. **100** [1984] No. 5581).

[14] Al-Rubaie, A. Z.; McWhinnie, W. R.; Granger, P.; Chapelle, S. (J. Organometal. Chem. **234** [1982] 287/98).

[15] Detty, M. R.; Murray, B. J. (U.S. 4365017 [1981/82] 1/16 from C.A. **98** [1983] No. 91042).

[16] Bender, S. L.; Detty, M. R.; Haley, N. F. (Tetrahedron Letters **23** [1982] 1531/4).

[17] Damerius, R.; Huppmann, P.; Lentz, D.; Seppelt, K. (J. Chem. Soc. Dalton Trans. **1984** 2821/6).

[18] Krebs, B. (Angew. Chem. **95** [1983] 113/34).

12 The System Boron-Polonium

Gert Heller
Institut für Anorganische und Analytische Chemie, Freie Universität Berlin
Berlin, Federal Republic of Germany

No distinct boron-polonium derivative has yet been prepared. However, the dissociation energy E_D has been calculated for **BPo** from the hyperbolic dependence of E_D of similar molecules with isostructural atoms on their atomic nuclear charges; a numerical value of 91 ± 19 kcal/mol was obtained [1].

Reference for 12:

[1] Cherkesov, A. I. (Zh. Neorgan. Khim. **26** [1981] 3181/5; Russ. J. Inorg. Chem. **26** [1981/82] 1703/5).

13 Carboranes

Thomas Onak

Department of Chemistry, California State University
Los Angeles, California, USA

13.1 Reviews and Nomenclature

For earlier work on carboranes see "Borverbindungen" 2, 1974, pp. 1/288; "Borverbindungen" 6, 1975, pp. 1/150; "Borverbindungen" 11, 1977, pp. 1/207; "Borverbindungen" 12, 1977, pp. 1/306; "Boron Compounds" 1st Suppl. Vol. 3, 1980, pp. 105/205; and "Boron Compounds" 2nd Suppl. Vol. 2, 1982, pp. 223/335. More recently, the general chemistry of the carboranes has been summarized in a number of accounts [1 to 4] and annual surveys [5 to 12]. Other reviews covering special aspects of carborane chemistry include: carborane-containing polymers [13 to 18], metallacarboranes [19 to 31], nomenclature rules [32 to 36], internuclear distances in carboranes [37], boron-11 NMR spectroscopy of carboranes [38], review of boron-substituted functional derivatives of $C_2B_{10}H_{12}$ [39], and carboranes viewed in terms of "hypercoordinate" carbon compounds [40].

closo-Carboranes of the type $C_2B_{n-2}H_n$ are discussed in the development of general structural relationships between polyboranes [41]. Various polyhedral arrangements for a 12-vertex system are considered, and the geometry of the $C_2B_{10}H_{12}$ isomers is intercorrelated with other 12-vertex molecules [42]. Additional reviews cover general synthetic methods for the carboranes [43, 44], structural comparisons between several carboranes and the higher borides [45], and carboranes with attached heterocyclic organic groups [46]. Abstracts of papers presented at a meeting and describing the use of carboranes for neutron capture therapy [47] have also appeared in the literature.

References for 13.1:

[1] Cavagnaro, D. M. (NTIS-PB-80-815558 [1980] 1/282; Govt. Rept. Announce. Index U.S. **80** No. 25 [1980] 5307 from C.A. **95** [1981] No. 5890).
[2] Matteson, D. S. (J. Organometal. Chem. **207** [1981] 13/65).
[3] Grimes, R. N. (Advan Inorg. Chem. Radiochem. **26** [1983] 55/117).
[4] Parry, R. W.; Kodama, G. (Boron Chemistry, Vol. 4, Pergamon, New York 1980).
[5] Leach, J. B. (Organometallic Chem. **9** [1981] 55/73).
[6] Leach, J. B. (Organometallic Chem. **10** [1982] 48/63).
[7] Leach, J. B. (Organometallic Chem. **11** [1983] 61/83).
[8] Morris, J. H. (Organometallic Chem. **12** [1984] 47/58).
[9] Hart, F. A.; Massey, A. G.; Harrison, P. G.; Holloway, J. H. (Ann. Rept. Progr. Chem. A **77** [1980/81] 3/156).
[10] Welch, A. J. (Ann. Rept. Progr. Chem. A **78** [1981/82] 19/73).

[11] Welch, A. J. (Ann. Rept. Progr. Chem. A **79** [1982/83] 19/89).
[12] Welch, A. J. (Ann. Rept. Progr. Chem. A **80** [1983/84] 19/59).
[13] Cavagnaro, D. M. (NTIS-PS-79-0950 [1979] 1/222; Govt. Rept. Announce. Index U.S. **79** No. 24 [1979] 168; C.A. **92** [1980] No. 147562).
[14] Cavagnaro, D. M. (NTIS-PS-80-814734 [1980] 1/254; Govt. Rept. Announce. Index U.S. **80** No. 24 [1980] 5109 from C.A. **94** [1981] No. 66789).
[15] Korshak, V. V.; Pavlova, S. A.; Gribkova, P. N.; Balykova, T. N. (Acta Polym. **32** [1981] 61/74 from C.A. **94** [1981] No. 140176).

[16] Peters, E. N. (Ind. Eng. Chem. Prod. Res. Develop. **23** [1984] 28/32).
[17] Bekasova, N. I. (Usp. Khim. **53** [1984] 107/34; C.A. **100** [1984] No. 86142).
[18] Allcock, H. R. (ACS Symp. Ser. No. 232 [1983] 49/67).
[19] Grimes, R. N. (in: Grimes, R. N., Metal Interactions with Boron Clusters, Plenum, New York 1982, pp. 269/319).
[20] Grimes, R. N. (Pure Appl. Chem. **54** [1982] 43/58).

[21] Grimes, R. N. (Accounts Chem. Res. **16** [1983] 22/6).
[22] Geiger, W. E. (in: Grimes, R. N., Metal. Interactions with Boron Clusters, Plenum, New York 1982, pp. 239/68).
[23] Bresadola, S., (in: Grimes, R. N., Metal Interactions with Boron Clusters, Plenum, New York 1982, pp. 173/237).
[24] Todd, L. J. (in: Grimes, R. N., Metal Interactions with Boron Clusters, Plenum, New York 1982, pp. 145/71).
[25] Hawthorne, M. F (Chem. Future Proc. 29th IUPAC Congr., Cologne, FRG, 1983 [1984], pp. 135/41; C.A. **101** [1984] No. 90996).
[26] Hawthorne, M. F. (Transition Metal Chem. Proc. Workshop, Bielefeld, FRG, 1980 [1981], pp. 299/306; C.A. **98** [1981] No. 89412).
[27] O'Neill, M. E.; Wade, K. (in: Grimes, R. N., Metal Interactions with Boron Clusters, Plenum, New York 1982, pp. 1/41).
[28] Greenwood, N. N. (ACS Symp. Ser. No. 232 [1983] 125/38).
[29] Siebert, W. (Transition Metal Chem. Proc. Workshop, Bielefeld, FRG, 1980 [1981], pp. 157/72).
[30] Bregadze, V. I. (Koord. Khim. **10** [1984] 666/72; C.A. **101** [1984] No. 171298).

[31] Kirillova, N. I.; Yanovskii, A. I.; Struchkov, Y. T. (Itogi Nauki Tekh. Ser. Kristallokhim. **15** [1981] 130/88; C.A. **96** [1982] No. 77628).
[32] Casey, J. B.; Evans, W. J.; Powell, W. H. (Inorg. Chem. **20** [1981] 1333/41).
[33] Casey, J. B.; Evans, W. J.; Powell, W. H. (Inorg. Chem. **20** [1981] 3556/61).
[34] Casey, J. B.; Evans, W. J.; Powell, W. H. (Inorg. Chem. **22** [1983] 2228/35).
[35] Casey, J. B.; Evans, W. J.; Powell, W. H. (Inorg. Chem. **22** [1983] 2236/45).
[36] Casey, J. B.; Evans, W. J.; Powell, W. H. (Inorg. Chem. **23** [1984] 4132/43).
[37] Mastryukov, V. S.; Dorofeeva, O. V.; Vilkov, L. B. (Usp. Khim. **49** [1980] 2377/88; Russ. Chem. Rev. **49** [1980] 1181/7).
[38] Siedle, A. R. (Ann. Rept. Nucl. Magn. Resonance Spectrosc. **12** [1982] 177/261).
[39] Kalinin, V. N. (Usp. Khim. **49** [1980] 2188/212; Russ. Chem. Rev. **49** [1980] 1084/96).
[40] Olah, G. A.; Prakash, G. K. S. (Chem. Brit. **19** [1983] 916/26).

[41] O'Neill, M. E.; Wade, K. (Polyhedron **3** [1984] 199/212).
[42] Favas, M. C.; Kepert, D. L. (Progr. Inorg. Chem. **28** [1981] 309/67).
[43] Grassberger, M.; Köster, R. (Houben-Weyl Methoden Org. Chem. 4th Ed. **13** Pt. 3c [1984] 92/214).
[44] Wrackmeyer, B.; Köster, R. (Houben-Weyl Methoden Org. Chem. 4th Ed. **13** Pt. 3c [1984] 377/611).
[45] Lipscomb, W. N. (J. Less-Common Metals **82** [1981] 1/20).
[46] Drygina, O. V.; Garnovskii, A. D. (Khim. Geterotsikl. Soedin. **1983** 579/91; C.A. **99** [1983] No. 105299).
[47] Fairchild, R. G.; Brownell, G. L. (Proc. 1st Intern. Symp. Neutron Capture Therapy, Cambridge, Mass., 1983, pp. 1/400; BNL-51730 [1983] 1/400; UC-48 [1983] 1/400; TIC-4500 [1983] 1/400).

13.2 Carboranes Containing One Boron Atom

For data on carboranes containing only one boron atom, see "Boron Compounds" 2nd Suppl. Vol. 2, 1982, p. 224. No new data on such species are available.

The reaction of tetrahydrofuran-borane, $C_4H_8O \cdot BH_3$, with the cobaltacycle $(\eta^5\text{-}C_5H_5)$-$[(C_6H_5)_3P]$ $CoC_4(C_6H_5)_4$ to give the *nido*-cobaltacarborane 3,4,5,6-$(C_6H_5)_4$-1,3,4,5,6-$(\eta^5\text{-}C_5H_5)$ CoC_4BH demonstrates a new route to carbon-rich metallacarboranes [1]. Reaction of $C_4H_8O \cdot BH_3$ with the alkylidyne-tungsten complex $W(\equiv CCH_3)(CO)_2(C_5H_5)$ yields $W_2[CH_3CB(H)C_2H_5]$-$(CO)_4(C_5H_5)_2$ with a CBW_2 tetrahedron containing a B-H-W bridging hydrogen [2]. Self-consistent charge MO calculations on the ferracarborane $C_4BH_5Fe(CO)_3$ predict that the most favorable structure of the compound should contain adjacent $Fe(CO)_3$ and BH units in the basal plane [3].

Various metal-carbon-boron containing "sandwich" compounds have been prepared in which the structures appear to obey carborane structure rules; these include: $(CO)_4V(C_5H_5BCH_3)$ [4], $(CO)_3Cr(C_5H_5BCH_3)$ [5], $(CO)_3Mn[(C_2H_5)C_4H_3BC_6H_5]Mn(CO)_3$ [6], $[(CO)_3Mn]_2(C_4H_4BC_6H_5)$ [7], derivatives of $(OC)_3Mn(C_5BH_6)$ [8], $(C_5H_5)Co[(C_6H_5)_4C_4BC_6H_5]$ [9], derivatives of $(C_4H_4)Co(C_5BH_6)$ [10], $(C_5H_5)Fe(2\text{-}CH_3\text{-}C_5H_4BC_6H_5)$ [6], $(C_6H_6)Ru(C_4H_4BC_6H_5)$ [7], $(CO)_3Ru(C_4H_4BC_6H_5)$ [7], and $(C_8H_{12})Pt[(C_6H_5)_4C_4BC_6H_5]$ [9].

References for 13.2:

[1] Palladino, D. B.; Fehlner, T. P. (Organometallics **2** [1983] 1692/3).

[2] Carriedo, G. A.; Elliot, G. P.; Howard, J. A. K.; Lewis, D. B.; Stone, F. G. A. (J. Chem. Soc. Chem. Commun. **1984** 1585/6).

[3] Brint, P.; Pelin, W. K.; Spalding, T. R. (J. Chem. Soc. Dalton Trans. **1981** 546/51).

[4] Herberich, G. E.; Boveleth, W.; Hessner, B.; Koch, W.; Raabe, E.; Schmitz, D. (J. Organometal. Chem. **265** [1984] 225/35).

[5] Herberich, G. E.; Soehnen, D. (J. Organometal. Chem. **254** [1983] 143/7).

[6] Herberich, G. E.; Hengesbach, J.; Huttner, G.; Frank, A.; Schubert, U. (J. Organometal. Chem. **246** [1983] 141/9).

[7] Herberich, G. E.; Hessner, B.; Boveleth, W.; Luethe, H.; Saive, R.; Zelenka, L. (Angew. Chem. **95** [1983] 1024/5; Angew. Chem. Intern. Ed. Engl. **22** [1983] 996).

[8] Herberich, G. E.; Hessner, B.; Kho, T. T. (J. Organometal. Chem. **197** [1980] 1/5).

[9] Herberich, G. E.; Buller, B.; Hessner, B.; Oschmann, W. (J. Organometal. Chem. **195** [1980] 253/9).

[10] Herberich, G. E.; Naithani, A. K. (J. Organometal. Chem. **241** [1983] 1/14).

13.3 Carboranes Containing Two Boron Atoms

For previous data on carboranes containing two boron atoms, see "Boron Compounds" 2nd Suppl. Vol. 2, 1982, pp. 224/6.

Molecular orbital calculations on $\mathbf{C_2B_2H_4}$ indicate that a tetrahedrane (*closo*-type of carborane) geometry for this molecule is less stable than the planar 1,3-diboretane structure [1]. NMR parameters of *nido*-hexaalkyl-2,3,4,5-tetracarbahexaboranes(6), ***nido*-$\mathbf{C_4B_2R_6}$**, have been determined by ^{11}B and ^{13}C NMR spectroscopy (Tables 13/1 to 13/3) [2].

References for 13.3 on pp. 157/8

Table 13/1
nido-2,3,4,5-C_4B_2-1-R^1-2-R^2-3-R^3-4-R^4-5-R^5-6-R^6 Compounds Studied by NMR Spectroscopy (see Tables 13/2 and 13/3) [2].

compound	R^1	R^2	R^3	R^4	R^5	R^6
I	CH_3	H	CH_3	CH_3	H	CH_3
II	CH_3	CH_3	H	CH_3	H	CH_3
III	CH_3	CH_3	CH_3	CH_3	CH_3	C_2H_5
IV	C_2H_5	CH_3	C_2H_5	C_2H_5	CH_3	C_2H_5
V	C_2H_5	CH_3	C_2H_5	CH_3	C_2H_5	C_2H_5
VI	C_2H_5	C_2H_5	C_2H_5	C_2H_5	C_2H_5	C_2H_5
VII	CH_3	CH_3	*i*-C_3H_7	*i*-C_3H_7	CH_3	*i*-C_3H_7
VIII	CH_3	CH_3	*i*-C_3H_7	CH_3	*i*-C_3H_7	*i*-C_3H_7

Table 13/2
Boron-11 NMR Data and Coupling Constants, $^1J(^{13}C\text{-}^{11}B)$ and $^1J(^{13}C\text{-}^1H)$, for *nido*-2,3,4,5-C_4B_2-1-R^1-2-R^2-3-R^3-4-R^4-5-R^5-6-R^6 [2].

compound	$\delta^{11}B$		$^1J(^{13}C\text{-}^{11}B)$			$^1J(^{13}C\text{-}^1H)$			
	B(1)	B(6)	B(1)R^1	B(6)R^6	B(6)$R^{2,5}$	R^1	$R^{2,5}$	$R^{3,5}$	R^6
I	−45.5	19.3	81.0	76.0	59.0	119.5	161.0	127.0	117.5
II	−46.2	19.3	—	—	—	—	—	—	—
III	−44.3	18.1	78.0	72.5	53.0	119.1	126.0	127.5	116.2 125.5
IV	−44.3	18.4	76.0	—	—	—	—	—	—
V	−44.3	18.4	76.0	—	—	—	—	—	—
VI	−44.4	18.6	72.0	—	—	117.0	126.0	126.0	116.0
VII	−44.9	19.9	—	—	—	—	—	—	—
VIII	−44.2	17.9	—	—	—	—	—	—	—

Table 13/3
Carbon-13 NMR Data for *nido*-2,3,4,5-C_4B_2-1-R^1-2-R^2-3-R^3-4-R^4-5-R^5-6-R^6 [2].

compound	C(2)	C(3)	C(4)	C(5)	R^1	R^2	R^3	R^4	R^5	R^6
I	87.3	109.2	109.2	87.3	−15.4	—	12.6	—	—	−5.9
II	—	102.5	—	—	—	14.5	—	11.7	—	—
III	93.5	106.3	106.3	93.5	−17.9	9.9	9.2	9.2	9.9	2.9 11.8
IV	93.2	111.1	111.1	93.2	−5.0 10.3	9.8	17.5 13.8	17.5 13.8	9.8	2.8 11.8
V	93.6	110.7	105.5	100.1	−5.0 10.3	9.8	17.5 13.0	8.9	18.6 15.3	2.8 12.2
VI	100.3	110.3	110.3	100.3	−4.3 10.5	18.6 16.0	17.5 14.4	17.5 14.4	18.6 16.0	3.2 12.4
VII	92.6	113.9	113.9	92.6	−15.0	—	—	—	—	12.3
VIII	93.5	114.5	104.3	103.3	−14.2	—	—	—	—	12.3

***nido*-2,3,4,5-$C_4B_2H_4$-1-CH_3-2-N(*i*-C_3H_7)$_2$**, **1**, a compound sensitive toward air and protic reagents, is prepared by treating the lithium salt of the borole dianion $Li_2[C_4H_4BN(i\text{-}C_3H_7)_2]$ with CH_3BBr_2. The ^1H NMR spectrum shows peaks at $\delta = 5.45$ ppm, J = 3.9 Hz, for H(3,4); = 3.37 ppm, J = 3.9 Hz, for H(2,5); = 3.50 ppm, J = 6.7 Hz, for CH of $CH(CH_3)_2$; = 1.16 ppm, J = 6.7 Hz, for CH_3 of $CH(CH_3)_2$; = −0.13 ppm for BCH_3; ^{11}B NMR: δ (in ppm) = +23 for BN; = −48 for BCH_3; ^{13}C NMR: δ (in ppm) = 98, J(C–H) = 180 Hz, for C(3,4); = 65 for C(2,5); = 47, J(C–H) = 132 Hz, for CH of $CH(CH_3)_2$; = 23, J(C–H) = 126 Hz, for CH_3 of $CH(CH_3)_2$. Upon treating $Li_2[C_4H_4BN(i\text{-}C_3H_7)_2]$ with $(i\text{-}C_3H_7)_2NBCl_2$ a bis-N(*i*-C_3H_7)$_2$ substituted C_4B_2 species is produced that does not have *nido*-cage geometry but instead is a fluxional 2,6-diborabicyclo[3.1.0]hex-3-ene derivative, **2** [3].

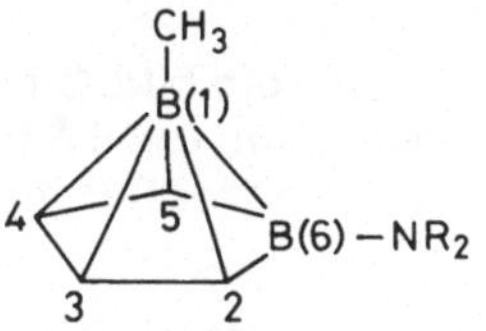

1

R = $CH(CH_3)_2$

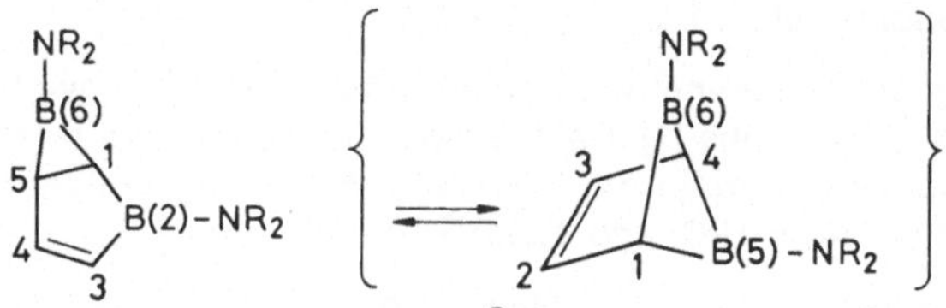

2

R = $CH(CH_3)_2$

The two-boron pentagonal pyramidal metallacarborane 1-(C_5H_5)-2,3-(C_2H_5)$_2$-4,6-(CH_3)$_2$-5-H-1,2,3,5-C_3CoB_2H, prepared from the reaction of $H_2(C_2H_5)_2C_3B_2(CH_3)_2$ with $(C_5H_5)Co(C_2H_4)_2$, has a melting point of 62°C [4]. Various other metal-carbon-boron-containing "sandwich" compounds have been prepared in which the structures appear to obey carborane structure rules; these include: nickel and iron complexes of $C_3B_2H_5$ derivatives [5, 6], derivatives of $(C_5H_5Co)_2(C_3B_2H_5)$ [4, 6, 7], $(C_5H_5)Co[C_4H_4B_2(CH_3)_2]$ [8], a number of compounds containing a C_4B_2M (M = Co, Rh, Ni, Pt) cage [9], $[(C_5H_5)_2FeBC_4H_4BFe(C_5H_5)_2]M(CO)_4$, M = Cr, Co, W [10], $[(C_5H_5)_2FeBC_4H_4BFe(C_5H_5)_2]M(CO)_3$, M = Fe, Ru, Os [10], $[(CH_3)_5C_5]Rh[C_4H_4B_2(CH_3)_2]$ [11], $[((CH_3)_5C_5)Rh]_2[C_4H_4B_2(CH_3)_2]^{2+}$ [11]. The photoelectron spectrum of $(C_5H_5)Co[(CH_3)BC_4H_4B(CH_3)]$ is reported [12].

References for 13.3:

[1] Krogh-Jespersen, K.; Cremer, D.; Dill, J. D.; Pople, J. A.; Schleyer, P. R. (J. Am. Chem. Soc. **103** [1981] 2589/94).

[2] Wrackmeyer, B. (Z. Naturforsch. **37b** [1982] 412/9).

[3] Herberich, G. E.; Ohst, H.; Mayer, H. (Angew. Chem. **96** [1984] 975; Angew. Chem. Intern. Ed. Engl. **23** [1984] 969/70).

[4] Siebert, W.; Edwin, J.; Pritzkow, H. (Angew. Chem. **94** [1982] 147; Angew. Chem. Intern. Ed. Engl. **21** [1982] 148/9).

[5] Kuhlmann, T.; Siebert, W. (Z. Naturforsch. **39b** [1984] 1046/9).

[6] Edwin, J.; Bochmann, M.; Boehm, M. C.; Brennan, D. E.; Geiger, W. E.; Krüger, C.; Pebler, J.; Pritzkow, H.; Siebert, W.; Swiridoff, W.; Wadepohl, H.; Weiss, J.; Zenneck, U. (J. Am. Chem. Soc. **105** [1983] 2582/98).

[7] Edwin, J.; Boehm, M. C.; Chester, N.; Hoffman, D. M.; Hoffmann, R.; Pritzkow, H.; Siebert, W.; Stumpf, K.; Wadepohl, H. (Organometallics **2** [1983] 1666/74).

[8] Herberich, G. E.; Hessner, B.; Beswethrick, S.; Howard, J. A. K.; Woodward, P. (J. Organometal. Chem. **192** [1980] 421/9).

[9] Herberich, G. E.; Hessner, B. (Chem. Ber. **115** [1982] 3115/27).

[10] Herberich, G. E.; Kucharska-Jansen, M. M. (J. Organometal. Chem. **243** [1983] 45/9).

[11] Herberich, G. E.; Hessner, B.; Huttner, G.; Zsolnai, L. (Angew. Chem. **93** [1981] 471/2; Angew. Chem. Intern. Ed. Engl. **20** [1981] 472/3).

[12] Boehm, M. C.; Eckert-Maksic, M.; Gleiter, R.; Herberich, G. E.; Hessner, B. (Chem. Ber. **115** [1982] 754/63).

13.4 Carboranes Containing Three Boron Atoms

For previous data on carboranes containing three boron atoms, see "Boron Compounds" 2nd Suppl. Vol. 2, 1982, pp. 226/30.

Optimized geometry parameters for $\mathbf{[CB_3H_4]^{3-}}$ have been derived from MNDO molecular orbital calculations [1]. Calculated distances (based on C_{3v} symmetry) are: 0.956 Å for B-B-B ring center to ring boron, 1.297 Å for B-B-B ring center to carbon, 1.657 Å for B-B, 1.195 Å for B-H, 1.611 Å for B-C, and 1.090 Å for C-H.

A 6-electron rule for explaining observed cage isomer stabilities of ***closo*-$\mathbf{C_2B_3H_5}$** isomers has been developed [2]. MNDO MO calculations on $C_2B_3H_5$ as well as other carboranes are used to assess the accuracy of A. J. Stone's theory of polyborane electronic structure and bonding [3]. For the boron atoms in 1,5-$C_2B_3H_5$, a cage "umbrella" angle of 77° is used in an empirical relationship (which includes the number of adjacent cage carbon atoms) to predict a $^1J(^{11}B\text{-}^1H)$ of 189 Hz which agrees well with an observed value of 188 Hz [4]. Aromatic solvent-induced ^1H NMR shifts are reported for 1,5-$C_2B_3H_5$ and correlated to PRDDO MO derived hydrogen charges [5]. The protonation of the 1,5-$C_2B_3H_5$ has been studied by the MNDO method; the calculated proton affinity for the compound is 144.2 kcal/mol and a carbon-protonated structure is reported [6]. 1,5-$C_2B_3H_5$ is converted to [1,5-$C_2B_3H_4$-2-]$_2$ in the presence of $PtBr_2$ [7].

NMR parameters of *closo*-pentaalkyl-1,5-dicarbapentaboranes(5), ***closo*-$\mathbf{C_2B_3R_5}$**, were determined by ^{10}B, ^{11}B, and ^{13}C NMR spectroscopy. In agreement with predictions from MO calculations $^1J(^{11}B\text{-}^{11}B)$ is <10 Hz. Together with these data the $\delta^{11}B$ and $\delta^{13}C$ values show that neither the structure nor the bonding situation in the polyhedron is significantly affected by alkyl substitution. For *closo*-1,5-C_2B_3-1,5-$(CH_3)_2$-2,3,4-$(C_2H_5)_3$: $\delta^{11}B = +13.8$ ppm; ^{13}C NMR resonances at δ (in ppm) = 97.6 for cage carbons, 8.4 for 1 and 5 attached CH_3 groups, 3.1 and 9.1 for CH_2 and CH_3 of ethyl groups; $^1J(^{13}C\text{-}^1H)$ (in Hz) = 117.5 for CH_2, 126.4 for CH_3 of ethyl group, 127.0 for 1 and 5 attached CH_3 groups, $^1J(^{13}C\text{-}^{11}B) = 20.0$ for BC(1,5), 78.0 for BCH_2, $^1J(^{13}C\text{-}^{13}C) = 53.0$ for C(1,5)CH_3. For *closo*-1,5-C_2B_3-1,2,3,4,5-$(C_2H_5)_5$: $\delta^{11}B = 13.8$ ppm; ^{13}C NMR resonances at δ (in ppm) = 106.3 for cage carbons, 4.1 and 9.8 for CH_2 and CH_3 of ethyl groups attached to cage carbons, 18.7 and 15.9 for CH_2 and CH_3 of ethyl groups attached to borons; $^1J(^{13}C\text{-}^1H)$ (in Hz) = 117.0 for B-CH_2, 126.4 for CH_3 of B ethyl group, 126.4 for C(1,5)-CH_2, 125.5 for CH_3 of ethyl groups attached to cage carbons, $^1J(^{13}C\text{-}^{11}B) = 75.0$ for BCH_2, $^1J(^{13}C\text{-}^{13}C) = 52.0$ for C(1,5)CH_2, 33.5 for B-CH_2CH_3, 33.5 for C-CH_2CH_3. For *closo*-1,5-(C_2B_3-1,5-$(C_2H_5)_2$-2,3,4-(*n*-$C_3H_7)_3$: $\delta^{11}B = +13.7$ ppm; ^{13}C NMR resonances at δ (in ppm) = 107.0 for cage carbons, 14.3 for B-CH_2, 19.4 for β-CH_2 of propyl group, 17.2 for CH_3 of propyl group, 18.7 for CH_2 of ethyl group, 15.6 for CH_3 of ethyl group; $^1J(^{13}C\text{-}^1H)$ (in Hz) = 117.0 for B-CH_2, 125.1 for the β-CH_2 of the propyl group, 126.0 for all other C-H; $^1J(^{13}C\text{-}^{11}B) = 75.0$ Hz for B-CH_2 [8].

closo-1,5-C_2B_3-1,2,3,4,5-$(C_2H_5)_5$, upon treatment with potassium and subsequently with iodine, yields the four-carbon carborane $(C_2H_5)_4C_4B_6(C_2H_5)_6$ [9]. Boron-substituted alkenyl derivatives of 1,5-$C_2B_3H_5$ (see Table 13/4 for ^{11}B NMR data; ^1H NMR and IR data are also reported) are prepared from the $(RC_2R')CO_2(CO)_6$ (R, R' = H, CH_3, C_2H_5) catalyzed reaction of alkynes (acetylene, 1-butyne, 2-butyne) with the parent carborane [10].

Table 13/4
Boron-11 NMR Data of Alkenyl Derivatives of 1,5-$C_2B_3H_5$ (chemical shifts in ppm, J values in Hz are given in parentheses) [10].

derivative	B(2)	B(3,4)
2-(ethenyl)-	5.9	8.5 (172)
2-(*trans*-1-but-1-enyl)-	5.8	8.5 (181)
2-(2-but-1-enyl)-	6.8	8.0 (190)
2-(*cis*-2-but-2-enyl)-	7.5	9.0 (190)
2,3,4-tris-(*cis*-2-but-2-enyl)-	18.5	18.5

nido-1,2,3-(η^5-$C_5(CH_3)_5$)$Co(CH_3)_2C_2B_3H_5$ is produced by the action of NaOH/CH_3CN on 1,2,3-[η^5-$C_5(CH_3)_5$]$Co(CH_3)_2C_2B_4H_4$ [11]. 1,7,2,3-[η^5-$C_5(CH_3)_5$]$_2Co_2(CH_3)_2C_2B_3H_3$ is formed from the reaction of Na[$(CH_3)_2C_2B_4H_5$], Li[$C_5(CH_3)_5$] ($C_5(CH_3)_5$ = pentamethylcyclopentadienyl), and $CoCl_2$ in cold tetrahydrofuran [11]. Reactions of the *nido*-cobaltacarborane anions [(η^5-C_5H_5)$Co(CH_3)_2C_2B_3H_4$]$^-$ and [(η^5-$C_5(CH_3)_5$)$Co(CH_3)_2C_2B_3H_4$]$^-$, respectively, with $HgCl_2$ in tetrahydrofuran give initially the unstable adducts [(η^5-C_5R_5)$Co(CH_3)_2C_2B_3H_4 \cdot HgCl_2$]$^-$ (R = H, CH_3), which lose Cl^- to form the isolable HgCl-bridged complexes μ-[(η^5-C_5R_5)-$Co(CH_3)_2C_2B_3H_4$]HgCl. The latter species undergo symmetrization to generate the bis-(cobaltacarboranyl)mercury complexes μ,μ'-[(η^5-C_5R_5)$Co(CH_3)_2C_2B_3H_4$]$_2$Hg [12]. Two-dimensional boron-11-boron-11 nuclear magnetic resonance spectroscopy confirms the B-B atom connectivities in μ-(4,5)-HgCl[1,2,3-($C_5(CH_3)_5Co(CH_3)_2C_2B_3H_4$] and in *nido*-1,2,3-[$C_6(CH_3)_6$]-$Fe(C_2H_5)_2C_2B_3H_5$ [13]. The bonding of 1,2,4-$C_2B_3H_5Fe(CO)_3$ was studied by self-consistent charge-type Hückel calculations [14]. Electrochemical data on 1,1,1-$(CO)_3$-1,2,6-$FeC_2B_3H_5$ are presented in [15]. Investigation of the bonding in $C_2B_3H_7Fe(CO)_3$ by self-consistent charge calculations provides a rationalization for the stability of the known compounds [16]. (η^6-C_8H_{10})Fe[$(C_2H_5)_2C_2B_3H_5$], characterized from spectroscopic data as a sandwich complex containing a planar carborane ligand, is formed from the reaction of (η^6-C_8H_{10})Fe-[$(C_2H_5)_2C_2B_4H_4$] with N,N,N',N'-tetramethyl-1,2-diaminoethane [17]. The four-carbon metallacarborane complexes 1-[η^6-$C_6(CH_3)_6$]Fe-4,5,7,8-$(CH_3)_4C_4B_3H_3$ and 1-[η^6-$(CH_3)C_6H_5$]Fe-4,5,7,8-$(CH_3)_4C_4B_3H_3$ are produced from the reaction of thermally generated iron atoms with pentaborane(9) and 2-butyne in the presence of toluene [18]. The metallacarborane [$(C_4H_4BC_6H_5)Rh$]$_2$-$(C_4H_4BC_6H_5)$ is prepared from $C_4H_4BC_6H_5$ and $Ru(C_6H_6)(C_6H_8)$ [19].

References for 13.4:

[1] Jemmis, E. D.; Pavankumar, P. N. V. (Proc. Indian Acad. Sci. Chem. Sci. **93** [1984] 479/89).
[2] Jemmis, E. D. (J. Am. Chem. Soc. **104** [1982] 7017/20).
[3] Brint, P.; Cronin, J. P.; Seward, E.; Whelan, T. (J. Chem. Soc. Dalton Trans. **1983** 975/80).
[4] Jarvis, W.; Abdou, Z. J.; Onak, T. (Polyhedron **2** [1983] 1067/70).
[5] Jarvis, W.; Inman, W.; Powell, B.; DiStefano, E. W.; Onak, T. (J. Magn. Resonance **43** [1981] 302/15).
[6] DeKock, R. L.; Jasperse, C. P. (Inorg. Chem. **22** [1983] 3843/8).
[7] Corcoran, E. W.; Sneddon, L. G. (J. Am. Chem. Soc. **106** [1984] 7793/800).
[8] Koester, R.; Wrackmeyer, B. (Z. Naturforsch. **36b** [1981] 704/7).
[9] Koester, R.; Seidel, G.; Wrackmeyer, B. (Angew. Chem. **96** [1984] 520/1; Angew. Chem. Intern. Ed. Engl. **23** [1984] 512/4).
[10] Wilczynski, R.; Sneddon, L. G. (Inorg. Chem. **21** [1982] 506/14).

[11] Finster, D. C.; Sinn, E.; Grimes, R. N. (J. Am. Chem. Soc. **103** [1981] 1399/407).
[12] Finster, D. C.; Grimes, R. N. (Inorg. Chem. **20** [1981] 863/71).
[13] Venable, T. L.; Hutton, W. C.; Grimes, R. N. (J. Am. Chem. Soc. **106** [1984] 29/37).
[14] Pelin, W. K.; Spalding, T. R.; Brint, R. P. (J. Chem. Res. S **1982** 120).
[15] Geiger, W. E.; Brennan, D. E. (Inorg. Chem. **21** [1982] 1963/6).
[16] Brint, P.; Pelin, W. K.; Spalding, T. R. (J. Chem. Soc. Dalton Trans. **1981** 546/51).
[17] Swisher, R. G.; Sinn, E.; Grimes, R. N. (Organometallics **2** [1983] 506/14).
[18] Micciche, R. P.; Briguglio, J. J.; Sneddon, L. G. (Organometallics **3** [1984] 1396/402).
[19] Herberich, G. E.; Hessner, B.; Boveleth, W.; Luethe, H.; Saive, R.; Zelenka, L. (Angew. Chem. **95** [1983] 1024/5; Angew. Chem. Intern. Ed. Engl. **22** [1983] 996).

13.5 Carboranes Containing Four Boron Atoms

For previous data on carboranes containing four boron atoms, see "Boron Compounds" 2nd Suppl. Vol. 2, 1982, pp. 230/8.

13.5.1 The $[CB_4H_5]^{3-}$ Anion

Optimized geometry parameters for $\mathbf{[CB_4H_5]^{3-}}$ have been derived from MNDO molecular orbital calculations [1]. Calculated distances (based on C_{4v} symmetry) are: 1.656 Å for B-B, 1.237 Å for B-H, 1.765 Å for B-C, and 1.194 Å for C-H [1].

Reference for 13.5.1:

[1] Jemmis, E. D.; Pavankumar, P. N. V. (Proc. Indian Acad. Sci. Chem. Sci. **93** [1984] 479/89).

13.5.2 $C_2B_4H_6$ and Derivatives

A rapid method for orbital localization is utilized for **1,2-**$\mathbf{C_2B_4H_6}$ which allows the rapid construction of localized molecular orbitals from eigenvector matrices and for a single-determinant wave function stemming from any quantum-chemical calculation [3].

The protonation of **1,6-**$\mathbf{C_2B_4H_6}$ has been studied by the MNDO method; the calculated proton affinity for the compound is 160.5 kcal/mol and the calculations predict B-B edge protonation. Ab initio calculations employing the 3-21G basis set indicate a proton affinity of 157 kcal/mol for 1,6-$C_2B_4H_6$ [1]. Gaussian-70 (STO-3G basis set) molecular orbital calculations and bond energy estimates are used to probe the changes that occur when B_4H_{10} reacts with ethylene or acetylene to form hydrogen and the *closo*-carboranes 1,2- and 1,6-$C_2B_4H_6$ via suggested *arachno* and *nido* precursors. The driving force for the stepwise cluster oxidation is provided by a progressive increase in the B-C bonding, which is accompanied by transfer of electronic charge to the C atoms [2].

MNDO MO calculations on $C_2B_4H_6$ as well as other carboranes are used to assess the accuracy of A. J. Stone's theory (1980) of their electronic structure and bonding [4]. CNDO calculations on 1,6-$C_2B_4H_6$ are reported in [5].

Aromatic solvent-induced ^{1}H NMR shifts are reported for 1,6-$C_2B_4H_6$ and correlated to PRDDO MO-derived hydrogen charges [6]. For the boron atoms in 1,2-$C_2B_4H_6$, cage "umbrella" angles of 86.6° (for B(3,5)) and 88.1° (for B(4,6)) are used in an empirical relationship, which includes the number of adjacent cage carbon atoms, to predict $^{1}J(^{11}B-^{1}H)$ values of 186 and 166 Hz, respectively; these values agree well with the observed data of 185 and 162 Hz, respectively; for the boron atoms in 1,6-$C_2B_4H_6$, a cage "umbrella" angle of 85.6° is used in an empirical relationship, which includes the number of adjacent cage carbon atoms, to predict a $^{1}J(^{11}B-^{1}H)$ of 186 Hz, which is exactly that observed experimentally [7]. From the IR and Raman spectra of 1,6-$C_2B_4H_6$ the following assignments have been made (frequencies in cm^{-1}): 3118 and 3107 (C–H stretching); 2667, 2664, and 2647 (B–H stretching); 1177, 1096, 799, 775, 474, and 447 (B–C stretching); 1261, 1159, and 986 (B–B stretching); 1295 and 1200 (C–H deformation); 1456, 1130, 993, 849, 794, and 744 (B–H deformation) [8]. 1,6-$C_2B_4H_6$ is converted to (1,6-$C_2B_4H_5$-2-)$_2$ in the presence of $PtBr_2$ [10].

Boron-substituted alkenyl derivatives of 1,6-$C_2B_4H_6$ (see Table 13/5 for ^{11}B NMR data; ^{1}H NMR and IR data are also reported) are prepared from the $(CH_3C_2CH_3)Co_2(CO)_6$-catalyzed reaction of 2-butyne with the parent carborane [9].

Table 13/5

Boron-11 NMR Data of Alkenyl Derivatives of 1,6-$C_2B_4H_6$ (chemical shifts in ppm, J values in Hz are in parentheses) [9].

derivative	B(2)	B(3)	B(4)	B(5)
2-(*cis*-2-but-2-enyl)-	−8.1	−17.2 (190)	−23.9 (193)	−17.2 (190)
2,3-bis(*cis*-2-but-2-enyl)-	−7.1	−7.1	−21.3 (187)	−21.3 (187)
2,4-bis(*cis*-2-but-2-enyl)-	−12.6	−15.2 (176)	−12.6	−15.2 (176)
2,3,4-tris(*cis*-2-but-2-enyl)-	−10.7	−5.4	−10.7	−18.8 (182)
2,3,4,5-tetrakis(*cis*-2-but-2-enyl)-	−11.0	−11.0	−11.0	−11.0

References for 13.5.2:

[1] DeKock, R. L.; Jasperse, C. P. (Inorg. Chem. **22** [1983] 3843/8).

[2] DeKock, R. L.; Fehlner, T. P.; Housecroft, C. E.; Lubben, T. V.; Wade, K. (Inorg. Chem. **21** [1982] 25/30).

[3] Perkins, P. G.; Stewart, J. J. P. (J. Chem. Soc. Faraday Trans. II **1982** 285/96).

[4] Brint, P.; Cronin, J. P.; Seward, E.; Whelan, T. (J. Chem. Soc. Dalton Trans. **1983** 975/80).

[5] Semenov, S. G. (Zh. Strukt. Khim. **22** [1981] 164/6; J. Struct. Chem. [USSR] **22** [1981] 776/8).

[6] Jarvis, W.; Inman, W.; Powell, B.; DiStefano, E. W.; Onak, T. (J. Magn. Resonance, **43** [1981] 302/15).

[7] Jarvis, W.; Abdou, Z. J.; Onak, T. (Polyhedron **2** [1983] 1067/70).

[8] Bragin, J.; Urevig, D. S.; Diem, M. (J. Raman Spectrosc. **12** [1982] 86/90).

[9] Wilczynski, R.; Sneddon, L. G. (Inorg. Chem. **21** [1982] 506/14).

[10] Corcoran, E. W.; Sneddon, L. G. (J. Am. Chem. Soc. **106** [1984] 7793/800).

13.5.3 $C_2B_4H_8$ and Derivatives

Extended Hückel MO calculations have been made on $\mathbf{C_4B_4H_8}$, and the results compared to those of related cage species [1]. Two-dimensional ^{11}B-^{11}B nuclear magnetic resonance spectroscopy confirms the B-B atom connectivities in Na[2,3-$C_2B_4H_5$-2,3-$(C_2H_5)_2$] [2].

References for 13.5.3 on p. 162

***nido*-2,3-$C_2B_4H_6$-2,3-$(C_2H_5)_2$** is prepared in 33 to 35% yield from a reaction of B_5H_9 with diethylacetylene in $(C_2H_5)_3N$. *nido*-2,3-$C_2B_4H_6$-2,3-$(C_2H_5)_2$ is a colorless, moderately air-sensitive liquid with a vapor pressure of 14 Torr at 24°C. Infrared absorptions for the compound are found at (frequencies in cm^{-1}, s=strong, m=medium, w=weak, v=very) 2976vs, 2940s, 2883s, 2600vs, 1952w, 1919w, 1468s, 1382m, 1073w, 998w, 946m, 924m, 848m, 799w, 770w, 724w, 661m [3].

Boron-substituted alkenyl derivatives of 2,3-$C_2B_4H_8$ (see Table 13/6 for ^{11}B NMR data; ^{1}H NMR and IR data are also reported) are prepared from the $(CH_3C_2CH_3)Co_2(CO)_6$-catalyzed reaction of 2-butyne with the parent carborane [4]. A method is presented for the determination of B-B connectivities in *nido*-2,3-$C_2B_4H_6$-2,3-$(C_2H_5)_2$ and its conjugate base, **[*nido*-2,3-$C_2B_4H_5$-2,3-$(C_2H_5)_2$]$^-$**, via two-dimensional J-correlated homonuclear (^{11}B-^{11}B) FT NMR spectroscopy [5].

Table 13/6
Boron-11 NMR Data of Alkenyl Derivatives of 2,3-$C_2B_4H_8$ (chemical shifts in ppm, J values in Hz are in parentheses) [4].

derivative	B(1)	B(4)	B(5)	B(6)
1-(*cis*-2-but-2-enyl)-	−41.1	−1.0 (153)	−1.0 (153)	−1.0 (153)
5-(*cis*-2-but-2-enyl)-	−51.9 (180)	−3.4 (158)	14.2	−3.4 (158)
4-(*cis*-2-but-2-enyl)-	−51.9 (180)	9.7	−2.4 (158)	−2.4 (158)

Extended Hückel molecular orbital calculations have been carried out on metal sandwich complexes derived from the *nido*-pentagonal bipyramidal **$[C_2B_4H_6]^{2-}$** ligands and the results contrasted with those obtained for related sandwich compounds derived from the cyclopentadienyl ligand [6].

The reaction of $Na[(CH_3)_2C_2B_4H_5]$, $Li[C_5(CH_3)_5]$ ($C_5(CH_3)_5$=pentamethylcyclopentadienyl) and $CoCl_2$ in cold tetrahydrofuran yields *closo*-1,2,3-$[\eta^5$-$C_5(CH_3)_5]Co(CH_3)_2C_2B_4H_4$, the triple-decker complex 1,7,2,3-$[\eta^5$-$C_5(CH_3)_5]_2Co_2(CH_3)_2C_2B_3H_3$, and $[\eta^5$-$C_5(CH_3)_5]_2Co_3(CH_3)_4C_4B_8H_7$ [7].

References for 13.5.3:

[1] Cox, D. N.; Mingos, D. M. P.; Hoffmann, R. (J. Chem. Soc. Dalton Trans. **1981** 1788/97).
[2] Venable, T. L.; Hutton, W. C.; Grimes, R. N. (J. Am. Chem. Soc. **106** [1984] 29/37).
[3] Maynard, R. B.; Borodinsky, L.; Grimes, R. N. (Inorg. Syn. **22** [1983] 211/4).
[4] Wilczynski, R.; Sneddon, L. G. (Inorg. Chem. **21** [1982] 506/14).
[5] Venable, T. L.; Hutton, W. C.; Grimes, R. N. (J. Am. Chem. Soc. **104** [1982] 4716/7).
[6] Calhorda, M. J.; Mingos, D. M. P. (J. Organometal. Chem. **229** [1982] 229/45).
[7] Finster, D. C.; Sinn, E.; Grimes, R. N. (J. Am. Chem. Soc. **103** [1981] 1399/407).

13.5.4 Metallacarboranes Containing Four Boron Atoms

closo-1,2,3-$SnC_2B_4H_4$-2-$Si(CH_3)_3$-3-R derivatives [R=$Si(CH_3)_3$, CH_3, or H] are prepared using the reaction of Na[*nido*-2,3-$C_2B_4H_5$-2-$Si(CH_3)_3$-3-R] with $SnCl_2$ [1]. The structure of the stannacarborane 1,2,3-$SnC_2B_4H_4$-2-$Si(CH_3)_3$-3-CH_3 has been determined by X-ray crystallography [2]. Reactions of $[C_8H_8]^{2-}$ and $[(C_2H_5)_2C_2B_4H_5]^-$ ions with $TiCl_3$ and VCl_3 give, as the major isolable products, $(\eta^8$-$C_8H_8)Ti[(C_2H_5)_2C_2B_4H_4]$ and $(\eta^8$-$C_8H_8)V[(C_2H_5)_2C_2B_4H_4]$, respectively. In the corresponding reaction with $CrCl_3$, no cyclooctatetraene complex is isolated; however,

$(\eta^7\text{-}C_7H_7)Cr[(C_2H_5)_2C_2B_4H_4]$ is obtained in low yield. $(\eta^8\text{-}C_8H_8)Ti[(C_2H_5)_2C_2B_4H_4]$ reacts with CH_3I/$AlCl_3$ to give the 5-iodo derivative, and with I_2 in C_6H_6 to give the 4,5-diiodo derivative [3, 4].

1,2,3-$[\eta^5\text{-}C_5(CH_3)_5]Co(CH_3)_2C_2B_4H_4$ is formed from the reaction of $Na[(CH_3)_2C_2B_4H_5]$, $Li[C_5(CH_3)_5]$ ($C_5(CH_3)_5$ = pentamethylcyclopentadienyl), and $CoCl_2$ in cold tetrahydrofuran. Degradation of 1,2,3-$[\eta^5\text{-}C_5(CH_3)_5]Co(CH_3)_2C_2B_4H_4$ in basic CH_3CN produces *nido*-1,2,3-$[\eta^5\text{-}C_5(CH_3)_5]Co(CH_3)_2C_2B_3H_5$ in 93% yield [5]. Electrochemical data on 1-C_5H_5-1,2,3-$CoC_2B_4H_6$, 1-C_5H_5-1,2,4-$CoC_2B_4H_6$, and 1-C_5H_5-1,2,3-$FeC_2B_4H_6$ are presented in [6]. Reaction of 2,3-$C_2B_4H_6$-2,3-$(C_2H_5)_2$ with NaH and subsequently with $FeCl_2$ yields $[2,3\text{-}C_2B_4H_4\text{-}2,3\text{-}(C_2H_5)_2]_2FeH_2$ [7]. Reactions of $(\eta^6\text{-}C_8H_{10})Fe(R_2C_2B_4H_4)$ (R = C_2H_5) with benzene or other arenes (over $AlCl_3$) form the corresponding $LFe(R_2C_2B_4H_4)$ (L = η^6-arene) species as air-stable crystalline solids. The reaction of $(\eta^6\text{-}C_8H_{10})Fe[(C_2H_5)_2C_2B_4H_4]$ with N,N,N',N'-tetramethyl-1,2-diaminoethane results in extraction of the apex boron and forms $(\eta^6\text{-}C_8H_{10})Fe[(C_2H_5)_2C_2B_3H_5]$ [8]. The syntheses of $(\eta^6\text{-}C_8H_{10})Fe[(C_2H_5)_2C_2B_4H_4]$, $(\eta^6\text{-}C_6H_6)Fe[(C_2H_5)_2C_2B_4H_4]$, and $(C_{16}H_{18})Fe[(C_2H_5)_2C_2B_4H_4]$ are reported in [9]. 1-$(\eta^6\text{-}C_6H_5CH_3)$-*closo*-1-Fe-2,3-$C_2B_4H_4$-2,3-$(C_2H_5)_2$ is formed from the reaction of thermally generated Fe atoms with toluene and *nido*-2,3-$C_2B_4H_4$-2,3-$(C_2H_5)_2$ under low-temperature conditions [10]. The (π-arene)ferracarborane sandwich complex 1-$[\eta^6\text{-}C_6(CH_3)_6]$Fe-2,3-$(CH_3)_2C_2B_4H_4$ is produced from the reaction of thermally generated iron atoms with pentaborane(9) and 2-butyne in the presence of toluene [11].

The air-stable *closo*-1-$Os(CO)_3$-2,3-$(Me_3Si)_2$-2,3-$C_2B_4H_4$ is prepared by treating $Os_3(CO)_{12}$ with either *closo*-$[(CH_3)_3Si]_2SnC_2B_4H_4$ or *nido*-$[(CH_3)_3Si]_2C_2B_4H_6$ [12]. Extended Hückel molecular orbital calculations on the *closo*-platinacarborane $(H_3P)_2Pt(C_2B_4H_6)$ have been carried out [13].

Various metal-carbon-boron containing "sandwich" compounds have been prepared in which the structures appear to obey carborane structure rules; these include: $[(C_5H_5Co)(H(C_2H_5)_2C_3B_2(CH_3)_2)]_2M$ (M = Zn, Fe, Co, Ni) [14], $[(C_5H_5Fe)((C_2H_5)_2C_2B_2(CH_3)_2S)]_2M$ (M = Fe, Co) [15], and various derivatives of $Pt(C_3B_2H_5)_2$ [16].

References for 13.5.4:

[1] Hosmane, N. S.; Sirmokadam, N. N.; Herber, R. H. (Organometallics **3** [1984] 1665/9).
[2] Cowley, A. H.; Galow, P.; Hosmane, N. S.; Jutzi, P.; Norman, N. C. (J. Chem. Soc. Chem. Commun. **1984** 1564/5).
[3] Swisher, R. G.; Sinn, E.; Grimes, R. N. (Organometallics **3** [1984] 599/605).
[4] Swisher, R. G.; Sinn, E.; Brewer, G. A.; Grimes, R. N. (J. Am. Chem. Soc. **105** [1983] 2079/80).
[5] Finster, D. C.; Sinn, E.; Grimes, R. N. (J. Am. Chem. Soc. **103** [1981] 1399/407).
[6] Geiger, W. E.; Brennan, D. E. (Inorg. Chem. **21** [1982] 1963/6).
[7] Maynard, R. B.; Grimes, R. N. (Inorg. Syn. **22** [1983] 215/8).
[8] Swisher, R. G.; Sinn, E.; Grimes, R. N. (Organometallics **2** [1983] 506/14).
[9] Maynard, R. B.; Swisher, R. G.; Grimes, R. N. (Organometallics **2** [1983] 500/5).
[10] Micciche, R. P.; Sneddon, L. G. (Organometallics **2** [1963] 674/8).

[11] Micciche, R. P.; Briguglio, J. J.; Sneddon, L. G. (Organometallics **3** [1984] 1396/402).
[12] Hosmane, N. S.; Sirmokadam, N. N. (Organometallics **3** [1984] 1119/21).
[13] Calhorda, M. J.; Mingos, D. M. P.; Welch, A. J. (J. Organometal. Chem. **228** [1982] 309/20).
[14] Siebert, W.; Edwin, J.; Wadepohl, H.; Pritzkow, H. (Angew. Chem. **94** [1982] 148; Angew. Chem. Intern. Ed. Engl. **21** [1982] 149/50).
[15] Siebert, W.; Böhle, C.; Krüger, C. (Angew. Chem. **92** [1980] 758/9; Angew. Chem. Intern. Ed. Engl. **19** [1980] 746/7).
[16] Wadepohl, H.; Siebert, W. (Z. Naturforsch. **39b** [1984] 50/5).

13.6 Carboranes Containing Five Boron Atoms

For previous data on carboranes containing five boron atoms, see "Boron Compounds" 2nd Suppl. Vol. 2, 1982, pp. 238/49.

13.6.1 CB_5H_7, CB_5H_9, 3,4-μ-$(CH_3)_3N$-CB_5H_{11}, and Related Compounds

Ab initio MO calculations (with full geometry optimization G-80 calculations) yield an octahedral structure for ***closo*-1-CB_5H_7**, in good agreement with the experimentally determined structure [1]. Optimized geometry parameters for $[CB_5H_6]^{3-}$ have been derived from MNDO molecular orbital calculations [2].

Protonation of **2-CB_5H_9** has been studied by the MNDO method; the calculated proton affinity for the compound is 165.5 kcal/mol and a carbon-protonated structure is reported [3]. Flash thermolysis (355°C) of 2-(*cis*-2-but-2-enyl)-B_5H_8 yields *nido*-2-CB_5H_7-2-CH_3-3-C_2H_5 (36%) and *nido*-2-CB_5H_7-2-CH_3-4-C_2H_5 (39%); 2-(2-propenyl)-B_5H_8 gives *nido*-2-CB_5H_7-2,3-$(CH_3)_2$ (28%) and *nido*-2-CB_5H_7-2,4-$(CH_3)_2$ (18%); 2-(1-*trans*-1-propenyl)-B_5H_8 gives *nido*-2-CB_5H_8-3-C_2H_5 (26%) and *nido*-2-CB_5H_8-4-C_2H_5 (16%); and 2-(ethenyl)-B_5H_8 gives *nido*-2-CB_5H_8-2-CH_3 (15%), *nido*-2-CB_5H_8-3-CH_3 (23%), and *nido*-2-CB_5H_8-4-CH_3 (13%). A mechanism to account for these conversions is discussed. ^{11}B NMR data are given in Table 13/7; ^{1}H NMR data and IR absorptions are also reported [4].

Table 13/7

Boron-11 NMR Data for *nido*-2-CB_5H_9 Derivatives (chemical shifts in ppm; J values in Hz are given in parentheses) [4].

derivative	B(1)	B(3)	B(4)	B(5)	B(6)
3-C_2H_5-	−49.9 (169)	28.4	−5.0 (140)	−6.2 (153)	16.6 (158)
4-C_2H_5-	−51.2 (167)	14.9 (160)	12.2	−4.1 (149)	14.9 (160)
2,4-$(CH_3)_2$-	−47.8 (162)	14.2 (171)	9.0	−4.7 (160)	14.2 (171)
2-CH_3-4-C_2H_5-	−48.4 (156)	13.7 (141)	11.7	−4.9 (172)	13.7 (141)
2-C_2H_5-3-CH_3-	−48.4 (156)	24.0	−6.8 (156)	−6.8 (156)	15.2 (156)

3,4-μ-$(CH_3)_3N$-CB_5H_{11}, **Fig.** 13-1, (^{11}B NMR data: δ (in ppm) = −12.4, J(B-H) = 115 Hz, for B(2,5); = −23.2, J(B-H) = 120 Hz, J(B-H bridging) = 30 Hz, for B(3,4); = −60.2, J(B-H) = 140 Hz, J(B-H bridging) = 55 Hz, for B(1)), is produced from the degradation of 6-$(CH_3)_3N$-6-CB_9H_{11} with KOH/CH_3OH [5].

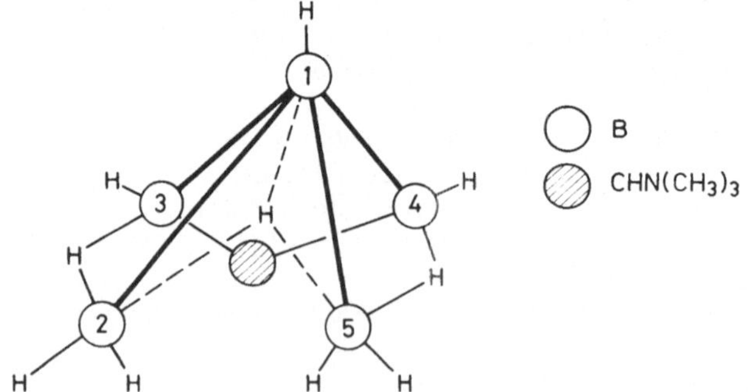

Fig. 13-1. The structure of 3,4-μ-$(CH_3)_3N$-CB_5H_{11}.

References for 13.6.1:

[1] Brint, P.; Healy, E. F.; Spalding, T. R.; Whelan, T. (J. Chem. Soc. Dalton Trans. **1981** 2515/22).
[2] Jemmis, E. D.; Pavankumar, P. N. V. (Proc. Indian Acad. Sci. Chem. Sci. **93** [1984] 479/89).
[3] DeKock, R. L.; Jasperse, C. P. (Inorg. Chem. **22** [1983] 3843/8).
[4] Wilczynski, R.; Sneddon, L. G. (Inorg. Chem. **20** [1981] 3955/62).
[5] Baše, K.; Štíbr, B.; Dolansky, J.; Duben, J. (Collection Czech. Chem. Commun. **46** [1981] 2345/53).

13.6.2 $C_2B_5H_7$ and Derivatives

A 6-electron rule for explaining observed cage isomer stabilities of ***closo*-$C_2B_5H_7$** isomers has been developed [1]. MNDO MO calculations on $C_2B_5H_7$ as well as other carboranes are used to assess the accuracy of A. J. Stone's theory (1980) of their electronic structure and bonding [2]. A coupling constant of 9.5 Hz for B(1)-B(5) of 2,4-$C_2B_5H_7$ is found [3]. From resolution-enhanced NMR spectra, the parameters shown in Table 13/8, pp. 166/7, have been derived for *closo*-2,4-$C_2B_5H_7$ and some of its derivatives [4]. Aromatic solvent-induced ^{1}H NMR shifts are reported for 2,4-$C_2H_5H_7$ and correlated to PRDDO MO-derived hydrogen charges [5]. For the boron atoms in 2,4-$C_2B_5H_7$, cage "umbrella" angles of 97.7° (for B(1,7)), 91.4° (for B(3)), and 90.2° (for B(5,6)) are used in an empirical relationship, which includes the number of adjacent cage carbon atoms, to predict $^1J(^{11}B\text{-}^1H)$ values of 182, 185, and 166 Hz, respectively; these values agree reasonably well with the observed values of 179, 184, and 168 Hz, respectively [6]. The protonation of the 2,4-$C_2B_5H_7$ has been studied by the MNDO method; the calculated proton affinity for the compound is 155.4 kcal/mol, and the calculations predict B-B-B face protonation [7].

Resolution-enhanced proton-decoupled ^{11}B NMR spectra of **5-X-2,4-$C_2B_5H_6$ (X=Cl, Br)** exhibit partially relaxed 1:1:1:1 quartets for B(5) and B(6) in each compound. Line shape analyses of the observed quartets, by comparison to theoretically derived curves, afford the following $^1J(^{11}B(5)\text{-}^{11}B(6))$ values: 30 Hz for X=Cl and 31 Hz for X=Br. In addition, $^1J(^{11}B(5)\text{-}^{11}B(6))$ values of 36 Hz for **1,5-Cl_2-2,4-$C_2B_5H_5$**, 42 Hz for **3,5-Cl_2-2,4-$C_2B_5H_5$**, 25 Hz for **[5-$(CH_3)_3$N-6-CH_3-2,4-$C_2B_5H_5$]$^+$**, and 27 Hz for **5-CH_3-6-Cl-2,4-$C_2B_5H_5$** were observed [4]. 5-Cl-2,4-$C_2B_5H_6$ reacts with Na[BH_4] to give the parent 2,4-$C_2B_5H_7$. Whereas the parent *closo*-2,4-$C_2B_5H_7$ does not react with $(CH_3)_3N$, both the 3-Cl-2,4-$C_2B_5H_6$ and 5-Cl-2,4-$C_2B_5H_6$ form 1:1 adducts with this amine. The ^{11}B NMR spectrum of **$(CH_3)_3N\cdot$5-Cl-2,4-$C_2B_5H_6$** shows peaks at δ (in ppm) = −19.4, J(B-H) = 179 Hz, 2B; = 6.3, J(B-H) = 157, 1B; = 17.4, 1B; = 2.3, J(B-H) = 130 Hz. The ^{1}H NMR spectrum of the compound exhibits resonances at δ (in ppm) = 5.68 (1HC), 7.63 (1HC), 0.36 (2HB), 4.57 (1HB), 3.77 (1HB), and 3.40 (($CH_3)_3N$). The adducts **$(CH_3)_3N\cdot$*z*-Cl-2,4-$C_2B_5H_6$** (*z* = 3 or 5) react with BCl_3 to yield the quaternary ammonium salts, [*z*-$(CH_3)_3$N-2,4-$C_2B_5H_6$][BCl_4]. The ^{11}B NMR spectrum of **[3-$(CH_3)_3$N-2,4-$C_2B_5H_6$][BCl_4]** shows peaks at δ (in ppm) = −17.8, J(B-H) = 184 Hz, for B(1,7); = 17.1 for B(3); = 3.9, J(B-H) = 160 Hz, for B(5,6); = 7.7 for [BCl_4]$^-$. ^{1}H NMR data for [3-$(CH_3)_3$N-2,4-$C_2B_5H_6$][BCl_4]: δ (in ppm) = 0.45 for HB(1,7), 5.85 for HC(2,4), 3.82 for HB(5,6), 3.40 for $(CH_3)_3N$. The ^{11}B NMR spectrum of **[5-$(CH_3)_3$N-2,4-$C_2B_5H_6$][BCl_4]** shows peaks at δ (in ppm) = −19.3, J(B-H) = 190 Hz, for B(1,7); = 6.3, J(B-H) = 183 Hz, for B(3); = 16.8 for B(5); = 1.7, J(B-H) = 156 Hz, for B(6); = 10.3 for [BCl_4]$^-$. ^{1}H NMR data for [5-$(CH_3)_3$N-2,4-$C_2B_5H_6$][BCl_4]: δ (in ppm) = 0.66 for HB(1,7), 5.86 for HC(2), 4.88 for HB(3), 6.27 for HC(4), 4.21 for HB(6), 3.41 for $(CH_3)_3N$. The reaction of $(CH_3)_3N$ with 1-Cl-2,4-$C_2B_5H_6$ proceeds with more difficulty than with the other two B-Cl isomers, but once formed, the 1:1 adduct combines with BCl_3 to give the rearrangement

References for 13.6.2 on p. 168

product [3-$(CH_3)_3N$-2,4-$C_2B_5H_6$][BCl_4]. The reaction of 5-Cl-2,4-$C_2B_5H_6$ with $(CH_3)_3P$ produces the **$(CH_3)_3P \cdot$5-Cl-2,4-$C_2B_5H_6$** adduct; this adduct reacts with BCl_3 to form **[5-$(CH_3)_3P$-2,4-$C_2B_5H_6$][BCl_4]** which exhibits ^{11}B NMR signals at δ (in ppm) = −18.9, J(B-H) = 189 Hz, for B(1,7); = 7 for B(3,6); = 3.8, J(P-B) = 158 Hz, for B(5). When 5-Cl-6-CH_3-$C_2B_5H_5$ is employed as the starting carborane in the $(CH_3)_3N/BCl_3$ reaction sequence, **[5-$(CH_3)_3N$-6-CH_3-2,4-$C_2B_5H_5$][BCl_4]** is produced. The ^{11}B NMR spectrum of [5-$(CH_3)_3N$-6-CH_3-2,4-$C_2B_5H_5$][BCl_4] shows signals at δ (in ppm) = −18.0, J(B-H) = 189 Hz, for B(1,7); = 5.8, J(B-H) = 200 Hz, for B(3); = 14.0 for B(5); = 9.9 for B(6). ^1H NMR data for [5-$(CH_3)_3N$-6-CH_3-2,4-$C_2B_5H_5$][BCl_4]: δ (in ppm) = 0.8 for HB(1,7), 5.43 for HC(2), 4.75 for HB(3), 6.00 for HC(4), 0.89 for CH_3B, 3.40 for $(CH_3)_3N$. **5-Cl-6-CH_3-2,4-$C_2B_5H_5$**, prepared by the reaction of 5-Cl-2,4-$C_2B_5H_6$ with $CH_3Cl/AlCl_3$, exhibits ^{11}B NMR resonances at δ (in ppm) = −19.0, J(B-H) = 180 Hz, for B(1,7); = 4.3, J(B-H) = 180 Hz, for B(3); = 12.0 for B(5); = 9.8 for B(6). ^1H NMR data for 5-Cl-6-CH_3-2,4-$C_2B_5H_5$: δ (in ppm) = 0.6 for HB(1,7), 5.1 for HC(2), 4.68 for HB(3), 5.3 for HC(4), 0.68 for CH_3B. A small quantity of **1-CH_3-5-Cl-2,4-$C_2B_5H_5$** is also produced from this modified Friedel-Crafts reaction [8]. The 1:1 amine-bromocarborane adduct **$(CH_3)_3N \cdot$5-Br-2,4-$C_2B_5H_6$** undergoes quantitative Br/Cl exchange with CH_2Cl_2, forming 5-Cl-2,4-$C_2B_5H_6$, via a presumed [5-$(CH_3)_3N$-2,4-$C_2B_5H_6$]$^+$ intermediate. The adduct $(CH_3)_3N \cdot$5-Br-2,4-$C_2B_5H_6$ exhibits ^{11}B NMR resonances at δ (in ppm) = −19.9, J(B-H) = 186 Hz, for B(1,7); = 0.8, J(B-H) = 176 Hz, for B(6); = 6.1, J(B-H) = 180 Hz, for B(3); = 16.9 for B(5). ^1H data: δ (in ppm) = 0.30, J(B-H) = 184 Hz, for H(1,7); = 3.74, J(B-H) = 173 Hz, for H(6); = 4.49, J(B-H) = 187 Hz, for H(3); = 5.26 and 7.18 for H(2 and 4); = 3.40 for CH_3 [9].

Photolysis of the small carborane 2,4-$C_2B_5H_7$ with hexafluoroacetone results in the formation of both **5-[$HO(CF_3)_2C$]-2,4-$C_2B_5H_6$** and **5-[$HOC(CF_3)_2C(CF_3)_2O$]-2,4-$C_2B_5H_6$**. The infrared spectrum of 5-[$HO(CF_3)_2C$]-2,4-$C_2B_5H_6$ displayed absorption bands at (frequencies in cm^{-1}) 3500w (HO stretching), 2600s, 1410m,sh, 1360s, 1260vs, 1140m, 1060m, 990m, 943m, 885s, 800m, 745m, 713m. The ^{11}B NMR spectrum of 5-[$HO(CF_3)_2C$]-2,4-$C_2B_5H_6$ shows signals at δ (in ppm) = 15.9 for B(5); = 2.8, J(B-H) = 184 Hz, for B(3); = −3.6, J(B-H) = 182 Hz, for B(6); = −20.2, J(BH) = 182 Hz, for B(1,7); ^1H NMR data: δ (in ppm) = 4.48 for H(2,4), 4.41 for OH, 4.24 for H(3), 3.61 for H(6), 0.55 for H(1,7); ^{19}F NMR datum: δ = −72.6 ppm. The ^{11}B NMR spectrum of 5-[$HOC(CF_3)_2C(CF_3)_2O$]-2,4-$C_2B_5H_6$ consists of peaks at δ (in ppm) = 15.9 for B(5); = 2.8, J(B-H) = 184 Hz, for B(3); = −3.3, J(B-H) = 170 Hz, for B(6); = −20.3, J(B-H) = 186 Hz, for B(1,7); ^1H NMR data: δ (in ppm) = 4.44 for H(2,4), 4.26 for OH, 4.24 for H(3), 3.61 for H(6), 0.58 for H(1,7); ^{19}F NMR data: δ = −72.6 and −70.9 ppm [10].

Table 13/8
$^1H\{^{11}B,^{10}B\}$ NMR Parameters for *closo*-2,4-$C_2B_5H_7$ and Derivatives [4].

compound	coupling constants, J(H-H)[a),b)] (in Hz)												
	1,3 and 3,7	1,5	1,6	2,3	2,4	2,5	2,6	3,4	4,5	4,6	5,6	5,7	6,7
2,4-$C_2B_5H_7$	1.13	0.57	0.57	6.66	4.00	1.51	8.89	6.66	8.89	1.51	2.75	0.57	0.57
5-CH_3-2,4-$C_2B_5H_6$	1.17	—	0.55	6.78	2.76	—	8.88	6.78	—	1.76	—	—	0.55
5,6-$(CH_3)_2$-2,4-$C_2B_5H_5$	1.17	—	—	6.90	3.03[d)]	—	—	6.90	—	—	—	—	—
5-C_2H_5-2,4-$C_2B_5H_6$[b)]	1.07	—	0.59	6.77	2.79	—	8.87	6.66	—	1.64	—	—	0.59
5,6-$(C_2H_5)_2$-2,4-$C_2B_5H_5$[c)]	1.10	—	—	6.69	?	—	—	6.69	—	—	—	—	—
5-Cl-6-CH_3-2,4-$C_2B_5H_5$	1.19	—	—	7.00	3.48	—	—	6.85	—	—	—	—	—
3-Cl-2,4-$C_2B_5H_6$	—	0.58	0.58	—	4.46	1.84	9.00	—	9.00	1.84	3.93	0.58	0.58
5-Cl-2,4-$C_2B_5H_6$	1.19	—	0.60	6.87	3.34	—	9.10	6.86	—	2.05	—	—	0.60

Table 13/8 (continued)

compound	coupling constants, $J(H\text{-}H)$[a),b)] (in Hz) 1,3 and 3,7	1,5	1,6	2,3	2,4	2,5	2,6	3,4	4,5	4,6	5,6	5,7	6,7
1,3-Cl_2-2,4-$C_2B_5H_5$	—	—	—	—	3.91	1.84	9.09	—	9.09	1.84	4.53	0.73	0.73
3,5-Cl_2-2,4-$C_2B_5H_5$	—	—	0.60	—	4.63	—	9.98	—	—	1.70	—	—	0.60
5,6-Cl_2-2,4-$C_2B_5H_5$	1.21	—	—	7.37	4.04[d)]	—	—	7.37	—	—	—	—	—

compound	chemical shifts, δ (in ppm)[a)] 1-BH	2-CH	3-BH	4-CH	5-BH	6-BH	7-BH
2,4-$C_2B_5H_7$	0.061	5.533	4.937	5.533	4.099	4.099	0.061
5-CH_3-2,4-$C_2B_5H_6$	0.158	5.326	4.827	5.135	0.699 (CH_3)	3.891	0.158
5,6-$(CH_3)_2$-2,4-$C_2B_5H_5$	0.221	5.000	4.715	5.000	0.620 (CH_3)	0.620 (CH_3)	0.221
5-C_2H_5-2,4-$C_2B_5H_6$	0.171	5.352	4.843	5.186	1.138 (CH_3) 1.274 (CH_2)	3.920	0.171
5,6-$(C_2H_5)_2$-2,4-$C_2B_5H_5$	0.264	5.050	4.757	5.050	1.057 (CH_3) 1.273 (CH_2)		0.264
5-Cl-6-CH_3-2,4-$C_2B_5H_5$	0.599	6.001	4.580	5.048	—	0.652 (CH_3)	0.599
3-Cl-2,4-$C_2B_5H_6$	0.548	5.383	—	5.383	3.983	3.983	0.548
5-Cl-2,4-$C_2B_5H_6$	0.550	5.428	4.764	5.423	—	3.891	0.550
1,3-Cl_2-2,4-$C_2B_5H_5$	—	5.513	—	5.513	4.057	4.057	0.445
3,5-Cl_2-2,4-$C_2B_5H_5$	0.972	5.167	—	5.243	—	3.746	0.972
5,6-Cl_2-2,4-$C_2B_5H_5$	0.972	4.890	4.421	4.890	—	—	0.972

a) Average estimated error for J: $<\pm 0.06$ Hz; for δ: 0.001 ppm error for $|\delta_i - \delta_j|$. – b) $J(CH_2\text{-}CH_3) = 7.813$ Hz for the monoethyl compound. – c) $J(CH_2\text{-}CH_3) = 7.911$ Hz for the diethyl compound. – d) Obtained from analysis of ^{13}C coupled ^1H resonances.

The gas-phase infrared spectrum of **3-CH_3-*closo*-2,4-$C_2B_5H_6$** exhibits strong absorption bands at 2620, 1350, 1290, 1280, and 1045 cm^{-1}. The rearrangement rates of 5- and 3-CH_3-*closo*-2,4-$C_2B_5H_6$ at 295°C to an equilibrium mixture of B-CH_3-*closo*-2,4-$C_2B_5H_6$ isomers have been measured and correlated to possible mechanisms. Of two plausible mechanistic routes (diamond-square-diamond, DSD, and triangle face rotation, TFR) the DSD path

$$5\text{-}CH_3\text{-}2,4\text{-}C_2B_5H_6 \underset{\underline{k}_b}{\overset{\underline{k}_a}{\rightleftharpoons}} 1\text{-}CH_3\text{-}2,4\text{-}C_2B_5H_6 \underset{\underline{k}_d}{\overset{\underline{k}_c}{\rightleftharpoons}} 3\text{-}CH_3\text{-}2,4\text{-}C_2B_5H_6$$

is in much better accord with the observed rate patterns than the TFR route. The best-fit rate constants (in $10^{-5}\,h^{-1}$) to the illustrated equation are $\underline{k}_a = 670$, $\underline{k}_b = 444$, $\underline{k}_c = 503$, $\underline{k}_d = 513$. ^1H NMR data for 1-CH_3-*closo*-2,4-$C_2H_5H_6$: δ (in ppm) = −0.47 for CH_3; = 5.58 for HC(2,4); = 4.94, J(B-H) = 190 Hz, for HB(3); = 4.10, J(B-H) = 163 Hz, for HB(5,6); = −0.03, J(B-H) = 178 Hz, for HB(7). ^1H NMR data for 3-CH_3-*closo*-2,4-$C_2B_5H_6$: δ (in ppm) = 1.02 for CH_3; = 5.25 for HC(2,4); = 4.03, J(B-H) = 168 Hz, for HB(5,6); = 0.18, J(B-H) = 176 Hz, for HB(1,7). ^1H NMR data for 5-CH_3-*closo*-2,4-$C_2B_5H_6$: δ (in ppm) = 0.75 for CH_3; = 5.36 for HC(2); = 5.19 for C(4); = 4.87, J(B-H) = 190 Hz, for HB(3); = 3.94, J(B-H) = 167 Hz, for HB(6); = 0.22, J(B-H) = 184 Hz, for HB(1,7) [11].

Boron-substituted alkenyl derivatives of 2,4-$C_2B_5H_7$ (see Table 13/9, p. 168, for ^{11}B NMR data; ^1H NMR and IR data are also reported) are prepared from the $(CH_3C_2CH_3)Co_2(CO)_6$-catalyzed reaction of 2-butyne with the parent carborane [12].

References for 13.6.2 on p. 168

Table 13/9

Boron-11 NMR Data for *cis*-2-But-2-enyl (= R) Derivatives of 2,4-$C_2B_5H_7$ (chemical shifts in ppm, J values in Hz are in parentheses) [12].

derivative	B(1)	B(3)	B(5)	B(6)	B(7)
1-R	−10.4	8.5 (178)	4.6 (178)	4.6 (178)	−27.8 (184)
5-R	−19.7 (178)	7.4 (180)	13.1	2.8 (189)	−19.7 (178)
3-R	−18.8 (178)	13.9	4.1 (173)	4.1 (173)	−18.9 (178)
5,6-R_2	−19.3 (178)	5.5 (173)	11.0	11.0	−19.3 (178)
1,5-R_2	−10.4	7.2 (178)	11.9	11.9	−25.9 (180)
1,3-R_2	−9.2	14.5	4.3 (173)	4.3 (173)	−24.5 (178)
1,7-R_2	−17.2	8.6 (160)	4.1 (160)	4.1 (160)	−17.2
3,5-R_2	−19.2 (176)	11.7	11.7	0.2 (173)	−19.2 (176)
1,5,6-R_3	−26.1 (178)	6.5 (189)	9.1	9.1	−26.1 (178)
1,5,7-R_3	−15.2	9.7 (178)	11.7	2.1 (173)	−15.2
1,3,5-R_3	−8.4	13.7	13.7	2.1 (173)	−22.9 (178)
3,5,6-R_3	−17.0 (178)	10.8	10.8	10.8	−17.0 (178)
1,5,6,7-R_4	−13.2	2.1 (173)	13.7	13.7	−13.2
1,3,5,7-R_4	−13.8	13.3	13.3	2.1 (167)	−13.8
1,3,5,6-R_4	−8.4	10.5	10.5	10.5	−22.5 (178)
1,3,5,6,7-R_5	−22.5	8.4	10.2	10.2	−22.5

Lithiation of 2,4-$C_2B_5H_7$ with LiC_4H_9 in a 40% solution of hexane in ether at room temperature gives a 95% yield of **Li_2[2,4-$C_2B_5H_5$]**; the latter compound reacts with $ClSi(CH_3)_3$ and $ClSi(CH_3)_2H$ to give **2,4-$C_2B_5H_5$-2,4-[$Si(CH_3)_3$]$_2$** and **2,4-$C_2B_5H_5$-2,4-[$Si(CH_3)_2H$]$_2$**, respectively. These two silanyl-carboranes react with chlorine or bromine to give corresponding halosilanylcarboranes, and with methanol or ethanol to give the corresponding alkoxylsilanylcarboranes. Li_2[2,4-$C_2B_5H_5$] reacts with deuterium oxide to give the C,C′-dideuterio derivative of 2,4-$C_2B_5H_7$ [13].

References for 13.6.2:

[1] Jemmis, E. D. (J. Am. Chem. Soc. **104** [1982] 7017/20).

[2] Brint, P.; Cronin, J. P.; Seward, E.; Whelan, T. (J. Chem. Soc. Dalton Trans. **1983** 975/80).

[3] Anderson, J. A.; Astheimer, R. J.; Odom, J. D.; Sneddon, L. G. (J. Am. Chem. Soc. **106** [1984] 2275/83).

[4] Nam, W.; Soltis, M.; Gordon, C.; Lee, S.; Onak, T. (J. Magn. Resonance **59** [1984] 399/405).

[5] Jarvis, W.; Inman, W.; Powell, B.; DiStefano, E. W.; Onak, T. (J. Magn. Resonance **43** [1981] 302/15).

[6] Jarvis, W.; Abdou, Z. J.; Onak, T. (Polyhedron **2** [1983] 1067/70).

[7] DeKock, R. L.; Jasperse, C. P. (Inorg. Chem. **22** [1983] 3843/8).

[8] Siwapinyoyos, G.; Onak, T. (Inorg. Chem. **21** [1982] 156/63).

[9] Fuller, K.; Onak, T. (J. Organometal. Chem. **249** [1983] C6/C8).

[10] Astheimer, R. J.; Sneddon, L. G. (Inorg. Chem. **23** [1984] 3207/12).

[11] Oh, B.; Onak, T. (Inorg. Chem. **21** [1982] 3150/4).

[12] Wilczynski, R.; Sneddon, L. G. (Inorg. Chem. **21** [1982] 506/14).

[13] Zhigach, A. F.; Petrunin, A. B.; Bekker, D. B. (Zh. Obshch. Khim. **50** [1980] 2727/32; J. Gen. Chem. [USSR] **50** [1980] 2206/11).

13.6.3 Metallacarboranes Containing Five Boron Atoms

Structural characterization of 1,2-$(CH_3)_2$-3,1,2-(η-C_5H_5)$CoC_2B_5H_5$ shows that the cage atoms adopt a closed dodecahedral-type geometry [1]. Reaction of *closo*-2,4-$C_2B_5H_5$-2,4-$(CH_3)_2$ with Co$[P(C_2H_5)_3]_4$ gives 4-$(C_2H_5)_3$P-1,7-$(CH_3)_2$-μ-4,8-Co(H)$[P(C_2H_5)_3]_2$-μ-H-μ-$P(C_2H_5)_2$-1,4,7-$CCoCB_5H_4$. In contrast, reaction of *closo*-2,4-$C_2B_5H_5$-2,4-$(CH_3)_2$ with Fe$[CNC(CH_3)_3]_5$ yields 4,4,4-$[(CH_3)_3CNC]_3$-1,7-$(CH_3)_2$-1,4,7-$CFeCB_5H_5$ [2]. The four-carbon metallacarborane complex 2-[η^6-$(CH_3)C_6H_5$]Fe-6,7,9,10-$(CH_3)_4C_4B_5H_5$ is produced from the reaction of thermally generated iron atoms with pentaborane(9) and 2-butyne in the presence of toluene [3]. The reaction of *closo*-2,4-$C_2B_5H_5$-2,4-$(CH_3)_2$ with Pt_2(μ-COD)$[P(C_2H_5)_3]_4$ (COD = cycloocta-1,5-diene) produces 4,4-$[(C_2H_5)_3P]_2$-1,7-$(CH_3)_2$-1,4,7-$CPtCB_5H_5$ and 1,1-$[(C_2H_5)_3P]_2$-6,6-$[(C_2H_5)_3P]_2$-4,5-$(CH_3)_2$-1,4,5,6-$PtC_2PtB_5H_5$ [4].

References to 13.6.3:

[1] Zimmerman, G. J.; Sneddon, L. G. (Acta Cryst. C **39** [1983] 856/8).
[2] Barker, G. K.; Garcia, M. P.; Green, M.; Stone, F. G. A.; Parge, H. E.; Welch, A. J. (J. Chem. Soc. Chem. Commun. **1982** 688/9).
[3] Micciche, R. P.; Briguglio, J. J.; Sneddon, L. G. (Organometallics **3** [1984] 1396/402).
[4] Barker, G. K.; Garcia, M. P.; Green, M.; Stone, F. G. A.; Welch, A. J. (J. Chem. Soc. Chem. Commun. **1982** 46/8).

13.7 Carboranes Containing Six Boron Atoms

For previous data on carboranes containing six boron atoms, see "Boron Compounds" 2nd Suppl. Vol. 2, 1982, pp. 249/52.

The LUMO (lowest unoccupied molecular orbital) of **1,6-$C_2B_6H_8$** is concentrated predominantly on the symmetry-related and four-connected boron atoms B(4) and B(8) (29% on each), and to a lesser extent on the carbon atoms C(1) and C(6) (11% on each). This molecular orbital lies at −8.5 eV [1]. A 6-electron rule for explaining *closo*-$C_2B_6H_8$ preference for a dodecahedron rather than a hexagonal bipyramid structure has been developed [2]. MNDO MO calculations on $C_2B_6H_8$ as well as other carboranes are used to assess the accuracy of A. J. Stone's theory (1980) of their electronic structure and bonding [3]. Aromatic solvent-induced ^{1}H NMR shifts are reported for **1,7-$C_2B_6H_8$** and correlated to PRDDO MO derived hydrogen charges [4]. For the boron atoms in 1,7-$C_2B_6H_8$, cage "umbrella" angles of 95.2° (for B(2,8)), 102° (for B(3,4)), and 104° (for B(5,6)) are used in an empirical relationship, which includes the number of adjacent cage carbon atoms, to predict $^1J(^{11}B\text{-}^1H)$ values of 166, 181, and 166 Hz, respectively; these values are only in fair agreement with the observed values of 168, 171, and 168 Hz, respectively [5].

The four-carbon carborane **$(C_2H_5)_4C_4B_6(C_2H_5)_6$** is obtained from the treatment of *closo*-1,5-C_2B_3-1,2,3,4,5-$(C_2H_5)_5$ with potassium and subsequently with iodine. The ^{11}B NMR spectrum of $(C_2H_5)_4C_4B_6(C_2H_5)_6$ displays signals at δ (in ppm) = −20.7 (intensity 2), −6.3 (1), 6.7 (2), 50.3 (1); ^{13}C NMR data: δ (in ppm) = −6.3 (intensity 1), 29.2 (2), 37.9 (1). The data are consistent with a skeletal structure, **Fig.** 13-**2**, p. 170, analogous to that of $B_{10}H_{14}$, and there is some evidence of fluxional behavior at room temperature [6].

1:2′-[1,5-$C_2B_3H_4$]$_2$, produced from the gas-phase pyrolysis of 1,5-$C_2B_3H_5$ at 400°C, exhibits ^{11}B NMR resonances at δ (in ppm) = 8.7, J(B-H) = 188 Hz, for B(3)′ and B(4)′; = 7.4 for B(2)′; = 5.6, J(B-H) = 190 Hz, for B(2,3,4); ^{1}H data: δ (in ppm) = 5.66 for HC(5); = 5.40, J(HCBH) =

References for 13.7 on pp. 170/1

2.0 Hz, for HC(1)′ and HC(5)′; =4.49, J(B-H)=188 Hz, for HB(3)′ and HB(4)′; =4.48, J(B-H)= 188 Hz, for HB(2,3,4). The infrared spectrum of the compound showed absorption bands at (frequencies in cm^{-1}) 2620vs, 1550vw, 1490m,sh, 1482m,sh, 1476m,sh, 1472s,sh, 1456s,sh, 1450s, 1409w,sh, 1245w, 1205w,sh, 1190w, 1105s, 1060m, 945w, 875w, 850w,sh, 805w [7].

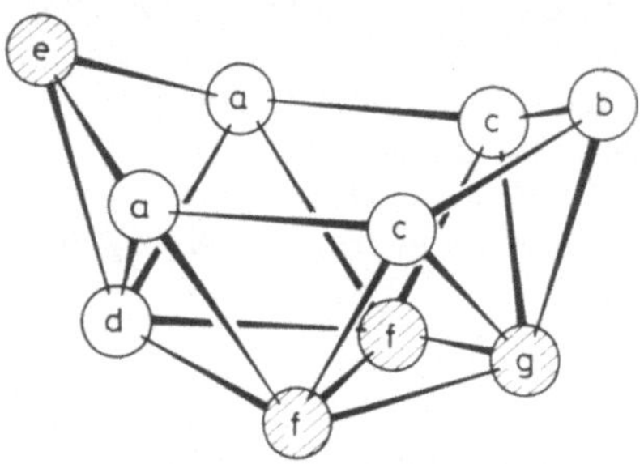

Fig. 13-2. Proposed skeletal structure of $(C_2H_5)_4C_4B_6(C_2H_5)_6$; open circles are boron atoms and filled circles are carbon atoms [6].

2:2′-[1,5-$C_2B_3H_4$]$_2$ is formed from 1,5-$C_2B_3H_5$ in the presence of $PtBr_2$ [8]. The exopolyhedral $J(^{11}B-^{11}B)$ for 2:2′-[1,5-$C_2B_3H_4$]$_2$ is found to be 137 Hz which corresponds to a 39% s-character for the coupled boron atoms [9].

The thiocarborane ***arachno*-4,6,8-$SC_2B_6H_{10}$** is produced from the action of Na_2SO_3 on $[7,9\text{-}C_2B_9H_{12}]^-$. The melting point of *arachno*-4,6,8-$SC_2B_6H_{10}$ is 104 to 105°C; ^{11}B NMR data: δ (in ppm)=7.60, J(B-H)=156 Hz (1B); =5.70, J(B-H)=168 Hz (2B); −21.68, J(B-H)=160 Hz (2B); =−35.60, J(B-H)=175 Hz (1B) [10].

The structures of 2,7,8,10,11- and 9,7,8,10,11-(η^5-C_5H_5)Co$(CH_3)_4C_4B_6H_6$, determined by single-crystal X-ray diffraction, both contain 11-vertex CoC_4B_6 cages having icosahedral-fragment (capped-pentagonal-antiprism) geometry with all four skeletal carbon atoms on the five-membered open face. In the 2,7,8,10,11 system (isomer I) the metal atom is located in the lower (CoB_4) belt adjacent to two cage C atoms, whereas in the 9,7,8,10,11 species (isomer II), the metal is on the open rim of the cage [11]. Extended Hückel molecular orbital calculations are reported for isomers of $Pt(PH_3)_2(C_2B_6H_8)$ [1].

Several metal-carbon-boron containing "sandwich" compounds have been prepared in which the structures appear to obey carborane structure rules; these include: the pentadecker compound $[C_5H_5Co((C_2H_5)_2(CH_3)C_3B_2(C_2H_5)_2)Ni]_2[H(C_2H_5)_2C_3]B_2(CH_3)_2$ with fused cages [12], and derivatives of the triple-decker $Co_2(C_2B_2SH_4)_3$ [13].

References for 13.7:

[1] Mingos. D. M. P.; Welch, A. J. (J. Chem. Soc. Dalton Trans. **1980** 1674/81).
[2] Jemmis, E. D. (J. Am. Chem. Soc. **104** [1982] 7017/20).
[3] Brint, P.; Cronin, J. P.; Seward, E.; Whelan, T. (J. Chem. Soc. Dalton Trans. **1983** 975/80).
[4] Jarvis, W.; Inman, W.; Powell, B.; DiStefano, E. W.; Onak, T. (J. Magn. Resonance **43** [1981] 302/15).
[5] Jarvis, W.; Abdou, Z. J.; Onak, T. (Polyhedron **2** [1983] 1067/70).
[6] Köster, R.; Seidel, G.; Wrackmeyer, B. (Angew. Chem. **96** [1984] 520/1; Angew. Chem. Intern. Ed. Engl. **23** [1984] 512/4).
[7] Astheimer, R. J.; Sneddon, L. G. (Inorg. Chem. **22** [1983] 1928/34).
[8] Corcoran, E. W.; Sneddon, L. G. (J. Am. Chem. Soc. **106** [1984] 7793/800).
[9] Anderson, J. A.; Astheimer, R. J.; Odom, J. D.; Sneddon, L. G. (J. Am. Chem. Soc. **106** [1984] 2275/83).
[10] Base, K.; Hermanek, S.; Hanousek, F. (J. Chem. Soc. Chem. Commun. **1984** 299/300).

[11] Maynard, R. B.; Sinn, E.; Grimes, R. N. (Inorg. Chem. **20** [1981] 3858/63).
[12] Whiteley, M. W.; Pritzkow, H.; Zenneck, U.; Siebert, W. (Angew. Chem. **94** [1982] 464; Angew. Chem. Intern. Ed. Engl. **21** [1982] 453/4).
[13] Siebert, W.; Schmidt, H.; Full, R. (Z. Naturforsch. **35b** [1980] 873/81).

13.8 Carboranes Containing Seven Boron Atoms

For previous data on carboranes containing seven boron atoms, see "Boron Compounds" 2nd Suppl. Vol. 2, 1982, pp. 252/6.

13.8.1 Carboranes

Tensor surface harmonic theory is used to describe the structure of $\mathbf{[C_2B_7H_9]^{2-}}$ as a *nido*-cluster with a nonplanar open face [1]. MNDO MO calculations on $C_2B_7H_9$ as well as other carboranes are used to assess the accuracy of A. J. Stone's theory (1980) of their electronic structure and bonding [2]. The effect of the carbon atoms in the capping positions of $C_2B_7H_9$, when compared to $[B_9H_9]^{2-}$, is to drain electronic charge from the prism ends of the tricapped trigonal prismatic cluster; this results in the lengthening of the triangle-edges of the prism, thus making the prism larger [3]. Aromatic solvent-induced ^1H NMR shifts are reported for 1,6-$C_2B_7H_9$ and correlated to PRDDO MO-derived hydrogen charges [4]. For the boron atoms in 1,6-$C_2B_7H_9$, cage "umbrella" angles of 105° (for B(2,3)), 106° (for B(4,5,7,9)), and 110° (for B(8)) are used in an empirical relationship, which includes the number of adjacent cage carbon atoms, to predict $^1J(^{11}B$-$^1H)$ values of 180, 165, and 165 Hz, respectively; these values agree reasonably well with the observed values of 176, 162, and 164 Hz, respectively [5].

***nido*-2,6-$\mathbf{C_2B_7H_{11}}$**, melting point of 90°C, is prepared from the action of $H_2CO/HCl/H_2O$ on $K[7,8\text{-}C_2B_9H_{12}]$. The 70.6 MHz ^{11}B NMR spectrum of *nido*-2,6-$C_2B_7H_{11}$ is composed of five doublets at 10.4, 4.7, −3.0, −5.2, and −55.2 ppm and one triplet at −28.3 ppm; the 220-MHz ^1H NMR spectrum shows two sharp 1:1 singlets at 2.81 and 1.95 ppm, a broad singlet of a hydrogen bridge at −0.87, and a number of overlapping signals of terminal B-H protons in the region of 3.72 to −0.15 ppm. The infrared spectrum of *nido*-2,6-$C_2B_7H_{11}$ consists of absorptions at (frequencies in cm^{-1}, s = strong, m = medium, w = weak, v = very, sh = shoulder, b = broad) 3048 w, 2982 w (cage C-H stretching), 2585 s, sh, 2505 s, 2460 s (terminal B-H stretching), 1880 vw (symmetric B-H-B stretching), and 1535 m, b (asymmetric B-H-B stretching) [6].

The ^{11}B NMR spectrum of **4,6-$\mathbf{C_2B_7H_{13}}$** shows signals at δ (in ppm) = −27.9, J(B-H) = 157 Hz, J(B-H bridging) = 40 Hz, for B(8); = 52.1, J(B-H) = 150 Hz, for B(3); −0.1, J(B-H) = 163 Hz, for B(5); = 0.5, J(B-H) = 152 Hz, J(B-H bridging) = 35 Hz, for B(7,9); = −17.3, J(B-H) = 165 Hz, for B(1,2). The STO-3G Mulliken atomic charges and overlap populations for 4,6-$C_2B_7H_{13}$ are given in **Fig.** 13-**3**, p. 172 [7].

The *nido*-carborane **1,2,8,10-$\mathbf{C_4B_7H_{11}}$**, **Fig.** 13-**4**, p. 172, produced from the gas-phase pyrolysis of 1,5-$C_2B_3H_5$ at 400°C, exhibits ^{11}B NMR resonances at δ (in ppm) = −14.1, J(B-H) = 150 Hz, for B(7,11); = −15.6, J(B-H) = 190 Hz, for B(9); = −16.6, J(B-H) = 171 Hz, for B(3,6 or 4,5); = −23.7, J = 172 Hz, for B(4,5 or 3,6); ^1H data: δ (in ppm) = 2.50, J(B-H) = 186 Hz, for HB(9); = 2.45, J(B-H) = 146 Hz, for HB(7,11); = 2.33, J(B-H) = 168 Hz, for HB(3,6 or 4,5); = 2.02, J(B-H) = 176 Hz, for H(B4,5 or 3,6); = 1.19 for HC(1,2,8,10); ^{13}C data: δ = 29.5 ppm, J(CH) = 161 Hz, J(C-B) = 38 Hz, for C(8,10); = 28.6 ppm for C(2); = −4.7, J(C-H) = 194 Hz, for C(1). The infrared spectrum of the compound showed absorption bands at (frequencies in cm^{-1}) 2615 vs, 2565 s, 1745 vw, 1078 m, 1042 w, br, 995 m, sh, 990 m, 968 m [8].

References for 13.8.1 on pp. 173/4

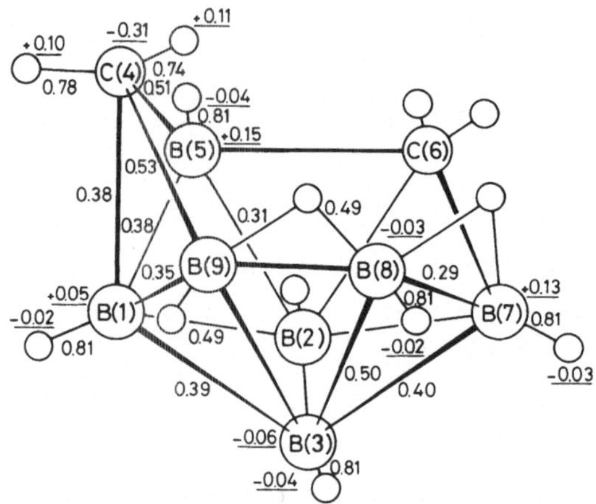

Fig. 13-3. STO-3G Mulliken atomic charges (underlined) and overlap populations of 4,6-$C_2B_7H_{13}$ [7].

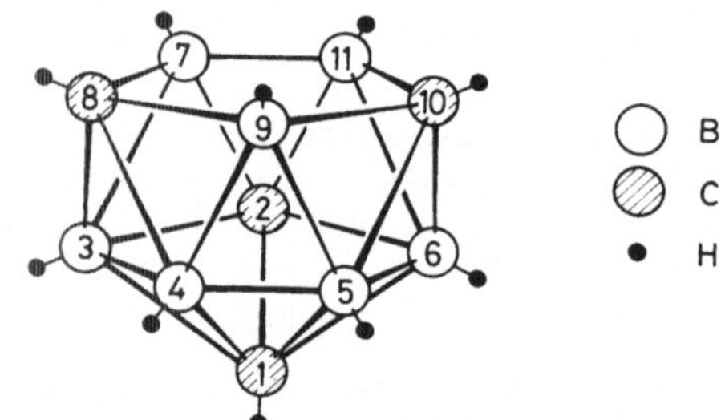

Fig. 13-4. Proposed structure of $C_4B_7H_{11}$ [8].

Two-dimensional boron-11-boron-11 nuclear magnetic resonance spectroscopy confirms many of the B-B atom connectivities in **$(CH_3)_4C_4B_7H_9$, (Fig.** 13-**5)** I; the ^{11}B NMR chemical shifts of the compound are: δ (in ppm) = 8.6 for B(9); = 3.1 for B(4); = 2.5 for B(6); = −3.9 for B(10); = −7.4 for B(1); = −11.0 for B(11); = −41.6 for B(5) [9]. $(CH_3)_4C_4B_7H_9$ is obtained from the controlled degradation of $(CH_3)_4C_4B_8H_8$ in 95% C_2H_5OH and air. Electrophilic bromination of $(CH_3)_4C_4B_7H_9$ produces **$(CH_3)_4C_4B_7H_8$-11-Br**, which is shown in an X-ray diffraction study to have an open-cage structure with a bridging -CH(CH_3)-group across the open face; $(CH_3)_4C_4B_7H_9$ can be described as an 11-vertex *arachno* cage. Boron-11 NMR data for $(CH_3)_4C_4B_7H_8$-11-Br: δ (in ppm) = 8.4, J(B-H) = 156 Hz (1B); = 0.5, J(B-H) = 166 Hz (2B); = −2.2, J(B-H) = 150 Hz (1B); = −7.3 (1B); −8.6, J(B-H) = 182 Hz (1B); = −41.7, J(B-H) = 156 Hz (1B). Deprotonation of $(CH_3)_4C_4B_7H_9$ with NaH in tetrahydrofuran generates the salt **Na[$(CH_3)_4C_4B_7H_8$]** in which the CH proton on the bridging group has been removed. Protonation of Na[$(CH_3)_4C_4B_7H_8$] gives a new isomer of $(CH_3)_4C_4B_7H_9$, II (see Fig. 13-5) [10]. ^{11}B NMR data for $(CH_3)_4C_4B_7H_9$, isomer I: δ (in ppm) = 8.7, J(B-H) = 128 Hz (1B); = 2.9, J(B-H) = 181 Hz (2B); = −3.3, J(B-H) = 156 Hz (1B); = −7.3, J(B-H) = 150 Hz (1B); −11.1, J(B-H) = 180 Hz (1B); = −41.7, J(B-H) = 156 Hz (1B). ^{11}B NMR data for Na[$(CH_3)_4C_4B_7H_8$]: δ (in ppm) = 7.6, J(B-H) = 116 Hz (1B); = 2.7, J(B-H) = 132 Hz (1B); = −4.3, J(B-H) = 147 Hz (2B); = −17.1, J(B-H) = 122 Hz (1B); = −26.8, J(B-H) = 146 Hz (2B). ^{11}B NMR data for $(CH_3)_4C_4B_7H_9$ (isomer II): δ (in ppm) = 4.4, J(B-H) = 147 Hz (3B); = −7.8, doublet (1B); = −10.5, doublet (1B); = −12.0, doublet (1B); = −40.3, J(B-H) = 156 Hz (1B). Partial mass spectroscopic data as well as proton NMR and infrared data are reported for $(CH_3)_4C_4B_7H_8$-11-Br as well as for the two isomers of $(CH_3)_4C_4B_7H_9$ [10].

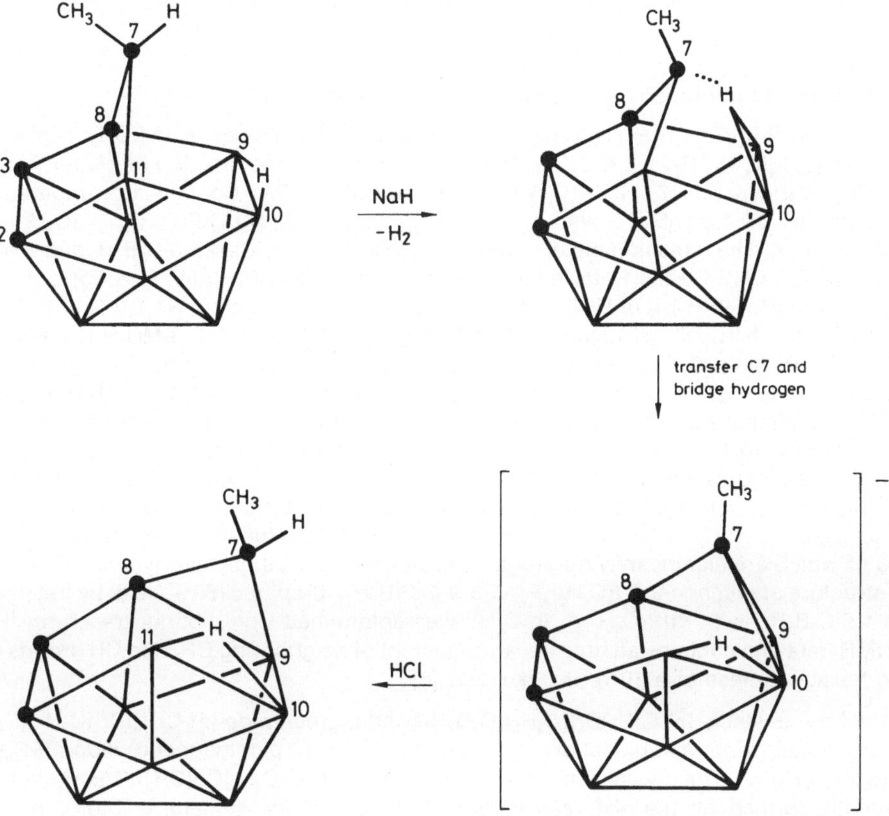

Fig. 13-5. Reaction scheme for the conversion of $(CH_3)_4C_4B_7H_9$, I, to its isomer II [10]. The CH_3 on C(7) is shown but all of the other carbon atoms (filled circles) also have CH_3 groups attached. Boron atoms, with attached terminal hydrogen atoms, are implied at all of the other vertices of the polyhedra.

2′:2-[1′,5′-$C_2B_3H_4$][1,6-$C_2B_4H_5$], formed from the copyrolysis of 1,5-$C_2B_3H_5$ and 1,6-$C_2B_4H_6$ at 400°C, exhibits an exopolyhedral B-B NMR coupling constant of 126 Hz [8].

References for 13.8.1:

[1] Stone, A. J.; Alderton, M. J. (Inorg. Chem. **21** [1982] 2297/302).
[2] Brint, P.; Cronin, J. P.; Seward, E.; Whelan, T. (J. Chem. Soc. Dalton Trans. **1983** 975/80).
[3] O'Neill, M. E.; Wade, K. (Polyhedron **2** [1983] 963/6).
[4] Jarvis, W.; Inman, W.; Powell, B.; DiStefano, E. W.; Onak, T. (J. Magn. Resonance **43** [1981] 302/15).
[5] Jarvis, W.; Abdou, Z. J.; Onak, T. (Polyhedron **2** [1983] 1067/70).
[6] Štíbr, B.; Plešek, J.; Heřmánek, S. (Inorg. Syn. **22** [1983] 237/9).
[7] Dolansky, J.; Heřmánek, S.; Zahradnik, R. (Collection Czech. Chem. Commun. **46** [1981] 2479/93).

[8] Astheimer, R. J.; Sneddon, L. G. (Inorg. Chem. **22** [1983] 1928/34).
[9] Venable, T. L.; Hutton, W. C.; Grimes, R. N. (J. Am. Chem. Soc. **106** [1984] 29/37).
[10] Finster, D. C.; Grimes, R. N. (J. Am. Chem. Soc. **103** [1981] 2675/83).

13.8.2 Metallacarboranes Containing Seven Boron Atoms

6,7-$[(C_5H_5)Ni]_2(\eta^5\text{-}1\text{-}CB_7H_8)$ is formed from the action of nickelocene on $[(CH_3)_4N][4\text{-}CB_8H_{13}]$ [1]. *closo*-[6,6-$[(C_2H_5)_3P]_2$-1,2,6-$C_2CoB_7H_9$] is produced from the reaction of *nido*-4,5-$C_2B_7H_{11}$ with $Co[P(C_2H_5)_3]_4$ [2]. [2-H-2,2-$[(C_2H_5)_3P]_2$-1,6,2-$C_2MB_7H_9$] (M = Co, Rh) are prepared from the reactions of 1,3-$C_2B_7H_{13}$ with $Co[P(C_2H_5)_3]_4$ and $[Rh(\eta\text{-}C_3H_5)(P(C_2H_5)_3)_2]$ (C_3H_5 = allyl), respectively [3]. The transition metal complexes *closo*-4,7-$(C_5H_5Co)_2$-2,3-$C_2B_7H_9$ and *arachno*-9,9-$[(C_6H_5)_3P]_2$-5,6,9-$C_2PtB_7H_{11}$ have been prepared from 4,5-$C_2B_7H_{11}$ [4]. Reactions of Na(1,3-$C_2B_7H_{12}$) with $[RhCl(PR_3)_2]_2$ or $RhCl(PR_3)_3$ (R = C_6H_5, p-$C_6H_4CH_3$) and of Na(1,3-R'_2-1,3-$C_2B_7H_{10}$) (R' = H, CH_3) with $IrCl[P(C_6H_5)_3]$ afford [*closo*-2,3-R'_2-6,6-$(PR_3)_2$-6-H-6,2,3-$MC_2B_7H_7$]. Reactions of Na(1-R-3-R'-1,3-$C_2B_7H_{10}$) (R = R' = H, R = C_6H_5, R' = H) with $RuHCl[P(C_6H_5)_3]_3$ yield the unsaturated complexes hyper-*closo*-2-R-3-R'-6,6-$[(C_6H_5)_3P]_2$-6,2,3-$RuC_2B_7H_7$. The latter compound is an effective catalyst for the homogeneous hydrogenation of terminal alkenes, and it reacts with CO to form *closo*-6,6-$(CO)_2$-6-$[(C_6H_5)_3P]$-6,2,3-$RuC_2B_7H_9$. The addition of excess $(C_6H_5)_3P$ to a CH_2Cl_2 solution of hyper-*closo*-2-R-3-R'-6,6-$[(C_6H_5)_3P]_2$-6,2,3-$RuC_2B_7H_7$ results in a thermochromic solution which exhibits a remarkable equilibrium between hyper- *closo*-6,6-$[P(C_2H_5)_3]_2$-6,2,3-$RuC_2B_7H_9$ and *closo*-6,6,6-$[(C_2H_5)_3P]_3$-6,2,3-$RuC_2B_7H_9$, the polyhedral structures of which are significantly different, as evidenced by multinuclear dynamic FT NMR [5]. The structure of *arachno*-9,9-$[(C_6H_5)_3P]_2$-5,6,9-$C_2PtB_7H_{11}$, prepared (80% yield) by treatment of *nido*-4,5-$C_2B_7H_{11}$ with $Pt[P(C_6H_5)_3]_4$ in C_6H_6, was determined from Fourier-transform ^1H and ^{11}B NMR data and contains an unusual arrangement of neighboring CH_2 and CH groups in the open hexagonal face of a 10-vertex skeleton [6].

A yellow, unstable $(\eta^5\text{-}C_5H_5)Cr[(C_2H_5)_4C_4B_7H_7]$ and a green *nido*-$(\eta^5\text{-}C_5H_5)Cr[(C_2H_5)_4C_4B_7H_7]$ are produced from air oxidation of solutions of red, paramagnetic *nido*-$(\eta^5\text{-}C_5H_5)Cr$-$[(C_2H_5)_4C_4B_8H_8]$ [7]. The crystal and molecular structure of $(\eta^5\text{-}C_5H_5)Co(CH_3)_4C_4B_7H_7$ (known as isomer III), formed by thermal rearrangement of one of its isomers at 140°C, has been determined by single-crystal X-ray diffraction. $(\eta^5\text{-}C_5H_5)Co(CH_3)_4C_4B_7H_7$, isomer III, has an open-cage geometry in which all four carbons and two boron atoms reside on a 6-membered open face, with one of the carbon atoms isolated from the other three. Lattice parameters of the monoclinic crystals are a = 25.944(8), b = 8.686(3), c = 15.410(5) Å, β = 108.47(3)°; D_{calc} = 1.27 g/cm³. Space group C2/c-C^6_{2h} (No. 15); Z = 8 [8].

References for 13.8.2:

[1] Štíbr, B.; Janoušek, Z.; Baše, K.; Plešek, J.; Solntsev, K. A.; Butman, L. A.; Kuznetsov, I. I.; Kuznetsov, N. T. (Collection Czech. Chem. Commun. **49** [1984] 1660/4).
[2] Barker, G. K.; Garcia, M. P.; Green, M.; Pain, G. N.; Stone, F. G. A.; Jones, S. K. R.; Welch, A. J. (J. Chem. Soc. Chem. Commun. **1981** 652/3).
[3] Barker, G. K.; Garcia, M. P.; Green, M.; Stone, F. G. A.; Bassett, J. M.; Welch, A. J. (J. Chem. Soc. Chem. Commun. **1981** 653/5).
[4] Štíbr, B.; Plešek, J.; Baše, K.; Zakharova, I. A. (Proc. Conf. Coord. Chem. **8** [1980] 423/5).
[5] Jung, C. W.; Baker, R. T.; Hawthorne, M. F. (J. Am. Chem. Soc. **103** [1981] 810/6).
[6] Štíbr, B.; Heřmánek, S.; Plešek, J.; Baše, K.; Zakharova, I. A. (Chem. Ind. [London] **1980** 468).
[7] Maynard, R. B.; Zhu-Ting, W.; Sinn, E.; Grimes, R. N. (Inorg. Chem. **22** [1983] 873/8).
[8] Maynard, R. B.; Sinn, E.; Grimes, R. N. (Inorg. Chem. **20** [1981] 1201/6).

13.9 Carboranes Containing Eight Boron Atoms

For previous data on carboranes containing eight boron atoms, see "Boron Compounds" 2nd Suppl. Vol. 2, 1982, pp. 257/63.

13.9.1 Compounds Containing Eight Boron Atoms and One Cage Carbon Atom

***arachno*-4-CB_8H_{14}** is formed from the reaction of Na[*nido*-6-CB_9H_{12}] with $FeCl_3/HCl/H_2O$; *arachno*-4-CB_8H_{14} shows ^{11}B resonances at δ (in ppm) = 17.0, J(B-H) = 160 Hz, for B(7); = −3.7, J(B-H) = 170 Hz, for B(1); = −6.3, J(B-H) = 160 Hz, J(B-H bridging) = 35 Hz, for (5,9); = −34.9, J(B-H) = 155 Hz, J(B-H bridging) = 55 Hz, for B(6,8); = −41.1, J(B-H) = 155 Hz, for B(2,3). ^1H NMR data: δ (in ppm) = 0.10 for equatorial CH, = −1.75 for axial CH, = −0.45 and −3.52 for the two hydrogen bridge types. The STO-3G Mulliken atomic charges and overlap populations of 4-CB_8H_{14} are shown in **Fig. 13-6**. 4-CB_8H_{14} reacts with K_2CO_3/H_2O, and subsequently with $[N(CH_3)_4]Cl$ to give **$[N(CH_3)_4][4\text{-}CB_8H_{13}]$**; ^{11}B NMR data: δ (in ppm) = 1.6 (2B), −6.5 (1B), −24.0 (1B), −32.8 (2B), −37.2 (2B). 4-CB_8H_{14} is dehydrogenated at 623 K to give 4-(or 7-)CB_8H_{12} [1, 2]. Treating Na[4-CB_8H_{13}] with nickelocene gives 6-η^5-C_5H_5Ni-η^5-1-CB_8H_9, which rearranges to 10-η^5-C_5H_5Ni-1-CB_8H_9. Reaction of $[N(CH_3)_4]$[4-CB_8H_{13}] with nickelocene in acetonitrile give the same products as the sodium salts and, in addition, 6,10-$(\eta^5\text{-}C_5H_5Ni)_2$-1-$CB_7H_8$. The reaction of 4-$CB_8H_{14}$ with $CoCl_2$ and cyclopentadiene affords the red-orange complex [2-η^5-C_5H_5Co-η^5-1-CB_8H_9]$^-$, isolated as the tetramethylammonium salt [3,4]. Reaction of 4-CB_8H_{14} with $CoCl_2/C_5H_6/KOH/C_2H_5OH$ yields $[N(CH_3)_4]$[2-(C_5H_5)Co-1-CB_8H_9] [4]. Reaction of Li[*arachno*-CB_8H_{13}] with $[(C_6H_5)_3P]_3IrCl$ gives [1,2,2-$[P(C_6H_5)_3]_3$-2-H-*closo*-2,10-$IrCB_8H_8$] [5]. The structure of μ-1,2-CH_3CO_2-2-H-2,10-$[(C_6H_5)_3P]_2$-closo-1,2-$IrCB_8H_7$ was determined by X-ray crystallography to have a *closo*-cluster geometry [6]. Crystals of 3-CH_3O-7-$(C_6H_5)_3P$-7-$P(C_6H_4)(C_6H_5)_2$-10-OH-*iso-nido*-7,10-$IrCB_8H_6$, with a C-B bond between (B_8) and a carbon of the C_6H_4, are hexagonal, space group $P6_1$-C_6^2 (No. 169) with a = 11.145(2) and c = 51.944(12) Å [7]. Thermal treatment of 9,9-$[(C_6H_5)_3P]_2$-6,9-$CPtB_8H_{11}$ at 250°C gives rise to an inner-sphere rearrangement resulting in the elimination of four hydrogen atoms and the coupling of the carborane ligand with the two

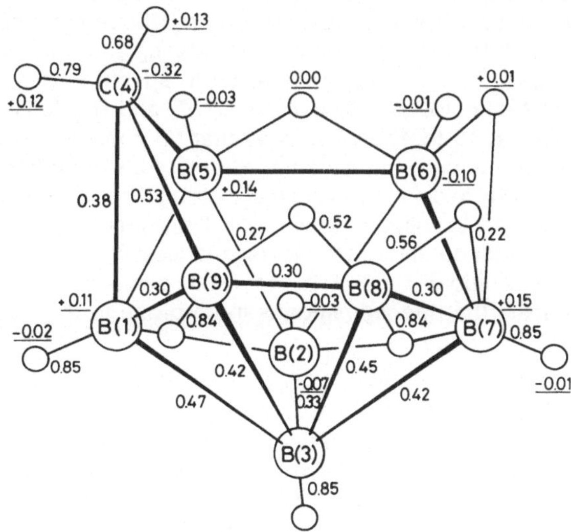

Fig. 13-6. STO-3G Mulliken atomic charges (underlined) and overlap populations of 4-CB_8H_{14} [2].

References for 13.9.1 on p. 176

phenyl groups of the phosphane ligands [8]. Heating 9,9-[$(C_6H_5)_3P]_2$-6,9-$CPtB_8H_{12}$ at 493 to 520 K results in the formation of 9,9-[$(C_6H_5)_2P]_2$-$(C_6H_4)_2$-6,9-$CPtB_8H_{10}$ involving two intramolecular $PtPC_2B$ rings [9].

References for 13.9.1:

[1] Baše, K.; Štíbr, B.; Dolanský, J.; Duben, J. (Collection Czech. Chem. Commun. **46** [1981] 2345/53).

[2] Dolanský, J.; Heřmánek, S.; Zahradnik, R. (Collection Czech. Chem. Commun. **46** [1981] 2479/93).

[3] Štíbr, B.; Janoušek, Z.; Baše, K.; Dolanský, J.; Heřmánek, S.; Solntsev, K. A.; Butman, L. A.; Kuznetsov, I. I.; Kuznetsov, N. T. (Polyhedron **1** [1982] 833/4).

[4] Štíbr, B.; Janoušek, Z.; Baše, K.; Plešek, J.; Solntsev, K. A.; Butman, L. A.; Kuznetsov, I. I.; Kuznetsov, N. T. (Collection Czech. Chem. Commun. **49** [1984] 1660/4).

[5] Alcock, N. W.; Taylor, J. G.; Wallbridge, M. G. H. (J. Chem. Soc. Chem. Commun. **1983** 1168/9).

[6] Crook, J. E.; Greenwood, N. N.; Kennedy, J. D.; McDonald, W. S. (J. Chem. Soc. Chem. Commun. **1983** 83/4).

[7] Crook, J. E.; Greenwood, N. N.; Kennedy, J. D.; McDonald, W. S. (J. Chem. Soc. Chem. Commun. **1981** 933/4).

[8] Kukina, G. A.; Sergienko, S. V.; Porai-Koshits, M. A.; Baše, K.; Zakharova, I. A. (Izv. Akad. Nauk SSSR Ser. Khim. **1981** 2838; Bull. Acad. Sci. USSR Div. Chem. Sci. **1981** 2369).

[9] Baše, K.; Štíbr, B.; Kukina, G. A.; Zakharova, I. A. (Proc. Conf. Coord. Chem. **8** [1980] 17/9; C.A. **95** [1981] No. 23933).

13.9.2 Compounds Containing Eight Boron Atoms and Two Cage Carbon Atoms

MNDO MO calculations on $\mathbf{C_2B_8H_{10}}$ as well as other carboranes are used to assess the accuracy of A. J. Stone's theory (1980) of their electronic structure and bonding [1]. Aromatic solvent-induced ^{1}H NMR shifts are reported for 1,6-$C_2B_8H_{10}$ and correlated to PRDDO MO-derived hydrogen charges [2]. The He(I) photoelectron spectrum of 1,10-$C_2B_8H_{10}$ exhibits bands at (in eV) 10.5, 11.4 (shoulder), 11.7, 12.6, 13.6, 14.5, and 16.0 [3]. A study of the Raman spectrum of 1,10-$C_2B_8H_{10}$ at 15 to 420 K indicates that on cooling to 15 K the broad background band increases in intensity and becomes structured. During heating the intensity of the background begins slowly to decrease so that at melting it has disappeared [4]. Reaction of 1,2-$C_2B_8H_{10}$ with LiC_4H_9 and subsequently with CH_3I produces **1,2-$\mathbf{C_2B_8H_8}$-1,2-$\mathbf{(CH_3)_2}$**; the ^{11}B NMR spectrum of the latter compound shows resonances at δ (in ppm) = 29.9, −6.6, −15.5, −21.7, and −24.6 in an area ratio of 1:1:2:2:2. The results of mono C-methylation of this same carborane system suggests that the 2-position is more prone to this substitution sequence than is the 1-position of this cage [5].

The compounds 9-(C_5H_5)-*nido*-7,8,9-$C_2NiB_8H_{11}$, 9-(C_5H_5)-μ-10,11-$(C_6H_5)_3$PAu-*nido*-7,8,9-$C_2NiB_8H_{10}$, and 1,3-$(C_5H_5)_2$-*closo*-1,2,3,4-$CrCCrCB_8H_{10}$ are prepared starting from *nido*-5,6-$C_2B_8H_{12}$, and their structures were determined by X-ray crystallography [6]. 1,1-[$(C_2H_5)_3P]_2$-*closo*-1,2,4-$CoC_2B_8H_{10}$ is prepared from the reaction of *nido*-5,6-$C_2B_8H_{12}$ with $Co[P(C_2H_5)_3]_4$ [7]. Electrophilic bromination of 1,2,3-$(C_5H_5)CoC_2B_8H_{10}$ by Br_2 in CH_2Cl_2 gives 1,2,3-$(C_5H_5)CoC_2B_8H_9$-10-Br and 1,2,3-$(C_5H_5)CoC_2B_8H_9$-4-Br. Further bromination yields 1,2,3-$(C_5H_5)CoC_2B_8H_5$-4,5,6,10,11-Br_5 [8]. Crystals of 9,9-[$P(C_6H_5)_3]_2$-7,8,9-$C_2RhB_8H_{11}$ are monoclinic, space group $P2_1/n$-C_{2h}^5 (No. 14), with a = 18.950(6), b = 12.057(3), c = 15.585(5) Å, and

$\beta = 97.80(3)°$; Z = 4 [9]. A crystal structure determination of $[P(C_6H_5)_3]_2Pt_2C_2B_8H_{10}$, prepared in 23% yield by treating $Na_2[C_2B_8H_{10}]$ with $[P(C_6H_5)_3]_2PtCl_2$ in tetrahydrofuran, indicates that the Pt atom in the cage complex is coordinated to two C atoms of the carborane as well as the two $P(C_6H_5)_3$ groups [10]. 9-H-9,9-$[(C_2H_5)_3P]_2$-μ-10,11-H-7,8,9-$C_2PtB_8H_{10}$ is prepared from the reaction of Pt_2(μ-cod)$[P(C_2H_5)_3]_4$ (cod = cycloocta-1,5-diene) with *nido*-5,6-$C_2B_8H_{12}$. Thermolysis of 9-H-9,9-$[(C_2H_5)_3P]_2$-μ-10,11-H-7,8,9-$C_2PtB_8H_{10}$ in toluene at 100°C yields 9-H-9,10-$[(C_2H_5)_3P]_2$-7,8,9-$C_2PtB_8H_9$ [11].

A conversion of the racemic (±)-5,6-$C_2B_8H_{12}$ to its levorotatory enantiomer is accomplished with the use of (+)-N-methylcamphidine [12]. From NMR studies on deuterio derivatives of 5,6-$C_2B_8H_{12}$, as well as other substituted derivatives of this same carborane, the ^{11}B NMR spectrum of the parent compound is assigned as follows: δ (in ppm) = 6.5 for B(7), 5.0 for B(1), 3.3 for B(8), −2.6 for B(3), −3.7 for B(9), −10.0 for B(10), −27.2 for B(2), −39.1 for B(4); ^{1}H NMR data: δ (in ppm) = 6.51 and 4.99 for the two CH hydrogens, = −2.4 for the bridging BHB hydrogens [12, 13].

A revised structure, **Fig. 13-7**, for the **$[6,9\text{-}C_2B_8H_{10}]^{2-}$** anion is suggested on the basis of NMR data: δ ^{11}B (in ppm) = −5.6 (4B), 8.6 (2B), −33.5 (2B); δ ^{1}H (in ppm) = 4.43 (CH). The $[6,9\text{-}C_2B_8H_{10}]^{2-}$ anion reacts with *cis*-Cl_2PtL_2 (L = $P(C_6H_5)_3$, $S(C_2H_5)_2$) and *cis*-Cl_2NiL' (L′ = 1,2-diaminocyclohexane) to produce square-planar *nido*-metalladicarbaboranes μ-6,9-PtL_2-6,9-$C_2B_8H_{10}$ and μ-6,9-NiL′-6,9-$C_2B_8H_{10}$, respectively, containing metal bridges linking both skeletal carbon atoms of the carborane ligand [14].

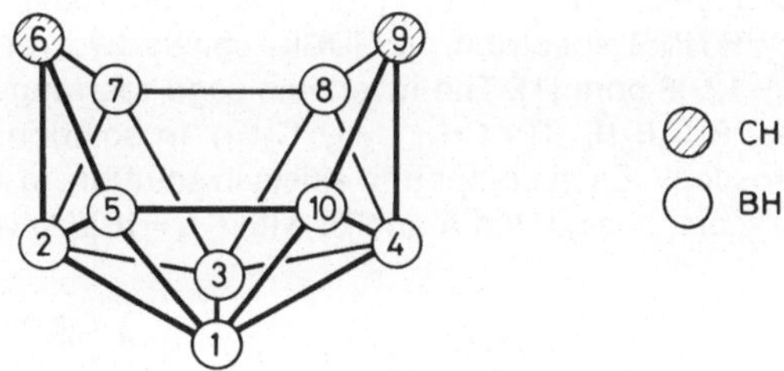

Fig. 13-7. Structure of the $[6,9\text{-}C_2B_8H_{10}]^{2-}$ anion [14].

μ-4′,5′-[1-(η-C_5H_5)Co-2,3-$(CH_3)_2C_2B_4H_3$]-6-[2′,3′-$(CH_3)_2C_2B_4H_5$], a coupled cage cobaltacarborane complex containing a three-center boron linkage is monoclinic, space group $P2_1/n\text{-}C_{2h}^5$ (No. 14); a = 14.092(4), b = 13.270(1), c = 9.988(2) Å; β = 103.06(2)°; Z = 4.

References for 13.9.2:

[1] Brint, P.; Cronin, J. P.; Seward, E.; Whelan, T. (J. Chem. Soc. Dalton Trans. **1983** 975/80).

[2] Jarvis, W.; Inman, W.; Powell, B.; DiStefano, E. W.; Onak, T. (J. Magn. Resonance **43** [1981] 302/15).

[3] Fehlner, T. P.; Wu, M.; Meneghelli, B. J.; Rudolph, R. W. (Inorg. Chem. **19** [1980] 49/54).

[4] Bukalov, S. S.; Leites, L. A. (Opt. Spektroskopiya **56** [1984] 10/2; Opt. Spectrosc. [USSR] **56** [1984] 6/7).

[5] Wong, E. H.; Nalband, G. T. (Inorg. Chim. Acta **53** [1981] L139/L140).

[6] Barker, G. K.; Godfrey, N. R.; Green, M.; Parge, H. E.; Stone, F. G. A.; Welch, A. J. (J. Chem. Soc. Chem. Commun. **1983** 277/9).

[7] Barker, G. K.; Garcia, M. P.; Green, M.; Pain, G. N.; Stone, F. G. A.; Jones, S. K. R.; Welch, A. J. (J. Chem. Soc. Chem. Commun. **1981** 652/3).

[8] Zakharkin, L. I.; Kanakhina, L. N.; Yanovskii, A. I.; Antonovich, V. A. (Koord. Khim. **7** [1981] 1692/8; Soviet J. Coord. Chem. **7** [1981] 850/5).

[9] Lu, P.; Knobler, C. B.; Hawthorne, M. F. (Acta Cryst. C **40** [1984] 1704/6).
[10] Kukina, G. A.; Porai-Koshits, M. A.; Sergienko, V. S.; Strouf, O.; Base, K.; Zakharova, I. A.; Štíbr, B. (Izv. Akad. Nauk SSSR Ser. Khim. **1980** 1686; C.A. **93** [1980] No. 239615).

[11] Barker, G. K.; Green, M.; Stone, F. G. A.; Wosley, W. C.; Welch, A. J. (J. Chem. Soc. Dalton Trans. **1983** 2063/9).
[12] Štíbr, B.; Plešek, J.; Zobacova, A.; (Polyhedron **1** [1982] 824/6).
[13] Štíbr, B.; Heřmánek, S.; Janoušek, Z.; Plzak, Z.; Dolanský, J.; Plešek, J. (Polyhedron **1** [1982] 822/4).
[14] Štíbr, B.; Janoušek, Z.; Baše, K.; Heřmánek, S.; Plešek, J.; Zakharova, I. A. (Collection Czech. Chem. Commun. **49** [1984] 1891/4).
[15] Borelli, A. J.; Plotkin, J. S.; Sneddon, L. G. (Inorg. Chem. **21** [1982] 1328/31).

13.9.3 Compounds Containing Eight Boron Atoms and Four or More Cage Carbon Atoms

2,3,7,8-$C_4B_8H_8$-2,3,7,8-$(C_2H_5)_4$ is formed by the oxidation of $[2,3\text{-}C_2B_4H_6\text{-}2,3\text{-}(C_2H_5)_2]_2FeH_2$ with O_2. 2,3,7,8-$C_4B_8H_8$-2,3,7,8-$(C_2H_5)_4$ is an air-stable solid melting at 57°C; the infrared spectrum of the compound exhibits major absorption bands at (frequencies in cm^{-1}) 2970, 2936, 2874, 2515, 1442, 1377, 1278, 1118, 1059, 1024, 1007, 950, 938, 870, 794, 766, 732, 713, 682, 645; the ^{11}B NMR spectrum, in $CDCl_3$, shows two resonances in a 3:1 area ratio at $\delta = -3.11$ and -12.08 ppm [1]. The reversible cage rearrangement of tetra-C-alkyltetracarbadodecaboranes, $\mathbf{R_4C_4B_8H_8}$ (R = CH_3, C_2H_5, C_3H_7), in solution was examined via ^{11}B, ^{13}C, and ^1H NMR spectroscopy. Each compound exists in solution, in a variety of solvents, as a mixture of two cage isomers, designated **A** and **B**, with K_{eq} = [B]/[A] values in $C_6H_5CH_3$ at 25°C of 0.56,

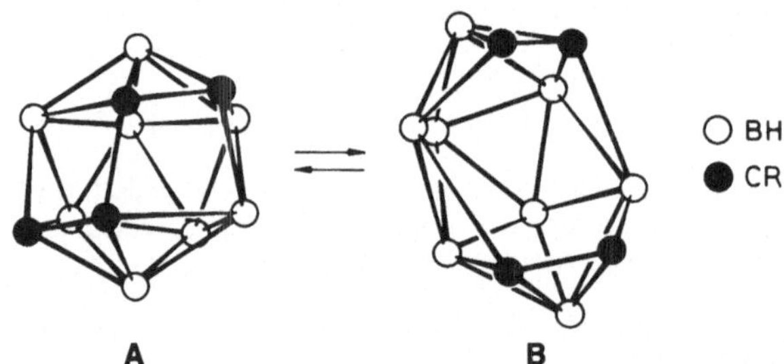

2.2, and 1.9 for R = CH_3, C_2H_5, C_3H_7, respectively. In the solid state, $R_4C_4B_8H_8$ (R = CH_3) exists as isomer **A**; when R = C_2H_5 the compound crystallizes as isomer **B**, which X-ray crystallography establishes as an open framework comprised of two pyramidal C_2B_4 units joined at their basal B-B edges. When R = C_3H_7 the compound contains both **A** and **B** isomers in the solid state. All three $R_4C_4B_8H_8$ (R = CH_3, C_2H_5, C_3H_7), when placed in solution, undergo changes in the [B]/[A] ratio until equilibrium is reached; however, the process is much more rapid in the case where R = CH_3 (minutes at ambient temperature) than in the higher homologs, which require hours. In each case, removal of solvent causes the compound to revert to its original solid-state cage isomer. From variable-temperature ^{11}B NMR measurements the values of ΔH and ΔS for the A→B conversion, which involves cleavage of a framework C-C bond, were found to be very small. The ΔH values range from 1.48 kcal/mol for R = CH_3 to 2.34 for R = C_3H_7, the slight increase attributed to an apparent steric effect [2]. The controlled degradation of $(CH_3)_4C_4B_8H_8$ in 95% C_2H_5OH in air gives $(CH_3)_4C_4B_7H_9$ [3].

nido-2,3-$C_2B_4H_6$-2,3-$[Si(CH_3)_3]_2$ undergoes thermolytic fusion at 210°C to give ***nido*-2,3,7,8-$C_4B_8H_{10}$-2,7-$[Si(CH_3)_3]_2$**, boiling point 208°C at 10 Torr; NMR data for *nido*-2,3,7,8-$C_4B_8H_{10}$-2,7-$[Si(CH_3)_3]_2$: δ ^{11}B (in ppm): =9.98, J(BH)=152 Hz, for B(4); =+8.86, J(BH)=156 Hz, for B(9); =6.15 for B(6,11); =−17.27, J(BH)=178 Hz, for B(5,10); =−20.85, J(BH)=181 Hz, for B(1,12). δ^{1}H (in ppm): =7.02 for cage CH; =5.66, 5.12, and 4.47 for terminal HB; =0.19 for $(CH_3)_3Si$. δ^{13}C: =105.44 ppm for cage CSi; =97.69 ppm for cage CH; =0.05 for $(CH_3)_3Si$. δ^{29}Si: =0.03 ppm for $Si(CH_3)_3$ [4].

The reaction of $[(C_2H_5)_4C_4B_8H_8]^{2-}$ with $CrCl_2$ and $Na[C_5H_5]$ (C_5H_5=cyclopentadienyl) yields primarily red, paramagnetic *nido*-(η^5-C_5H_5)$Cr[(C_2H_5)_4C_4B_8H_8]$ (isomer I). Column chromatography of the product mixture under N_2 gives purple II, an unstable species isomeric with I. Air oxidation of solutions of I over prolonged periods of time gave yellow, unstable *nido*-(η^5-C_5H_5)$Cr[(C_2H_5)_4C_4B_7H_7]$ [5]. An X-ray diffraction study of $[\eta^5$-$C_5(CH_3)_5]_2Co_3(CH_3)_4C_4B_8H_7$, formed from the reaction of $Na[(CH_3)_2C_2B_4H_5]$, $Li[C_5(CH_3)_5]$ ($C_5(CH_3)_5$=pentamethylcyclopentadienyl), and $CoCl_2$ in cold tetrahydrofuran, disclosed a structure consisting of two identical (η^5-$C_5(CH_3)_5$)$Co(CH_3)_2C_2B_4H_3$ units face-coordinated to a third Co atom, with a direct B-B bond (1.758(5) Å) between the two ligands; the linked B atoms have no terminal H atoms in the solid-state structure. Structural parameters are: a=8.555(2), b=12.599(2), c=16.268(9) Å, α=102.76(4)°, β=92.71(3)°, γ=99.15(4)°; Z=2; space group $P\bar{1}$-C_i^1 (No. 2) [6].

2,2′-[1,6-$C_2B_4H_5$]$_2$ is formed from 1,6-$C_2B_4H_6$ in the presence of $PtBr_2$; ^{11}B NMR data for (1,6-$C_2B_4H_5$-2-)$_2$: δ (in ppm)=−13.8, J(B-H)=202 Hz, for B(4); =−15.3 for B(2); =−15.6, J(B-H)=196 Hz, for B(3,5); ^{1}H NMR data: δ=2.74 ppm for CH; =2.14 ppm for BH [7].

The X-ray diffraction of the purple, mixed-spin diiron "wedged" metallacarborane, $[(CH_3)_2C_2B_4H_4]_2[Fe^{II}$ (low-spin)$][Fe^{II}$ (high-spin)$]\cdot 2L$ [L=thf=tetrahydrofuran or 2L=dme=$(CH_3O)_2C_2H_4$], established it as a bimetallic complex having one Fe atom sandwiched between two $[(CH_3)_2C_2B_4H_4]^{2-}$ ligands with the other Fe in a wedging location, coordinated to the complex via 4 Fe-B interactions. Structural parameters of the monoclinic crystals of the 1,2-dimethoxyethane (=dme) complex are: a=9.320(7), b=23.39(2), c=9.546(5) Å, β=103.53(5)°. Space group $P2_1/n$-C_{2h}^5 (No. 14); Z=4. D_{calc}=1.328 g/cm^3 [8]. Conversion of the red diamagnetic sandwich complexes $(R_2C_2B_4H_4)_2FeH_2$ or $(R_2C_2B_4H_4)_2CoH$ to the corresponding $R_4C_4B_8H_8$ carborane (R=alkyl) via oxidative fusion of the formal $[R_2C_2B_4H_4]^{2-}$ ligands has been studied in detail. The reaction is intramolecular with respect to the ligands [9]. The formation of $R_4C_4B_8H_8$ (R=alkyl) via oxidative fusion of the formal $[R_2C_2B_4H_4]^{2-}$ ligands of the sandwich complexes $(R_2C_2B_4H_4)_2FeH_2$ or $(R_2C_2B_4H_4)_2CoH$ has been studied; in tetrahydrofuran (THF) slow conversion of $(R_2C_2B_4H_4)_2FeH_2$ to a purple, paramagnetic diiron complex, formulated as $(R_2C_2B_4H_4)_2Fe_2(thf)_2$, is observed. This species on exposure to O_2 rapidly forms $R_4C_4B_8H_8$, and hence is an intermediate in the oxidative fusion of the monoiron complex. Disproportionation of $(R_2C_2B_4H_4)_2FeH_2$ to $R_4C_4B_8H_8$ and $(R_2C_2B_4H_4)_2Fe_2(thf)_2$ is catalyzed by traces of $FeCl_3$ in THF. At $FeCl_3/(R_2C_2B_4H_4)_2FeH_2$ molar ratios of 2.0 or greater in THF, $R_4C_4B_8H_8$ is produced essentially quantitatively [9]. $FeH_2(2,3$-$C_2B_4H_4$-2,3-$(CH_3)_2)_2$ reacts with $Co[P(C_2H_5)_3]_4$, (μ-cod)$Pt_2[P(C_2H_5)_3]_4$ (cod=1,5-cyclooctadiene) and η^5-C_5H_5Fe(cod) to afford $FeM[(CH_3)_4C_4B_8H_8][P(C_2H_5)_3]_2$ (M=Co, Pt) and η^5-$C_5H_5Fe_2[(CH_3)_4C_4B_8H_8]$, respectively, each of which features a metal-metal bond [10].

The preparation of the green paramagnetic bis(carbaboranyl) triple-decker sandwich compound $[(H(C_2H_5)_2C_3B_3(CH_3)_3)Ni]_2[H(C_2H_5)_2C_3B_2(CH_3)_2]$ has been reported [11].

References for 13.9.3:

[1] Maynard, R. B.; Grimes, R. N. (Inorg. Syn. **22** [1983] 215/8).
[2] Venable, T. L.; Maynard, R. B.; Grimes, R. N. (J. Am. Chem. Soc. **106** [1984] 6187/93).

[3] Finster, D. C.; Grimes, R. N. (J. Am. Chem. Soc. **103** [1981] 2675/83).
[4] Hosmane, N. S.; Dehghan, M.; Davies, S. (J. Am. Chem. Soc. **106** [1984] 6435/6).
[5] Maynard, R. B.; Zhu-Ting, W.; Sinn, E.; Grimes, R. N. (Inorg. Chem. **22** [1983] 873/8).
[6] Finster, D. C.; Sinn, E.; Grimes, R. N. (J. Am. Chem. Soc. **103** [1981] 1399/407).
[7] Corcoran, E. W.; Sneddon, L. G. (J. Am. Chem. Soc. **106** [1984] 7793/800).
[8] Grimes, R. N.; Maynard, R. B.; Sinn, E.; Brewer, G. A.; Long, G. J. (J. Am. Chem. Soc. **104** [1982] 5987/92).
[9] Maynard, R. B.; Grimes, R. N. (J. Am. Chem. Soc. **104** [1982] 593/6).
[10] Barker, G. K.; Garcia, M. P.; Green, M.; Stone, F. G. A.; Welch, A. J. (J. Chem. Soc. Dalton Trans. **1982** 1679/86).

[11] Kuhlmann, T.; Pritzkow, H.; Zenneck, U.; Siebert, W. (Angew. Chem. **96** [1984] 994/5; Angew. Chem. Intern. Ed. Engl. **23** [1984] 956).

13.10 Carboranes Containing Nine Boron Atoms

For previous data on carboranes containing nine boron atoms, see "Boron Compounds" 2nd Suppl. Vol. 2, 1982, pp. 263/75.

13.10.1 Monocarbon Cage Compounds Containing Nine Boron Atoms

The infrared spectra of **M[CB_9H_{10}]** (M = Li, Na, K, Rb, Cs) are displayed in [1]. The *nido*-6-CB_9H_{13} carborane structure has been discussed in relationship to isoelectronic platinapolyborane compounds [2]. Reversed-phase ion-pair liquid chromatographic separations involving the heteroborane anion $[CB_9H_{12}]^-$ have been carried out [3].

9-L-6-CB_9H_{13} (L = $(CH_3)_2S$, CH_3CN, $P(C_6H_5)_3$) is formed from the reaction of $[(CH_3)_4N]$[*nido*-6-CB_9H_{12}] with L/HCl. Degradation of 6-$(CH_3)_3N$-6-CB_9H_{11} with KOH in CH_3OH yields 3,4-μ-$(CH_3)_3N$-CB_5H_{11}. Treating 6-$(CH_3)_3N$-6-CB_9H_{11} with Na in liquid NH_3 gives, after treatment with $[N(CH_3)_4]Cl$, $[N(CH_3)_4]$[*nido*-6-CB_9H_{12}]; the ^{11}B NMR spectrum of the latter compound exhibits resonances at δ (in ppm) = 1.7 (2B), −2.6 (1B), −4.3 (2B), −12.6 (2B), −40.7 (1B), −38.6 (1B). The reaction of Na[*nido*-6-CB_9H_{12}] with $FeCl_3$ and HCl in H_2O yields *arachno*-4-CB_8H_{14} [4].

$[N(CH_3)_4][1$-(η^5-C_5H_5)-1,2-$FeCB_9H_{10}]$ is formed from $[N(CH_3)_4]$[*nido*-6-CB_9H_{12}] and KOH, C_5H_6, $FeCl_2$, $[N(CH_3)_4]Cl$; 2,3-(η^5-$C_5H_5)_2$-2,3-Co_2-1-CB_9H_{10} is formed from $[N(CH_3)_4]$[*nido*-6-CB_9H_{12}] and KOH, C_5H_6, $CoCl_2$, $[N(CH_3)_4]Cl$ [4].

The salt Na[1,7-CPB_9H_{10}-7-CH_3], when treated with $[(C_6H_5)_3P]_3RhCl$, gives 2,2-$[(C_6H_5)_3P]_2$-2,1,7-$RhCPB_9H_{10}$-7-CH_3. Analogous treatment of Na[1,7-CPB_9H_{10}-7-CH_3] with $[(C_6H_5)_3P]_3NiBr_2$ gives 2-$(C_6H_5)_3P$-2-Br-2,1,7-$NiCPB_9H_{10}$-7-CH_3 [5].

References for 13.10.1:

[1] Myasoedov, S. F.; Tsimerinova, T. V.; Solntsev, K. A.; Kuznetsov, N. T. (Koord. Khim. **10** [1984] 1174/7; C.A. **102** [1985] No. 46003).
[2] Baše, K.; Štíbr, B.; Zakharova, I. A. (Syn. Reactiv. Inorg. Metal-Org. Chem. **10** [1980] 509/14).
[3] Plzak, Z.; Plešek, J.; Štíbr, B. (J. Chromatogr. **212** [1981] 283/93).

[4] Baše, K.; Štíbr, B.; Dolanský, J.; Duben, J. (Collection Czech. Chem. Commun. **46** [1981] 2345/53).
[5] Zakharkin, L. I.; Zhigareva, G. G. (Zh. Obshch. Khim. **52** [1982] 2802/3; J. Gen. Chem. [USSR] **52** [1982] 2471).

13.10.2 $C_2B_9H_{13}$, $[C_2B_9H_{12}]^-$, $[C_2B_9H_{11}]^{2-}$, and Related Compounds

The acid dissociation constants of $\mathbf{C_2B_9H_{13}}$, as potentiometrically and spectrophotometrically determined, are pKa(1) = 2.98 and pKa(2) = 14.25 [1]. Treatment of $[7,9\text{-}C_2B_9H_{12}]^-$ with $NaNO_2$ in dilute HCl gives *arachno*-6,8-$C_2B_7H_{13}$ whereas treatment of $[7,9\text{-}C_2B_9H_{12}]^-$ with Na_2SO_3 provides the thiocarborane *arachno*-4,6,8-$SC_2B_6H_{10}$ [2].

A rapid high-yield conversion of 1,12-$C_2B_{10}H_{12}$ to [K(18-crown-6)][*nido*-2,9-$C_2B_9H_{12}$] in the presence of 18-crown-6-ether and KOH is presently the best method to prepare the ***[nido*-2,9-$\mathbf{C_2B_9H_{12}]^-}$** ion. [K(18-crown-6)][*nido*-2,9-$C_2B_9H_{12}$] has been used to prepare $[NH(C_2H_5)_3]$[*nido*-2,9-$C_2B_9H_{12}$], 2-(η^5-C_5H_5)-*closo*-2,1,12-$CoC_2B_9H_{11}$, and 2,2-$[P(C_6H_5)_3]_2$-2-H-*closo*-2,1,12-$RhC_2B_9H_{11}$. The ^{11}B NMR spectrum of [K(18-crown-6)][*nido*-2,9-$C_2B_9H_{12}$] exhibits resonances at δ (in ppm) = −13.10, J(B-H) = 140 Hz (2B); = −19.23, J(B-H) = 148 Hz (2B); = −21.58, J(B-H) = 148 Hz (2B); = −28.54, J(B-H) = 139 Hz (2B); = −42.94, J(B-H) = 156 Hz (1B) [3]. In an electrolyte containing Ni ions, the $[C_2B_9H_{12}]^-$ ion is adsorbed on the cathode surface with subsequent autocatalytic decomposition to elemental boron [4]. The Raman spectrum of two isomers of $Cs[C_2B_9H_{12}]$ shows a bridging BHB band at ca. 2080 cm^{-1} for one of the isomers and bands at 2125, 1912, 1840, and 1770 cm^{-1} for the other isomer [5]. Toxicity tests on $K[C_2B_9H_{12}]$ and $K[C_2B_9H_{11}\text{-}6\text{-}NH_2]$ have been conducted [6]. Reversed-phase ion-pair liquid chromatographic separations of the heteroborane anions $[7,8\text{-}C_2B_9H_{12}]^-$, $[7,8\text{-}C_2B_9H_{11}\text{-X-R}]^-$ (X-R = 5-SH, 5-$CH(CH_3)_2$, 5-I, 5-Cl, 9-Cl, 9-I, 9-OH), $[7,8\text{-}C_2B_9H_{10}\text{-X,X'-}R_2]^-$ (X,X'-R_2 = 5,6-I_2, 5,6-Cl_2, 9,11-I_2, 9,11-Cl_2), $[7,9\text{-}C_2B_9H_{12}]^-$, and $[7,9\text{-}C_2B_9H_{11}\text{-X-R}]^-$ (X-R = 10-OH, 10-OCH_3) have been carried out [7]. Treatment of $K[7,8\text{-}C_2B_9H_{12}]$ with aqueous formaldehyde in the presence of HCl gave 57 to 75% of *nido*-2,6-$C_2B_7H_{11}$ [8]. Treatment of $[NH(CH_3)_3][C_2B_9H_{12}]$ with 40% NaOH, heating to remove $(CH_3)_3N$, and subsequent addition of $CoCl_2$ and then Cs_2SO_4, yields $Cs_2[(C_2B_9H_{11})_2Co_2(C_2B_8H_{10})]$, $Cs_3[(C_2B_9H_{11})_2Co_3(C_2B_8H_{10})_2]$, and $Cs_4[(C_2B_9H_{11})_2Co_4(C_2B_8H_{10})_3]$ [9].

Electrolysis of $K[7,8\text{-}C_2B_9H_{12}]$ in $(CH_3)_2SO$ on a cobalt anode yields $M[Co(C_2B_9H_{11})_2]$ (M = Cs, $N(CH_3)_4$). Similarly, electrolysis of $K[7,8\text{-}C_2B_9H_{12}]$ on a nickel anode gives a mixture of $[N(CH_3)_4][Ni(C_2B_9H_{11})_2]$ and $Ni(C_2B_9H_{11})_2$ [10]. Electrolysis of $K[C_2B_9H_{12}]$ in a 0.1N NaBr solution in $(CH_3)_2SO$ with an Fe anode, followed by treatment with CsCl, yields $Cs[Fe(C_2B_9H_{11})_2]$ [11].

7,8-$\mathbf{C_2B_9H_{11}}$-9-$\mathbf{S(CH_3)_2}$, melting point 147 to 148°C, is produced from the reaction of $K[7,8\text{-}C_2B_9H_{12}]$ with $(CH_3)_2SO$ and aqueous H_2SO_4. The ^{11}B NMR spectrum of the dimethyl sulfide compound exhibits one singlet at −3.2 ppm and eight doublets centered at −2.6, −9.7, −13.4, −14.8, −19.3, −22.1, −25.8, −34.2 ppm; the ^{1}H NMR spectrum exhibits two carborane-CH signals at 2.22 and 2.70 ppm, respectively, and a B-H-B bridging hydrogen at −3.37 ppm [12].

The structure of **7,8-$\mathbf{C_2B_9H_{11}}$-5-$\mathbf{SO(CH_3)_2}$**, prepared from the action of $(CH_3)_2SO$ on $[1,2\text{-}C_2B_{10}H_{11}\text{-}9\text{-}IC_6H_5]BF_4$, has been established by X-ray crystallography. Crystals of 7,8-$C_2B_9H_{11}$-5-$SO(CH_3)_2$, **Fig. 13-8**, p. 182, are monoclinic, space group $P2_1/a\text{-}C^5_{2h}$ (No. 14) with a = 10.099(5), b = 10.073(4), c = 11.852(5) Å, β = 105.92(4)°; Z = 4, D_{calc} = 1.206 g/cm^3 [13, 14].

Crystals of $\mathbf{[N(CH_3)_4][7,8\text{-}C_2B_9H_{11}\text{-}7\text{-}C_6H_5]}$ are monoclinic, space group $P2_1\text{-}C^2_2$ (No. 4), with a = 7.292(1), b = 10.401(2), c = 12.488(2) Å, and β = 101.67(1)°; D_{calc} = 1.020 g/cm^3 for Z = 2; the atomic parameters and bond lengths and angles are given [15]. Crystals of $Cs[7,8\text{-}C_2B_9H_9\text{-}9,10,11\text{-}(CH_3)_3]$ are triclinic with a = 11.069(3), b = 11.339(3), c = 11.402(3) Å, α = 100.97(2)°,

References for 13.10.2 on pp. 182/3

β=103.55(2)°, and γ=107.22(2)°; Z=4; space group P$\bar{1}$-C_i^1 (No. 2); the framework of the carborane anion is essentially that of an icosahedron with one vertex missing and with both carbon atoms occupying adjacent positions in the open face. All three boron atoms of the open face have CH_3 substituents with the middle boron also having an *exo* terminal hydrogen atom. Selected bond distances (in Å): 1.58 for C(7)-C(8), 1.59 for C(7)-B(11), 1.93 for B(9)-B(10), 1.84 for B(5)-B(10), 1.75 for B(3)-C(7) [16]. Degradation of 1,2-$C_2B_{10}H_{10}$-1-CH_3-2-R (R=C_4H_9, C_5H_{11}, C_7H_{15}, $C_{11}H_{23}$, $C_{16}H_{33}$) to [N(CH_3)$_4$][7,8-$C_2B_9H_{10}$-7-CH_3-8-R] is accomplished by the action of KOH in C_2H_5OH followed by the addition of [N(CH_3)$_4$]Cl. [N(CH_3)$_4$][7,8-$C_2B_9H_{10}$-7-CH_3-8-(CH_2)$_3$OC_6H_2-2′,4′,5′-Cl_3] is formed by the action of piperidine on 1,2-$C_2B_{10}H_{10}$-1-CH_3-2-(CH_2)$_3$OC_6H_2-2′,4′,5′-Cl_3 followed by the addition of [N(CH_3)$_4$]Cl [17]. Benzylation of KNa[7,8-$C_2B_9H_{11}$] and KNa[7,8-(CH_3)$_2$-7,8-$C_2B_9H_9$] has been carried out with $C_6H_5CH_2Cl$, *p*-Br$C_6H_4CH_2$Br, and *o*-, *m*-, *p*-$CH_3C_6H_4CH_2$Br in liquid NH_3. Thus, treating KNa[7,8-$C_2B_9H_{11}$] with $C_6H_5CH_2Cl$ followed by treatment with CsBr gives a 90% yield of Cs[7,8-$C_2B_9H_{11}$-9-$CH_2C_6H_5$]. The ^{11}B NMR spectrum of Cs[7,8-$C_2B_9H_{11}$-9-$CH_2C_6H_5$] consists of resonances at δ (in ppm) = −38.7 for B(1); −33.0 for B(10); −21.0 for B(4); −16.7 for B(2,3,5,6); −11.1 for B(11); 0.9 for B(9) [18].

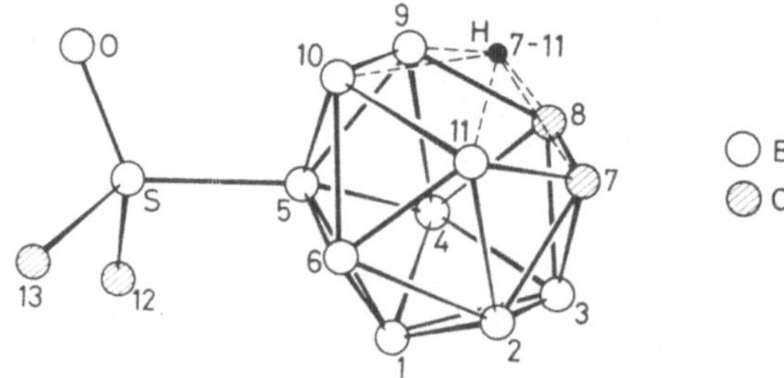

Fig. 13-8. Molecular structure of 7,8-$C_2B_9H_{11}$-5-SO(CH_3)$_2$ [13, 14].

Treating [NH(CH_3)$_3$][7,8-$C_2B_9H_{12}$] with NaH gives **[7,8-$C_2B_9H_{11}$]$^{2-}$** [19]. Tensor surface harmonic theory is used to describe the structure and bonding in [$C_2B_9H_{11}$]$^{2-}$ [20]. From a variety of experimental evidence it is suggested that both the electron donor abilities and the steric requirements of the carborane unit [$C_2B_9H_{11}$]$^{2-}$ (as a ligand in transition metal complexes) are roughly equivalent to those of the η^5-pentamethylcyclopentadiene, and are clearly greater than those of unsubstituted η^5-cyclopentadiene [21].

References for 13.10.2:

[1] Truba, N. A.; Nabivanets, B. I. (Zh. Obshch. Khim. **49** [1979] 1333/6; J. Gen. Chem. [USSR] **49** [1979] 1171/3).

[2] Baše, K.; Heřmanek, S.; Hanoušek, F. (J. Chem. Soc. Chem. Commun. **1984** 299/300).

[3] Busby, D. C.; Hawthorne, M. F. (Inorg. Chem. **21** [1982] 4101/3).

[4] Sadakov, G. A.; Ezikyan, A. Ya.; Kukoz, F. I. (Elektrokhimiya **16** [1980] 1837/40; Soviet Electrochem. **16** [1980] 1507/10).

[5] Leites, L. A.; Bukalov, S. S.; Vinogradova, L. E.; Kalinin, V. N.; Kobel'kova, N. I.; Zakharkin, L. I. (Izv. Akad. Nauk SSSR Ser. Khim. **1984** 954; Bull. Acad. Sci. USSR Div. Chem. Sci. **1984** 880).

[6] Spryshkova, R. A.; Karaseva, L. I.; Bratsev, V. A.; Serebryakov, N. G. (Med. Radiol. **26** [1981] 62/4; C.A. **95** [1981] No. 108411).

[7] Plzak, Z.; Plešek, J.; Štíbr, B. (J. Chromatogr. **212** [1981] 283/93).

[8] Štíbr, B.; Plešek, J.; Heřmánek, S. (Inorg. Syn. **22** [1983] 237/9).

[9] Volkov, V. V.; Dvurechenskaya, S. Ya. (Izv. Akad. Nauk SSSR Ser. Khim. **1981** 2356/9; Bull. Acad. Sci. USSR Div. Chem. Sci. **30** [1981] 1940/3).
[10] Erdman, A. A.; Zubreichuk, Z. P.; Shirokii, V. L.; Maier, N. A.; Ol'dekop, Y. A. (Vestsi Akad. Navuk Belarusk. SSR Ser. Khim. Navuk **1984** No. 4, pp. 86/8; C.A. **101** [1984] No. 192180).

[11] Shirokii, V. L.; Erdman, A. A.; Zubreichuk, Z. P.; Maier, N. A.; Ol'dekop, Y. A. (Zh. Obshch. Khim. **53** [1983] 951/2; C.A. **99** [1983] No. 53923).
[12] Plešek, J.; Janoušek, Z.; Heřmánek, S. (Inorg. Syn. **22** [1983] 239/41).
[13] Grushin, V. V.; Tolstaya, T. P.; Yanovskii, A. I.; Struchkov, Yu. T. (Dokl. Akad. Nauk SSSR **276** [1984] 1136/9; Dokl. Chem. Proc. Acad. Sci. USSR **274/279** [1984] 205/7).
[14] Grushin, V. V.; Tolstaya, T. P.; Yanovskii, A. I.; Struchkov, Yu. T. (Izv. Akad. Nauk SSSR Ser. Khim. **1984** 855/61; Bull. Acad. Sci. USSR Div. Chem. Sci. **1983** 788/93).
[15] Yanovskii, A. I.; Struchkov, Yu. T.; Kalinin, V. N.; Zakharkin, L. I. (Zh. Strukt. Khim. **23** [1982] 77/80; J. Struct. Chem. [USSR] **23** [1982] 232/5).
[16] Antipin, M. Yu.; Struchkov, Yu. T.; Kirillova, N. I.; Knyazev, S. P.; Brattsev, V. A.; Stanko, V. I. (Cryst. Struct. Commun. **9** [1980] 599/603).
[17] Totani, T.; Aono, K.; Yamamoto, K.; Tawara, K. (J. Med. Chem. **24** [1981] 1492/9).
[18] Zakharkin, L. I.; Zhigareva, G. G.; Antonovich, V. A. (Zh. Obshch. Khim. **50** [1980] 1026/31; J. Gen. Chem. [USSR] **50** [1980] 820/5).
[19] Behnken, P. E.; Hawthorne, M. F. (Inorg. Chem. **23** [1984] 3420/3).
[20] Stone, A. J.; Alderton, M. J. (Inorg. Chem. **21** [1982] 2297/302).

[21] Hanusa, T. P. (Polyhedron **1** [1982] 663/5).

13.10.3 Additional Carboranes with Nine Boron Atoms

For the boron atoms in **2,3-$C_2B_9H_{11}$** cage "umbrella" angles of 116° (for B(1)), 111° (for B(4,5,6,7)), 112° (for B(8,9)), and 118° (for B(10,11)) are used in an empirical relationship, which includes the number of adjacent cage carbon atoms, to predict $^1J(^{11}B-^1H)$ values of 176, 165, 165, and 149 Hz, respectively; these values are compared to the observed values of 171, 167, 172, and 150 Hz, respectively [1]. Aromatic solvent-induced ^1H NMR shifts are reported for 2,3-$C_2B_9H_{11}$ and correlated to PRDDO MO-derived hydrogen charges [2]. Several carboranylporphyrins (carboranyl = $C_2B_9H_{11}$) have been prepared and characterized [3].

1-Methyl-1-[(carboranyl)methylene]-3,3,5,5-tetrachlorocyclotriphosphazene anion (in which carboranyl = $C_2B_9H_{11}$) reacts with $Rh[P(C_6H_5)_3]_3Cl$ to form a metallacarboranylphosphazene derivative. The dianion, 1-methyl-1-[(carboranyl)methylene]-3,3,5,5-tetrachlorocyclotriphosphazene (in which carboranyl = $C_2B_9H_{10}$) reacts with $M(CO)_6$ (M = Mo, W) to yield bismetallacarboranyl-phosphazene derivatives [4, 5]. The application of thin layer and of medium-pressure liquid chromatography on silica gel is described for the separation and purification of various carborane anions of the type $[C_2B_9H_{11}R]^-$, in which R = H, C_6H_5, *p*-NH_2-C_6H_4, *p*-COOH-C_6H_4, and CH_2O-6-(1-oxa-2-hydroxymethyl-3-hydroxy-cyclohex-4-ene), and of B-iodo derivatives of $[C_2B_9H_{11}R]^-$ (R = C_6H_5, -$CH_2CH_2CH_2C(=\overset{+}{N}H_2)(COCH_3)$) [6].

2:2′,1′:2″-[1,5-$C_2B_3H_4$][1′,5′-$C_2B_3H_3$][1″,5″-$C_2B_3H_4$], produced from the gas-phase pyrolysis of 1,5-$C_2B_3H_5$ at 400°C, exhibits ^{11}B NMR resonances at δ (in ppm) = 10.2 for B(2′); = 9.0 for B(2); = 8.6, J(B-H) = 190 Hz, for B(3″) and B(4″); = 7.3 for B(2″); = 4.2, J(B-H) = 185 Hz, for B(3′,4′); = 2.3, J(B-H) = 185 Hz, for B(3), B(4); ^1H data: δ (in ppm) = 5.80 for HC(5′); = 5.58, J(HCBH) = 2.2 Hz, for HC(1) and HC(5); = 5.30, J(HCBH) = 2.2 Hz, for HC(1″), HC(5″). The

References for 13.10.3 on p. 184

infrared spectrum of the compound showed absorption bands at (frequencies in cm^{-1}) 2625 m, sh, 2615 s, 1480 m, br, 1080 m, vbr [7].

2:2′,3′:1″-[1,5-$C_2B_3H_4$][1′,5′-$C_2B_3H_3$][1″,5″-$C_2B_3H_4$], also produced from the gas-phase pyrolysis of 1,5-$C_2B_3H_5$ at 400°C, exhibits ^{11}B NMR resonances at δ (in ppm) = 14.7 for B(2′); = 9.1 for B(2); = 7.5, J(B-H) = 187 Hz, for B(4′); = 6.0 for B(3′); = 5.4, J(B-H) = 187 Hz, for B(2″,3″,4″); = 2.4, J(B-H) = 187 Hz, for B(3), B(4); ^{1}H data: δ (in ppm) = 5.72, J(HCBH) = 1.9 Hz, for HC(1) and HC(5); = 5.68, for HC(5″); = 5.54, J(HCBH) = 2.0 Hz, for HC(1′), HC(5′). The infrared spectrum of the compound showed absorption bands at (frequencies in cm^{-1}) 2602 s, 1480 m, br, 1460 m, sh, 1453 m, 1105 s, 1069 m, sh [7].

References for 13.10.3:

[1] Jarvis, W.; Abdou, Z. J.; Onak, T. (Polyhedron **2** [1983] 1067/70).

[2] Jarvis, W.; Inman, W.; Powell, B.; DiStefano, E. W.; Onak, T. (J. Magn. Resonance **43** [1981] 302/15).

[3] Haushalter, R. C.; Butler, W. M.; Rudolph, R. W. (J. Am. Chem. Soc. **103** [1981] 2620/7).

[4] Allcock, H. R.; Scopelianos, A. G.; Whittle, R. R.; Tollefson, N. M. (J. Am. Chem. Soc. **105** [1983] 1316/21).

[5] Allcock, H. R. (Preprints Am. Chem. Soc. Div. Petrol. Chem. **27** [1982] 652/3; C. A. **101** [1984] No. 73189).

[6] Serino, A. J.; Hawthorne, M. F. (J. Chromatogr. **291** [1984] 384/8).

[7] Astheimer, R. J.; Sneddon, L. G. (Inorg. Chem. **22** [1983] 1928/34).

13.10.4 Metallacarboranes and Related Compounds Containing Nine Boron Atoms

The stannacarborane $(C_{10}H_8N_2)Sn(CH_3)_2C_2B_9H_9$ has been prepared and the structure determined by X-ray crystallography [1]. The high Mößbauer shift for $SnC_2B_9H_{11}$ indicates that the unshared electron pair found in the 5s orbital of Sn does not participate in bonding; Sn bonds are formed as a result of 5p-electrons and some 5d-electrons. The large quadrupole splitting indicated that in the formation of Sn-B and Sn-C bonds $5p_z$ and $5p_{x,y}$-electrons participate unequally [2].

$(C_2B_9H_{11})ML_3X_2$ (M = Cr, Co; L = $H_2NCH_2CH_2NH_2$; X = Cl, Br) were prepared by the reaction of $K[C_2B_9H_{12}]$ with ML_3X_2 [3].

The salt $[N(P(C_6H_5)_3)_2][W(\equiv CC_6H_5(CH_3))(CO)_2(\eta^5\text{-}1,2\text{-}C_2B_9H_9(CH_3)_2)]$ reacts with $(C_6H_5)_3PMCl$ (M = Rh, Au) to give metallacarborane products with tungsten-rhodium and tungsten-gold bonds [4].

Bonding in 3,3,3-$(CO)_3$-3,1,2-$MnC_2B_9H_{11}^-$ has been studied by the SCCC method (self-consistent charge and configuration method), and the results compared with those of $(C_5H_5)Mn(CO)_3$. The calculations suggest that while the bonding in metal-carborane complexes and analogous metal-cyclopentadienyl complexes is formally similar, the $[C_2B_9H_{11}]^{2-}$ dianion should be regarded primarily as a σ-electron donor and only in a secondary manner as a π-electron donor in carbonyl complexes as well as in complexes without carbonyl groups [5].

[2,3-$C_2B_4H_4$-2,3-$(C_2H_5)_2$]-2-Co[B_5H_{10}] is formed from $CoCl_2$, tetrahydrofuran, Na[2,3-$C_2B_4H_5$-2,3-C_2H_5]$_2$, and Na[B_5H_9] [6]. Electrochemical data on 1-C_5H_5-1,2,3-$CoC_2B_9H_{11}$, and 1-C_5H_5-1,2,3-$FeC_2B_9H_{11}$ are presented [7].

Reaction of $Na_2[C_2B_9H_{11}]$ with $FeCl_2$ and 6,6-dimethylfulvene produces $[CH_2{=}C(CH_3)C_5H_4]$-$Fe[C_2B_9H_{11}]$; the chemistry of the organic ligand of the latter compound has been studied [8]. η^5-$C_5H_5FeC_2B_9H_{11}$ has been used as an example of a simple generalization of the Mingos/Wade equation [9]. Reaction of bis(arene)iron(II) salts (arene = mesitylene, $C_6(CH_3)_6$) or $[(C_6H_6)RuCl_2]_2$ with Tl[3,1,2-$TlC_2B_9H_{11}$] in tetrahydrofuran produces neutral, air-stable π-(arene)$MC_2B_9H_{11}$ (M = Fe, Ru) complexes in low or moderate yields [10]. (C_5H_5)Fe[1,2-$C_2B_9H_{10}$-1-CHCHCH(C_6H_5)], with an allyl cation attached to the formally negatively charged carborane moiety, is prepared by the reaction of (C_5H_5)Fe(1,2-$C_2B_9H_{10}$-1-CHO) with C_6H_5CH=CHMgBr [11]. (C_5H_5)Fe(1,2-$C_2B_9H_{10}$-1-C≡CH) is formed by the action of $FeCl_2$, CH_3OH, KOH, and C_5H_6 on 1,2-$C_2B_{10}H_{11}$-1-C≡CH. Reduction of (C_5H_5)Fe(1,2-$C_2B_9H_{10}$-1-C≡CH) with $Na[BH_4]$ yields, after treatment with $[N(CH_3)_4]Br$, $[N(CH_3)_4][(C_5H_5)$Fe(1,2-$C_2B_9H_{10}$-1-C≡CH)]; protonation of the latter compound yields the zwitterionic (C_5H_5)Fe(1,2-$C_2B_9H_{10}^-$-1-C^+=CH_2) [12, 13]. The syntheses of several 1-substituted functional derivatives of 3-(η^5-C_5H_5)Fe^{III}(1,2-$C_2B_9H_{10}$-1-R) (R = CH_2OH, CHO, CO_2H, CH_2CO_2H, $CH(OC_2H_5)_2$) starting with 1,2-$C_2B_{10}H_{11}$-1-R are reported [14]. 3-(η^5-C_5H_5)-3,1,2-$FeC_2B_9H_{10}^-$-1-CH_2^+ is prepared by treating Cs[3-(η^5-C_5H_5)-3,1,2-$FeC_2B_9H_{10}$-1-CH_2OH] with HCl [15]. The synthesis and structure of 3-η^5-$C_5H_5Fe^{II}$-η^5-1,2-$C_2B_9H_{10}^-$-1-$C^+H(C_6H_5)$ are also reported [16]. Reaction of 2,3-$C_2B_9H_9$-2,3-$(CH_3)_2$ with (η^5-C_5H_5)Fe(C_8H_{12}) (C_8H_{12} = 1,5-cyclooctadiene) in $C_6H_5CH_3$, C_6H_6, *o*-xylene, or with naphthalene affords *closo*-1-(η^6-L)-2,4-$(CH_3)_2$-1,2,4-$FeC_2B_9H_9$ (L = arene). The structure of the compound with L = $C_6H_5CH_3$ was determined by X-ray crystallographic analysis. Reaction of *closo*-1-(η^6-L)-2,4-$(CH_3)_2$-1,2,4-$FeC_2B_9H_9$ (L = naphthalene) with CO or with $P[(OCH_3)_3]$ gives *closo*-1,1,1-$(CO)_3$-2,4-$(CH_3)_2$-1,2,4-$FeC_2B_9H_9$ or *closo*-1,1,1-$[(CH_3O)_3P]_3$-2,4-$(CH_3)_2$-1,2,4-$FeC_2B_9H_9$, respectively [17]. Mercuration of (C_5H_5)Fe(1,2-$C_2B_9H_{11}$) with $Hg(OOCCF_3)_2$ in F_3CCOOH/CH_2Cl_2 followed by treatment with NaCl and H_2O yields (C_5H_5)Fe(1,2-$C_2B_9H_{10}$-9-HgCl); the latter compound, when treated with Br_2 in CH_2Cl_2, gives (C_5H_5)Fe(1,2-$C_2B_9H_{10}$-9-Br) [18].

$[7,8\text{-}C_2B_9H_{11}]^{2-}$ reacts with $[(CO)_3RuCl_2]_2$ to give *closo*-3,3,3-$(CO)_3$-3,1,2-$RuC_2B_9H_{11}$; the chemistry of the latter compound is described [19].

Dynamic ^1H and proton-decoupled ^{31}P Fourier transform NMR spectra of a series of 12-vertex *closo*-phosphanometallacarboranes of the type [L_2HM(carb)] (L = $P(C_6H_5)_3$, $P(C_2H_5)_3$, or $P(CH_3)_2C_6H_5$; M = Rh, Ir; carb = 1,2-, 1,7-, or 1,12-$C_2B_9H_{10}$R; R = H, CH_3, C_6H_5, or C_4H_9), and 2,2-$[(C_6H_5)_3P]_2$-2,2-H_2-2,1,7-$RuC_2B_9H_{11}$, 2,2-$[(C_6H_5)_3P]_2$-2-CO-2,1,7-$RuC_2B_9H_{11}$, 3,3-$[(C_6H_5)_3P]_2$-4-C_5H_5N-3,1,2-$RhC_2B_9H_{10}$, and 3,3-$[(C_6H_5)_3P]_2$-3-H-4-C_5H_5N-3,1,2-$RuC_2B_9H_{10}$ indicate that the metal vertex undergoes hindered rotation with respect to the 5-membered face of the carborane cage, with free energies of activation varying from 8.4 to 17.5 kcal/mol [20]. Isomeric rhodacarboranes, i.e., *closo*-3,3-$[(C_6H_5)_3P]_2$-3-H-3,1,2-$RhC_2B_9H_{11}$, *closo*-2,2-$[(C_6H_5)_3P]_2$-2-H-2,1,7-$RhC_2B_9H_{11}$, and *closo*-2,2-$[(C_6H_5)_3P]_2$-2-H-2,1,12-$RhC_2B_9H_{11}$ react with potassium tri-*s*-butylhydroborate in tetrahydrofuran to produce anionic rhodacarboranes $[((C_6H_5)_3P)_2$-$RhC_2B_9H_{11}]^-$ [21]. The synthesis and X-ray crystal structure of $[Rh(P(C_6H_5)_3)_2]$[7,8-μ-$(CH_2)_3$-10-$((CH_2)_2C(O)O(CH_2)_3CH_3)$-7,8-$C_2B_9H_9$] are reported, and a mechanism for its formation from exo-*nido*-$[Rh(P(C_6H_5)_3)_2]$[7,8-μ-$(CH_2)_3$-7,8-$C_2B_9H_{10}$] is proposed based on deuterium-labelling studies. The insertion of 1-butyl acrylate into the B(10)-H bond of Cs[7,8-μ-$(CH_2)_3$-7,8-$C_2B_9H_{10}$] occurs if a catalytic amount of exo-*nido*-$[(P(C_6H_5)_3)_2Rh]$[7,8-μ-$(CH_2)_3$-7,8-$C_2B_9H_{10}$] is present, giving Cs[7,8-μ-$(CH_2)_3$-10-$((CH_2)_2C(O)O(CH_2)_3CH_3)$-7,8-$C_2B_9H_9$] [22]. $[HP(C_6H_5)_3]$-[*closo*-3-$(C_6H_5)_3$P-3,3-Br_2-3,1,2-$RhC_2B_9H_{11}$] is obtained in low yield from the reaction of *closo*-3,3-$((C_6H_5)_3P)_2$-3-H-3,1,2-$RhC_2B_9H_{11}$ and BBr_3; the rhodacarborane anion is pseudo-octahedral about the Rh atom. The iodo analog of the anion, [*closo*-3-$(C_6H_5)_3$P-3,3-I_2-3,1,2-$RhC_2B_9H_{11}]^-$, is isolated as the $[N(C_4H_9)_4]^+$ salt, in high yield, from the reaction of *closo*-3-$(C_6H_5)_3$P-3,3-(O_2NO)-3,1,2-$RhC_2B_9H_{11}$, $[N(C_4H_9)_4]I$, and NaI. The chloro analog of [*closo*-3-$(C_6H_5)_3$P-3,3-I_2-3,1,2-$RhC_2B_9H_{11}]^-$ is prepared as the $[N(C_2H_5)_4]^+$ salt from the reaction of the 16-electron complex $[N(C_2H_5)_4]$[*closo*-3-$(C_6H_5)_3$P-3,1,2-$RhC_2B_9H_{11}$] with CH_2Cl_2 [23]. The *nido*

References for 13.10.4 on pp. 187/8

and *closo* geometrical forms of $LRhC_2B_9H_{10}RR'$ (L = $[(C_6H_5)_3P]_2$, R and/or R' = H, CH_3, C_4H_9, C_6H_5, *ortho*-xylyl, are correlated to the nature of R and R' [24]. The reactions of two isomeric rhodacarborane anions, both the 3,1,2- and 2,1,7-isomers of $[(C_6H_5)_3P(CO)RhC_2B_9H_{11}]^-$, with benzonitrile N-oxide have been studied [25]. The synthesis and X-ray molecular structure of the homogeneous hydrogenation catalyst *closo*-1,3-[μ-(η^2-3-CH_2=$CHCH_2CH_2$)]-3-H-3-$(C_6H_5)_3P$-3,1,2-$RhC_2B_9H_{10}$ have been reported [26]. Rhodacarborane clusters, *closo*-3,3-$[(C_6H_5)_3P]_2$-3-H-3,1,2-$RhC_2B_9H_{11}$, *closo*-2,2-$[(C_6H_5)_3P]_2$-2-H-2,1,7-$RhC_2B_9H_{11}$, *closo*-2,2-$[(C_6H_5)_3P]_2$-2-H-2,1,12-$RhC_2B_9H_{11}$, and exo-*nido*-$[(C_6H_5)_3P]_2$Rh-μ-7,8-$C_2B_9H_{10}$-7,8-$(CH_3)_2$, have been examined as catalyst precursors in the hydrogenation of 1-butyl acrylate [27]. Reaction of *closo*-3,3-$[P(C_6H_5)_3]_2$-3-H-3,1,2-$RhC_2B_9H_{11}$ with H_2SO_4 or HNO_3 affords *closo*-3,3-$[P(C_6H_5)_3]_2$-3-HSO_4-3,1,2-$RhC_2B_9H_{11}$ and *closo*-3,3-$[P(C_6H_5)_3]_2$-3,3-(O_2NO)-3,1,2-$RhC_2B_9H_{11}$, respectively. *closo*-3,3-$[P(C_6H_5)_3]_2$-3,3-(O_2NO)-3,1,2-$RhC_2B_9H_{11}$ is also prepared from HNO_3 and the dimeric metallacarborane [*closo*-Rh($P(C_6H_5)_3$)($C_2B_9H_{11}$)$]_2$ or from NO_2/N_2O_4 and *closo*-3,3-$[P(C_6H_5)_3]_2$-3-H-3,1,2-$RhC_2B_9H_{11}$; additional compounds of this type with varying substituents on Rh are also reported [28]. 3,3-$[(C_6H_5)_3P]_2$-3-H-*closo*-3,1,2-$RhC_2B_9H_9$-1-R-2-R' (R, R' = -$CH_2C_6H_4CH_2$-), exo-*nido*-$[(C_6H_5)_2P]$Rh(7,8-$C_2B_9H_{10}$-7-R-8-R') (R = CH_3, R' = C_6H_5; R = R' = CH_3; R, R' = -CH_2-CH_2CH_2-), and the salt $[(C_6H_5)_3P]$Rh[7,8-$C_2B_9H_{11}$-7-R-] (R = 1'-(*closo*-1',2'-$C_2B_{10}H_{11}$)) are prepared from the reaction of the corresponding carborane anions [*nido*-7,8-$C_2B_9H_{10}$-7-R-8-R'$]^-$ with $RhCl[P(C_6H_5)_3]_3$ [29].

Similarly, 3,3-$[(C_6H_5)_3P]_2$-3-H-*closo*-3,1,2-$RhC_2B_9H_9$-1-R-2-R' (R = R' = H; R = R' = D; R = H and R' = C_6H_5, CH_3, and C_4H_9) is prepared from the reaction of $RhCl[P(C_6H_5)_3]_3$ with the corresponding anions [*nido*-7,8-$C_2B_9H_{10}$-7-R-8-R'$]^-$; 2,2-$[(C_6H_5)_3P]_2$-2-H-*closo*-2,1,7-RhC_2-B_9H_{10}-1-R (R = H, C_6H_5, and CH_3) is formed from the reaction of $RhCl[P(C_6H_5)_3]_3$ with the corresponding anions [*nido*-7,9-$C_2B_9H_{11}$-7-R$]^-$; and 2,2-$[(C_6H_5)_3P]_2$-2-H-*closo*-2,1,12-RhC_2-B_9H_{11} is prepared from the reaction of $RhCl[P(C_6H_5)_3]_3$ with [*nido*-2,9-$C_2B_9H_{12}]^-$ [30]. The carborane moieties of a series of rhodacarboranes, *closo*-$[(C_6H_5)_3P]_2$Rh(H)[η^5-(*nido*-$C_2B_9H_9RR'$)] (e.g., R = R' = H, D, CH_3) are replaced easily by a series of [*nido*-$C_2B_9H_{12}]^-$ ligands using thermal reactions. The ease of displacement of the *nido*-carborane anion in the rhodacarborane was 7,8-di->7,8-mono->7,8-un->7,9-un->2,9-unsubstituted. Kinetic studies of the reaction of 3,3-$[(C_6H_5)_3P]_2$-3-H-*closo*-3,1,2-$RhC_2B_9H_{10}$-1-CH_3 with [*nido*-7,8-$C_2B_9H_{12}]^-$ and [*nido*-2,9-$C_2B_9H_{12}]^-$ showed no anion concentration dependence and a common first-order rate constant for the two reactions; however, for [*nido*-7,9-$C_2B_9H_{12}]^-$, anion dependence was observed [31]. Various rhodacarboranes such as [*closo*-3,3-$[(C_6H_5)_3P]_2$-3-H-3,1,2-$RhC_2B_9H_{11}$], [*closo*-2,2-$[(C_6H_5)_3P]_2$-2-H-2,1,7-$RhC_2B_9H_{11}$], [*closo*-2,2-$[(C_6H_5)_3P]_2$-2-H-2,1,12-$RhC_2B_9H_{11}$], have been employed as catalyst precursors for 1-hexene isomerization [32]. In contrast to the observed equilibrium between the *closo* and exo-*nido* tautomers of the bis(triphenylphosphane)rhodacarborane formally derived from $[(C_6H_5)_3P]_2Rh^+$ and the [*nido*-7,8-μ-(1',2'-$CH_2C_6H_4CH_2$-)-7,8-$C_2B_9H_{10}]^-$ ion, the crystalline solid obtained from this system was determined to be *closo*-1,2-μ-(1',2'-$CH_2C_6H_4CH_2$-)-3,3-$[(C_6H_5)_3P]_2$-3-H-3,1,2-$RhC_2B_9H_9$. However, when one of the $(C_6H_5)_3P$ groups is replaced by $P(C_6H_{11})_3$ (= tris-cyclohexylphosphane) a *nido*-rhodacarborane with the rhodium exo to the C_2B_9 cage is the preferred isomer in the solid. Reaction of this and related exo-*nido* complexes with two-electron donor ligands resulted in displacement of Rh from the carborane cage to generate an ion pair [33]. Synthesis of [*closo*-1,3-μ-2,3-μ-[1,2-μ-(η^2-3,4-$CH_2CH_2C(CH_3)$=$CHCH_2CH_2CH_2$)]-3-H-3-$(C_6H_5)_3P$-3,1,2-Rh-$C_2B_9H_9$], a metallacarborane hydrogenation catalyst, is accomplished by reacting $[(C_6H_5)_3P]_3$-RhCl with Cs[7,8-bis(butenyl)-7,8-$C_2B_9H_{10}$] [34].

closo-2,2-$[(C_6H_5)_3P]_2$-2-H-8-(R)-2,1,8-$IrC_2B_9H_{10}$ (R = CH_3, C_6H_5) are prepared in 33 and 84% yields, respectively, by heating *closo*-3,3-$[(C_6H_5)_3P]_2$-1-(R)-3-H-3,1,2-$IrC_2B_9H_{10}$ in toluene at the reflux temperature. The crystal structure of *closo*-2,2-$[(C_6H_5)_3P]_2$-2-H-8-C_6H_5-2,1,8-$IrC_2B_9H_{10}$ has been determined. The adduct *closo*-2,2-$[(C_6H_5)_3P]_2$-2-H-8-C_6H_5-2,1,8-$IrC_2B_9H_{10}$ · ½CH_2Cl_2

crystallizes in the triclinic system with space group $P\bar{1}$-C_i^1 (No. 2); Z=2. Lattice parameters are a=11.722(4), b=19.994(7), c=9.923(4) Å, α=100.17(3)°, β=107.59(3)°, γ=86.40(3)°; D_{calc}= 1.47, D_{meas}=1.48 g/cm^3 [35].

Extended Hückel molecular orbital calculations on the *closo*-platinacarborane $(H_3P)_2Pt(C_2B_9H_{11})$ have been carried out [36].

References for 13.10.4:

[1] Cowley, A. H.; Galow, P.; Hosmane, N. S.; Jutzi, P.; Norman, N. C. (J. Chem. Soc. Chem. Commun. **1984** 1564/5).

[2] Varnek, V. A.; Dvurechenskaya, S. Ya.; Volkov, V. V.; Mazalov, L. N. (Zh. Strukt. Khim. **21** [1980] 185/7; C.A. **93** [1980] No. 104247).

[3] Shevchenko, Yu. N.; Yashina, N. I.; Svitsyn, R. A.; Egorova, N. V. (Zh. Obshch. Khim. **51** [1981] 1258/63; J. Gen. Chem. [USSR] **51** [1981] 1065/70).

[4] Green, M.; Howard, J. A. K.; James, A. P.; Nunn, C. M.; Stone, F. G. A. (J. Chem. Soc. Chem. Commun. **1984** 1113/4).

[5] Brown, D. A.; Fanning, M. O.; Fitzpatrick, N. J. (Inorg. Chem. **19** [1980] 1822/3).

[6] Borodinsky, L.; Grimes, R. N. (Inorg. Chem. **21** [1982] 1921/7).

[7] Geiger, W. E.; Brennan, D. E. (Inorg. Chem. **21** [1982] 1963/6).

[8] Zakharkin, L. I.; Kobak, V. V. (Izv. Akad. Nauk SSSR Ser. Khim. **1983** 2611/5; Bull. Acad. Sci. USSR Div. Chem. Sci. **1983** 2348/51).

[9] Nishimura, E. K. (Inorg. Nucl. Chem. Letters **17** [1981] 269/71).

[10] Hanusa, T. P.; Huffman, J. C.; Todd, L. J. (Polyhedron **1** [1982] 77/82).

[11] Zakharkin, L. I.; Kobak, V. V.; Antonovich, V. A. (Izv. Akad. Nauk SSSR Ser. Khim. **1983** 2809/14; Bull. Acad. Sci. USSR Div. Chem. Sci. **1983** 2521/6).

[12] Zakharkin, L. I.; Kobak, V. V. (J. Organometal. Chem. **270** [1984] 229/35).

[13] Zakharkin, L. I.; Kobak, V. V. (Izv. Akad. Nauk SSSR Ser. Khim. **1982** 1919; Bull. Acad. Sci. USSR Div. Chem. Sci. **1982** 1711).

[14] Zakharkin, L. I.; Kobak, V. V.; Yanovskii, A. I.; Struchkov, Yu. T. (J. Organometal. Chem. **228** [1982] 119/33).

[15] Zakharkin, L. I.; Kobak, V. V. (Izv. Akad. Nauk SSSR Ser. Khim. **1980** 2671/2; C.A. **94** [1981] No. 103523).

[16] Zakharkin, L. I.; Kobak, V. V.; Yanovsky, A. I.; Struchkov, Yu. T. (Tetrahedron **38** [1982] 3515/25).

[17] Garcia, M. P.; Green, M.; Stone, F. G. A.; Somerville, R. G.; Welch, A. J. (J. Chem. Soc. Chem. Commun. **1981** 871/2).

[18] Zakharkin, L. I.; Kobak, V. V.; Antonovich, V. A. (Zh. Obshch. Khim. **53** [1983] 2153; J. Gen. Chem. [USSR] **53** [1983] 1942).

[19] Behnken, P. E.; Hawthorne, M. F. (Inorg. Chem. **23** [1984] 3420/3).

[20] Marder, T. B.; Baker, R. T.; Long, J. A.; Doi, J. A.; Hawthorne, M. F. (J. Am. Chem. Soc. **103** [1981] 2988/94).

[21] Walker, J. A.; Knobler, C. B.; Hawthorne, M. F. (J. Am. Chem. Soc. **105** [1983] 3368/9).

[22] Hewes, J. D.; Kreimendahl, C. W.; Marder, T. B.; Hawthorne, M. F. (J. Am. Chem. Soc. **106** [1984] 5757/9).

[23] Zheng, L.; Baker, R. T.; Knobler, C. B.; Walker, J. A.; Hawthorne, M. F. (Inorg. Chem. **22** [1983] 3350/5).

[24] Long, J. A.; Marder, T. B.; Behnken, P. E.; Hawthorne, M. F. (Preprints Am. Chem. Soc. Div. Petrol. Chem. **25** [1980] 724/9; C.A. **97** [1982] No. 144988).

[25] Walker, J. A.; Knobler, C. B.; Hawthorne, M. F. (J. Am. Chem. Soc. **105** [1983] 3370/1).
[26] Delaney, M. S.; Knobler, C. B.; Hawthorne, M. F. (Inorg. Chem. **20** [1981] 1341/7).
[27] Behnken, P. E.; Busby, D. C.; Delaney, M. S.; King, R. E.; Kreimendahl, C. W.; Marder, T. B.; Wilczynski, J. J.; Hawthorne, M. F. (J. Am. Chem. Soc. **106** [1984] 7444/50).
[28] Kalb, W. C.; Demidowicz, Z.; Speckman, D. M.; Knobler, C.; Teller, R. G.; Hawthorne, M. F. (Inorg. Chem. **21** [1982] 4027/36).
[29] Long, J. A.; Marder, T. B.; Behnken, P. E.; Hawthorne, M. F. (J. Am. Chem. Soc. **106** [1984] 2979/89).
[30] Baker, R. T.; Delaney, M. S.; King, R. E.; Knobler, C. B.; Long, J. A.; Marder, T. B.; Paxson, T. E.; Teller, R. G.; Hawthorne, M. F. (J. Am. Chem. Soc. **106** [1984] 2965/78).

[31] Long, J. A.; Marder, T. B.; Hawthorne, M. F. (J. Am. Chem. Soc. **106** [1984] 3004/10).
[32] Behnken, P. E.; Belmont, J. A.; Busby, D. C.; Delaney, M. S.; King, R. E.; Kreimendahl, C. W.; Marder, T. B.; Wilczynski, J. J.; Hawthorne, M. F. (J. Am. Chem. Soc. **106** [1984] 3011/25).
[33] Knobler, C. B.; Marder, T. B.; Mizusawa, E. A.; Teller, R. G.; Long, J. A.; Behnken, P. E.; Hawthorne, M. F. (J. Am. Chem. Soc. **106** [1984] 2990/3004).
[34] Delaney, M. S.; Teller, R. G.; Hawthorne, M. F. (J. Chem. Soc. Chem. Commun. **1981** 235/6).
[35] Doi, J. A.; Mizusawa, E. A.; Knobler, C. B.; Hawthorne, M. F. (Inorg. Chem. **23** [1984] 1482/4).
[36] Calhorda, M. J.; Mingos, D. M. P.; Welch, A. J. (J. Organometal. Chem. **228** [1982] 309/20).

13.11 Carboranes Containing Ten Boron Atoms

For previous data on carboranes containing ten boron atoms, see "Boron Compounds" 2nd Suppl. Vol. 2, 1982, pp. 275/326.

13.11.1 $[CB_{10}H_{13}]^-$ and Derivatives

Reversed-phase ion-pair liquid chromatographic separations involving the heteroborane anion $\mathbf{[CB_{10}H_{13}]^-}$ have been carried out [1]. Treatment of **7-$CB_{10}H_{12}$-7-$N(CH_3)_3$** with $(CH_3)_3N \cdot BH_3$ at 180 to 200°C proceeds by splitting off one CH_3 group and inserting one boron vertex to obtain 1-$CB_{11}H_{11}$-1-$NH(CH_3)_2$ [2]. Treatment of *nido*-H_3N-$CB_{10}H_{12}$ with $RhCl[P(C_6H_5)_3]_3$ in the presence of $[N(C_4H_9)_4]OH$ gives $[N(C_4H_9)_4]$[*closo*-2,2-$[(C_6H_5)_3P]_2$-2-H-1-(NH_2)-2,1-$RhCB_{10}H_{10}$] which, on heating in CH_3OH, yields the dimeric salt $[N(C_4H_9)_4][\{(C_6H_5)_3PRhCB_{10}H_{10}NH_2\}_2H]$ [3].

References for 13.11.1:

[1] Plzak, Z.; Plešek, J.; Štíbr, B. (J. Chromatogr. **212** [1981] 283/93).
[2] Plešek, J.; Jelinek, T.; Drdakova, E.; Heřmánek, S.; Štíbr, B. (Collection Czech. Chem. Commun. **49** [1984] 1559/62).
[3] Walker, J. A.; O'Con, C. A.; Zheng, L.; Knobler, C. B.; Hawthorne, M. F. (J. Chem. Soc. Chem. Commun. **1983** 803/4).

13.11.2 1,2-$C_2B_{10}H_{12}$ and Derivatives

13.11.2.1 1,2-$C_2B_{10}H_{12}$

The ^{11}B quadrupolar coupling constants for *closo*-1,2-$C_2B_{10}H_{12}$ are determined to be (in kHz) 917 for B(9,12), 801 for B(8,10), 1109 for B(4,5,7,11), and 1451 for B(3,6) [1]. Two-dimensional boron-11-boron-11 nuclear magnetic resonance spectroscopy confirms many of the B-B atom connectivities in 1,2-$C_2B_{10}H_{12}$ [2]. For the boron atoms in 1,2-$C_2B_{10}H_{12}$ cage "umbrella" angles of 114° (for B(3,6)), 116° (for B(4,5,7,11)), 117° (for B(8,10)), and 117° (for B(9,12)) are used in an empirical relationship, which includes the number of adjacent cage carbon atoms, to predict $^1J(^{11}B{-}^1H)$ values of 177, 165, 151, and 151 Hz, respectively; these values agree well with the observed values of 178, 164, 151, and 151 Hz, respectively [3].

Proton NMR second moments and spin-lattice relaxation times T_1 and $T_{1\varrho}$ have been measured for 1,2-$C_2B_{10}H_{12}$ over the temperature range 320>T>77 K. In addition to the known phase transition at 263 K, the results reveal a phase transition at 200 K. In phase III (T<200 K) the motion of the B-C cage is restricted while the protons exchange position. From the nonexponential decays of the proton magnetization it follows that $T_{1\varrho}$>22 ms. The activation energy of the exchange mechanism is 6.0±0.2 kcal/mol. In phase II (263>T>200 K), T_1(B) is short (approximately 100μs) and the decays of the magnetization exponential. Dipolar interactions of the second kind contribute significantly to $T_{1\varrho}$(H) in this phase. The B-C cage is involved in an anisotropic reorientation, probably about one of the 2-fold axes of the molecule. The activation energy of this motion is 5.5±0.1 kcal/mol. In phase I (T>263 K), the molecule executes an isotropic reorientation with an activation energy of 4.3±0.1 kcal/mol [4]. Aromatic solvent-induced ^{1}H NMR shifts are reported for 1,2-$C_2B_{10}H_{12}$ and correlated to PRDDO MO-derived hydrogen charges [5].

Short-lived absorption spectra of a methanolic solution of 1,2-$C_2B_{10}H_{12}$, under the action of a pulse of accelerated electrons, have been recorded over the whole spectral range from the UV to the IR spectra [6]. Electron resonance capture mass spectra of the radical anion of 1,2-$C_2B_{10}H_{12}$ in the gas phase confirm the existence of long-lived negative ions in this carborane [7].

The temperature-dependent 1,2-$C_2B_{10}H_{12}$ Raman spectrum at 11 to 300 K shows that solid 1,2-$C_2B_{10}H_{12}$ exists in four phases, two of them being orientationally disordered; phase IV is an ordered crystal; the pattern obtained at 11 to 165 K depends on the rate of cooling, the metastable phase III sometimes emerging [8]. The disappearance of the Raman spectrum background for solid 1,2-$C_2B_{10}H_{12}$, just before melting, is accompanied by a reduction in the intensity of Rayleigh scattering [9].

The use of 1,2-$C_2B_{10}H_{12}$ as a plastic crystal allows for the production of tetrahydrofuran radicals by α-irradiation at 77 K of tetrahydrofuran imbedded in the carborane [10]. An analytical procedure for the determination of boron in 1,2-$C_2B_{10}H_{12}$ derivatives has been developed [11]. Microdetermination of boron in 1,2-$C_2B_{10}H_{12}$ was carried out by Fourier-transform NMR [12]. 1,2-$C_2B_{10}H_{12}$ has been used in connection with neutron dosimetry measurements [13].

References for 13.11.2.1:

[1] Leffler, A. J. (J. Chem. Phys. **81** [1984] 2574/6).
[2] Venable, T. L.; Hutton, W. C.; Grimes, R. N. (J. Am. Chem. Soc. **106** [1984] 29/37).
[3] Jarvis, W.; Abdou, Z. J.; Onak, T. (Polyhedron **2** [1983] 1067/70).
[4] Reynhardt, E. C.; Watton, A.; Petch, H. E. (J. Magn. Resonance **46** [1982] 453/68).
[5] Jarvis, W.; Inman, W.; Powell, B.; DiStefano, E. W.; Onak, T. (J. Magn. Resonance **43** [1981] 302/15).

[6] Zimina, G. M.; Vannikov, A. V.; Stanko, V. I. (Khim. Vysokikh Energ. **16** [1982] 335/9; High Energy Chem. [USSR] **16** [1982] 263/6; C.A. **97** [1982] No. 101576).
[7] Mazunov, V. A.; Nekrasov, Y. S.; Khvostenko, V. I.; Stanko, V. I. (Izv. Akad. Nauk SSSR Ser. Khim. **1983** 223/5; Bull. Acad. Sci. USSR Div. Chem. Sci. **1983** 197/9).
[8] Bukalov, S. S.; Leites, L. A. (Chem. Phys. Letters **87** [1982] 327/9).
[9] Bukalov, S. S.; Leites, L. A. (Opt. Spektroskopiya **56** [1984] 10/2; Opt. Spectrosc. [USSR] **56** [1984] 6/7).
[10] Tabata, M.; Murakami, S.; Sohma, J. (Chem. Phys. Letters **75** [1980] 311/4).

[11] Volodina, M. A.; Buyanovskaya, A. G. (Dokl. Akad. Nauk SSSR **255** [1980] 1120/3; C.A. **94** [1981] No. 131690).
[12] Kasler, F.; Tierney, M. (Mikrochim. Acta **2** [1981] 301/12).
[13] Tsuruta, T.; Juto, N. (J. Nucl. Sci. Technol. **21** [1984] 871/6).

13.11.2.2 Halogen Derivatives of 1,2-$C_2B_{10}H_{12}$

The crystal structure of the adduct **1,2-$C_2B_{10}H_2$-3,4,5,6,7,8,9,10,11,12-Cl_{10}·$(CH_3)_2SO$** indicates that there is a cyclic association of two molecules of $(CH_3)_2SO$ with two molecules of the carborane; the molecules are linked by hydrogen bonds of the type C-C-H···O···H-C-C-H···O···H-. The crystals are monoclinic. At −120°C, a = 9.180(9), b = 14.76(1), c = 15.8(1) Å; β = 90.67(8)°; D_{calc} = 1.757 g/cm³; Z = 4; space group P2$_1$/n-C^5_{2h} (No. 14) [1]. Solid complexes of boron 1,2-$C_2B_{10}H_2$-3,4,5,6,7,8,9,10,11,12-Cl_{10} with some oxygen- and nitrogen-containing bases (dioxane, dimethylformamide, dimethyl sulfoxide, triphenylphosphanoxide, pyridine, trimethylamine, triethylamine) were investigated by IR and Raman spectroscopy. Complexes containing C-H··O bonds are characterized by a relative C-H stretching frequency shift up to 12%, and a half-width of the C-H stretching band up to 220 cm^{-1}. C-H··N bonds, with $(CH_3)_3N$ for example, are stronger with a relative shift of approximately 18% and a half-width of the C-H stretching band of about 500 cm^{-1}. $(C_2H_5)_3N$ complexes, however, form an NH^+··C^- proton transfer hydrogen-bond while pyridine can give either C-H··N or C^-··^+HN hydrogen-bonded adducts depending on the solvent and temperature. The CH··N ⇄ C^-··^+HN equilibrium appears to be shifted towards ion-pair formation at considerably smaller enthalpy values as compared to the O-H··N⇄O^-··^+HN system. C-H and N-H stretching frequencies are correlated with the acidity of the donor and the basicity of acceptor molecules [2]. **1,2-$C_2B_{10}H$-3,4,5,6,7,8,9,10,11,12-Cl_{10}-1-R** compounds are prepared from the action of 1,2-$C_2B_{10}H_2$-3,4,5,6,7,8,9,10,11,12-Cl_{10} with LiC_4H_9 followed by the addition of the bromo or epoxy derivatives of R (R = CH_2CH_2COOH, $CH_2CHOHCH_2OCH_2CH{=}CH_2$) [3].

A molecular and crystal structure of **[1,2-$C_2B_{10}H_{11}$-9-IC_6H_5]I** indicate that the carboranyl group has the expected icosahedral configuration, but slightly compressed in the direction of the carbon atoms. The monoclinic crystals have cell parameters a = 13.234(2), b = 13.248(2), c = 21.546(4) Å; γ = 109.25(3)°; D_{calc} = 0.89 g/cm³; Z = 8; space group P2$_1$/b-C^5_{2h} (No. 14) [4].

References for 13.11.2.2:

[1] Yanovskii, A. I.; Struchkov, Yu. T.; Vinogradova, L. E.; Leites, L. A. (Izv. Akad. Nauk SSSR Ser. Khim. **31** [1982] 2257/61; Bull. Acad. Sci. USSR Div. Chem. Sci. **31** [1982] 1988/91).
[2] Leites, L. A.; Vinogradova, L. E.; Belloc, J.; Novak, A. (J. Mol. Struct. **100** [1983] 379/92).
[3] Gabel, D.; Walczyna, R. (Z. Naturforsch. **37c** [1982] 1038/9).
[4] Ionov, V. M.; Subbotin, M. Yu.; Grushin, V. V.; Tolstaya, T. P.; Lisichkina, I. N.; Aslanov, L. A. (Zh. Strukt. Khim. **24** [1983] 139/43; J. Struct. Chem. [USSR] **24** [1983] 638/41).

13.11.2.3 Sulfur Derivatives of $1,2\text{-}C_2B_{10}H_{12}$

Crystals of **$1,2\text{-}C_2B_{10}H_{10}\text{-}9,12\text{-}S_2CH(CH_3)_2$**, with a BBSCS five-membered ring, are monoclinic, space group C2/c-C_{2h}^6 (No. 15), with a = 22.331(4), b = 9.326(1), c = 13.138(1) Å; β = 105.39(1)°; Z = 8. The cage B-C bonds range from 1.692 to 1.720 Å, B-B bonds from 1.739 to 1.802 Å; C-H bonds 0.97 Å; B-H bonds from 1.06 to 1.12 Å; C-S bonds average 1.86 Å; B-S bonds average 1.85 Å [1].

Degradation of **$[NH_4]_2[1,2\text{-}C_2B_{10}H_{10}\text{-}1,2\text{-}(S)_2]$** with C_2H_5OH in the presence of $BrCH_2CH_2Br$, followed by the addition of $[N(CH_3)_4]Cl$, yields $[N(CH_3)_4]_2[(CH_2S)_2C_2B_9H_{10}]_2$ containing a 10-membered C_2SCSC_2SCS ring [2].

References for 13.11.2.3:

[1] Subrtova, V.; Linek, A.; Hašek, J. (Acta Cryst. B **36** [1980] 858/61).
[2] Teixidor, F.; Rudolph, R. W. (J. Organometal. Chem. **241** [1983] 301/12).

13.11.2.4 Nitrogen and Phosphorus Derivatives of $1,2\text{-}C_2B_{10}H_{12}$

The reaction of **$1,2\text{-}C_2B_{10}H_{11}\text{-}3\text{-}NH_2$** with Na in NH_3 and subsequently with CH_2I_2, CHI_3 and CI_4 gives mixtures of $1,2\text{-}C_2B_{10}H_{12}$, $1,7\text{-}C_2B_{10}H_{11}\text{-}2\text{-}NH_2$, and $1,7\text{-}C_2B_{10}H_{10}\text{-}2\text{-}NH_2\text{-}6\text{-}CH_3$; in the case of CI_4 a small amount of $1,7\text{-}C_2B_{10}H_{10}\text{-}B\text{-}NH_2\text{-}B'\text{-}I$ is also formed [1]. Reaction of $1,2\text{-}C_2\text{-}B_{10}H_{11}\text{-}3\text{-}NH_2$ with maleic anhydride and acetic anhydride yields $1,2\text{-}C_2B_{10}H_{11}\text{-}3\text{-}N(\text{-}COCH{=}CHCO\text{-})$, melting point 159 to 160°C [2]. Reaction of $1,2\text{-}C_2B_{10}H_{10}\text{-}1\text{-}R\text{-}2\text{-}R'$ (R = CH_3, C_6H_5, $CH_2{=}C(CH_3)$ or $(CH_3)_2CH$, R' = $N{=}NC_6H_5$) with $CH_3Re(CO)_5$ at 100°C gives $1,2\text{-}C_2B_{10}H_9\text{-}1\text{-}R\text{-}2\text{-}R'\text{-}B\text{-}Re(CO)_4$ [3]. The crystal structure of $1,2\text{-}C_2B_{10}H_9\text{-}1\text{-}(CH_3)_2CH\text{-}2\text{-}N_2C_6H_5\text{-}7\text{-}Re(CO)_4$ has been determined; crystals of the compound are monoclinic with two independent molecules in the unit cell. Lattice parameters at −120°C are a = 11.355(5), b = 19.163(6), c = 20.542(7) Å; β = 94.56(8)°; space group $P2_1/c\text{-}C_{2h}^5$ (No. 14); Z = 8 [4].

$1,2\text{-}C_2B_{10}H_{10}\text{-}1\text{-}PRR'\text{-}2\text{-}PR''R'''$ (R = R' = C_6H_5, R = R' = $N(CH_3)_2$, R'' = R''' = $N(CH_3)_2$; R = R'' = $N(CH_3)_2$, R' = R''' = F; R = R' = C_6H_5, R'' = $N(CH_3)_2$, R''' = F) reacts with PF_5 to give bis-PF_5 adducts; infrared data are given for the adducts. The adduct **$1,2\text{-}C_2B_{10}H_{10}\text{-}1,2\text{-}[PFN(CH_3)_2]_2\cdot 2PF_5$** in CH_3CN solution exists in the ionic form $[1,2\text{-}C_2B_{10}H_{10}]\text{-}1,2\text{-}[PN(CH_3)_2][PF_6]_2$ containing dicoordinate P. Thermal decomposition of this adduct gives $1,2\text{-}C_2B_{10}H_{10}\text{-}1,2\text{-}(PF_2)_2$; the latter compound exhibits a phosphorus NMR signal at 171 ppm [5]. The *cis*-chelating ligands *closo*-$1,2\text{-}C_2B_{10}H_{10}\text{-}1\text{-}P(C_6H_5)_2\text{-}2\text{-}PRR'$ (R = R' = C_6H_5, $N(CH_3)_2$, F; R = $N(CH_3)_2$, R' = F) and *closo*-$1,2\text{-}C_2B_{10}H_{10}\text{-}1\text{-}P\{N(CH_3)_2\}_2\text{-}2\text{-}PRR'$ (R = R' = C_6F_5; R = $N(CH_3)_2$, R' = F) form chelates with platinum(II) of the type $[PtCl_2L]$ (L = ligand) [6]. ESR data indicate that the $1,2\text{-}C_2B_{10}H_{11}\text{-}1\text{-}P(CH_3)(OC_2H_5)[OC(CH_3)_3]$ radical has a tetrahedral structure about the phosphorus atom with the unpaired electron localized in the carborane ring [7]. Crystallographic data, bond lengths, and bond angles are reported for $1,2\text{-}C_2B_{10}H_{10}\text{-}1\text{-}CH_3\text{-}2\text{-}P(O)Cl$; crystals of the compound are orthorhombic, space group Pbca-D_{2h}^{15} (No. 61); a = 21.687(2), b = 15.000(1), c = 13.505(1) Å. Lengths of the B-B (average 1.769 Å) and B-C (average 1.736 Å) bonds in the cages are normal while the C-C bonds (1.703 and 1.722 Å) are slightly longer than in other known $1,2\text{-}C_2B_{10}H_{12}$ derivatives [8].

References for 13.11.2.4:

[1] Stanko, V. I.; Gol'tyapin, Yu. V. (Zh. Obshch. Khim. **52** [1982] 78/80; J. Gen. Chem. [USSR] **52** [1982] 72/3).

[2] Sergeev, V. A.; Danilova, M. P.; Shitikov, V. K.; Kalinin, V. N.; Zakharkin, L. I. (Izv. Akad. Nauk SSSR Ser. Khim. **1981** 226/7; C.A. **95** [1981] No. 62286).

[3] Kalinin, V. N.; Usatov, A. V.; Zakharkin, L. I. (Zh. Obshch. Khim. **51** [1981] 2151/2; J. Gen. Chem. [USSR] **51** [1981] 1853/4).

[4] Yanovskii, A. I.; Struchkov, Yu. T.; Kalinin, V. N.; Usatov, A. V.; Zakharkin, L. I. (Koord. Khim. **8** [1982] 240/4; C.A. **97** [1982] No. 39078).

[5] Colburn, C. B.; Hill, W. E.; Silva-Trivino, L. M.; Verma, R. D. (J. Fluorine Chem. **23** [1983] 261/6).

[6] Hill, W. E.; Rackley, B. G.; Silva-Trivino, L. M. (Inorg. Chim. Acta **75** [1983] 51/6).

[7] Tumanskii, B. L.; Degtyarev, A. N.; Bubnov, N. N.; Solodovnikov, S. P.; Bregadze, V. I.; Godovikov, N. N.; Kabachnik, M. I. (Izv. Akad. Nauk SSSR Ser. Khim. **1980** 2627/30; C.A. **94** [1981] No. 102438).

[8] Furmanova, N. G.; Struchkov, Yu. T.; Degtyarev, A. N.; Bregadze, V. I.; Godovikov, N. N.; Kabachnik, M. I. (Izv. Akad. Nauk SSSR Ser. Khim. **1980** 845/8; C.A. **93** [1980] No. 113814).

13.11.2.5 1,2-$C_2B_{10}H_{12}$ Derivatives Having a Substituent with a Carbon Atom Attached to the Carborane Cage

The zero-field splitting parameters of the triplet EPR spectra of the carboranylcarbene, **1,2-$C_2B_{10}H_{11}$-1-CH_3**, and its 2-CH_3 derivative indicate little delocalization of the π-spin density onto the carboranyl moiety [1]. Electron resonance capture mass spectra of the radical anion of 1,2-$C_2B_{10}H_{11}$-1-CH_3 in the gas phase confirms the existence of long-lived negative ions in this carborane [2].

Li[*closo*-1,2-$C_2B_{10}H_{10}$-1-CH_3]·[$CH_3N\{CH_2CH_2N(CH_3)_2\}_2$], prepared by successive treatment of 1-CH_3-*closo*-1,2-$C_2B_{10}H_{11}$ with LiC_4H_9 and $CH_3N\{CH_2CH_2N(CH_3)_2\}_2$, crystallizes in the monoclinic space group $P2_1/m$-C^2_{2h} (No. 11) with a = 8.242(1), b = 13.522(1), c = 10.799(1) Å; β = 108.75(1)°; Z = 2, D_{calc} = 0.983 g/cm³. The compound has a terminal Li atom singly bonded to the six-coordinate C atom of the carborane. An infrared spectrum in Nujol mull is reported [3].

Electrophilic alkylation of 1,2-$C_2B_{10}H_{12}$ with $(CH_3)_2CHX$ (X = Cl, Br) in the presence of $AlCl_3$ in CS_2 gives a mixture of 8- and 9-*n*-C_3H_7- and 8- and 9-*i*-C_3H_7 derivatives of the carborane. Additionally, 1,2-$C_2B_{10}H_{11}$-9-$CH(CH_3)_2$ undergoes alkyl migration and redistribution, in the presence of $AlCl_3$, to give an equilibrium mixture of 27% 1,2-$C_2B_{10}H_{12}$, 52% 1,2-$C_2B_{10}H_{11}$-*B*-C_3H_7, and 21% 1,2-$C_2B_{10}H_{10}$-*B,B'*-$(C_3H_7)_2$, in which the 1,2-$C_2B_{10}H_{11}$-*B*-C_3H_7 portion consists of 36% 4-*i*-C_3H_7-, 20% 8-*i*-C_3H_7-, 4% 8-*n*-C_3H_7-, 20% 9-*i*-C_3H_7-, and 20% of other *n*-C_3H_7 derivatives of 1,2-$C_2B_{10}H_{12}$ [4]. The phase equilibria in 1,2-$C_2B_{10}H_{11}$-1-$CH(CH_3)_2$, 1,2-$C_2B_{10}H_{11}$-1-C_4H_9, and 1,2-$C_2B_{10}H_{11}$-1-C_6H_{11} were determined by differential scanning colorimetry [5]. A melting point diagram for the binary system 1,2-$C_2B_{10}H_{11}$-1-$CH(CH_3)_2$/$(C_2H_5)_3NBH_3$ is of the eutectic type [6]. The kinetics of the reaction between Li[1,2-$C_2B_{10}H_{11}$-1-$CH(CH_3)_2$] and C_2H_5Br and C_2H_5I indicate that the Li[1,2-$C_2B_{10}H_{11}$-1-$CH(CH_3)_2$] is monomeric in an ether-benzene solution; the melting point of 1,2-$C_2B_{10}H_{10}$-1-C_2H_5-2-$CH(CH_3)_2$ is 37.5 to 38.5°C [7].

The boiling points of various alkyl carboranes, 1,2-$C_2B_{10}H_{11}$-1-R, have been extrapolated to be 298.05°C for R = *n*-C_4H_9, 290.7°C for R = *i*-C_4H_9, 313.6°C for R = C_5H_{11}, 327.7°C for R = C_6H_{13} [8]. The heat of formation for 1,2-$C_2B_{10}H_{11}$-1-$C(CH_3)_3$ is found to be −313.8 ± 16.2 kJ/mol in the crystalline state and −230.8 ± 17.5 kJ/mol in the gaseous state at 298.15 K [9].

CNDO/2 MO calculations on 1,2-$C_2B_{10}H_{12}$ with a C-attached allyl anion suggest that the negative charge on the allyl group carbon atoms increases in the order 2' > 3' > 1' [10].

1,2-$C_2B_{10}H_{11}$-1-$CH_2CH{=}CH_2$, prepared from 1,2-$C_2B_{10}H_{12}$, NaH, $CH_2{=}CHCH_2Br$, and NaI in $CH_3OCH_2CH_2OCH_3$, rearranges to **1,2-$C_2B_{10}H_{11}$-1-$CH{=}CHCH_3$** upon heating with $NaOC(CH_3)_3$ in $HOC(CH_3)_3$ [11]. Reaction of 1,2-$C_2B_{10}H_{10}$-1,2-$(CH_2CH{=}CH_2)_2$ with $WCl_6/Sn(CH_3)_4$ or $WOCl_4/Sn(CH_3)_4$ produces the dihydrobenzocarborane **1,2-$C_2B_{10}H_{10}$-(-1-$CH_2CH{=}CH$-CH_2-2-)**; the melting point of the compound is 116 to 118°C [12].

1,2-$C_2B_{10}H_{10}$-1,2-$(CH_2Cl)_2$ reacts with Mg to form **1,2-$C_2B_{10}H_{10}$-1,2-(-CH_2CH_2-)** having a four-membered C_4 ring [14].

1,2-$C_2B_{10}H_{12}$ reacts with carboethoxycarbene, :$CHCO_2C_2H_5$, to form all four possible products of B-H insertion, **1,2-$C_2B_{10}H_{11}$-*x*-$CH_2CO_2(C_2H_5)$** (*x*=3, 4, 8, or 9) [15]. The ^{1}H NMR spectrum of 1,2-$C_2B_{10}H_{10}$-1-$CH(CH_2COR)_2$-2-R′ shows a signal at δ=4.12 ppm (cage CH) for R=$C(CH_3)_3$, R′=H; partial NMR data are also reported for compounds where R=$C(CH_3)_3$, R′=CH_3; R=$C(CH_3)_3$, R′=C_6H_5; R=R′=C_6H_5. The dipole moment of the compound where R=R′=C_6H_5 is 4.97 D [16].

The reaction of 1,2-$C_2B_{10}H_{10}$-1,2-$(CH_2OH)_2$ with PCl_5 results in a spiro phosphorus derivative, [1,2-$C_2B_{10}H_{10}$-1,2-$(CH_2O\text{-})_2]_2PCl$, melting point 137 to 138°C, and 1,2-$C_2B_{10}H_{10}$-1,2-$(CH_2\text{-})_2O$ [17]. Hydrolyzing 1,2-$C_2B_{10}H_{10}$-1,2-(-$CH_2OPClOCH_2$-), prepared by cyclizing 1,2-$C_2B_{10}H_{10}$-1,2-$(CH_2OH)_2$ with PCl_3, yields 1,2-$C_2B_{10}H_{10}$-1,2-(-$CH_2OP(H)(O)OCH_2$-), boiling point 95 to 100°C at 0.0001 Torr. Treatment of the latter compound with $(C_2H_5)_2NH$ gives 1,2-$C_2B_{10}H_{10}$-1,2-[-$CH_2OP\{N(C_2H_5)_2\}(O)OCH_2$-]. The infrared spectrum of the latter compound shows absorption bands at (frequencies in cm^{-1}) 2560 (B-H stretching), 1460 (B-C), 1220 (P=O), 1030 and 1090 (C-O-P) [18].

The reaction of Li[1,2-$C_2B_{10}H_{10}$-1-R] (R=CH_3, $CH(CH_3)_2$, C_6H_5) with $ClCH_2(C_2H_3S)$, where C_2H_3S contains a C_2S three-membered ring, produces **1,2-$C_2B_{10}H_{10}$-1-R-2-$CH_2(C_2H_3S)$**. The latter compound reacts with $C_6H_5NHCH_3$ to give aminothiols, with SO_2Cl_2 to give dichloro disulfides, and with H_2O_2 to give sulfonic acids [19]. Electrophilic ring cleavage of 1,2-$C_2B_{10}H_{10}$-1-R-2-$CH_2(CHSCH_2)$ (in which $CHSCH_2$ is a 3-membered C_2S ring, R=CH_3, $CH(CH_3)_2$, C_6H_5) with Cl_2, Br_2, or I_2 produces the disulfides **[1,2-$C_2B_{10}H_{10}$-1-R-2-$CH_2CH(S\text{-})CH_2X]_2$**, X=Cl, Br, I (except R=CH_3, X=Cl). Ring cleavage of 1,2-$C_2B_{10}H_{10}$-1-R-2-$CH_2(CHSCH_2)$ (in which $CHSCH_2$ is a 3-membered C_2S ring, R=$CH(CH_3)_2$, C_6H_5) with CH_3COX (X=Cl, Br, I) produces **1,2-$C_2B_{10}H_{10}$-1-R-2-$CH_2CH(CH_2X)SCOCH_3$**; melting points, in °C, for the latter compounds: 71 to 72 for R=C_6H_5, X=Cl; 74 to 75 for R=C_6H_5, X=Br; 86 to 87 for R=$CH(CH_3)_2$, X=I; 90 to 91 for R=C_6H_5, X=I [20]. 1,2-$C_2B_{10}H_{10}$-1-R-2-$CH_2(CHSCH_2)$ (in which R=$CH(CH_3)_2$, C_6H_5 and in which $CHSCH_2$ represents a three-membered C_2S ring) reacts with C_6H_5SH, C_2H_5OH, and $NaSC_6H_5$ to yield **1,2-$C_2B_{10}H_{10}$-1-R-2-$CH_2CH(SH)CH_2SC_6H_5$**; the boiling point of 1,2-$C_2B_{10}H_{10}$-1-$CH(CH_3)_2$-2-$CH_2CH(SH)CH_2SC_6H_5$ is 218 to 220°C; the melting point of 1,2-$C_2B_{10}H_{10}$-1-C_6H_5-2-$CH_2CH(SH)CH_2SC_6H_5$ is 68 to 69°C. Reaction of 1,2-$C_2B_{10}H_{10}$-1-R-2-$CH_2(CHSCH_2)$ (in which R=$CH(CH_3)_2$, C_6H_5 and in which $CHSCH_2$ represents a three-membered C_2S ring) with KSH in C_2H_5OH yields **1,2-$C_2B_{10}H_{10}$-1-R-2-$CH_2CH(SH)CH_2SH$**; the melting point of 1,2-$C_2B_{10}H_{10}$-1-$CH(CH_3)_2$-2-$CH_2CH(SH)CH_2SH$ is 46 to 47°C; the melting point of 1,2-$C_2B_{10}H_{10}$-1-C_6H_5-2-$CH_2CH(SH)CH_2SH$ is 64 to 65°C [21]. 1,2-$C_2B_{10}H_{10}$-1-R-2-CH_2-$CHSCH_2$ (in which $CHSCH_2$ is a three-membered C_2S ring) reacts with $C_6H_5NH_2$ to produce 1,2-$C_2B_{10}H_{10}$-1-R-2-$CH_2CH(SH)$-$CH_2NHC_6H_5$, melting point 124 to 125°C for R=$CH(CH_3)_2$, 136 to 137°C for R=C_6H_5. Treatment of 1,2-$C_2B_{10}H_{10}$-1-R-2-$CH_2CH(SH)CH_2NHC_6H_5$ with aldehydes and ketones gives rise to thiazolidine and 1,3-dithiolane derivatives. **1,2-$C_2B_{10}H_{10}$-1-R-2-CH_2-$CHSCH_2$** (in which $CHSCH_2$ is a three-membered C_2S ring) reacts with KSH in C_2H_5OH and subsequently with ketones to yield cyclic thioacetals. 1,2-$C_2B_{10}H_{10}$-1,2-$(CH_2\text{-}CHSCH_2)_2$ (in which $CHSCH_2$ is a three-membered C_2S ring) reacts with KSH in C_2H_5OH to form **1,2-$C_2B_{10}H_{10}$-1,2-$[CH_2CH(SH)CH_2SH]_2$**, melting point 115 to 116°C. By an analogous procedure 1,2-$C_2B_{10}H_{11}$-1-CH_2-$CHSCH_2$ (in which $CHSCH_2$ is a three-membered C_2S ring) yields 1,2-$C_2B_{10}H_{11}$-1-$CH_2CH(SH)CH_2SH$, melting point 79 to 80°C [22].

 References for 13.11.2.5 on pp. 196/8

Carboranyl-substituted dithiols react with MCl_2 (M = Co, Ni) and $[(C_2H_5)_4N]Br$ to give four-coordinate dithiolato metal complexes [22]. Reaction of $1,2\text{-}C_2B_{10}H_{11}\text{-}1\text{-}CH_2N(C_2H_5)_2$ with $CH_3Re(CO)_5$ at 100°C gives $1,2\text{-}C_2B_{10}H_{10}\text{-}1\text{-}CH_2N(C_2H_5)_2\text{-}B\text{-}Re(CO)_4$ [23].

Crystals of $1,2\text{-}C_2B_{10}H_{10}\text{-}1\text{-}CH_2C_6H_5\text{-}2\text{-}MgCH_3$ are monoclinic, space group $P2_1/n\text{-}C^5_{2h}$ (No. 14) with a = 20.44(1), b = 7.580(5), c = 22.07(1) Å; β = 108.79(4)°; D_{calc} = 1.842, D_{meas} = 1.793 g/cm³; Z = 8 [13]. $1,2\text{-}C_2B_{10}H_{10}\text{-}1,2\text{-}(CH_2MgBr)_2$ reacts with $RPCl_2$ (R = CH_3, C_6H_5) to yield $1,2\text{-}C_2B_{10}H_{10}\text{-}1,2\text{-}(\text{-}CH_2P(R)CH_2\text{-})$ containing a five-membered C_4P ring; the melting point is 62 to 63°C for the compound in which R = CH_3, 138 to 140°C for the compound in which R = C_6H_5 [24].

1-Methyl-1-(2-propynyl)-3,3,5,5-tetrachlorocyclotriphosphazene reacts with bis(acetonitrile)-decaborane to yield 1-methyl-1-[(carboranyl)methylene]-3,3,5,5-tetrachlorocyclotriphosphazene (where carboranyl = $1,2\text{-}C_2B_{10}H_{11}\text{-}1\text{-}$). 1-Methyl-1-[(carboranyl)methylene]-3,3,5,5-tetrachlorocyclotriphosphazene (where carboranyl = $1,2\text{-}C_2B_{10}H_{11}\text{-}1\text{-}$) is converted by a base such as $C_5H_{10}NH$ to the *nido*-carboranyl anions 1-methyl-1-[(carboranyl)methylene]-3,3,5,5-tetrachlorocyclotriphosphazene (where carboranyl = $C_2B_9H_{11}$ or $C_2B_9H_{10}$) [26]. CH_3AsI_2 and $C_6H_5AsCl_2$ react with $1,2\text{-}C_2B_{10}H_{10}\text{-}1,2\text{-}(CH_2MgBr)_2$ to form $1,2\text{-}C_2B_{10}H_{10}\text{-}1,2\text{-}(\text{-}CH_2As(R)CH_2\text{-})$ containing a five-membered C_4As ring; the melting point is 77 to 78°C for the compound in which R = CH_3, 128 to 129°C for the compound in which R = C_6H_5. On treatment of $1,2\text{-}C_2B_{10}H_{10}\text{-}1,2\text{-}(\text{-}CH_2E(R)CH_2\text{-})$ (E = P, As) with $M(CO)_6$ (M = Cr, W, Mo) compounds of the type $1,2\text{-}C_2B_{10}H_{10}\text{-}1,2\text{-}(\text{-}CH_2E[M(CO)_5](R)CH_2\text{-})$ are formed. Crystals of $1,2\text{-}C_2B_{10}H_{10}\text{-}1,2\text{-}(\text{-}CH_2P(CH_3)CH_2\text{-})$ are monoclinic, space group $P2_1/c\text{-}C^5_{2h}$ (No. 14) with a = 7.8764(8), b = 22.339(2), c = 8.9124(9) Å; β = 99.359(9)°; Z = 4; D_{calc} = 1.195 g/cm³. The cage bond lengths (average) are 1.769(9) Å for B-B, 1.712(9) Å for C-B and 1.656(3) Å for C-C [24]. $1,2\text{-}C_2B_{10}H_{10}\text{-}1\text{-}CH_2As(CH_3)_2\text{-}2\text{-}Li$ reacts with $(C_6H_5CN)_2MCl_2$ (M = Pd, Pt) to give bis-chelate complexes $(1,2\text{-}C_2B_{10}H_{10}\text{-}1\text{-}CH_2As(CH_3)_2\text{-}2\text{-})_2M$ in which both As atoms are bonded to the metal [27].

$(CH_3C_5H_4N)_2Rh(H)(Cl)[1,2\text{-}C_2B_{10}H_{10}\text{-}1\text{-}CH_2P(C_6H_5)_2]$ ($CH_3C_5H_4N$ = 4-methylpyridine) is prepared by reacting $1,2\text{-}C_2B_{10}H_{11}\text{-}1\text{-}CH_2P(C_6H_5)_2$ with $(L_2RhCl)_2$ (L = CO, cyclohexene, or ½ cyclooctadiene) in the presence of 4-methylpyridine. Treating $1,2\text{-}C_2B_{10}H_{11}\text{-}1\text{-}CH_2P(C_6H_5)_2$ with $[Rh(CO)_2Cl]_2$ gives $(1,2\text{-}C_2B_{10}H_{11}\text{-}1\text{-}CH_2PPh_2)_2Rh(CO)Cl$. Reaction of $1,2\text{-}C_2B_{10}H_{11}\text{-}1\text{-}CH_2P(C_6H_5)_2$ with $(C_8H_{12}IrCl)_2$ (C_8H_{12} = 1,5-cyclooctadiene) in hexane or C_2H_5OH gives $C_8H_{12}Ir(H)(Cl)[1,2\text{-}C_2B_{10}H_{10}\text{-}1\text{-}CH_2P(C_6H_5)_2]$ [25].

Crystals of $1,2\text{-}C_2B_{10}H_{10}\text{-}1,2\text{-}[\text{-}CH_2Si(CH_3)_2OSi(CH_3)_2CH_2\text{-}]_2$ are monoclinic, space group $P2_1/m\text{-}C^2_{2h}$ (No. 11), a = 7.609(4), b = 14.834(6), c = 8.190(4) Å; β = 103.00(4)°; Z = 2. The seven-membered C_4SiOSi ring has the chair conformation, the oxygen and the carbon atoms of the carborane cage being on the opposite sides of the $C(H_2)Si\cdots SiC(H_2)$ plane [28]. The thermal decomposition of $1,2\text{-}C_2B_{10}H_{11}\text{-}1\text{-}CH_2Si(CH_3)_2OOC(CH_3)_3$ has been reported [29]. The reactivity of $1,2\text{-}C_2B_{10}H_{11}\text{-}1\text{-}CH_2M(CH_3)_3$ toward alkaline solvolysis in CH_3OH decreases in the order M = Sn, Si, Ge [30].

$1,2\text{-}C_2B_{10}H_{10}\text{-}1,2\text{-}(CH_2MgBr)_2$, prepared from $1,2\text{-}C_2B_{10}H_{10}\text{-}1,2\text{-}(CH_2Br)_2$ and Mg, reacts with $[(C_6H_5)_3]_2PMCl_2$ (M = Pt, Pd, Ni) to give $1,2\text{-}C_2B_{10}H_{10}\text{-}1,2\text{-}(CH_2\text{-})_2M[P(C_6H_5)_3]_2$ containing a five-membered C_4M ring. Analogous treatment of $1,2\text{-}C_2B_{10}H_{10}\text{-}1,2\text{-}(CH_2MgBr)_2$ with $(C_5H_5)_2MCl_2$ (M = Ti, Zr) yields $1,2\text{-}C_2B_{10}H_{10}\text{-}1,2\text{-}(CH_2\text{-})_2M(C_5H_5)_2$. Reaction of $1,2\text{-}C_2B_{10}H_{10}\text{-}1,2\text{-}(CH_2MgBr)_2$ with MCl_4 (M = Si, Sn, Ge) provides the spiro-heterocyclic compounds $[1,2\text{-}C_2B_{10}H_{10}\text{-}1,2\text{-}(CH_2\text{-})_2]_2M$ [31].

$1,2\text{-}C_2B_{10}H_{10}\text{-}1\text{-}HgCH_3\text{-}2\text{-}CH_2OCH_2CH_3$, which crystallizes in the monoclinic system in space group $P2_1/n\text{-}C^5_{2h}$ (No. 14), has a melting point of 117 to 118°C. Lattice constants at −120°C are a = 7.394(1), b = 12.844(3), c = 15.087(4) Å; β = 96.30(2)°; Z = 4; D_{calc} = 1.944 g/cm³ [32].

Stable carboranylmethyl complexes of Fe and Ni, i.e., $(1,2\text{-}C_2B_{10}H_{10}\text{-}1\text{-}CH_3\text{-}2\text{-}CH_2\text{-})Fe(CO)_2(C_5H_5)$ and $(1,2\text{-}C_2B_{10}H_{10}\text{-}1\text{-}CH_3\text{-}2\text{-}CH_2\text{-})Ni(PPh_3)(C_5H_5)$, are obtained by reaction of the corresponding carboranylmethylmagnesium chlorides with halogen complexes of Fe and Ni. $(1,2\text{-}C_2B_{10}H_{10}\text{-}1\text{-}CH_3\text{-}2\text{-}CH_2\text{-})Fe(CO)_2(C_5H_5)$, when treated with Br_2 or $HgCl_2$, undergoes rearrangement with migration of the carboranylmethyl group from the Fe atom to the carbonyl group [33].

Isomerization of $1,2\text{-}C_2B_{10}H_{11}\text{-}3\text{-}CH{=}CH_2$ in the gas phase at 560 to 580°C gives a mixture of $1,7\text{-}C_2B_{10}H_{11}\text{-}2\text{-}CH{=}CH_2$ and $1,7\text{-}C_2B_{10}H_{11}\text{-}4\text{-}CH{=}CH_2$. Isomerization of $1,2\text{-}C_2B_{10}H_{11}\text{-}3\text{-}C{\equiv}CH$ in the gas phase at 560 to 580°C gives a mixture of $1,7\text{-}C_2B_{10}H_{11}\text{-}2\text{-}C{\equiv}CH$ and $1,7\text{-}C_2B_{10}H_{11}\text{-}4\text{-}C{\equiv}CH$ [34]. From infrared spectra of $1,2\text{-}C_2B_{10}H_{11}\text{-}x\text{-}C{\equiv}CH$ ($x=1, 3, 9$) it is concluded that the ethynyl C-H bond has greater proton-donating capacity than the carborane C-H bond. Introduction of the ethynyl group into the carborane ring causes no change in the proton-donating capacity of the carborane C-H bond in comparison to the unsubstituted carborane. The carboranyl group increases the proton-donating capacity of acetylenic C-H bonds in comparison with $C_4H_9C{\equiv}CH$; this effect decreases in the series $x=1>3>9$ [35].

Thermal decarboxylation of $1,2\text{-}C_2B_{10}H_{11}\text{-}1\text{-}COOCu\cdot bpy$ (where bpy = 2,2′-bipyridyl) at approximately 100°C yields $1,2\text{-}C_2B_{10}H_{10}\text{-}1,2\text{-}Cu\cdot bpy$ containing a three-membered CuC_2 ring [36]. Reaction of $CoCl_2\cdot bpy$ with $1,2\text{-}C_2B_{10}H_{10}\text{-}1\text{-}COOLi\text{-}2\text{-}Li$ produces $1,2\text{-}C_2B_{10}H_{10}\text{-}1,2\text{-}(\text{-}COOCo\text{-})$ containing a five-membered C_3OCo ring [37]. Heating $1,2\text{-}C_2B_{10}H_{10}\text{-}1,2\text{-}(\text{-}COONiOOC\text{-})\cdot bpy$ at 174°C produces $1,2\text{-}C_2B_{10}H_{10}\text{-}1,2\text{-}Ni\cdot bpy$ in which it is proposed that there is a three-membered NiCC ring [38].

The carboranylcyclopentadiene $1,2\text{-}C_2B_{10}H_{10}\text{-}1\text{-}C_6H_5\text{-}2\text{-}C_5H_3\text{-}3',4'\text{-}(C_6H_5)_2$, melting point 170 to 172°C, is prepared by refluxing $1,2\text{-}C_2B_{10}H_{10}\text{-}1\text{-}C_6H_5\text{-}2\text{-}CH(CH_2COC_6H_5)_2$ in Zn/CH_3COOH followed by dehydration of the intermediate product with HCl/C_2H_5OH [39].

The dipole moment of $1,2\text{-}C_2B_{10}H_{11}\text{-}1\text{-}C_6H_5$ is 4.65 D [16]. $1,2\text{-}C_2B_{10}H_{11}\text{-}1\text{-}C_6H_5$ with tritium in the 2-position is prepared by treating the corresponding $1,2\text{-}C_2B_{10}H_{10}\text{-}1\text{-}C_6H_5\text{-}2\text{-}Li$ with tritiated water. Using previously described methods, the tritiated $1,2\text{-}C_2B_{10}H_{11}\text{-}1\text{-}C_6H_5$ was converted to the corresponding tritiated diazonium cation $1,2\text{-}C_2B_{10}H_{11}\text{-}1\text{-}(C_6H_4\text{-}4'\text{-}N_2)$ suitable for azo coupling reactions [40]. The mono-nitration of the phenyl ring of $1,2\text{-}C_2B_{10}H_{11}\text{-}1\text{-}C_6H_5$ produces a 4% *ortho*, 26% *meta*, and 70% *para* isomer mixture. No dinitro derivatives were detected [41]. The synthesis of $1,2\text{-}C_2B_{10}H_{10}\text{-}1,2\text{-}R_2$ (where $R=C_6H_4\text{-}4'\text{-}C_6H_5$, $C_6H_4\text{-}4\text{-}OC_6H_4\text{-}4'\text{-}OCH_3\text{-}$, $C_6H_4\text{-}4\text{-}SC_6H_4\text{-}4'\text{-}OCH_3$, $C_6H_4\text{-}4\text{-}C_6H_4\text{-}4'\text{-}OCH_3$) is briefly described [42]. $1,2\text{-}C_2B_{10}H_{10}\text{-}1,2\text{-}(C_6H_5)_2$ dimerizes at 260 to 230°C with the formation of stable B-B bonds, probably between the B atoms farthest removed from the C atoms in the carborane icosahedra [43].

Crystallographic data have been collected for $1,2\text{-}C_2B_{10}H_{10}\text{-}1\text{-}C_6H_5\text{-}2\text{-}HgX$ (X = Br, I). Lattice constants for the bromide are a = 7.753(2), b = 10.800(4), c = 18.029(9) Å, γ = 90.73(3)°; and for the iodide a = 7.782(2), b = 10.945(2), c = 18.141(6) Å, γ = 90.08(2)°. In both cases Z = 4, space group $P2_1/n\text{-}C^5_{2h}$ (No. 14) [44].

Treatment of $[1,2\text{-}C_2B_{10}H_{10}\text{-}1\text{-}R\text{-}2\text{-}C_5H_2O\text{-}2',6'\text{-}(R')_2][ClO_4]$ (where $R=C(CH_3)_3$, C_6H_5, and $R'=CH_3$, C_6H_5) with NH_3, CH_3NH_2, $C_6H_5NH_2$, N_2H_4, or $C_6H_5NNH_2$ yields a variety of nitrogen heterocycles with a carboranyl substituent [45].

The rate constants for *B*-metallation of $1,2\text{-}C_2B_{10}H_{10}\text{-}1,2\text{-}(CH_3)_2$, $1,2\text{-}C_2B_{10}H_{11}\text{-}1\text{-}CH_3$, and $1,2\text{-}C_2B_{10}H_{12}$ by $Tl(OOCCF_3)_3$ decrease in the stated order [46]. $1,2\text{-}C_2B_{10}H_{10}\text{-}x\text{-}R\text{-}2\text{-}LnCl_2$ (in which $x=1$, $R=CH_3$, Ln = Tm; $x=1$, $R=C_6H_5$, Ln = Tm, Yb; $x=9$, R = H, Ln = Tm, Yb), as tetrahydrofuran adducts, are obtained from the treatment of $(1,2\text{-}C_2B_{10}H_{10}\text{-}x\text{-}R\text{-}2\text{-})_2Hg$ with $LnCl_3$ in tetrahydrofuran solution [47]. Reaction of *closo*-$1\text{-}CH_3\text{-}1,2\text{-}C_2B_{10}H_{11}$ with two equivalents of $Co(P(C_2H_5)_3)_4$ in toluene at room temperature yields $1\text{-}CH_3\text{-}4\text{-}(C_2H_5)_3P\text{-}\mu(4,6 \text{ or } 4,7)\text{-}[Co(P(C_2H_5)_3)_2\text{-}\mu\text{-}(H)_2]\text{-}4,1,2\text{-}CoC_2B_{10}H_{10}$, the structure of which was determined by X-ray crystallography. The orthorhombic crystals have lattice parameters a = 16.7723(16), b = 21.662(2), c = 19.149(3) Å; Z = 8; D_{calc} = 1.206 g/cm³; space group $Pbca\text{-}D^{15}_{2h}$ (No. 61) [48].

References for 13.11.2.5 on pp. 196/8

References for 13.11.2.5:

[1] Hutton, R. S.; Roth, H. D.; Chari, S. (J. Phys. Chem. **85** [1981] 753/4).
[2] Mazunov, V. A.; Nekrasov, Yu. S.; Khvostenko, V. I.; Stanko, V. I. (Izv. Akad. Nauk SSSR Ser. Khim. **1983** 223/5; Bull. Acad. Sci. USSR Div. Chem. Sci. **1983** 197/9).
[3] Clegg, W.; Brown, D. A.; Bryan, S. J.; Wade, K. (Polyhedron **3** [1984] 307/11).
[4] Zakharkin, L. I.; Kovredov, A. I.; Ol'shevskaya, V. A.; Vitt, S. V. (Izv. Akad. Nauk SSSR Ser. Khim. **1983** 1680; Bull. Acad. Sci. USSR Div. Chem. Sci. **1983** 1528/9).
[5] Chistov, S. F.; Sobolev, E. S.; Skorokhodov, I. I.; Larikov, E. I. (Zh. Obshch. Khim. **53** [1983] 733/5; J. Gen. Chem. [USSR] **53** [1983] 638/40).
[6] Sobolev, E. S.; Chistov, S. F.; Larikov, E. I.; Skorokhodov, I. I. (Zh. Obshch. Khim. **53** [1983] 1847/9; J. Gen. Chem. [USSR] **53** [1983] 1662/4).
[7] Savel'eva, I. S.; Kanakhina, L. N.; Kazantsev, A. V.; Zakharkin, L. I. (Izv. Akad. Nauk SSSR Ser. Khim. **1981** 2211/5; Bull. Acad. Sci. USSR Div. Chem. Sci. **1981** 1811/5).
[8] Shul'tse, P.; Varushchenko, R. M.; Gal'chenko, G. L.; Klimova, T. V.; Stanko, V. I. (Zh. Obshch. Khim. **50** [1980] 1818/25; J. Gen. Chem. [USSR] **50** [1980] 1482/8).
[9] Shul'tse, P.; Gal'chenko, G. L.; Pavlovich, V. K.; Miroshnichenko, E. A.; Lebedev, Yu. A.; Stanko, V. I.; Klimova, T. V. (Vestn. Mosk. Univ. Ser. II Khim. **36** [1981] 450/4; Moscow Univ. Chem. Bull. **36** No. 5 [1981] 27/31; C.A. **95** [1981] No. 226687).
[10] Minaev, B. F.; Kazantsev, A. V.; Kozhakov, B. E. (Zh. Strukt. Khim. **22** [1981] 166/8; J. Struct. Chem. [USSR] **22** [1981] 778/80).

[11] Plešek, J.; Štíbr, B.; Drdakova, E.; Plzak, Z.; Heřmánek, S. (Chem. Ind. [London] **1982** 778/9).
[12] Vdovin, V. M.; Bespalova, N. B.; Bovina, M. A.; Kalinin, V. N.; Zakharkin, L. I. (Izv. Akad. Nauk SSSR Ser. Khim. **1984** 474; Bull. Acad. Sci. USSR Div. Chem. Sci. **1984** 441).
[13] Furmanova, N. G.; Struchkov, Yu. T.; Kalinin, V. N.; Finkel'shtein, B. Ya.; Zakharkin, L. I. (Zh. Strukt. Khim. **21** No. 5 [1980] 96/100; C.A. **94** [1981] No. 112826).
[14] Zakharkin, L. I.; Kovredov, A. I.; Meiramov, M. G. (Zh. Obshch. Khim. **51** [1981] 238/9; C.A. **95** [1981] No. 25183).
[15] Zheng, G.; Jones, M. (J. Am. Chem. Soc. **105** [1983] 6487/8).
[16] Dorofeenko, G. N.; Drygina, O. V.; Gruntfest, M. G.; Osipov, O. A. (Zh. Obshch. Khim. **51** [1981] 1570/5; J. Gen. Chem. [USSR] **51** [1981] 1332/6).
[17] Kazantsev, A. V.; Meiramov, M. G.; Kovredov, A. I.; Zakharkin, L. I. (Izv. Akad. Nauk SSSR Ser. Khim. **1982** 1603/5; Bull. Acad. Sci. USSR Div. Chem. Sci. **1982** 1425/7).
[18] Nifant'ev, E. E.; Prokof'eva, T. Yu.; Magdeeva, R. K.; Sklyarskii, L. S.; Galstyan, K. D. (Zh. Obshch. Khim. **52** [1982] 217/8; J. Gen, Chem. [USSR] **52** [1982] 197/8).
[19] Shustova, T. V. (Izv. Akad. Nauk Kaz.SSR Ser. Khim. **1982** No. 2, pp. 62/4; C.A. **97** [1982] No. 6360).
[20] Zakharkin, L. I.; Shustova, T. V.; Kazantsev, A. V. (Zh. Obshch. Khim. **51** [1981] 1071/8; J. Gen. Chem. [USSR] **51** [1981] 893/9).

[21] Kazantsev, A. V.; Shustova, T. V.; Zakharkin, L. I. (Izv. Akad. Nauk SSSR Ser. Khim. **1982** 2134/7; Bull. Acad. Sci. USSR Div. Chem. Sci. **1982** 1884/7).
[22] Kazantsev, A. V.; Shustova, T. V.; Antonovich, V. A.; Zakharkin, L. I. (Zh. Obshch. Khim. **54** [1984] 579/83; J. Gen. Chem. [USSR] **54** [1984] 511/4).
[23] Kalinin, V. N.; Usatov, A. V.; Zakharkin, L. I. (Zh. Obshch. Khim. **51** [1981] 2151/2; J. Gen. Chem. [USSR] **51** [1981] 1853/4).
[24] Zakharkin, L. I.; Meiramov, M. G.; Antonovich, V. A.; Kazantsev, A. V.; Yanovskii, A. I.; Struchkov, Yu. T. (Zh. Obshch. Khim. **53** [1983] 90/7; J. Gen. Chem. [USSR] **53** [1983] 73/9).

[25] Kalinin, V. N.; Usatov, A. V.; Zakharkin, L. I. (Zh. Obshch. Khim. **53** [1983] 945/6; J. Gen. Chem. [USSR] **53** [1983] 833/4).

[26] Allcock, H. R.; Scopelianos, A. G.; Whittle, R. R.; Tollefson, N. M. (J. Am. Chem. Soc. **105** [1983] 1316/21).

[27] Zakharkin, L. I.; Kovredov, A. I.; Savel'eva, I. S.; Meiramov, M. T. (Izv. Akad. Nauk SSSR Ser. Khim. **1980** 2645/6; C.A. **94** [1981] No. 175242).

[28] Yanovsky, A. I.; Struchkov, Yu. T.; Kalinin, V. N.; Ismailov, B. A.; Myakushev, V. D. (Cryst. Struct. Commun. **10** [1981] 817/21).

[29] Yablokov, V. A.; Sluchevskaya, N. P.; Yablokova, N. V.; Kalinin, V. N.; Izmailov, B. A. (Zh. Obshch. Khim. **52** [1982] 893/6; J. Gen. Chem. [USSR] **52** [1982] 777/9).

[30] Stanko, V. I.; Klimova, T. V. (Dokl. Akad. Nauk SSSR **257** [1981] 891/5; Dokl. Chem. Proc. Acad. Sci. USSR **256/261** [1980] 132/5).

[31] Zakharkin, L. I.; Shemyakin, N. F. (Izv. Akad. Nauk SSSR Ser. Khim. **1981** 1856/9; Bull. Acad. Sci. USSR Div. Chem. Sci. **1982** 1525/7).

[32] Yanovskii, A. I.; Struchkov, Yu. T.; Kalinin, V. N.; Zurlova, O. M.; Zakharkin, L. I. (Zh. Strukt. Khim. **22** [1981] 120/4; J. Struct. Chem. [USSR] **22** [1981] 406/9).

[33] Kovredov, A. I.; Meiramov, M. G.; Kazantsev, A. V.; Zakharkin, L. I. (Zh. Obshch. Khim. **51** [1981] 854/9; J. Gen. Chem. [USSR] **51** [1981] 708/12).

[34] Kalinin, V. N.; Kobel'kova, N. I.; Zakharkin, L. I. (Izv. Akad. Nauk SSSR Ser. Khim. **1982** 1661/2; Bull. Acad. Sci. USSR Div. Chem. Sci. **1982** 1479/81).

[35] Vinogradova, L. E.; Leites, L. A.; Kovredov, A. I.; Ol'shevskaya, V. A.; Zakharkin, L. I. (Izv. Akad. Nauk SSSR Ser. Khim. **1982** 1663/4; Bull. Acad. Sci. USSR Div. Chem. Sci. **1983** 1481/2).

[36] Ol'dekop, Yu. A.; Maier, N. A.; Erdman, A. A.; Prokopovich, V. P. (Vestsi Akad. Navuk Belarusk.SSR Ser. Khim. Navuk **1981** No. 2, pp. 71/5; C.A. **96** [1982] No. 20149).

[37] Ol'dekop, Yu. A.; Maier, N. A.; Erdman, A. A.; Prokopovich, V. P. (Zh. Obshch. Khim. **52** [1982] 2256/9; J. Gen. Chem. [USSR] **52** [1982] 2008/10).

[38] Ol'dekop, Yu. A.; Maier, N. A.; Erdman, A. A.; Prokopovich, V. P. (Dokl. Akad. Nauk SSSR **257** [1981] 647/51; Dokl. Chem. Proc. Acad. Sci. [USSR] **256/261** [1981] 118/20).

[39] Drygina, O. V.; Garnovskii, A. D.; Kolodyazhnyi, Y. V.; Gruntfest, M. G.; Kazantsev, A. V. (Zh. Obshch. Khim. **53** [1983] 1066/9; J. Gen. Chem. [USSR] **53** [1983] 942/5).

[40] Mizusawa, E.; Dahlman, H. L.; Bennett, S. J.; Goldenberg, D. M.; Hawthorne, M. F. (Proc. Natl. Acad. Sci. [U.S.A.] **79** [1982] 3011/4).

[41] Goldenberg, D. M.; Sharkey, R. M.; Primus, F. J.; Mizusawa, E.; Hawthorne, M. F. (Proc. Natl. Acad. Sci. [U.S.A.] **81** [1984] 560/3).

[42] Teplyakov, M. M.; Khotina, I. A.; Gelashvili, T. L.; Korshak, V. V. (Dokl. Akad. Nauk SSSR **271** [1983] 874/7; Dokl. Chem. Proc. Acad. Sci. [USSR] **268/273** [1983] 252/4).

[43] Sidorenko, V. I.; Vasyukova, N. I.; Kuzaev, A. I.; Nekrasov, Yu. S.; Valetskii, P. M.; Vinogradova, S. V.; Korshak, V. V. (Vysokomol. Soedin. B **22** [1980] 884/5; C.A. **94** [1981] No. 122221).

[44] Medvedev, A. V.; Pakhomov, V. I. (Koord. Khim. **8** [1982] 627/30; C.A. **97** [1982] No. 39059).

[45] Drygina, O. V.; Dorofeenko, G. N.; Okhlobystin, O. Y. (Khim. Geterotsikl. Soedin. **1981** 454/8; C.A. **95** [1981] No. 97873).

[46] Usyatinskii, A. Y.; Bregadze, V. I.; Shcherbina, T. M.; Godovikov, N. N. (Izv. Akad. Nauk SSSR Ser. Khim. **1981** 1428/31; Bull. Acad. Sci. USSR Div. Chem. Sci. **1981** 1148/50).

[47] Suleimanov, G. Z.; Koval'chuk, N. A.; Bregadze, V. I.; Kurbanov, T. K.; Agaeva, R. A.; Godovikov, N. N.; Beletskaya, I. P. (Zh. Org. Khim. **19** [1983] 2258/63; Russ. J. Org. Chem. **19** [1983] 1969/73).

[48] Barker, G. K.; Garcia, M. P.; Green, M.; Stone, F. G. A.; Welch, A. J. (J. Chem. Soc. Chem. Commun. **1983** 137/9).

13.11.2.6 Si, Sn, Ge, Hg, Tl, Ni, and Ir Derivatives of 1,2-$C_2B_{10}H_{12}$

Crystals of 1,2-$C_2B_{10}H_{11}$-1-$Si(CH_3)_3$ are monoclinic. Lattice constants at −120°C are a = 6.9491(8), b = 17.400(2), c = 11.259(1) Å; β = 92.66(2)°; Z = 4; space group $P2_1/n$-C^5_{2h} (No. 14); D_{calc} = 1.06 g/cm³. Crystals of 1,2-$C_2B_{10}H_{11}$-1-$Ge(CH_3)_3$ are monoclinic, a = 10.668(9), b = 6.868(6), c = 19.93(2) Å; β = 92.02°; Z = 4; D_{meas} = 1.21, D_{calc} = 1.19 g/cm³; space group $P2_1/n$-C^5_{2h} (No. 14) [1]. The reactivity of 1,2-$C_2B_{10}H_{11}$-1-$M(CH_3)_3$ toward alkaline solvolysis in CH_3OH decreases in the order M = Si, Sn, Ge [2].

Infrared and Raman spectra of mercury derivatives of 1,2-$C_2B_{10}H_{12}$ are reported [3]. Refluxing 1,2-$C_2B_{10}H_{11}$-9-$Tl(OOCCF_3)_2$ with Hg in dimethylformamide yields 1,2-$C_2B_{10}H_{11}$-9-$HgOOCCF_3$ and (1,2-$C_2B_{10}H_{11}$-9-$)_2$Hg [4]. The reduction of 1,2-$C_2B_{10}H_{11}$-9-$Tl(SCN)_2$ with $K_2[C_8H_8]$ gives 1,2-$C_2B_{10}H_{12}$ [5].

The preparation of (1,2-$C_2B_{10}H_{11}$-1-)$Ni(C_5H_5)\cdot P(C_6H_5)_3$ has been reported [6]. Complexes of the type *cis*- and *trans*-Ni(X)(1,2-$C_2B_{10}H_{10}$-1-$P[N(CH_3)_2]_2$-2-$P(C_6H_5)_2$) (X = Cl, Br, I, SCN) were prepared and characterized by IR and Raman spectroscopy [7].

The addition reactions of d^8 carborane-iridium(I) complexes of the general formula *trans*-[Ir-(1,2-$C_2B_{10}H_{10}$-2-R-1-)(CO)L_2], where L = $P(C_6H_5)_3$, $P(CH_3)(C_6H_5)_2$, and R = H, CH_3, C_6H_5, with HCl, HBr, Cl_2, and Br_2 have been studied both in the solid state and in solution using a variety of solvents. The HX adducts of the general formula [Ir(H)(X)(1,2-$C_2B_{10}H_{11}$-1-)(CO)L_2] (X = Cl, Br, I) reductively eliminate the carborane molecule 1,2-$C_2B_{10}H_{12}$ [8].

References for 13.11.2.6:

[1] Kirillova, N. I.; Klimova, T. V.; Struchkov, Yu. T.; Stanko, V. I. (Izv. Akad. Nauk SSSR Ser. Khim. **1981** 600/4; Bull. Acad. Sci. USSR Div. Chem. Sci. **30** [1981] 442/6).
[2] Stanko, V. I.; Klimova, T. V. (Dokl. Akad. Nauk SSSR **257** [1981] 891/5; Dokl. Chem. Proc. Acad. Sci. USSR **256/261** [1980] 132/5).
[3] Leites, L. A.; Vinogradova, L. E.; Bukalov, S. S.; Kampel, V. T.; Bregadze, V. I. (Izv. Akad. Nauk SSSR Ser. Khim. **1981** 2035/43; Bull. Acad. Sci. USSR Div. Chem. Sci. **1981** 1670/6).
[4] Bregadze, V. I.; Usyatinskii, A. Ya.; Godovikov, N. N. (Izv. Akad. Nauk SSSR Ser. Khim. **1983** 1903/5; Bull. Acad. Sci. USSR Div. Chem. Sci. **1983** 1726/8).
[5] Usyatinskii, A. Y.; Todres, Z. V.; Shcherbina, T. M.; Bregadze, V. I.; Godovikov, N. N. (Izv. Akad. Nauk SSSR Ser. Khim. **1983** No. 1, pp. 1640/2; Bull. Acad. Sci. USSR Div. Chem. Sci. **1983** 1488/90).
[6] Ol'dekop, Yu. A.; Maier, N. A.; Erdman, A. A.; Prokopovich, V. P. (Vestsi Akad. Navuk Belarusk. SSR Ser. Khim. Navuk **1983** 114/6; C.A. **98** [1983] No. 160775).
[7] Conteras, J. G.; Pena, J.; Silva-Trivino, L. M.; (Inorg. Chim. Acta **87** [1984] 181/4).
[8] Longato, B.; Morandini, F.; Bresadola, S. (Inorg. Chim. Acta **39** [1980] 27/34).

13.11.2.7 Additional Derivatives of 1,2-$C_2B_{10}H_{12}$

Toxicity tests on 1,2-$C_2B_{10}H_{11}$-1-R (R = SH, COOH, CH_2COOH), 1,2-$C_2B_{10}H_{10}$-1-SH-2-CH_3, 1,2-$C_2B_{10}H_{10}$-1,2-$(SH)_2$, and 1,2-$C_2B_{10}H_{11}$-3-NH_2 have been conducted [1]. Accumulation of 1,2-$C_2B_{10}H_{11}$-?-CH_2COOH, 1,2-$C_2B_{10}H_{11}$-?-aminosulfonic acid, and a mercaptan derivative of 1,2-

$C_2B_{10}H_{12}$ in animal tissues for neutron-capture therapy has been studied [2]. 1,2-$C_2B_{10}H_{11}$-1-C_6H_4-4'-N_2^+ ion, with tritium-enrichment in the 2-position, has been used in connection with tumor localization of boron-10-labelled antibodies to carcinoembryonic antigens in the GW-39 tumor model system [3]. The preparation of the peptide, H-Tyr-Gly-Gly-Car-Leu-OH (Car = 1,2-$C_2B_{10}H_{11}$-1-alanine residue), has been described [4]. The pharmacological properties of analogs of an angiotensin containing carboranylalanine, 1,2-$C_2B_{10}H_{11}$-1-CH_2-CH(NH_3^+)CO_2^-, have been reported [5]. Lipophilic, steric, electronic, and enzyme resistant characteristics of carboranylalanine (carboranyl = 1,2-$C_2B_{10}H_{11}$-1-) have been described [6]. Several carboranylporphyrins (carboranyl = 1,2-$C_2B_{10}H_{11}$-) have been prepared and characterized [7]. An alanine derivative of 1,2-$C_2B_{10}H_{12}$ has been used for the synthesis of enkephalin-like peptides [8]. Various carborane derivatives of estrogens (where the carborane is 1,2-$C_2B_{10}H_{11}$-) have been synthesized and some of their biochemical and biological properties evaluated [9]. The synthesis of a carboranylestradiol and carboranyltestosterone (carboranyl = 1,2-$C_2B_{10}H_{11}$-1-) has been described and their estrogenic and receptor-binding activities studied [10]. A method for the trace analysis for steroidal carboranes, containing a 1,2-$C_2B_{10}H_{11}$-1- unit, in biological matrices is described that consists of gas chromatography combined with flame ionization detection and microwave-induced plasma emission detection [11].

In addition to those compounds mentioned above in this section, and in Sections 13.11.2.2 to 13.11.2.6, a supplemental list of 1,2-$C_2B_{10}H_{12}$ derivatives, along with reagents used to synthesize each one, and physical properties of each product, is given in Table 13/10, pp. 200/11.

References for 13.11.2.7:

[1] Spryshkova, R. A.; Karaseva, L. I.; Bratsev, V. A.; Serebryakov, N. G. (Med. Radiol. **26** [1981] 62/4; C.A. **95** [1981] No. 108411).

[2] Spryshkova, R. A.; Bratsev, V. A.; Sherman, T. L.; Stanko, V. I. (Med. Radiol. **26** [1981] 51/5; C.A. **97** [1982] No. 158872).

[3] Goldenberg, D. M.; Sharkey, R. M.; Primus, F. J.; Mizusawa, E.; Hawthorne, M. F. (Proc. Natl. Acad. Sci. [U.S.A.] **81** [1984] 560/3).

[4] Fauchère, J.-L.; Leukart, O.; Eberle, A.; Schwyzer, R. (Helv. Chim. Acta **62** [1979] 1385/95).

[5] Escher, E.; Guillemette, G.; Leukart, O.; Regoli, D. (Eur. J. Pharmacol. **66** [1980] 267/72).

[6] Fauchère, J.-L.; Do, K. Q.; Jow, P. Y. C.; Hansch, C. (Experientia **36** [1980] 1203/4).

[7] Haushalter, R. C.; Butler, W. M.; Rudolph, R. W. (J. Am. Chem. Soc. **103** [1981] 2620/7).

[8] Schwyzer, R.; Do, K. Q.; Eberle, A. N.; Fauchère, J.-L. (Helv. Chim. Acta **64** [1981] 2078/83).

[9] Sweet, F. (Steroids **37** [1981] 223/38).

[10] Sweet, F.; Samant, B. R. (Syn. Appl. Isot. Labeled Compounds, Proc. Intern. Symp., Kansas City, Mo., 1982 [1983], pp. 175/80; C.A. **98** [1983] No. 157193).

(references continued on p. 212)

Table 13/10

$1,2\text{-}C_2B_{10}H_{12}$ Derivatives; Synthesis and Physical Properties (m.p. = melting point in °C; b.p. = boiling point in °C; IR = infrared spectrum reported; NMR = nuclear magnetic resonance spectrum reported; MS = mass spectrum reported).

compound (product)	reagents used to synthesize compound listed in the first column	properties of product	Ref.
$1,2\text{-}C_2B_{10}H_{11}\text{-}9\text{-}Cl$	$1,2\text{-}C_2B_{10}H_{11}\text{-}9\text{-}HgCl$ (350°C)		[12]
$1,2\text{-}C_2B_{10}H_{10}\text{-}1\text{-}SH\text{-}9\text{-}Cl$	$1,2\text{-}C_2B_{10}H_{11}\text{-}9\text{-}Cl$, NaH, S	^1H, ^{11}B NMR	[13]
$1,2\text{-}C_2B_{10}H_{10}\text{-}9\text{-}SH\text{-}12\text{-}Cl$	$1,2\text{-}C_2B_{10}H_{11}\text{-}9\text{-}Cl$, NaH, S	^1H, ^{11}B NMR	[13]
$1,2\text{-}C_2B_{10}H_{11}\text{-}9\text{-}Br$	$1,2\text{-}C_2B_{10}H_{12}$, C_6H_5Br, $AlCl_3$		[14]
$1,2\text{-}C_2B_{10}H_{10}\text{-}1\text{-}SH\text{-}9\text{-}Br$	$1,2\text{-}C_2B_{10}H_{11}\text{-}9\text{-}Br$, NaH, S	^1H, ^{11}B NMR	[13]
$1,2\text{-}C_2B_{10}H_{10}\text{-}9\text{-}SH\text{-}12\text{-}Br$	$1,2\text{-}C_2B_{10}H_{11}\text{-}9\text{-}Br$, NaH, S	^1H, ^{11}B NMR	[13]
$1,2\text{-}C_2B_{10}H_{11}\text{-}9\text{-}I$	$1,2\text{-}C_2B_{10}H_{12}$, $C_6H_5I(OCOCF_3)_2$, I_2	m.p. 120 to 121	[15]
$1,2\text{-}C_2B_{10}H_{11}\text{-}9\text{-}I$	$[1,2\text{-}C_2B_{10}H_{11}\text{-}9\text{-}IC_6H_5]BF_4$, $(C_6H_5)_3P$	m.p. 118 to 120	[16]
$1,2\text{-}C_2B_{10}H_{11}\text{-}9\text{-}ICl_2$	$1,2\text{-}C_2B_{10}H_{11}\text{-}9\text{-}I$, Cl_2, CCl_4		[17]
$1,2\text{-}C_2B_{10}H_{11}\text{-}9\text{-}I(OCOCF_3)_2$	$1,2\text{-}C_2B_{10}H_{11}\text{-}9\text{-}I$, CF_3COOAg, CH_3CN	m.p. 96 to 97	[17]
$[1,2\text{-}C_2B_{10}H_{11}\text{-}9\text{-}IC_6H_5][BF_4]$	$1,2\text{-}C_2B_{10}H_{11}\text{-}9\text{-}I(OCOCF_3)_2$, C_6H_6, H_2SO_4, CH_3COOH, $(CH_3CO)_2O$, $NH_4[BF_4]$		[17]
$[1,2\text{-}C_2B_{10}H_{11}\text{-}9\text{-}IC_6H_5][BF_4]$	$1,2\text{-}C_2B_{10}H_{11}\text{-}9\text{-}I$, C_6H_6, H_2SO_4, $K_2S_2O_8$, $(CH_3CO)_2O$, $NH_4[BF_4]$		[17]
$1,2\text{-}C_2B_{10}H_{10}\text{-}1\text{-}SH\text{-}9\text{-}I$	$1,2\text{-}C_2B_{10}H_{11}\text{-}9\text{-}I$, NaH, S	^1H, ^{11}B NMR	[13]
$1,2\text{-}C_2B_{10}H_{10}\text{-}9\text{-}SH\text{-}12\text{-}I$	$1,2\text{-}C_2B_{10}H_{11}\text{-}9\text{-}Cl$, NaH, S	^1H, ^{11}B NMR	[13]
$1,2\text{-}C_2B_{10}H_{10}\text{-}1,2\text{-}I_2$	$(1,2\text{-}C_2B_{10}H_{10})Pd$, ICl		[18]
$1,2\text{-}C_2B_{10}H_{10}\text{-}9,12\text{-}I_2$	$1,2\text{-}C_2B_{10}H_{12}$, $C_6H_5I(OCOCF_3)_2$, I_2	m.p. 192 to 193	[15]
$1,2\text{-}C_2B_{10}H_{11}\text{-}1\text{-}SH$	$1,2\text{-}C_2B_{10}H_{12}$, NaH, S	^1H, ^{11}B NMR	[13]

$1,2\text{-}C_2B_{10}H_{10}\text{-}1\text{-}SH\text{-}2\text{-}CH_3$	$1,2\text{-}C_2B_{10}H_{11}\text{-}1\text{-}CH_3$, NaH, S	^1H, ^{11}B NMR	[13]
$1,2\text{-}C_2B_{10}H_{11}\text{-}9\text{-}SH$	$(1,2\text{-}C_2B_{10}H_{11}\text{-}9\text{-})_2S_2$, Zn, HCl	m.p. 201 to 202	[19]
$1,2\text{-}C_2B_{10}H_{11}\text{-}9\text{-}SH$	$1,2\text{-}C_2B_{10}H_{12}$, S, $AlCl_3$	m.p. 203 to 205, ^1H, ^{11}B NMR	[20]
$1,2\text{-}C_2B_{10}H_{10}\text{-}1\text{-}CH_3\text{-}9\text{-}SH$	$1,2\text{-}C_2B_{10}H_{11}\text{-}1\text{-}CH_3$, S, $AlCl_3$	^1H, ^{11}B NMR	[13]
$1,2\text{-}C_2B_{10}H_{10}\text{-}1\text{-}CH_3\text{-}12\text{-}SH$	$1,2\text{-}C_2B_{10}H_{11}\text{-}1\text{-}CH_3$, S, $AlCl_3$	^1H, ^{11}B NMR	[13]
$1,2\text{-}C_2B_{10}H_{11}\text{-}9\text{-}SCl$	$1,2\text{-}C_2B_{10}H_{11}\text{-}9\text{-}SH$, Cl_2		[21]
$1,2\text{-}C_2B_{10}H_{11}\text{-}9\text{-}SCN$	$1,2\text{-}C_2B_{10}H_{11}\text{-}9\text{-}SCl$, KCN	m.p. 81 to 81.5	[21]
$1,2\text{-}C_2B_{10}H_9\text{-}1\text{-}R\text{-}2\text{-}R'\text{-}9\text{-}SCN$	$1,2\text{-}C_2B_{10}H_9\text{-}1\text{-}R\text{-}2\text{-}R'\text{-}9\text{-}Tl(SCN)_2/h\nu$	m.p. 64 for $R = R' = H$ m.p. 57 to 59 for $R = R' = CH_3$	[22]
$1,2\text{-}C_2B_{10}H_{10}\text{-}9,12\text{-}(SH)_2$	$1,2\text{-}C_2B_{10}H_{12}$, (a) S_2Cl_2, $AlCl_3$; (b) Zn, CH_3COOH, HCl	m.p. 235 to 236, ^{11}B NMR	[23]
$1,2\text{-}C_2B_{10}H_{10}\text{-}9,12\text{-}(SCH_3)_2$	$1,2\text{-}C_2B_{10}H_{10}\text{-}9,12\text{-}(SH)_2$, NaOH, CH_3I	^{11}B NMR	[23]
$1,2\text{-}C_2B_{10}H_8\text{-}8,9,10,12\text{-}(SH)_4$	$1,2\text{-}C_2B_{10}H_{12}$, (a) S_2Cl_2, $AlCl_3$; (b) $Na[BH_4]$	^{11}B NMR	[23]
$1,2\text{-}C_2B_{10}H_{11}\text{-}9\text{-}SeH$	$(1,2\text{-}C_2B_{10}H_{11}\text{-}9\text{-}Se)_2$, Zn, HCl, CH_3COOH	m.p. 150, IR, ^1H NMR	[24]
$1,2\text{-}C_2B_{10}H_{10}\text{-}9,12\text{-}(SeH)_2$	$1,2\text{-}C_2B_{10}H_{12}$, (a) Se_2Cl_2, $AlCl_3$; (b) $Na[BH_4]$	^{11}B NMR	[23]
$1,2\text{-}C_2B_{10}H_{10}\text{-}9,12\text{-}(SeCH_3)_2$	$1,2\text{-}C_2B_{10}H_{10}\text{-}9,12\text{-}(SeH)_2$, NaOH, CH_3I	^{11}B NMR	[23]
$1,2\text{-}C_2B_{10}H_{11}\text{-}9\text{-}TeCl_3$	$1,2\text{-}C_2B_{10}H_{12}/TeCl_4$, $AlCl_3$, CH_2Cl_2		[25]
$1,2\text{-}C_2B_{10}H_{11}\text{-}9\text{-}TeCH_3$	$(1,2\text{-}C_2B_{10}H_{11}\text{-}9\text{-})_2Te_2$, $Na[BH_4]$, CH_3I	m.p. 86 to 87	[25]
$1,2\text{-}C_2B_{10}H_9\text{-}1\text{-}CH(CH_3)_2\text{-}2\text{-}N{=}N{=}C_6H_5\text{-}B\text{-}Mn(CO)_4$	$1,2\text{-}C_2B_{10}H_{10}\text{-}1\text{-}CH(CH_3)_2\text{-}2\text{-}N{=}N{=}C_6H_5$, $CH_3Mn(CO)_5$	m.p. 138 to 138.5	[26]
$1,2\text{-}C_2B_{10}H_{11}\text{-}9\text{-}N_3$	$[1,2\text{-}C_2B_{10}H_{11}\text{-}9\text{-}IC_6H_5][BF_4]$, $[N(C_4H_9)_4]N_3$	m.p. 83 to 84, IR	[27]
$1,2\text{-}C_2B_{10}H_{11}\text{-}3\text{-}NO_2$	$1,2\text{-}C_2B_{10}H_{11}\text{-}3\text{-}NH_2$, CrO_3, H_2SO_4, CH_3COOH	m.p. 323 to 325	[28]

References for 13.11.2.7 on pp. 199, 212/4

Table 13/10 (continued)

compound (product)	reagents used to synthesize compound listed in the first column	properties of product	Ref.
1,2-$C_2B_{10}H_{11}$-9-$AsCl_2$	(1,2-$C_2B_{10}H_{11}$-9-)$_2$Hg, $AsCl_3$	m.p. 74	[29]
1,2-$C_2B_{10}H_{10}$-1-$CH(CH_3)_2$-2-$AsCl_2$	1,2-$C_2B_{10}H_{10}$-1-$CH(CH_3)_2$-2-Li, $ClAs[N(C_2H_5)_2]_2$, HCl	m.p. 30 to 31	[29]
1,2-$C_2B_{10}H_{10}$-1-$C(CH_3)$=CH_2-2-$AsCl_2$	1,2-$C_2B_{10}H_{10}$-1-$C(CH_3)$=CH_2-2-Li, $ClAs[N(C_2H_5)_2]_2$, HCl	b.p. 148 to 150 (1 Torr)	[29]
1,2-$C_2B_{10}H_{10}$-1-$CH(CH_3)_2$-2-AsO	1,2-$C_2B_{10}H_{10}$-1-$CH(CH_3)_2$-2-$AsCl_2$, H_2O	m.p. 244 to 245	[29]
1,2-$C_2B_{10}H_{10}$-1-$CH(CH_3)_2$-2-AsO_3H_2	1,2-$C_2B_{10}H_{10}$-1-$CH(CH_3)_2$-2-$AsCl_2$, Cl_2, H_2O		[29]
1,2-$C_2B_{10}H_{11}$-9-CH_3	1,2-$C_2B_{10}H_{12}$, CH_3Br, $AlCl_3$	m.p. 163 to 165, NMR, MS	[14]
1,2-$C_2B_{10}H_{10}$-9,12-$(CH_3)_2$	1,2-$C_2B_{10}H_{12}$, C_2H_5Br, $AlCl_3$	m.p. 157 to 158, NMR	[14]
1,2-$C_2B_{10}H_{11}$-9-C_2H_5	1,2-$C_2B_{10}H_{12}$, C_2H_5Br, $AlCl_3$	m.p. 14, NMR, MS	[14]
1,2-$C_2B_{10}H_{11}$-9-C_2H_5	1,2-$C_2B_{10}H_{11}$-9-Br(or I), C_2H_5MgX (X = halogen), $[(C_6H_5)_3P]_2MCl_2$ or $[(C_6H_5)_3P]_4M$ (M = Pd, Ni)	b.p. 74 at 1 Torr	[30]
1,2-$C_2B_{10}H_{11}$-9-C_2H_5	1,2-$C_2B_{10}H_{11}$-9-I, C_2H_5MgX (X = halogen), $[(C_6H_5)_3P]_2PdCl_2$ or $[(C_6H_5)_3P]_4Pd$	b.p. 74 to 75 at 1 Torr	[31]
1,2-$C_2B_{10}H_{10}$-9,12-$(C_2H_5)_2$	1,2-$C_2B_{10}H_{12}$, C_2H_5Br, $AlCl_3$	b.p. 95 at 13 Pa, NMR	[14]
1,2-$C_2B_{10}H_8$-8,9,10,12-$(C_2H_5)_4$	1,2-$C_2B_{10}H_{12}$, C_2H_5Br, $AlCl_3$	b.p. 115 at 13 Pa, NMR	[14]
1,2-$C_2B_{10}H_7$-4,8,9,10,12-$(C_2H_5)_5$	1,2-$C_2B_{10}H_{12}$, C_2H_5Br, $AlCl_3$	b.p. 132 at 13 Pa	[14]
1,2-$C_2B_{10}H_6$-4,?,8,9,10,12-$(C_2H_5)_6$	1,2-$C_2B_{10}H_{12}$, C_2H_5Br, $AlCl_3$	m.p. 25 to 27	[14]
1,2-$C_2B_{10}H_5$-4,5,7,8,9,10,12-$(C_2H_5)_7$	1,2-$C_2B_{10}H_{12}$, C_2H_5Br, $AlCl_3$	b.p. 158 at 13 Pa	[14]
1,2-$C_2B_{10}H_{11}$-9-C_3H_7	1,2-$C_2B_{10}H_{11}$-9-I, C_3H_7MgBr, $[(C_6H_5)_3P]_2PdCl_2$	m.p. 42 to 43	[32]
1,2-$C_2B_{10}H_{11}$-9-$CH(CH_3)_2$	1,2-$C_2B_{10}H_{12}$, $(CH_3)_2CHOH$, H_2SO_4	b.p. 86 at 13 Pa, NMR	[14]
1,2-$C_2B_{10}H_{11}$-9-$CH(CH_3)_2$	1,2-$C_2B_{10}H_{11}$-9-I, $(CH_3)_2CHMgX$ (X = halogen), $[(C_6H_5)_3P]_2PdCl_2$ or $[(C_6H_5)_3P]_4Pd$	b.p. 97 to 98 at 1 Torr	[31]

1,2-$C_2B_{10}H_{10}$-*B*,*B'*-$[CH(CH_3)_2]_2$	1,2-$C_2B_{10}H_{12}$, $(CH_3)_2CHOH$, H_2SO_4	NMR	[14]
1,2-$C_2B_{10}H_{11}$-9-*n*-C_4H_9	1,2-$C_2B_{10}H_{11}$-9-I, *n*-C_4H_9MgX (X = halogen), $[(C_6H_5)_3P]_2PdCl_2$ or $[(C_6H_5)_3P]_4Pd$	b.p. 110 to 111 at 1 Torr	[31]
1,2-$C_2B_{10}H_{11}$-9-C_4H_9	1,2-$C_2B_{10}H_{11}$-9-Br (or I), C_4H_9MgX (X = halogen), $[(C_6H_5)_3P]_2MCl_2$ or $[(C_6H_5)_3P]_4M$ (M = Pd, Ni)	b.p. 110 at 1 Torr	[30]
1,2-$C_2B_{10}H_{11}$-9-$CH_2CH(CH_3)_2$	1,2-$C_2B_{10}H_{11}$-9-I, $(CH_3)_3CCHCH_2MgX$ (X = halogen), $[(C_6H_5)_3P]_2PdCl_2$ or $[(C_6H_5)_3P]_4Pd$	b.p. 108 to 109 at 1 Torr	[31]
1,2-$C_2B_{10}H_{11}$-1-C_5H_{11}	1,2-$C_2B_{10}H_{11}$-1-CHN_2, C_4H_{10}/hv	IR, ^1H NMR	[33]
1,2-$C_2B_{10}H_{10}$-1,2-$(C_5H_{11})_2$	Li[1,2-$C_2B_{10}H_{10}$], *n*-$C_5H_{11}Br$		[34]
1-(1',2'-$C_2B_{10}H_{11}$-1'-)C_4H_8-2-CH_3	1,2-$C_2B_{10}H_{11}$-1-CHN_2, C_4H_{10}/hv	IR, ^1H NMR	[33]
1'-(1,2-$C_2B_{10}H_{11}$-1-)C_3H_3-2',3'-$(CH_3)_2$ (C_3H_3 = cyclopropyl group) (syn-*cis* and anti-*cis* isomers)	1,2-$C_2B_{10}H_{11}$-1-CHN_2, cis-2-butene/hv	IR, ^1H NMR	[33]
1'-(1,2-$C_2B_{10}H_{11}$-1-)C_3H_3-2',3'-$(CH_3)_2$ (C_3H_3 = cyclopropyl group) (*trans* isomer)	1,2-$C_2B_{10}H_{11}$-1-CHN_2, *trans*-2-butene/hv	IR, ^1H NMR	[33]
1'-(1,2-$C_2B_{10}H_{11}$-1-)C_3H_3-2',2'-$(CH_3)_2$ (C_3H_3 = cyclopropyl group)	1,2-$C_2B_{10}H_{11}$-1-CHN_2, $(CH_3)_2C{=}CH_2$/hv	IR, ^1H NMR	[33]
1'-(1,2-$C_2B_{10}H_{10}$-1-CH_3-2-)C_3H_3-2',3'-$(CH_3)_2$ (C_3H_3 = cyclopropyl group) (anti-*cis* isomer)	1,2-$C_2B_{10}H_{10}$-1-CH_3-2-CHN_2, *cis*-2-butene/hv	IR, ^1H NMR	[33]
1,2-$C_2B_{10}H_{11}$-9-$CH_2CH{=}CH_2$	1,2-$C_2B_{10}H_{11}$-9-Br(or I), $CH_2{=}CHCH_2MgX$ (X = halogen), $[(C_6H_5)_3P]_2MCl_2$ or $[(C_6H_5)_3P]_4M$ (M = Pd, Ni)	b.p. 85 at 1 Torr, ^1H NMR	[30, 31]
1,2-$C_2B_{10}H_{10}$-1-R-2-$CH_2C({=}CH_2)SO_2N(CH_3)C_6H_5$	1,2-$C_2B_{10}H_{10}$-1-R-2-$CH_2CH(CH_2X)SO_2Cl$, $C_6H_5NH(CH_3)$, X = Cl, Br	m.p. 131 to 132 for R = CH_3, 157 to 158 for R = $CH(CH_3)_2$, 147 to 148 for R = C_6H_5	[35]

References for 13.11.2.7 on pp. 199, 212/4

Table 13/10 (continued)

compound (product)	reagents used to synthesize compound listed in the first column	properties of product	Ref.
1,2-$C_2B_{10}H_{11}$-1-$CH_2C{\equiv}CH$	1,2-$C_2B_{10}H_{11}$-1-CH=C=CH_2, C_4H_9Li	m.p. 79 to 80	[36]
1,2-$C_2B_{10}H_{10}$-1-$CH_2C{\equiv}CH$-2-CH_3	1,2-$C_2B_{10}H_{10}$-1-CH=C=CH_2-2-CH_3, C_4H_9Li	b.p. 104 at 1.3 hPa	[36]
1,2-$C_2B_{10}H_{11}$-9-$CH_2C_6H_5$	1,2-$C_2B_{10}H_{11}$-9-Br(or I), $C_6H_5CH_2MgX$ (X = halogen), $[(C_6H_5)_3P]_2MCl_2$ or $[(C_6H_5)_3P]_4M$ (M = Pd, Ni)	m.p. 102 to 103	[30, 31]
1,2-$C_2B_{10}H_{10}$-1,2-$(CH_2C_6H_4$-4'-$R)_2$	$M_2[1,2\text{-}C_2B_{10}H_{10}]$, p-R-$C_6H_4CH_2X$ M = Li, Na; R = CH_3 when X = Br; R = CH_3O when X = Cl	m.p. 148 to 150, R = CH_3; 129 to 130, R = CH_3O	[37]
1,2-$C_2B_{10}H_{10}$-1,2-$(CH_2C_6H_4$-4'-$OH)_2$	1,2-$C_2B_{10}H_{10}$-1,2-$(CH_2C_6H_4$-4'-$OCH_3)_2$, HI/100°C	m.p. 227 to 228	[37]
1,2-$C_2B_{10}H_{10}$-1,2-$(CH_2C_6H_4$-4'-$COOH)_2$	1,2-$C_2B_{10}H_{10}$-1,2-$(CH_2C_6H_4$-4'-$CH_3)_2$, CrO_3, CH_3COOH, H_2SO_4	m.p. 331 to 333	[37]
1,2-$C_2B_{10}H_{11}$-1-CH_2Cl	1,2-$C_2B_{10}H_{11}$-1-CH_2OH, PCl_5		[38]
1,2-$C_2B_{10}H_{10}$-1,2-$(CH_2Br)_2$	1,2-$C_2B_{10}H_{10}$-1,2-$(CH_2OH)_2$, Br_2, $(C_6H_5)_3P$		[38]
1,2-$C_2B_{10}H_{11}$-3-$CHBrCH_2Br$	1,2-$C_2B_{10}H_{11}$-3-CH=CH_2, Br_2	b.p. 173 to 174 at 1 Torr	[39]
1,2-$C_2B_{10}H_{11}$-1-CH_2OR	1,2-$C_2B_{10}H_{11}$-1-CH_2OH, RBr, $[(C_2H_5)_3N(CH_2C_6H_5)]Cl$, NaOH, R = C_4H_9, $C_6H_5CH_2$	MS	[40]
1,2-$C_2B_{10}H_{10}$-1,2-$(CH_2OR)_2$	1,2-$C_2B_{10}H_{10}$-1,2-$(CH_2OH)_2$, RBr, $[(C_2H_5)_3N(CH_2C_6H_5)]Cl$, NaOH, R = C_4H_9, $C_6H_5CH_2$ (m.p. for R = $C_6H_5CH_2$ 80 to 80.5)	MS	[40]
1,2-$C_2B_{10}H_{10}$-1-CH_2OH-2-CH_2Br	1,2-$C_2B_{10}H_{10}$-1,2-$(CH_2OH)_2$, Br_2, $(C_6H_5)_3P$		[38]
1,2-$C_2B_{10}H_{11}$-1-$CH_2OC(O)R$	1,2-$C_2B_{10}H_{11}$-1-CH_2OH, RCOCl	b.p. 64 to 65 at 0.02 Torr for R = C_2H_5, m.p. 56 to 57 for R = $CH(CH_3)_2$, 93 to 94 for R = C_6H_5	[41]

$1,2\text{-}C_2B_{10}H_{10}\text{-}1,2\text{-}[CH_2OC(O)R]_2$	$1,2\text{-}C_2B_{10}H_{10}\text{-}1,2\text{-}(CH_2OH)_2$, RCOCl	m.p. 28.5 to 29.5 for $R=C_2H_5$, b.p. 106 to 109 at 0.02 Torr for $R=CH(CH_3)_2$ m.p. 111.5 to 112 for $R=C_6H_5$	[41]
$1,2\text{-}C_2B_{10}H_{10}\text{-}1,2\text{-}(CH_2O\text{-})_2SO$	$1,2\text{-}C_2B_{10}H_{10}\text{-}1,2\text{-}(CH_2OH)_2$, $SOCl_2$	m.p. 52 to 53	[38]
$1,2\text{-}C_2B_{10}H_{11}\text{-}1\text{-}CH_2OCONHP(O)(OC_2H_5)_2$	$1,2\text{-}C_2B_{10}H_{11}\text{-}1\text{-}CH_2OH$, $(C_2H_5O)_2P(O)N{=}C{=}O$	m.p. 66 to 67 IR	[42]
$1,2\text{-}C_2B_{10}H_{11}\text{-}1\text{-}CH_2XPRR'Y$	$B_{10}H_{14}$, $HCCCH_2SPRR'X$, $C_6H_5N(CH_3)_2$	m.p. 82 to 84 for $R=CH_3$, $R'=C_2H_5O$, $X=S$, $Y=O$; 61 to 62 for $R=C_6H_5$, $R'=C_2H_5O$, $X=S$, $Y=O$; 87 to 89 for $R=R'=C_6H_5$, $X=Y=S$; 56 to 58 for $R=CH_3$, $R'=C_2H_5O$, $X=Y=O$; 52 to 54 for $R=CH_3$, $R'=C_2H_5O$, $X=O$, $Y=S$	[43]
$1,2\text{-}C_2B_{10}H_{11}\text{-}1\text{-}CH_2SP(O)(OC_2H_5)(CH_3)$	$HC{\equiv}CCH_2SP(O)(OC_2H_5)(CH_3)$, $B_{10}H_{14}$, $C_6H_5N(CH_3)_2$	m.p. 82 to 84	[44]
$1,2\text{-}C_2B_{10}H_{11}\text{-}1\text{-}CH_2SP(O)(OC_2H_5)(C_6H_5)$	$HC{\equiv}CCH_2SP(O)(OC_2H_5)(C_6H_5)$, $B_{10}H_{14}$, $C_6H_5N(CH_3)_2$	m.p. 61 to 62	[44]
$1,2\text{-}C_2B_{10}H_{11}\text{-}1\text{-}CH_2OSiRR'Cl$	$1,2\text{-}C_2B_{10}H_{11}\text{-}1\text{-}CH_2OH$, Cl_2SiRR'	b.p. 127 to 128 at 2 Torr for $R=R'=CH_3$; b.p. 110 to 113 at 2 Torr for $R=CH_3$, $R'=Cl$; b.p. 113 to 114 at 2 Torr for $R=R'=Cl$	[45]
$1,2\text{-}C_2B_{10}H_{10}\text{-}1,2\text{-}[CH_2OSi(CH_3)_2CH_2Cl]_2$	$1,2\text{-}C_2B_{10}H_{10}\text{-}1,2\text{-}(CH_2OH)_2$, $ClSi(CH_3)_2CH_2Cl$	b.p. 212 to 215 at 3 Torr	[45]
$1,2\text{-}C_2B_{10}H_{10}\text{-}1,2\text{-}[\text{-}CH_2OSi(CH_3)(Cl)OCH_2\text{-}]_2$	$1,2\text{-}C_2B_{10}H_{10}\text{-}1,2\text{-}(CH_2OH)_2$, Cl_3SiCH_3	m.p. 116 to 117	[45]
$1,2\text{-}C_2B_{10}H_{10}\text{-}1,2\text{-}[\text{-}CH_2OSiCl_2OCH_2\text{-}]_2$	$1,2\text{-}C_2B_{10}H_{10}\text{-}1,2\text{-}(CH_2OH)_2$, $SiCl_4$	m.p. 79 to 80	[45]

References for 13.11.2.7 on pp. 199, 212/4

Table 13/10 (continued)

compound (product)	reagents used to synthesize compound listed in the first column	properties of product	Ref.
1,2-$C_2B_{10}H_{11}$-1-$CH_2OSi(CH_3)_2H$	1,2-$C_2B_{10}H_{11}$-1-CH_2OH, $HSi(CH_3)_2NHSi(CH_3)_2H$	^1H NMR	[45]
1,2-$C_2B_{10}H_{10}$-1,2-$[CH_2OSi(CH_3)_2CH{=}CH_2]_2$	1,2-$C_2B_{10}H_{10}$-1,2-$(CH_2OH)_2$, $ClSi(CH_3)_2CH{=}CH_2$	b.p. 155 to 156 at 1 Torr	[45]
1,2-$C_2B_{10}H_{10}$-1-R-2-$CH_2Si(CH_3)_2OR'$	1,2-$C_2B_{10}H_{10}$-1-R-2-Li, $ClCH_2Si(CH_3)_2OR'$	b.p. (1 Torr) 132 to 134 for R = H, R' = CH_3; 154 to 156 for R = H, R' = C_2H_5; 135 to 137 for R = CH_3, R' = CH_3	[46]
1,2-$C_2B_{10}H_{10}$-1-R-2-$CH_2Si(CH_3)(OR')_2$	1,2-$C_2B_{10}H_{10}$-1-R-2-Li, $ClCH_2Si(CH_3)(OR')_2$	b.p. (1 Torr) 135 to 137 for R = H, R' = CH_3; 150 to 152 for R = H, R' = C_2H_5; 140 to 143 for R = CH_3, R' = C_2H_5; 183 to 185 for R = C_6H_5, R' = C_2H_5	[46]
1,2-$C_2B_{10}H_{10}$-1-R-2-$CH_2Si(OR')_3$	1,2-$C_2B_{10}H_{10}$-1-R-2-Li, $ClCH_2Si(OR')_3$	b.p. (1 Torr) 146 to 148 for R = H, R' = CH_3; 154 to 156 for R = H, R' = C_2H_5; 142 to 144 for R = CH_3, R' = C_2H_5; 143 to 145 for R = $CH_2{=}C(CH_3)$, R' = C_2H_5; 183 to 186 for R = C_6H_5, R' = C_2H_5	[46]
1,2-$C_2B_{10}H_{10}$-1,2-$[CH_2Si(CH_3)_2(OCH_3)]_2$	$Li_2[1,2-C_2B_{10}H_{10}]$, $ClCH_2Si(CH_3)_2OCH_3$	b.p. 177 to 180 at 1 Torr	[46]
1,2-$C_2B_{10}H_{10}$-1-$CH_2Si(CH_3)_2OCH_3$-2-$Si(CH_3)_2CH_2Cl$	$Li_2[1,2-C_2B_{10}H_{10}]$, $ClCH_2Si(CH_3)_2OCH_3$	b.p. 173 to 175 at 1 Torr	[46]
1,2-$C_2B_{10}H_{10}$-1,2-(-$CH_2Si(CH_3)_2OSi(CH_3)_2OCH_2$-)	$Li_2[1,2-C_2B_{10}H_{10}]$, $ClCH_2Si(CH_3)_2OCH_3$	m.p. 167	[46]
1,2-$C_2B_{10}H_{10}$-1-$CH_2Si(CH_3)_3$-2-$CH_2Si(CH_3)_2OCH_3$	$Li[1,2-C_2B_{10}H_{10}-1-CH_2Si(CH_3)_3]$, $ClCH_2Si(CH_3)_2OCH_3$	m.p. 62	[46]

1,2-$C_2B_{10}H_{10}$-1-CH_3-2-$CH_2Si(CH_3)_2C_4H_9$	1,2-$C_2B_{10}H_{10}$-1-CH_3-2-$CH_2Si(CH_3)_2OCH_3$, LiC_4H_9		[46]
1,2-$C_2B_{10}H_{11}$-1-$CH_2Si(CH_3)_2Cl$	1,2-$C_2B_{10}H_{11}$-1-$CH_2Si(CH_3)_2OCH_3$, PCl_3	b.p. 115 to 117 at 1 Torr	[46]
1,2-$C_2B_{10}H_{11}$-1-$CH_2Si(CH_3)Cl_2$	1,2-$C_2B_{10}H_{11}$-1-$CH_2Si(CH_3)(OC_2H_5)_2$, PCl_3	b.p. 120 to 123 at 1 Torr	[46]
1,2-$C_2B_{10}H_{11}$-9-$CH_2Si(CH_3)_3$	1,2-$C_2B_{10}H_{11}$-9-I, $(CH_3)_3SiCH_2MgCl$, $[(C_6H_5)_3P]_2PdCl_2$	m.p. 52 to 53	[47]
1,2-$C_2B_{10}H_{11}$-1-$CH_2CH_2OSiCl_3$	1,2-$C_2B_{10}H_{11}$-1-CH_2CH_2OH, $SiCl_4$	b.p. 135 to 137 at 1 Torr	[45]
1,2-$C_2B_{10}H_{10}$-1-CH_3-2-$CH_2CH(OC_2H_5)_2$	Li[1,2-$C_2B_{10}H_{10}$-1-CH_3], $BrCH_2CH(OC_2H_5)_2$	^1H NMR	[7]
1,2-$C_2B_{10}H_{11}$-1-CH_2CH_2OH	1,2-$C_2B_{10}H_{11}$-1-$CH_2CH_2OC(O)CH_3$, CH_3OH		[7]
1,2-$C_2B_{10}H_{11}$-1-CH_2CHO	1,2-$C_2B_{10}H_{11}$-1-CH_2CH_2OH, pyridinium chlorochromate	IR, ^1H NMR	[7]
1,2-$C_2B_{10}H_{10}$-1-CH_3-2-CH_2CHO	1,2-$C_2B_{10}H_{10}$-1-CH_3-2-$CH_2CH(OC_2H_5)_2$, HCl, CH_3COOH	IR	[7]
1,2-$C_2B_{10}H_{11}$-1-CH_2COOH	Na[1,2-$C_2B_{10}H_{11}$], Na[$OOCCH_2Br$]		[7]
1,2-$C_2B_{10}H_{11}$-1-CH_2COCl	1,2-$C_2B_{10}H_{11}$-1-CH_2COOH, PCl_5	IR, ^1H NMR	[7]
1,2-$C_2B_{10}H_{10}$-1-CH_3-2-$(CH_2)_3I$	1,2-$C_2B_{10}H_{10}$-1-CH_3-2-Li, $Br(CH_2)_3Cl$, NaI	m.p. 33	[34]
1,2-$C_2B_{10}H_{10}$-1-CH_3-2-$(CH_2)_3OC_6H_2$-2′,4′,5′-Cl_3	1,2-$C_2B_{10}H_{10}$-1-CH_3-2-$(CH_2)_3I$, HOC_6H_2-2,4,5-Cl_3	m.p. 124	[34]
1,2-$C_2B_{10}H_{10}$-1-CH=CH_2-2-CH_3	1,2-$C_2B_{10}H_{11}$-1-CH=CH_2, (a) C_4H_9Li; (b) CH_3I	m.p. 164 to 166	[33]
1,2-$C_2B_{10}H_{11}$-9-CH=CHC_6H_5	1,2-$C_2B_{10}H_{11}$-9-$Tl(OCOCF_3)_2$, $Pd(OCOCH_3)_2$, C_6H_5CH=CH		[48]
1,2-$C_2B_{10}H_{11}$-1-C(Cl)=CH_2	1,2-$C_2B_{10}H_{11}$-1-C≡CH, HCl, $AlCl_3$	^1H NMR	[49]
1,2-$C_2B_{10}H_{11}$-1-C(Br)=C(Br)H	1,2-$C_2B_{10}H_{11}$-1-C≡CH, Br_2	^1H NMR	[49]
1,2-$C_2B_{10}H_{11}$-1-CH=C(I)H	1,2-$C_2B_{10}H_{11}$-1-C≡CH, LiI, CH_3COOH	^1H NMR	[49]
1,2-$C_2B_{10}H_{11}$-1-CH=CH(SC_6H_5)	1,2-$C_2B_{10}H_{11}$-1-C≡CH, C_6H_5SH, KOH	^1H NMR	[49]
1,2-$C_2B_{10}H_{11}$-1-CH=C=CH_2	Cu[1,2-$C_2B_{10}H_{11}$], HC≡CCH_2X (X = Cl, Br)	b.p. 95 at 1.3 hPa	[50]
1,2-$C_2B_{10}H_{10}$-1-CH=C=CH_2-2-CH_3	Cu[1,2-$C_2B_{10}H_{10}$-2-CH_3], HC≡CCH_2X (X = Cl, Br)	m.p. 30 to 31	[50]

References for 13.11.2.7 on pp. 199, 212/4

Table 13/10 (continued)

compound (product)	reagents used to synthesize compound listed in the first column	properties of product	Ref.
1,2-$C_2B_{10}H_{11}$-9-C≡CH	1,2-$C_2B_{10}H_{11}$-9-C≡CSi$(CH_3)_3$, KOH, CH_3OH	m.p. 146 to 147	[51]
1,2-$C_2B_{10}H_{11}$-1-C≡CCH_3	1,2-$C_2B_{10}H_{11}$-1-CH=C=CH_2, [$(CH_3)_3$CO]K	m.p. 119 to 120	[36]
1,2-$C_2B_{10}H_{10}$-1-C≡CCH_3-2-CH_3	1,2-$C_2B_{10}H_{10}$-1-CH=C=CH_2-2-CH_3, [$(CH_3)_3$CO]K	m.p. 91 to 92	[36]
1,2-$C_2B_{10}H_{11}$-1-C≡CFe$(CO)_2(C_5H_5)$	1,2-$C_2B_{10}H_{11}$-1-C≡CLi, $(C_5H_5)Fe(CO)_2Br$	IR	[52]
1,2-$C_2B_{10}H_{11}$-9-C_6H_5	1,2-$C_2B_{10}H_{11}$-9-Br(or I), C_6H_5MgX (X = halogen), [$(C_6H_5)_3P]_2MCl_2$ or [$(C_6H_5)_3P]_4$M (M = Pd, Ni)	m.p. 143 to 143.5	[30, 31]
1,2-$C_2B_{10}H_{11}$-9-(C_6H_4-3′-F)	1,2-$C_2B_{10}H_{11}$-9-I, 3-F-C_6H_4MgBr, [$(C_6H_5)_3P]_2PdCl_2$	m.p. 132 to 134	[53]
1,2-$C_2B_{10}H_{11}$-9-(C_6H_4-4′-F)	1,2-$C_2B_{10}H_{11}$-9-I, 4-F-C_6H_4MgBr, [$(C_6H_5)_3P]_2PdCl_2$	m.p. 96 to 97	[53]
1,2-$C_2B_{10}H_{10}$-1-C_6H_5-2-CH$(CH_2COC_6H_5)_2$	1,2-$C_2B_{10}H_{10}$-1-C_6H_5-2-[4-H-pyran-2,6-$(C_6H_5)_2$]-4′-, $HClO_4$ (4-H-pyran = C_5H_3O)	m.p. 154 to 156	[54]
1,2-$C_2B_{10}H_{10}$-1-C_6H_5-2-CH_2COCH=CH-CH=CHC_6H_5	1,2-$C_2B_{10}H_{10}$-1-C_6H_5-2-CH_2COCH_3, C_6H_5CH=CHCHO, $B(OH)_3$	m.p. 111 to 112, IR	[55]
1,2-$C_2B_{10}H_{10}$-1-C_6H_5-2-$COCH_3$	1,2-$C_2B_{10}H_{10}$-1-C_6H_5-2-YbI, Cl$COCH_3$	m.p. 102 to 103	[56]
1,2-$C_2B_{10}H_9$-1-C_6H_5-2-$CH_2N(CH_3)_2$-*B*-Mn$(CO)_4$	1,2-$C_2B_{10}H_{10}$-1-C_6H_5-2-$CH_2N(CH_3)_2$, $CH_3Mn(CO)_5$	m.p. 154 to 155	[26]
1,2-$C_2B_{10}H_{11}$-9-C_6H_4-3′-CH_3	1,2-$C_2B_{10}H_{11}$-9-I, 3-CH_3-C_6H_4MgX (X = halogen), [$(C_6H_5)_3P]_2PdCl_2$ or [$(C_6H_5)_3P]_4$Pd	m.p. 140 to 141	[31]
1,2-$C_2B_{10}H_{11}$-9-C_6H_5-4′-CH_3	1,2-$C_2B_{10}H_{11}$-9-I, 4-CH_3-C_6H_5MgX (X = halogen), [$(C_6H_5)_3P]_2PdCl_2$ or [$(C_6H_5)_3P]_4$Pd	m.p. 129 to 129.5	[31]
1,2-$C_2B_{10}H_{10}$-1-C_6H_4-4′-N$(CH_2OH)_2$	1,2-$C_2B_{10}H_{10}$-1-C_6H_4-4′-NH_2, CH_2O	IR	[57]
1,2-$C_2B_{10}H_{10}$-1,2-(C_6H_4-3′(and 4′)-$NO_2)_2$	1,2-$C_2B_{10}H_{10}$-1,2-$(C_6H_5)_2$, HNO_3, H_2SO_4	m.p. 120 to 147	[57]
1,2-$C_2B_{10}H_{10}$-1,2-(C_6H_4-3′(and 4′)-$NH_2)_2$	1,2-$C_2B_{10}H_{10}$-1,2-(C_6H_4-3′(and 4′)-$NO_2)_2$, Ni, H_2		[57]

1,2-$C_2B_{10}H_{10}$-1-R-2-$C_6H_5Cr(CO)_3$	1,2-$C_2B_{10}H_{10}$-1-R-2-Li, $ClC_6H_5Cr(CO)_3$	m.p. 140 for R = CH_3; m.p. 150 to 151 for R = C_6H_5	[58]
1,2-$C_2B_{10}H_{10}$-1,2-$[C_6H_5Cr(CO)_3]_2$	$Li_2[1,2-C_2B_{10}H_{10}]$, $ClC_6H_5Cr(CO)_3$		[58]
1,2-$C_2B_{10}H_{11}$-1-CHO	1,2-$C_2B_{10}H_{11}$-1-CH=CH_2, O_3	IR	[7]
1,2-$C_2B_{10}H_{11}$-1-CHO	1,2-$C_2B_{10}H_{11}$-1-CH=CH_2, O_3	m.p. 207 to 209	[33]
1,2-$C_2B_{10}H_{10}$-1-CHO-2-CH_3	1,2-$C_2B_{10}H_{10}$-1-CH=CH_2-2-CH_3, O_3	m.p. 223 to 225	[33]
1,2-$C_2B_{10}H_{11}$-9-COOH	1,2-$C_2B_{10}H_{11}$-9-R, CrO_3, H_2SO_4, CH_3COOH (R = CH_3, C_2H_5, C_3H_7)	m.p. 239 to 240 (with decomposition)	[59]
1,2-$C_2B_{10}H_{11}$-9-COOH	1,2-$C_2B_{10}H_{11}$-9-R, CrO_3 (R = CH_3, C_2H_5, C_3H_7)	m.p. 239 to 240	[32]
1,2-$C_2B_{10}H_{10}$-1-CH_3-2-COC_6H_2-2',4',6'-$(CH_3)_3$	1,2-$C_2B_{10}H_{10}$-1-CH_3-2-Li, 2,4,6-$(CH_3)_3C_6H_2COCl$	m.p. 120 to 121	[60]
1,2-$C_2B_{10}H_{10}$-1-R-2-COCH=CHR'	1,2-$C_2B_{10}H_{10}$-1-R-2-$COCH_3$, R'CHO, $B(OH)_3$	m.p. 85 to 86 for R = CH_3, R' = C_6H_5; 142 to 144 for R = C_6H_5, R' = C_6H_4-4'-OCH_3; 141 to 152 for R = C_6H_5, R' = C_6H_4-4'-$N(CH_3)_2$; 133 to 135 for R = $CH(CH_3)_2$, R' = C_6H_5; 128 to 130 for R = C_6H_5, R' = CH=CHC_6H_5; 94 to 95 for R = $CH(CH_3)_2$, R' = C_6H_4-4'-OCH_3; 132 to 133 for R = R' = C_6H_5; 137 to 139 for R = $CH(CH_3)_2$, R' = CH=CHC_6H_5	[55]
1,2-$C_2B_{10}H_{11}$-9-$COFe(CO)_2(C_5H_5)$	1,2-$C_2B_{10}H_{11}$-9-COOH, $SOCl_2$, $NaFe(CO)_2(C_5H_5)$	m.p. 132 to 133	[32]
1,2-$C_2B_{10}H_{11}$-1-CH=$NNHS(O)_2C_6H_4CH_3$	1,2-$C_2B_{10}H_{11}$-1-CHO, $CH_3C_6H_4S(O)_2NHNH_2$	m.p. 195 to 196	[33]

References for 13.11.2.7 on pp. 199, 212/4

Table 13/10 (continued)

compound (product)	reagents used to synthesize compound listed in the first column	properties of product	Ref.
1,2-$C_2B_{10}H_{10}$-1-CH=NNHS$(O)_2C_6H_4CH_3$-2-CH_3	1,2-$C_2B_{10}H_{10}$-CHO-2-CH_3, $CH_3C_6H_4S(O)_2NHNH_2$	m.p. 165 to 166	[33]
1,2-$C_2B_{10}H_{10}$-1-CHN_2-2-CH_3	1,2-$C_2B_{10}H_{10}$-1-CH=NNHS(O)$OC_6H_4CH_3$-2-CH_3, NaH, 70 to 90°C	IR, ^{1}H NMR	[33]
1,2-$C_2B_{10}H_{10}$-1-R-2 (H, O, R′, R′)	1,2-$C_2B_{10}H_{11}$-1-Li, 2,6-di-*t*-butylpyrylium perchlorate	R = H, R′ = $(CH_3)_3C$, m.p. 119 to 121; R = H, R′ = C_6H_5, m.p. 212 to 215; R = CH_3, R′ = $(CH_3)_3C$, m.p. 121 to 123	[61]
1,2-$C_2B_{10}H_{10}$-1-R-2-CH$[CH_2COC(CH_3)_3]_2$	1,2-$C_2B_{10}H_{10}$-1-R-2-[4-H-pyran-2,6-$(C(CH_3)_3)_2$]-$HClO_4$ (4-H-pyran = C_5H_3O)	m.p. 185 to 187 for R = H; m.p. 141 to 143 for R = CH_3; 107 to 109 for R = C_6H_5	[54]
1,2-$C_2B_{10}H_{10}$-1-C_6H_5-2-$Si(CH_3)_3$	1,2-$C_2B_{10}H_{10}$-1-C_6H_5-2-YbI, $ClSi(CH_3)_3$	m.p. 103 to 104	[56]
1,2-$C_2B_{10}H_{10}$-1-$Si(CH_3)_3$-2-$CH_2Si(CH_3)_2OCH_3$	Li[1,2-$C_2B_{10}H_{10}$-1-$Si(CH_3)_3$], $ClCH_2Si(CH_3)_2OCH_3$	m.p. 113 to 115	[46]
1,2-$C_2B_{10}H_{11}$-9-$SnCl_3$	1,2-$C_2B_{10}H_{11}$-9-HgCl, $SnCl_2$	m.p. 249 to 249, ^{1}H NMR	[62]
1,2-$C_2B_{10}H_9$-1,2-$(CH_3)_2$-9-$SnCl_3$	1,2-$C_2B_{10}H_9$-1,2-$(CH_3)_2$-9-HgCl, $SnCl_2$	m.p. 124 to 126, ^{1}H NMR	[62]
1,2-$C_2B_{10}H_{11}$-9-MgCl·$C_{12}H_8N_2$ ($C_{12}H_8N_2$ = 1,10-phenanthroline)	1,2-$C_2B_{10}H_{11}$-9-MgCl, $C_{12}H_8N_2$	m.p. 158 to 159	[63]
1,2-$C_2B_{10}H_{11}$-9-Tl(C_4H_9)Cl	1,2-$C_2B_{10}H_{11}$-9-Tl$(OCOCF_3)_2$, C_4H_9MgCl	m.p. 168	[22]
1,2-$C_2B_{10}H_{11}$-9-Tl(C_2H_5)Br	1,2-$C_2B_{10}H_{11}$-9-Tl$(OCOCF_3)_2$, C_2H_5MgBr	m.p. 306 to 308	[22]
1,2-$C_2B_{10}H_9$-1-R-2-R′-9-Tl$(SCN)_2$	1,2-$C_2B_{10}H_9$-1-R-2-R′-9-Tl$(OCOCF_3)_2$, KSCN	m.p. 158 (with decomposition) for R = R′ = H; m.p. 146 (with decomposition) for R = R′ = CH_3	[22]

1,2-$C_2B_{10}H_{10}$-1-CH_3-2-$LaCl_2$	1,2-$C_2B_{10}H_{10}$-1-CH_3-2-Li, $LaCl_3$	m.p. 133 to 135	[64]
1,2-$C_2B_{10}H_{10}$-1-C_6H_5-2-LnI (Ln = Sm, Eu, Yb)	1,2-$C_2B_{10}H_{10}$-1-C_6H_5-2-Li, LnI_2		[56]
1,2-$C_2B_{10}H_{11}$-9-$Fe(CO)_2(C_5H_5)$	1,2-$C_2B_{10}H_{11}$-9-$COFe(CO)_2(C_5H_5)$, 140°C	m.p. 109 to 110	[32]

References for 13.11.2.7 on pp. 199, 212/4

References for 13.11.2.7: (continued from p. 199)

[11] Krull, I. S.; Jordan, S. W.; Kahl, S.; Smith, S. B. (J. Chromatogr. Sci. **20** [1982] 489/98).
[12] Zakharkin, L. I.; Pisareva, I. V. (Izv. Akad. Nauk SSSR Ser. Khim. **1983** 1158/61; Bull. Acad. Sci. USSR Div. Chem. Sci. **1983** 1046/8).
[13] Plešek, J.; Heřmánek, S. (Collection Czech. Chem. Commun. **46** [1981] 687/92).
[14] Plešek, J.; Plzak, Z.; Stuchlik, J.; Heřmánek, S. (Collection Czech. Chem. Commun. **46** [1981] 1748/63).
[15] Merkushev, E. B.; Simakhina, N. D.; Grigor'ev, M. G. (Izv. Akad. Nauk SSSR Ser. Khim. **1980** 2649; C.A. **94** [1981] No. 103456).
[16] Grushin, V. V.; Tolstaya, T. P.; Lisichkina, I. N.; Grishin, Yu. K.; Shcherbina, T. M.; Kampel, V. Ts.; Bregadze, V. I.; Godovikov, N. N. (Izv. Akad. Nauk SSSR Ser. Khim. **1983** 472/4; Bull. Acad. Sci. USSR Div. Chem. Sci. **1983** 429/31).
[17] Grushin, V. V.; Tolstaya, T. P.; Lisichkina, I. N. (Dokl. Akad. Nauk SSSR **261** [1981] 99/102; Dokl. Chem. Proc. Acad. Sci. USSR. **256/261** [1981] 456/9).
[18] Ol'dekop, Yu. A.; Maier, N. A.; Erdman, A. A.; Prokopovich, V. P. (Vestsi Akad. Navuk Belarusk.SSR Ser. Khim Navuk **1982** 72/6; C.A. **98** [1983] No. 198405).
[19] Zakharkin, L. I.; Pisareva, I. V. (Izv. Akad. Nauk SSSR Ser. Khim. **1981** 2794/6; Bull. Acad. Sci. USSR Div. Chem. Sci. **1981** 2328/30).
[20] Plešek, J.; Janoušek, Z.; Heřmánek, S. (Inorg. Syn. **22** [1983] 241/3).

[21] Zakharkin, L. I.; Pisareva, I. V. (Izv. Akad. Nauk SSSR Ser. Khim. **1984** 396/402; Bull. Acad. Sci. USSR Div. Chem. Sci. **1984** 355/61).
[22] Bregadze, V. I.; Usyatinskii, A. Ya.; Godovikov, N. N. (Izv. Akad. Nauk SSSR Ser. Khim. **1981** 398/401; Bull. Acad. Sci. USSR Div. Chem. Sci. **30** [1981] 315/8).
[23] Zakharkin, L. I.; Pisareva, I. V. (J. Organometal. Chem. **267** [1984] 73/9).
[24] Zakharkin, L. I.; Pisareva, I. V. (Izv. Akad. Nauk SSSR Ser. Khim. **1982** 718/9; Bull. Acad. Sci. USSR Div. Chem. Sci. **1982** 644).
[25] Zakharkin, L. I.; Pisareva, I. V. (Izv. Akad. Nauk SSSR Ser. Khim. **1984** 472/3; Bull. Acad. Sci. USSR Div. Chem. Sci. **1984** 438).
[26] Kalinin, V. N.; Usatov, A. V.; Popello, I. A.; Zakharkin, L. I. (Izv. Akad. Nauk SSSR Ser. Khim. **1982** 1433; Bull. Acad. Sci. USSR Div. Chem. Sci. **1982** 1281).
[27] Grushin, V. V.; Tolstaya, T. P.; Lisichkina, I. N. (Izv. Akad. Nauk SSSR Ser. Khim. **1982** 2633; Bull. Acad. Sci. USSR Div. Chem. Sci. **1982** 2329).
[28] Zakharkin, L. I.; Zhigareva, G. G.; Litonina, E. I. (Zh. Obshch. Khim. **52** [1982] 2367; J. Gen. Chem. [USSR] **52** [1982] 2106).
[29] Zakharkin, L. I.; Pisareva, I. V. (Zh. Obshch. Khim. **51** [1981] 1280/8; J. Gen. Chem. [USSR] **51** [1981] 1084/91).
[30] Zakharkin, L. I.; Kovredov, A. I.; Ol'shevskaya, V. A.; Shaugumbekova, Z. S. (Izv. Akad. Nauk SSSR Ser. Khim. **1980** 1691; C.A. **94** [1981] No. 65740).

[31] Zakharkin, L. I.; Kovredov, A. I.; Ol'shevskaya, V. A.; Shaugumbekova, Z. S. (J. Organometal. Chem. **226** [1982] 217/22).
[32] Zakharkin, L. I.; Kovredov, A. I.; Olshevskaya, V. A.; Antonovich, V. A. (J. Organometal. Chem. **267** [1984] 81/91).
[33] Chari, S. L.; Chiang, S. H.; Jones, M. (J. Am. Chem. Soc. **104** [1982] 3138/45).
[34] Totani, T.; Aono, K.; Yamamoto, K.; Tawara, K. (J. Med. Chem. **24** [1981] 1492/9).
[35] Zakharkin, L. I.; Shustova, T. V.; Kazantsev, A. V. (Zh. Obshch. Khim. **51** [1981] 1071/8; J. Gen. Chem. [USSR] **51** [1981] 893/9).
[36] Zakharkin, L. I.; Kovredov, A. I.; Shaugumbekova, Z. S.; Vinogradova, L. E.; Leites, L. A. (Zh. Obshch. Khim. **51** [1981] 1575/82; J. Gen. Chem. [USSR] **51** [1981] 1337/42).

[37] Zakharkin, L. I.; Kalinin, V. N.; Shitikov, V. K.; Sergeev, V. A. (Zh. Obshch. Khim. **51** [1981] 593/5; J. Gen. Chem. [USSR] **51** [1981] 468/70).
[38] Kazantsev, A. V.; Meiramov, M. G.; Kovredov, A. I.; Zakharkin, L. I. (Izv. Akad. Nauk SSSR Ser. Khim. **1982** 1603/5; Bull. Acad. Sci. USSR Div. Chem. Sci. **1982** 1425/7).
[39] Kalinin, V. N.; Kobel'kova, N. I.; Zakharkin, L. I. (Izv. Akad. Nauk SSSR Ser. Khim. **1982** 1661/2; Bull. Acad. Sci. USSR Div. Chem. Sci. **1982** 1479/81).
[40] Churkina, L. A.; Zvereva, T. D.; Shingel, I. A.; Ol'dekop, Yu. A. (Vestsi Akad. Navuk Belarusk. SSR Ser. Khim. Navuk **1982** 59/66; C.A. **97** [1982] No. 6359).

[41] Churkina, L. A.; Zvereva, T. D.; Shingel, I. A.; Ol'dekop, Yu. A. (Vestsi Akad. Navuk Belarusk. SSR Ser. Khim. Navuk **1983** 56/64; C.A. **98** [1983] No. 160774).
[42] Kolesova, V. A.; Tagirova, N. A.; Strepikheev, Yu. A.; Kalinin, V. N.; Virin, L. I.; Popova, I. Yu. (Zh. Obshch. Khim. **53** [1983] 239/40; J. Gen. Chem. [USSR] **53** [1983] 210/1).
[43] Rys, E. G.; Godovikov, N. N.; Kabachnik, M. I. (Izv. Akad. Nauk SSSR Ser. Khim. **1983** 2640/4; Bull. Acad. Sci. USSR Div. Chem. Sci. **1983** 2372/5).
[44] Kabachnik, M. I.; Godovikov, N. N.; Rys, E. G. (Izv. Akad. Nauk SSSR Ser. Khim. **1980** 1455/6; C.A. **94** [1981] No. 15805).
[45] Kalinin, V. N.; Izmailov, B. A.; Petrushkina, E. A.; Myakushev, V. D.; Zhdanov, A. A.; Zakharkin, L. I. (Izv. Akad. Nauk SSSR Ser. Khim. **1980** 2406/9; C.A. **94** [1981] No. 65751).
[46] Kalinin, V. N.; Izmailov, B. A.; Kazantsev, A. A.; Myakushev, V. D.; Zhdanov, A. A.; Zakharkin, L. I. (J. Organometal. Chem. **216** [1981] 295/320).
[47] Zakharkin, L. I.; Kovredov, A. I.; Savel'eva, I. S.; Safronova, E. V. (Zh. Obshch. Khim. **51** [1981] 2383; J. Gen. Chem. [USSR] **51** [1981] 2056).
[48] Usyatinskii, A. Yu.; Ryabov, A. D.; Bregadze, V. I.; Shcherbina, T. M.; Godovikov, N. N. (Izv. Akad. Nauk SSSR Ser. Khim. **1982** 1598/603; Bull. Acad. Sci. USSR Div. Chem. Sci. **31** [1982] 1420/4).
[49] Zakharkin, L. I.; Kovredov, A. I.; Ol'shevskaya, V. A. (Zh. Obshch. Khim. **52** [1982] 1911/8; J. Gen. Chem. [USSR] **52** [1982] 1694/700).
[50] Zakharkin, L. I.; Kovredov, A. I.; Shaugumbekova, Z. S.; Vinogradova, L. E.; Leites, L. A. (Zh. Obshch. Khim. **51** [1981] 1582/6; J. Gen. Chem. [USSR] **51** [1981] 1343/6).

[51] Zakharkin, L. I.; Kovredov, A. I.; Ol'shevskaya, V. A. (Zh. Obshch. Khim. **51** [1981] 2807/8; J. Gen. Chem. [USSR] **51** [1981] 2056).
[52] Zakharkin, L. I.; Kovredov, A. I.; Ol'shevskaya, V. A. (Izv. Akad. Nauk SSSR Ser. Khim. **1982** 673/5; Bull. Acad. Sci. USSR Div. Chem. Sci. **1982** 599/602).
[53] Zakharkin, L. I.; Kovredov, A. I.; Ol'shevskaya, V. A. (Izv. Akad. Nauk SSSR Ser. Khim. **1981** 2159/61; Bull. Acad. Sci. USSR Div. Chem. Sci. **1982** 1775/7).
[54] Drygina, O. V.; Dorofeenko, G. N.; Okhlobystin, O. Yu. (Zh. Obshch. Khim. **51** [1981] 868/75; J. Gen. Chem. [USSR] **51** [1981] 720/6).
[55] Kazantsev, A. V.; Ibraev, M. I. (Zh. Obshch. Khim. **51** [1981] 2501/3; J. Gen. Chem. [USSR] **51** [1981] 2155/7).
[56] Suleimanov, G. Z.; Bregadze, V. I.; Koval'chuk, N. A.; Khalilov, K. S.; Beletskaya, I. P. (J. Organometal. Chem. **255** [1983] C5/C7).
[57] Sergeev, V. A.; Kalinin, V. N.; Shitikov, V. K.; Svetogorov, Y. E.; Chizhova, N. V.; Danilova, M. P.; Chimishkin, A. L.; Zakharkin, L. I. (Zh. Obshch. Khim. **51** [1981] 863/8; (J. Gen. Chem. [USSR] **51** [1981] 716/20).
[58] Zakharkin, L. I.; Zhigareva, G. G. (Zh. Obshch. Khim. **53** [1983] 953/4; J. Gen. Chem. [USSR] **53** [1983] 841/2).
[59] Zakharkin, L. I.; Kovredov, A. I.; Ol'shevskaya, V. A. (Zh. Obshch. Khim. **53** [1983] 1431/2; J. Gen. Chem. [USSR] **53** [1983] 1287).

[60] Zakharkin, L. I.; L'vov, A. I.; Zhigareva, G. G. (Zh. Obshch. Khim. **53** [1983] 683/9; J. Gen. Chem. [USSR] **53** [1983] 594/9).

[61] Drygina, O. V.; Panov, V. B.; Okhlobystin, O. Yu. (Khim. Geterotsikl. Soedin. **1980** 185/8; Chem. Heterocyclic Compounds [USSR] **1980** 130/3).

[62] Bregadze, V. I.; Dzhashiashvili, T. K.; Sadzhaya, D. N.; Petriashvili, M. V.; Ponomareva, O. B.; Shcherbina, T. M.; Kampel, V. Ts.; Kukushkina, L. B.; Rochev, V. Ya.; Godovikov, N. N. (Izv. Akad. Nauk SSSR Ser. Khim. **1983** 907/12; Bull. Acad. Sci. USSR Div. Chem. Sci. **32** [1983] 824/7).

[63] Zakharkin, L. I.; Pisareva, I. V.; Vasil'eva, N. S. (Zh. Obshch. Khim. **52** [1982] 711; J. Gen. Chem. [USSR] **52** [1982] 618).

[64] Bregadze, V. I.; Koval'chuk, N. A.; Godovikov, N. N.; Suleimanov, G. Z.; Beletskaya, I. P. (J. Organometal. Chem. **241** [1983] C13/C15).

13.11.3 1,7-$C_2B_{10}H_{12}$ and Derivatives

A second phase transition in solid **1,7-$C_2B_{10}H_{12}$** in the vicinity of 165 K has been deduced from Raman and NMR T_1 studies [1]. Short-lived absorption spectra of a methanolic solution of 1,7-$C_2B_{10}H_{12}$, under the action of a pulse of accelerated electrons, have been recorded over the whole spectral range from the UV to the IR spectra [2]. Two-dimensional ^{11}B-^{11}B nuclear magnetic resonance spectroscopy confirms many of the B-B atom connectivities in 1,7-$C_2B_{10}H_{12}$ [3]. Aromatic solvent-induced ^1H NMR shifts are reported for 1,7-$C_2B_{10}H_{12}$ and correlated to PRDDO MO-derived hydrogen charges [4]. For the boron atoms in 1,7-$C_2B_{10}H_{12}$ cage "umbrella" angles of 114° (for B(2,3)), 116° (for B(4,6,8,11)), 116° (for B(5,12)), and 117° (for B(9,10)) are used in an empirical relationship, which includes the number of adjacent cage carbon atoms, to predict $^1J(^{11}B^1H)$ values of 177, 165, 165, and 151 Hz, respectively; these values agree well with the observed values of 178, 165, 162, and 151 Hz, respectively [5].

Solid boron-bonded complexes 1,7-$C_2B_{10}H_2$-2,3,4,5,6,8,9,10,11,12-Cl_{10} with some oxygen- and nitrogen-containing bases (dioxane, dimethyl sulfoxide, triphenylphosphanoxide, pyridine, trimethylamine, triethylamine) were investigated by IR and Raman spectroscopy. Complexes containing C-H··O bonds are characterized by a relative CH stretching frequency shift up to 12%, and a half-width of the CH stretching band up to 220 cm^{-1}. C-H··N bonds, with $N(CH_3)_3$ for example, are stronger with a relative shift of approximately 18% and a half-width of the CH stretching band of about 500 cm^{-1}. $N(C_2H_5)_3$ complexes, however, form an $\overset{+}{N}H\cdots C^-$ proton transfer hydrogen-bond while pyridine can give either C-H···N or $C^-\cdots H\overset{+}{N}$ hydrogen-bonded adducts depending on the solvent and temperature. The CH··N $\rightleftharpoons$ $C^-\cdots H\overset{+}{N}$ equilibrium appears to be shifted towards ion-pair formation at considerably smaller enthalpy values compared to the O-H··N $\rightleftharpoons$ $O^-\cdots H\overset{+}{N}$ system. CH and NH stretching frequencies are correlated with the acidity of the donor and the basicity of acceptor molecules [6]. Microdetermination of boron in 1,7-$C_2B_{10}H_{11}$-9-Br was carried out by Fourier-transform NMR [7].

The reduction of 1,7-$C_2B_{10}H_{11}$-9-SCN with $K_2[C_8H_8]$ proceeds with SC cleavage to give **1,7-$C_2B_{10}H_{11}$-9-SH** and **(1,7-$C_2B_{10}H_{11}$-9-S-)$_2$** [8]. Crystals of **1,7-$C_2B_{10}H_{11}$-9-SO_2CH_3** are orthorhombic, space group Pnma-D_{2h}^{16} (No. 62), with a=13.440(2), b=8.539(2), c=10.525(3) Å; Z=4; mean values of lengths in the borane cage are B-C=1.705 and B-B=1.749 Å [9].

1,7-$C_2B_{10}H_{11}$-2-NH_2 and **1,7-$C_2B_{10}H_{10}$-2-NH_2-6-CH_3** are formed from the reaction of 1,2-$C_2B_{10}H_{11}$-3-NH_2 with Na/NH_3 and subsequently with CH_2I_2, CHI_3, and CI_4; in the case of CI_4 a small amount of 1,7-$C_2B_{10}H_{10}$-*B*-NH_2-*B*-I is also formed [10].

ESR data indicate that the 1,7-$C_2B_{10}H_{11}$-1-$P(CH_3)(OC_2H_5)[OC(CH_3)_3]$ radical has a trigonal bipyramid structure about the phosphorus atom with the carborane ring in the apical position [11].

Electrophilic alkylation of 1,7-$C_2B_{10}H_{12}$ with $(CH_3)_2CHX$ (X = Cl, Br) in the presence of $AlCl_3$ in CS_2 gives a mixture of 9-*n*-C_3H_7 and 9-*i*-C_3H_7 derivatives of the carborane. Additionally, 1,7-$C_2B_{10}H_{11}$-9-$CH(CH_3)_2$ undergoes alkyl migration and redistribution, in the presence of $AlCl_3$, to give an equilibrium mixture of 9% 1,7-$C_2B_{10}H_{12}$, 82% 1,7-$C_2B_{10}H_{11}$-*B*-C_3H_7, and 9% 1,7-$C_2B_{10}H_{10}$-*B*,*B'*-$(C_3H_7)_2$, in which the 1,7-$C_2B_{10}H_{11}$-*B*-C_3H_7 portion consists of 9% 4- and 5-C_3H_7, 84% 9-*i*-C_3H_7, and 7% 9-*n*-C_3H_7 derivatives of 1,7-$C_2B_{10}H_{12}$ [12]. The boiling points of various alkyl carboranes, 1,7-$C_2B_{10}H_{11}$-1-R, have been extrapolated to be 264.33°C for R = *n*-C_4H_9, 258.9°C for R = *i*-C_4H_9, 281.91°C for R = C_5H_{11}, 298.6°C for R = C_6H_{13} [13].

1,7-$C_2B_{10}H_{11}$-2-CH=CH_2 and **1,7-$C_2B_{10}H_{11}$-4-CH=CH_2** are obtained from the isomerization of 1,2-$C_2B_{10}H_{11}$-3-CH=CH_2 in the gas phase at 560 to 580°C. Similarly, 1,7-$C_2B_{10}H_{11}$-2-C≡CH and 1,7-$C_2B_{10}H_{11}$-4-C≡CH are obtained from the isomerization of 1,2-$C_2B_{10}H_{11}$-3-C≡CH in the gas phase at 560 to 580°C [14]. From infrared spectra of 1,7-$C_2B_{10}H_{11}$-*x*-C≡CH (*x*=1, 2, 9) it is concluded that the ethynyl C-H bond has greater proton-donating capacity than the carborane C-H bond. Introduction of the ethynyl group into the carborane ring causes no change in the proton-donating capacity of the carborane C-H bond in comparison to the unsubstituted carborane. The carboranyl group increases the proton-donating capacity of acetylenic C-H bonds in comparison with C_4H_9C≡CH; this effect decreases in the series *x*=1>2>9 [15].

1,7-$C_2B_{10}H_{11}$-7-C_6H_5 is formed from the reductive-elimination reaction of the 6-coordinate Ir^{III} complex (1,7-$C_2B_{10}H_{10}$-7-C_6H_5-1-)Ir(H)(Cl)(CO)$[P(C_6H_5)_3]_2$ [16]. 1,7-$C_2B_{10}H_{11}$-1-C(Cl)=NOH, melting point 108 to 110°C, is obtained by passing Cl_2 into a $CHCl_3$ solution of 1,7-$C_2B_{10}H_{11}$-1-CH=NOH. Upon treating 1,7-$C_2B_{10}H_{11}$-1-C(Cl)=NOH with $N(C_2H_5)_3$ the nitrile oxide, 1,7-$C_2B_{10}H_{11}$-1-CNO, is obtained. 1,7-$C_2B_{10}H_{11}$-1-CNO reacts with acetylenes and olefins to yield the respective isooxazoles and isooxalines, respectively [17]. The phase equilibrium in 1,7-$C_2B_{10}H_{11}$-1-$CH(OOCCH_3)_2$ was determined by differential scanning colorimetry [18]. Crystals of 1,7-$C_2B_{10}H_9$-1-$CH_2N(CH_3)_2$-7-C_6H_5-4-$Re(CO)_4$ are monoclinic, space group $P2_1/n$-C^5_{2h} (No. 14), with a = 9.31(1), b = 18.22(2), c = 13.82(2) Å; β = 104.6(1)°; Z = 4 [19].

Hydrolysis of 1,7-$C_2B_{10}H_{10}$-1,7-$[CH_2Si(CH_3)_2Cl]_2$ yields 1,7-$C_2B_{10}H_{10}$-1,7-$[CH_2Si(CH_3)_2OH]_2$ [20]. Crystals of 1,7-$C_2B_{10}H_{10}$-1,7-$[CH_2Si(CH_3)_2OH]_2$ are triclinic, with space group $P\bar{1}$-C^1_i (No. 2), a = 7.131(6), b = 11.47(1), c = 12.26(1) Å; α = 108.33(9)°, β = 97.11(8)°, γ = 94.33(8)°; Z = 2, D_{calc} = 1.142 g/cm³. The symmetry of the molecule is C_{2v} [21]. Initial reaction rates are reported for the condensation reactions of 1,7-$C_2B_{10}H_{10}$-1,7-$(CHO)_2$ with various aniline derivatives [22]. 1,7-$C_2B_{10}H_{10}$-1-$Si(CH_3)_2R$-7-$Si(CH_3)R'R''$ (R = CH_3O, C_2H_5O, R' = CH_3, R'' = H; R = H, R' = $F_3CCH_2CH_2$, R'' = CH_3O; R = C_2H_5O, R' = CH_3, R'' = Cl) were prepared by treating 1,7-$C_2B_{10}H_{10}$-1,7-$[Si(CH_3)_2R]_2$ with 1,7-$Li_2C_2B_{10}H_{10}$ to give 1,7-$C_2B_{10}H_{10}$-1-$Si(CH_3)_2R$-7-Li which was then treated with $ClSi(CH_3)R'R''$. The boiling points of 1,7-$C_2B_{10}H_{10}$-1-$Si(CH_3)_2R$-7-$Si(CH_3)R'R''$ are (at 0.1 Torr) 86°C for R = OCH_3, R' = CH_3, R'' = H; 90°C for R = C_2H_5O, R' = CH_3, R'' = H; 99°C for R = H, R' = $CF_3CH_2CH_2$, R'' = CH_3O; 105°C for R = C_2H_5O, R' = CH_3, R'' = Cl [23].

J(^{199}Hg-^{11}B) values for 1,7-$C_2B_{10}H_9$-1,7-R_2-9-HgR' are (in Hz): 1366 for R = H, R' = CH_3; 1170 for R = H, R' = 1,7-$C_2B_{10}H_9$-1,2-R_2-9-; 1180 for R = CH_3, R' = 1,7-$C_2B_{10}H_9$-1,2-R_2-9- [24]. Infrared and Raman spectra of mercury derivatives of 1,7-$C_2B_{10}H_{12}$ are reported [25]. Compounds with B-Hg-Ge or Ge-Hg-B-B-Hg-Ge chains in which the B atoms are members of a C_2B_{10} carborane cage are prepared by treating the digermane $(C_6F_5)_3GeGe(C_2H_5)_3$ with *B*-mercurated carboranes, 1,7-$C_2B_{10}H_{11}$HgX (X = Cl, $OOCCF_3$) and 1,7-$C_2B_{10}H_{10}(HgOOCCF_3)_2$. The prepared compounds undergo oxidative insertion with $Pt[P(C_6H_5)_3]_n$ (n = 3,4) to give the chains B-Hg-Pt-Ge and Ge-Pt-Hg-B-B-Hg-Ge. Treatment of $HGe(C_6F_5)_2Ge(C_6F_5)_2H$ with 1,7-$C_2B_{10}H_{11}$-9-$HgCH_3$ results in a compound with a B-Hg-Ge-Ge-Hg-B chain containing two carborane cages at the ends of the chain [26].

References for 13.11.3 on pp. 216, 225/7

Crystals of $1,7\text{-}C_2B_{10}H_9\text{-}1,7\text{-}(CH_3)_2\text{-}9\text{-}Tl(CF_3COO)_2 \cdot bpy$ (bpy = 2,2′-bipyridyl) are monoclinic, lattice parameters at −120°C are a = 14.151(7), b = 10.190(6), c = 23.91(1) Å; β = 97.97(4)°; D_{calc} = 1.620 g/cm^3; Z = 4; space group $P2_1/n\text{-}C_{2h}^5$ (No. 14) [27]. Treatment of $1,7\text{-}C_2B_{10}H_9\text{-}1,7\text{-}(CH_3)_2\text{-}9\text{-}Tl(OOCCF_3)_2$ with $K_2[C_8H_8]$ gives $1,7\text{-}C_2B_{10}H_{12}$ and $(1,7\text{-}C_2B_{10}H_{11}\text{-}9\text{-})_2$ [8]. $1,7\text{-}C_2B_{10}H_9\text{-}1\text{-}R\text{-}7\text{-}R'\text{-}9\text{-}TlSn(OOCCH_2COCH_3)_2Br_2$ (R = R′ = H, CH_3) is obtained by treatment of $1,7\text{-}C_2B_{10}H_9\text{-}1\text{-}R\text{-}7\text{-}R'\text{-}9\text{-}TlBr_2$ with $Sn(OOCCH_2COCH_3)_2$ [28]. The rate constant for *B*-metallation of $1,7\text{-}C_2B_{10}H_{10}\text{-}1,7\text{-}(CH_3)_2$ by $Tl(OOCCF_3)_3$ is greater than of $1,7\text{-}C_2B_{10}H_{12}$ [29].

A carborane-containing 3,6-di-*t*-butyl-*o*-semiquinolate complex of tin, $[3,6\text{-}(t\text{-}C_4H_9)_2\text{-}C_6H_2O_2]SnCl_2(1,7\text{-}C_2B_{10}H_{11}\text{-}1\text{-})$, has been prepared, where the carborane unit is $1,7\text{-}C_2B_{10}H_{11}\text{-}1\text{-}$, and its ESR and NMR studied [30].

Stable carboranylmethyl complexes of Fe and Ni, i.e., $(1,7\text{-}C_2B_{10}H_{10}\text{-}1\text{-}CH_3\text{-}7\text{-}CH_2\text{-})\text{-}Fe(CO)_2(C_5H_5)$ and $(1,7\text{-}C_2B_{10}H_{10}\text{-}1\text{-}CH_3\text{-}7\text{-}CH_2\text{-})Ni\{P(C_6H_5)_3\}(C_5H_5)$, are obtained by reaction of the corresponding carboranylmethylmagnesium chlorides with halogen complexes of Fe and Ni [31]. Treatment of $(1,7\text{-}C_2B_{10}H_{10}\text{-}1\text{-}CH_3\text{-}7\text{-})Fe(CO)_2(C_5H_5)$ with Br_2 results in electrophilic bromination of the carboranyl group at positions 9 and 10 [31].

Reactions of $[Ir(H)_2(\sigma\text{-}1,7\text{-}C_2B_{10}H_{10}\text{-}7\text{-}C_6H_5)(CO)(RCN)(P(C_6H_5)_3)]$ (R = CH_3, C_6H_5) with activated olefins and acetylenes give hydridoalkyl- and hydridoalkenyliridium(III) derivatives, respectively. In addition, the reductive elimination of the alkane or alkene molecule from the obtained hydrido complexes is also examined [32]. $1,7\text{-}C_2B_{10}H_{10}\text{-}1\text{-}C_6H_5\text{-}7\text{-}Ir(CO)(C_6H_5CN)\text{-}[P(C_6H_5)_3]$ is found to be an effective catalyst for the homogeneous hydrogenation of terminal olefins and acetylenes [33]. The addition reactions of d^8 carborane-iridium(I) complexes of general formula *trans*-$[Ir\text{-}(1,7\text{-}C_2B_{10}H_{10}\text{-}7\text{-}R\text{-}1\text{-})(CO)L_2]$, where L = $P(C_6H_5)_3$, $P(CH_3)(C_6H_5)_2$, and R = H, CH_3, C_6H_5, with HCl, HBr, Cl_2, and Br_2 have been studied both in the solid state and in solution using a variety of solvents. The HX adducts of general formula $[Ir(H)(X)(1,7\text{-}C_2B_{10}H_{11}\text{-}1\text{-})(CO)L_2]$ (X = Cl, Br, I) reductively eliminate the carborane molecule $1,7\text{-}C_2B_{10}H_{12}$ [34]. The complex $(1,7\text{-}C_2B_{10}H_{10}\text{-}7\text{-}C_6H_5\text{-}1\text{-})Ir(H)(CO)(C_6H_5CN)(\sigma\text{-}CHCH_2C(O)OC(O))$ (where σ = sigma and a bond exists between the first carbon of the last group and the last carbon of the last group in parenthesis) undergoes a reductive elimination reaction with succinic anhydride [35].

In addition to those compounds mentioned above in this section, a supplemental list of $1,7\text{-}C_2B_{10}H_{12}$ derivatives, along with reagents used to synthesize each one, and physical properties of each product, is given in Table 13/11, pp. 217/24.

References for 13.11.3:

[1] Bukalov, S. S.; Leites, L. A.; Blumenfeld, A. L.; Fedin, E. I. (J. Raman Spectrosc. **14** [1983] 210/1).

[2] Zimina, G. M.; Vannikov, A. V.; Stanko, V. I. (Khim. Vysokikh Energ. **16** [1982] 335/9; High Energy Chem. [USSR] **16** [1982] 263/6; C.A. **97** [1982] No. 101576).

[3] Venable, T. L.; Hutton, W. C.; Grimes, R. N. (J. Am. Chem. Soc. **106** [1984] 29/37).

[4] Jarvis, W.; Inman, W.; Powell, B.; DiStefano, E. W.; Onak, T. (J. Magn. Resonance **43** [1981] 302/15).

[5] Jarvis, W.; Abdou, Z. J.; Onak, T. (Polyhedron **2** [1983] 1067/70).

[6] Leites, L. A.; Vinogradova, L. E.; Belloc, J.; Novak, A. (J. Mol. Struct. **100** [1983] 379/92).

[7] Kasler, F.; Tierney, M. (Mikrochim. Acta **2** [1981] 301/12).

[8] Usyatinskii, A. Ya.; Todres, Z. V.; Shcherbina, T. M.; Bregadze, V. I.; Godovikov, N. N. (Izv. Akad. Nauk SSSR Ser. Khim. **1983** 1640/2; Bull. Acad. Sci. USSR Div. Chem. Sci. **32** [1983] 1488/90).

[9] Maly, K.; Petrina, A.; Petricek, V.; Hummel, L.; Linek, A. (Acta Cryst. B **36** [1980] 181/3).

[10] Stanko, V. I.; Gol'tyapin, Yu. V. (Zh. Obshch. Khim. **52** [1982] 78/80; J. Gen. Chem. [USSR] **52** [1982] 72/3).

(continued on p. 225)

Table 13/11

1,7-$C_2B_{10}H_{12}$ Derivatives; Synthesis and Physical Properties (m.p. = melting point in °C; b.p. = boiling point in °C; IR = infrared spectrum reported; NMR = nuclear magnetic resonance spectrum reported; MS = mass spectrum reported).

compound (product)	reagents used to synthesize compound listed in the first column	properties of product	Ref.
1,7-$C_2B_{10}H_{11}$-9-Cl	1,7-$C_2B_{10}H_{11}$-9-HgCl/350°C	—	[36]
1,7-$C_2B_{10}H_{10}$-1-SH-9-Cl	1,7-$C_2B_{10}H_{11}$-9-Cl, NaH, S	^1H, ^{11}B NMR	[37]
1,7-$C_2B_{10}H_{10}$-9-SH-10-Cl	1,7-$C_2B_{10}H_{11}$-10-Cl, S, $AlCl_3$	^1H, ^{11}B NMR	[37]
1,7-$C_2B_{10}H_{11}$-9-Br	[(1,7-$C_2B_{10}H_{11}$-9-)$_2$Br][BF_4], $P(C_6H_5)_3$	m.p. 171 to 172	[38]
1,7-$C_2B_{10}H_{10}$-1-SH-9-Br	1,7-$C_2B_{10}H_{11}$-9-Br, NaH, S	^1H, ^{11}B NMR	[37]
1,7-$C_2B_{10}H_{10}$-9-SH-10-Br	1,7-$C_2B_{10}H_{11}$-10-Br, S, $AlCl_3$	^1H, ^{11}B NMR	[37]
1,7-$C_2B_{10}H_{11}$-9-I	1,7-$C_2B_{10}H_{12}$, $C_6H_5I(OOCCF_3)_2$, I_2	m.p. 111 to 112	[39]
1,7-$C_2B_{10}H_{11}$-9-I	[1,7-$C_2B_{10}H_{11}$-9-IC_6H_5][BF_4], $P(C_6H_5)_3$	m.p. 107 to 108	[38]
1,7-$C_2B_{10}H_{10}$-9-SH-10-I	1,7-$C_2B_{10}H_{11}$-10-I, S, $AlCl_3$	^1H, ^{11}B NMR	[37]
1,7-$C_2B_{10}H_{11}$-9-ICl_2	1,7-$C_2B_{10}H_{11}$-9-I, Cl_2, CCl_4	—	[40]
1,7-$C_2B_{10}H_{11}$-9-$I(OOCCF_3)_2$	1,7-$C_2B_{10}H_{11}$-9-I, CF_3COOAg, CH_3CN	—	[40]
[1,7-$C_2B_{10}H_{11}$-9-IC_6H_5]BF_4	1,7-$C_2B_{10}H_{11}$-9-$I(OOCCF_3)_2$, C_6H_6, H_2SO_4, CH_3COOH, $(CH_3CO)_2O$, $NH_4[BF_4]$		[40]
[1,7-$C_2B_{10}H_{11}$-9-IC_6H_5]BF_4	1,7-$C_2B_{10}H_{11}$-9-I, C_6H_6, H_2SO_4, $K_2S_2O_8$, $(CH_3CO)_2O$, $NH_4[BF_4]$		[40]
1,7-$C_2B_{10}H_{11}$-9-OH	1,7-$C_2B_{10}H_{11}$-9-$Tl(OOCCF_3)_2$, CrO_3, H_2SO_4	m.p. 326	[41]
1,7-$C_2B_{10}H_{11}$-9-OH	1,7-$C_2B_{10}H_{11}$-9-$OOCC_6H_5$, $LiAlH_4$	m.p. 325 to 326	[42]
1,7-$C_2B_{10}H_{11}$-9-$OOCCH_3$	[1,7-$C_2B_{10}H_{11}$-9-IC_6H_5][BF_4], CH_3COONa	m.p. 103.5 to 104.5	[42]

References for 13.11.3 on pp. 216, 225/7

Table 13/11 (continued)

compound (product)	reagents used to synthesize compound listed in the first column	properties of product	Ref.
$1,7\text{-}C_2B_{10}H_{11}\text{-}9\text{-}OOCC_6H_5$	$[1,7\text{-}C_2B_{10}H_{11}\text{-}9\text{-}IC_6H_5]BF_4$, C_6H_5COONa		[42]
$1,7\text{-}C_2B_{10}H_{11}\text{-}1\text{-}SH$	$1,7\text{-}C_2B_{10}H_{12}$, NaH, S	^1H, ^{11}B	[37]
$1,7\text{-}C_2B_{10}H_{10}\text{-}9,10\text{-}(SH)_2$	$1,7\text{-}C_2B_{10}H_{12}$, (a) S_2Cl_2, $AlCl_3$; (b) $Na[BH_4]$	m.p. 193 to 194, ^{11}B NMR	[43]
$1,7\text{-}C_2B_{10}H_{10}\text{-}1\text{-}CH_3\text{-}9\text{-}SH$	$(1,7\text{-}C_2B_{10}H_{10}\text{-}1\text{-}CH_3\text{-}9\text{-})_2S_2$, Zn, HCl	m.p. 113	[44]
$1,7\text{-}C_2B_{10}H_9\text{-}1,2\text{-}(CH_3)_2\text{-}9\text{-}SH$	$(1,7\text{-}C_2B_{10}H_9\text{-}1,2\text{-}(CH_3)_2\text{-}9\text{-})_2S_2$, Zn, HCl	m.p. 78 to 79	[44]
$1,7\text{-}C_2B_{10}H_{11}\text{-}9\text{-}SCl$	$1,7\text{-}C_2B_{10}H_{11}\text{-}9\text{-}SH$, SO_2Cl_2, or Cl_2		[45]
$1,7\text{-}C_2B_{10}H_{11}\text{-}9\text{-}SP(O)(OC_2H_5)_2$	$1,7\text{-}C_2B_{10}H_{11}\text{-}9\text{-}SCl$, $P(OC_2H_5)_3$	m.p. 69 to 70	[45]
$1,7\text{-}C_2B_{10}H_{11}\text{-}9\text{-}SO_2C_6H_5$	$1,7\text{-}C_2B_{10}H_{11}\text{-}9\text{-}SO_2Cl$, C_6H_6, $AlCl_3$	m.p. 162 to 163	[45]
$1,7\text{-}C_2B_{10}H_{11}\text{-}9\text{-}SC_6H_5$	$1,7\text{-}C_2B_{10}H_{11}\text{-}9\text{-}SCl$, C_6H_6, $FeCl_3$	m.p. 82	[45]
$1,7\text{-}C_2B_{10}H_{10}\text{-}9,10\text{-}(SCH_3)_2$	$1,7\text{-}C_2B_{10}H_{10}\text{-}9,10\text{-}(SH)_2$, NaOH, CH_3I	^{11}B NMR	[43]
$1,7\text{-}C_2B_{10}H_{11}\text{-}9\text{-}SCN$	$[1,7\text{-}C_2B_{10}H_{11}\text{-}9\text{-}IC_6H_5][BF_4]$, NCSNa	m.p. 59.5 to 60.5	[42]
$1,7\text{-}C_2B_{10}H_{11}\text{-}9\text{-}SCN$	$1,7\text{-}C_2B_{10}H_{11}\text{-}9\text{-}Tl(SCN)_2/h\nu$	m.p. 48	[41]
$1,7\text{-}C_2B_{10}H_{11}\text{-}9\text{-}SCN$	$1,7\text{-}C_2B_{10}H_{11}\text{-}9\text{-}SCl$, KCN	m.p. 59 to 59.5	[45]
$1,7\text{-}C_2B_{10}H_{11}\text{-}9\text{-}SeH$	$(1,7\text{-}C_2B_{10}H_{11}\text{-}9\text{-}Se\text{-})_2$, Zn, HCl, CH_3COOH	m.p. 121 to 122, IR, NMR	[46]
$1,7\text{-}C_2B_{10}H_{10}\text{-}9,10\text{-}(SeH)_2$	$1,7\text{-}C_2B_{10}H_{12}$, (a) Se_2Cl_2, $AlCl_3$; (b) $Na[BH_4]$	^{11}B NMR	[43]
$1,7\text{-}C_2B_{10}H_{10}\text{-}9,10\text{-}(SeCH_3)_2$	$1,7\text{-}C_2B_{10}H_{10}\text{-}9,10\text{-}(SeH)_2$, NaOH, NaOH, CH_3I	^{11}B NMR	[43]
$1,7\text{-}C_2B_{10}H_{11}\text{-}9\text{-}TeCl_3$	$1,7\text{-}C_2B_{10}H_{12}$, $TeCl_4$, $AlCl_3$, CH_2Cl_2		[47]
$1,7\text{-}C_2B_{10}H_{11}\text{-}9\text{-}TeCH_3$	$(1,7\text{-}C_2B_{10}H_{11}\text{-}9\text{-})_2Te_2$, $Na[BH_4]$, CH_3I	m.p. 48 to 50	[47]
$1,7\text{-}C_2B_{10}H_{11}\text{-}9\text{-}NH_2$	$1,7\text{-}C_2B_{10}H_{11}\text{-}9\text{-}N{=}P(C_6H_5)_3$, CH_3COOH, HBr		[42]
$1,7\text{-}C_2B_{10}H_{11}\text{-}9\text{-}NHCH_3$	$1,7\text{-}C_2B_{10}H_{11}\text{-}9\text{-}NHCHO$, $LiAlH_4$	^1H NMR	[48]

$1,7\text{-}C_2B_{10}H_{11}\text{-}9\text{-}N(CH_3)_2$	$1,7\text{-}C_2B_{10}H_{11}\text{-}9\text{-}NH_2$, $Na[BH_4]$, H_2SO_4, CH_2O	m.p. 185, ^{1}H NMR	[48]
$1,7\text{-}C_2B_{10}H_{11}\text{-}9\text{-}NHCHO$	$1,7\text{-}C_2B_{10}H_{11}\text{-}9\text{-}NH_2$, $HCOOC_2H_5$	IR	[48]
$1,7\text{-}C_2B_{10}H_9\text{-}1\text{-}CH_3\text{-}7\text{-}N{=}N{=}C_6H_5\text{-}B\text{-}Mn(CO)_4$	$1,7\text{-}C_2B_{10}H_9\text{-}1\text{-}CH_3\text{-}7\text{-}N{=}N{=}C_6H_5\text{-}B\text{-}Mn(CO)_4$	m.p. 93 to 95	[49]
$1,7\text{-}C_2B_{10}H_{11}\text{-}9\text{-}N_3$	$[1,7\text{-}C_2B_{10}H_{11}\text{-}9\text{-}IC_6H_5][BF_4]$, NaN_3	m.p. 96 to 97	[42]
$1,7\text{-}C_2B_{10}H_{11}\text{-}N{=}P(C_6H_5)_3$	$1,7\text{-}C_2B_{10}H_{11}\text{-}9\text{-}N_3$, $P(C_6H_5)_3$	m.p. 174 to 174.5	[42]
$1,7\text{-}C_2B_{10}H_{11}\text{-}9\text{-}NC$	$1,7\text{-}C_2B_{10}H_{11}\text{-}9\text{-}NHCHO$, $POCl_3$, C_5H_5N	IR	[48]
$1,7\text{-}C_2B_{10}H_{11}\text{-}9\text{-}NCS$	$[1,7\text{-}C_2B_{10}H_{11}\text{-}9\text{-}IC_6H_5][BF_4]$, NCSNa	m.p. 158 to 159	[42]
$[1,7\text{-}C_2B_{10}H_{11}\text{-}9\text{-}P(C_6H_5)_3][BF_4]$	$[(1,7\text{-}C_2B_{10}H_{11}\text{-}9\text{-})_2Br][BF_4]$, $P(C_6H_5)_3$	m.p. 222 to 224	[38]
$1,7\text{-}C_2B_{10}H_{11}\text{-}9\text{-}AsCl_2$	$(1,7\text{-}C_2B_{10}H_{11}\text{-}9\text{-})_2Hg$, $AsCl_3$	m.p. 78	[50]
$1,7\text{-}C_2B_{10}H_{11}\text{-}9\text{-}AsO_2H$	$1,7\text{-}C_2B_{10}H_{11}\text{-}9\text{-}AsCl_2$, Cl_2, H_2O	decomp. temp. 230	[50]
$1,7\text{-}C_2B_{10}H_{10}\text{-}1\text{-}CH_3\text{-}7\text{-}AsCl_2$	$1,7\text{-}C_2B_{10}H_{10}\text{-}1\text{-}CH_3\text{-}7\text{-}Li$, $ClAs[N(C_2H_5)_2]_2$, HCl	b.p. 100 at 1 Torr	[50]
$1,7\text{-}C_2B_{10}H_{10}\text{-}1\text{-}CH_3\text{-}7\text{-}AsO$	$1,7\text{-}C_2B_{10}H_{10}\text{-}1\text{-}CH_3\text{-}7\text{-}AsCl_2$, H_2O	m.p. 218 to 220	[50]
$1,7\text{-}C_2B_{10}H_{11}\text{-}9\text{-}CH_3$	$1,7\text{-}C_2B_{10}H_{11}\text{-}9\text{-}I$, CH_3MgBr, $[P(C_6H_5)_3]_2PdCl_2$	m.p. 145 to 146	[51]
$1,7\text{-}C_2B_{10}H_{11}\text{-}9\text{-}C_2H_5$	$1,7\text{-}C_2B_{10}H_{11}\text{-}9\text{-}Br$(or I), C_2H_5MgX (X = halogen), $[P(C_6H_5)_3]_2MCl_2$, or $[P(C_6H_5)_3]_4M$ (M = Pd, Ni)	b.p. 71 at 1 Torr	[52, 53]
$1,7\text{-}C_2B_{10}H_{11}\text{-}9\text{-}CH(CH_3)_2$	$1,7\text{-}C_2B_{10}H_{11}\text{-}9\text{-}I$, $(CH_3)_2CHMgX$ (X = halogen), $[P(C_6H_5)_3]_2PdCl_2$, or $[P(C_6H_5)_3]_4Pd$	b.p. 71 to 72 at 1 Torr	[52]
$1,7\text{-}C_2B_{10}H_{11}\text{-}9\text{-}C_4H_9$	$1,7\text{-}C_2B_{10}H_{11}\text{-}9\text{-}Br$(or I), C_4H_9MgX (X = halogen), $[P(C_6H_5)_3]_2MCl_2$, or $[P(C_6H_5)_3]_4M$ (M = Pd, Ni)	b.p. 82 to 83 at 1 Torr	[52, 53]
$1,7\text{-}C_2B_{10}H_{11}\text{-}9\text{-}CH_2CH(CH_3)_2$	$1,7\text{-}C_2B_{10}H_{11}\text{-}9\text{-}I$, $(CH_3)_2CHCH_2MgX$ (X = halogen), $[P(C_6H_5)_3]_2PdCl_2$, or $[P(C_6H_5)_3]_4Pd$	b.p. 80 to 81 at 1 Torr	[52]
$1,7\text{-}C_2B_{10}H_{11}\text{-}9\text{-}CH_2CH{=}CH_2$	$1,7\text{-}C_2B_{10}H_{11}\text{-}9\text{-}Br$(or I), $CH_2{=}CHCH_2MgX$ (X = halogen), $[P(C_6H_5)_3]_2MCl_2$, or $[P(C_6H_5)_3]_4M$ (M = Pd, Ni)	b.p. 78 to 79 at 1 Torr	[52, 53]

References for 13.11.3 on pp. 216, 225/7

Table 13/11 (continued)

compound (product)	reagents used to synthesize compound listed in the first column	properties of product	Ref.
$1,7-C_2B_{10}H_{11}-1-CH_2C{\equiv}CH$	$1,7-C_2B_{10}H_{11}-1-CH{=}C{=}CH_2$, C_4H_9Li	b.p. 65 at 1.3 hPa	[54]
$1,7-C_2B_{10}H_{10}-1-CH_2C{\equiv}CH-7-CH_3$	$1,7-C_2B_{10}H_{10}-1-CH{=}C{=}CH_2-7-CH_3$, C_4H_9Li	^1H NMR	[54]
$1,7-C_2B_{10}H_{11}-1-CH_2Cl$	$1,7-C_2B_{10}H_{11}-1-CH_2OH$, PCl_5	m.p. 91 to 92	[55]
$1,7-C_2B_{10}H_{11}-2-CHClCH_3$	$1,7-C_2B_{10}H_{11}-2-CH{=}CH_2$, HCl, $AlCl_3$	b.p. 89 to 91 at 1 Torr	[14]
$1,7-C_2B_{10}H_{11}-1-CH_2Br$	$1,7-C_2B_{10}H_{11}-1-CH_2OH$, $P(C_6H_5)_3Br_2$	m.p. 31 to 32	[55]
$1,7-C_2B_{10}H_{10}-1,7-(CH_2Br)_2$	$1,7-C_2B_{10}H_{10}-1,7-(CH_2OH)_2$, $P(C_6H_5)_3Br_2$	b.p. 141 at 2.7 hPa	[55]
$1,7-C_2B_{10}H_{11}-1-CH_2OH$	$1,7-C_2B_{10}H_{12}$, C_4H_9Li, CH_2O, H^+	m.p. 214 to 215	[55]
$1,7-C_2B_{10}H_{10}-1,7-(CH_2OH)_2$	$1,7-C_2B_{10}H_{12}$, C_4H_9Li, CH_2O, H^+	m.p. 193 to 195	[55]
$1,7-C_2B_{10}H_{11}-1-CH_2OR$	$1,7-C_2B_{10}H_{11}-1-CH_2OH$, RBr, $[(C_2H_5)_3N(CH_2C_6H_5)]Cl$, NaOH	$R = C_4H_9$, $C_6H_5CH_2$, MS	[56]
$1,7-C_2B_{10}H_{10}-1,7-(CH_2OR)_2$	$1,7-C_2B_{10}H_{10}-1,7-(CH_2OH)_2$, RBr, $[(C_2H_5)_3N(CH_2C_6H_5)]Cl$, NaOH, $R = C_4H_9$, $C_6H_5CH_2$	m.p. for $R = C_6H_5CH_2$ 79 to 80, MS	[56]
$1,7-C_2B_{10}H_{11}-1-CH_2OC(O)R$	$1,7-C_2B_{10}H_{11}-1-CH_2OH$, RCOCl	b.p. 56 to 58.5 at 0.03 Torr for $R = C_2H_5$, m.p. 22.5 to 23 for $R = CH(CH_3)_2$, 97 to 98 for $R = C_6H_5$	[57]
$1,7-C_2B_{10}H_{10}-1,2-[CH_2OC(O)R]_2$	$1,7-C_2B_{10}H_{10}-1,2-(CH_2OH)_2$, RCOCl	m.p. 28.5 to 30.5 for $R = C_2H_5$, b.p. 111 to 113 at 0.02 Torr for $R = CH(CH_3)_2$, m.p. 109.5 to 110 for $R = C_6H_5$	[57]
$1,7-C_2B_{10}H_{11}-2-CH[CH_2COC(CH_3)_3]_2$	$1,7-C_2B_{10}H_{10}-2-[4-H-pyran-2,6-(C(CH_3)_3)_2]$, $HClO_4$ (4-H-pyran = C_5H_3O)	m.p. 109 to 111	[58]

$1,7\text{-}C_2B_{10}H_{10}\text{-}1\text{-}CH_3\text{-}7\text{-}CH_2HgCl$	$(1,7\text{-}C_2B_{10}H_{10}\text{-}1\text{-}CH_3\text{-}7\text{-}CH_2)_2Hg$, $HgCl_2$	m.p. 126 to 127	[55]
$[(1,7\text{-}C_2B_{10}H_{11}\text{-}1\text{-})C(OH)(CH_3)C_5H_4]Fe(C_5H_5)$	$Li[1,7\text{-}C_2B_{10}H_{11}]$, $(CH_3COC_5H_4)Fe(C_5H_5)$	m.p. 143 to 145	[59]
$1,7\text{-}C_2B_{10}H_{11}\text{-}9\text{-}CH_2C_6H_5$	$1,7\text{-}C_2B_{10}H_{11}\text{-}9\text{-}I$, $C_6H_5CH_2MgX$ (X = halogen), $[(C_6H_5)_3P]_2PdCl_2$, $[(C_6H_5)_3P]_4Pd$	m.p. 102 to 103	[52]
$1,7\text{-}C_2B_{10}H_{10}\text{-}1,7\text{-}(CH_2C_6H_4\text{-}4'\text{-}R)_2$	$M_2[1,7\text{-}C_2B_{10}H_{10}]$, $p\text{-}R\text{-}C_6H_4CH_2X$ (M = Li, Na; R = CH_3 when X = Br; R = CH_3O when X = Cl)	m.p. 58 to 59, R = CH_3; 66 to 67, R = CH_3O	[60]
$1,7\text{-}C_2B_{10}H_{10}\text{-}1,7\text{-}(CH_2C_6H_4\text{-}4'\text{-}COOH)_2$	$1,7\text{-}C_2B_{10}H_{10}\text{-}1,7\text{-}(CH_2C_6H_4\text{-}4'\text{-}CH_3)_2$, CrO_3, CH_3COOH, H_2SO_4	m.p. 321 to 323	[60]
$1,7\text{-}C_2B_{10}H_{10}\text{-}1,7\text{-}(CH_2C_6H_4\text{-}4'\text{-}OH)_2$	$1,7\text{-}C_2B_{10}H_{10}\text{-}1,7\text{-}(CH_2C_6H_4\text{-}4'\text{-}OCH_3)_2$, HI, 100°C	m.p. 200 to 203	[60]
$1,7\text{-}C_2B_{10}H_{10}\text{-}1,7\text{-}[CH_2Si(CH_3)_2Cl]_2$	$1,2\text{-}C_2B_{10}H_{10}\text{-}1,2\text{-}[CH_2Si(CH_3)_2Cl]_2$, 280 to 320°C	b.p. 151 to 153°C; ^1H NMR	[61]
$(1,7\text{-}C_2B_{10}H_{10}\text{-}1\text{-}CH_3\text{-}7\text{-})Si(CH_3)C_3H_6$	$Li[1,7\text{-}C_2B_{10}H_{10}\text{-}1\text{-}CH_3]$, $Cl(CH_3)SiC_3H_6$ (where SiC_3H_6 is a silacyclobutane ring)	b.p. 141 to 142 at 5 Torr	[62]
$1,7\text{-}C_2B_{10}H_{11}\text{-}1\text{-}CH_2Si(CH_3)_2OCH_3$	$Li[1,7\text{-}C_2B_{10}H_{11}]$, $ClCH_2Si(CH_3)_2OCH_3$	b.p. 115 to 117 at 1 Torr	[20]
$1,7\text{-}C_2B_{10}H_{11}\text{-}1\text{-}Si(CH_3)_2CH_2Cl$	$Li[1,7\text{-}C_2B_{10}H_{11}]$, $ClCH_2Si(CH_3)_2OCH_3$		[20]
$1,7\text{-}C_2B_{10}H_{10}\text{-}1,7\text{-}[CH_2Si(CH_3)_2OCH_3]_2$	$Li_2[1,7\text{-}C_2B_{10}H_{10}]$, $ClCH_2Si(CH_3)_2OCH_3$	b.p. 168 to 170 at 1 Torr	[20]
$1,7\text{-}C_2B_{10}H_{10}\text{-}1,7\text{-}[Si(CH_3)_2CH_2Cl]_2$	$Li_2[1,7\text{-}C_2B_{10}H_{10}]$, $ClCH_2Si(CH_3)_2OCH_3$		[20]
$1,7\text{-}C_2B_{10}H_{11}\text{-}1\text{-}CH_2Si(CH_3)_2Cl$	$1,7\text{-}C_2B_{10}H_{11}\text{-}1\text{-}CH_2Si(CH_3)_2OCH_3$, PCl_3	m.p. 34	[20]
$1,7\text{-}C_2B_{10}H_{11}\text{-}1\text{-}CH_2Si(CH_3)Cl_2$	$1,7\text{-}C_2B_{10}H_{11}\text{-}1\text{-}CH_2Si(CH_3)(OCH_3)_2$, PCl_3	b.p. 119 to 122 at 1 Torr	[20]
$1,7\text{-}C_2B_{10}H_{11}\text{-}9\text{-}CH_2Si(CH_3)_3$	$1,7\text{-}C_2B_{10}H_{11}\text{-}9\text{-}I$, $(CH_3)_3SiCH_2MgCl$, $[P(C_6H_5)_3]_2PdCl_2$	b.p. 99 at 2 h Pa	[63]
$1,7\text{-}C_2B_{10}H_{11}\text{-}1\text{-}CH{=}CH(SC_6H_5)$	$1,7\text{-}C_2B_{10}H_{11}\text{-}1\text{-}C{\equiv}CH$, C_6H_5SH, KOH	^1H NMR	[64]
$1,7\text{-}C_2B_{10}H_{11}\text{-}1\text{-}C(Cl){=}CH_2$	$1,7\text{-}C_2B_{10}H_{11}\text{-}1\text{-}C{\equiv}CH$, HCl, $AlCl_3$	^1H NMR	[64]
$1,7\text{-}C_2B_{10}H_{11}\text{-}1\text{-}C(Br){=}C(Br)H$	$1,7\text{-}C_2B_{10}H_{11}\text{-}1\text{-}C{\equiv}CH$, Br_2	^1H NMR	[64]
$1,7\text{-}C_2B_{10}H_{11}\text{-}1\text{-}CH{=}C(I)H$	$1,7\text{-}C_2B_{10}H_{11}\text{-}1\text{-}C{\equiv}CH$, LiI, CH_3COOH	^1H NMR	[64]
$1,7\text{-}C_2B_{10}H_{11}\text{-}1\text{-}CH{=}C{=}CH_2$	$Cu[1,7\text{-}C_2B_{10}H_{11}]$, $HC{\equiv}CCH_2X$ (X = Cl, Br)	b.p. 75 at 1.3 hPa	[65]
$1,7\text{-}C_2B_{10}H_{10}\text{-}1\text{-}CH{=}C{=}CH_2\text{-}7\text{-}CH_3$	$Cu[1,7\text{-}C_2B_{10}H_{11}\text{-}7\text{-}CH_3]$, $HC{\equiv}CCH_2X$ (X = Cl, Br)	b.p. 84 at 1.3 hPa	[65]

References for 13.11.3 on pp. 216, 225/7

Table 13/11 (continued)

compound (product)	reagents used to synthesize compound listed in the first column	properties of product	Ref.
$1,7\text{-}C_2B_{10}H_{11}\text{-}9\text{-}C{\equiv}CH$	$1,7\text{-}C_2B_{10}H_{11}\text{-}9\text{-}C{\equiv}CSi(CH_3)_3$, KOH, CH_3OH	m.p. 120 to 121	[66]
$1,7\text{-}C_2B_{10}H_{11}\text{-}1\text{-}C{\equiv}CCH_3$	$1,7\text{-}C_2B_{10}H_{11}\text{-}1\text{-}CH{=}C{=}CH_2$, $[(CH_3)_3CO]K$	m.p. 121 to 122	[54]
$1,7\text{-}C_2B_{10}H_{10}\text{-}1\text{-}C{\equiv}CCH_3\text{-}7\text{-}CH_3$	$1,7\text{-}C_2B_{10}H_{10}\text{-}1\text{-}CH{=}C{=}CH_2\text{-}7\text{-}CH_3$, $[(CH_3)_3CO]K$	^{1}H NMR	[54]
$1,7\text{-}C_2B_{10}H_{11}\text{-}9\text{-}C{\equiv}CR$	$1,7\text{-}C_2B_{10}H_{11}\text{-}9\text{-}I$, $RC{\equiv}CMgBr$, Pd	b.p. 150 at 1 Torr for R = C_6H_5, m.p. 59 to 60 for R = $Si(CH_3)_3$	[66]
$1,7\text{-}C_2B_{10}H_{11}\text{-}1\text{-}C{\equiv}CCu$	$1,7\text{-}C_2B_{10}H_{11}\text{-}1\text{-}C{\equiv}CH$, $[Cu(NH_3)_2]Cl$, NH_4OH, H_2O, CH_3OH	IR	[67]
$1,7\text{-}C_2B_{10}H_{11}\text{-}1\text{-}C{\equiv}CAg$	$1,7\text{-}C_2B_{10}H_{11}\text{-}1\text{-}C{\equiv}CH$, $[Ag(NH_3)_2]NO_3$, NH_4OH, H_2O, CH_3OH	IR	[67]
$1,7\text{-}C_2B_{10}H_{11}\text{-}1\text{-}C{\equiv}CFe(CO)_2(C_5H_5)$	$1,7\text{-}C_2B_{10}H_{11}\text{-}1\text{-}C{\equiv}CLi$, $(C_5H_5)Fe(CO)_2Br$	IR	[67]
$1,7\text{-}C_2B_{10}H_{11}\text{-}9\text{-}C_6H_5$	$1,7\text{-}C_2B_{10}H_{11}\text{-}9\text{-}I$, C_6H_5MgX (X = halogen), $[(C_6H_5)_3P]_2PdCl_2$, or $[P(C_6H_5)_3]_4Pd$	m.p. 69 to 70	[52]
$1,7\text{-}C_2B_{10}H_{11}\text{-}9\text{-}(C_6H_4\text{-}3'\text{-}F)$	$1,7\text{-}C_2B_{10}H_{11}\text{-}9\text{-}I$, $3\text{-}F\text{-}C_6H_4MgBr$, $[(C_6H_5)_3P]_2PdCl_2$	m.p. 89 to 91	[68]
$1,7\text{-}C_2B_{10}H_{11}\text{-}9\text{-}(C_6H_4\text{-}4'\text{-}F)$	$1,7\text{-}C_2B_{10}H_{11}\text{-}9\text{-}I$, $4\text{-}F\text{-}C_6H_4MgBr$, $[(C_6H_5)_3P]_2PdCl_2$	m.p. 58 to 59	[68]
$1,7\text{-}C_2B_{10}H_9\text{-}1\text{-}C_6H_5\text{-}7\text{-}CH_2N(CH_3)_2\text{-}\mathit{B}\text{-}Mn(CO)_4$	$1,7\text{-}C_2B_{10}H_{10}\text{-}1\text{-}C_6H_5\text{-}7\text{-}CH_2N(CH_3)_2$, $CH_3Mn(CO)_5$	m.p. 148 to 149	[49]
$1,7\text{-}C_2B_{10}H_{11}\text{-}9\text{-}C_6H_4\text{-}3'\text{-}CH_3$	$1,7\text{-}C_2B_{10}H_{11}\text{-}9\text{-}I$, $3\text{-}CH_3\text{-}C_6H_4MgX$ (X = halogen), $[(C_6H_5)_3P]_2PdCl_2$, or $[P(C_6H_5)_3]_4Pd$	m.p. 78 to 79	[52]
$1,7\text{-}C_2B_{10}H_{11}\text{-}9\text{-}C_6H_4\text{-}4'\text{-}CH_3$	$1,7\text{-}C_2B_{10}H_{11}\text{-}9\text{-}I$, $4\text{-}CH_3\text{-}C_6H_5MgX$ (X = halogen), $[(C_6H_5)_3P]_2PdCl_2$, or $[P(C_6H_5)_3]_4Pd$	m.p. 73 to 74	[52]
$1,7\text{-}C_2B_{10}H_{10}\text{-}1,7\text{-}(C_6H_4\text{-}4'\text{-}NO_2)_2$	$1,7\text{-}C_2B_{10}H_{10}\text{-}1,7\text{-}(C_6H_5)_2$, HNO_3, H_2SO_4	m.p. 120 to 147	[69]
$1,7\text{-}C_2B_{10}H_{10}\text{-}1,7\text{-}(C_6H_4\text{-}4'\text{-}NH_2)_2$	$1,7\text{-}C_2B_{10}H_{10}\text{-}1,7\text{-}(C_6H_4\text{-}4'\text{-}NO_2)_2$, Pd, $BaSO_4$, H_2		[69]

$1,7\text{-}C_2B_{10}H_{10}\text{-}1,7\text{-}[C_6H_5Cr(CO)_3]_2$	$Li_2[1,7\text{-}C_2B_{10}H_{10}]$, $ClC_6H_5Cr(CO)_3$		[70]
$1,7\text{-}C_2B_{10}H_{11}\text{-}1\text{-}CHN_2$	$1,7\text{-}C_2B_{10}H_{11}\text{-}1\text{-}CH{=}NNHS(O)_2C_6H_4CH_3$, NaH, 70 to 90°C		[71]
$1,7\text{-}C_2B_{10}H_{11}\text{-}1\text{-}CH{=}NNHS(O)_2C_6H_4CH_3$	$1,7\text{-}C_2B_{10}H_{11}\text{-}1\text{-}CHO$, $CH_3C_6H_4S(O)_2NHNH_2$	m.p. 150 to 154, IR, ^1H NMR	[71]
$1,7\text{-}C_2B_{10}H_{11}\text{-}1\text{-}CHO$	$1,7\text{-}C_2B_{10}H_{11}\text{-}1\text{-}CH{=}CH_2$, O_3	IR	[71]
$1,7\text{-}C_2B_{10}H_{10}\text{-}1,7\text{-}(COCH{=}CHR)_2$	$1,7\text{-}C_2B_{10}H_{10}\text{-}1,7\text{-}(COCH_3)_2$, RCHO, $B(OH)_3$	m.p. 117 to 118 for $R = C_6H_5$, 124 to 125 for $R = C_6H_4\text{-}4'\text{-}OCH_3$, 135 to 136 for $R = C_6H_4\text{-}4'\text{-}F$	[72]
$1,7\text{-}C_2B_{10}H_{11}\text{-}9\text{-}COOH$	$1,7\text{-}C_2B_{10}H_{11}\text{-}9\text{-}R$, CrO_3, H_2SO_4, CH_3COOH ($R = CH_3$, C_2H_5, C_3H_7)	m.p. 220 to 222	[51, 73]
$1,7\text{-}C_2B_{10}H_{10}\text{-}1,7\text{-}(CO_2R)_2$	$1,7\text{-}C_2B_{10}H_{10}\text{-}1,7\text{-}(COCl)_2$, ROH, $N(C_2H_5)_3$	b.p. 142 to 143 at 1 Torr for $R = CH_2CH{=}CH_2$; b.p. 154 to 155 at 1 Torr for $R = CH_2C{\equiv}CH$	[74]
$1,7\text{-}C_2B_{10}H_{11}\text{-}9\text{-}COFe(CO)_2(C_5H_5)$	$1,7\text{-}C_2B_{10}H_{11}\text{-}9\text{-}COOH$, $SOCl_2$, $NaFe(CO)_2(C_5H_5)$	m.p. 114 to 115	[51]
$1,7\text{-}C_2B_{10}H_{11}\text{-}9\text{-}MgCl \cdot C_{12}H_8N_2$	$1,7\text{-}C_2B_{10}H_{11}\text{-}9\text{-}MgCl$, $C_{12}H_8N_2$ ($C_{12}H_8N_2$ = 1,10-phenanthroline)		[75]
$1,7\text{-}C_2B_{10}H_{11}\text{-}9\text{-}Tl(C_6H_5)Br$	$1,7\text{-}C_2B_{10}H_{11}\text{-}9\text{-}Tl(OOCCF_3)_2$, C_6H_5MgBr		[76]
$1,7\text{-}C_2B_{10}H_9\text{-}1,7\text{-}(CH_3)_2\text{-}9\text{-}Tl(C_6H_5)Br$	$1,7\text{-}C_2B_{10}H_9\text{-}1,7\text{-}(CH_3)_2\text{-}9\text{-}Tl(OOCCF_3)_2$, C_6H_5MgBr		[76]
$1,7\text{-}C_2B_{10}H_{11}\text{-}9\text{-}Tl(SCN)_2$	$1,7\text{-}C_2B_{10}H_{11}\text{-}9\text{-}Tl(OOCCF_3)_2$, KSCN	m.p. 186 (with decomposition)	[41]
$1,7\text{-}C_2B_{10}H_{11}\text{-}9\text{-}Tl(C_4H_9)Cl$	$1,7\text{-}C_2B_{10}H_{11}\text{-}9\text{-}Tl(OOCCF_3)_2$, C_4H_9MgCl	m.p. decomposition >150	[41]
$1,7\text{-}C_2B_{10}H_{11}\text{-}9\text{-}Tl(C_2H_5)Br$	$1,7\text{-}C_2B_{10}H_{11}\text{-}9\text{-}Tl(OOCCF_3)_2$, C_2H_5MgBr	m.p. decomposition >160	[41]
$1,7\text{-}C_2B_{10}H_{11}\text{-}9\text{-}Tl(CH_3)I$	$1,7\text{-}C_2B_{10}H_{11}\text{-}9\text{-}Tl(OOCCF_3)_2$, CH_3MgI	m.p. decomposition >215	[41]
$(1,7\text{-}C_2B_{10}H_{11}\text{-}9\text{-})_2Hg[Sn(acetylacetonate)_2]_2$	$(1,7\text{-}C_2B_{10}H_{11}\text{-}9\text{-})_2Hg$, excess $Sn(acetylacetonate)_2$	m.p. 138 to 144	[77]

References for 13.11.3 on pp. 216, 225/7

Table 13/11 (continued)

compound (product)	reagents used to synthesize compound listed in the first column	properties of product	Ref.
1,7-$C_2B_{10}H_{11}$-9-$SnCl_3$	1,7-$C_2B_{10}H_{11}$-9-HgCl, $SnCl_2$	m.p. 133 to 134, ^{1}H NMR	[78]
1,7-$C_2B_{10}H_9$-1,7-$(CH_3)_2$-9-$SnCl_3$	1,7-$C_2B_{10}H_9$-1,7-$(CH_3)_2$-9-HgCl, $SnCl_2$	m.p. 82 to 83, ^{1}H NMR	[78]
1,7-$C_2B_{10}H_{11}$-9-$Fe(CO)_2(C_5H_5)$	1,7-$C_2B_{10}H_{11}$-9-$COFe(CO)_2(C_5H_5)$, 140°C	m.p. 70 to 72	[51]

References for 13.11.3: (continued from p. 216)

[11] Tumanskii, B. L.; Degtyarev, A. N.; Bubnov, N. N.; Solodovnikov, S. P.; Bregadze, V. I.; Godovikov, N. N.; Kabachnik, M. I. (Izv. Akad. Nauk SSSR Ser. Khim. **1980** 2627/30; C.A. **94** [1981] No. 102438).

[12] Zakharkin, L. I.; Kovredov, A. I.; Ol'shevskaya, V. A.; Vitt, S. V. (Izv. Akad. Nauk SSSR Ser. Khim. **1983** 1680; Bull. Acad. Sci. USSR Div. Chem. Sci. **32** [1983] 1528/9).

[13] Shul'tse, P.; Varushchenko, R. M.; Gal'chenko, G. L.; Klimova, T. V.; Stanko, V. I. (Zh. Obshch. Khim. **50** [1980] 1818/25; J. Gen. Chem. [USSR] **50** [1980] 1482/8).

[14] Kalinin, V. N.; Kobel'kova, N. I.; Zakharkin, L. I. (Izv. Akad. Nauk SSSR Ser. Khim. **1982** 1661/2; Bull. Acad. Sci. USSR Div. Chem. Sci. **31** [1982] 1479/81).

[15] Vinogradova, L. E.; Leites, L. A.; Kovredov, A. I.; Ol'shevskaya, V. A.; Zakharkin, L. I. (Izv. Akad. Nauk SSSR Ser. Khim. **1982** 1663/4; Bull. Acad. Sci. USSR Div. Chem. Sci. **32** [1983] 1481/2).

[16] Basato, M.; Morandini, F.; Longato, B.; Bresadola, S. (Inorg. Chem. **23** [1984] 649/53).

[17] Kalinin, V. N.; Astakhin, A. V.; Kazantsev, A. V.; Zakharkin, L. I, (Zh. Obshch. Khim. **52** [1982] 1932/3; J. Gen. Chem. [USSR] **52** [1982] 1714/5).

[18] Chistov, S. F.; Sobolev, E. S.; Skorokhodov, I. I.; Larikov, E. I. (Zh. Obshch. Khim. **53** [1983] 733/5; J. Gen. Chem. [USSR] **53** [1983] 638/40).

[19] Yanovskii, A. I.; Struchkov, Yu. T.; Kalinin, V. N.; Usatov, A. V.; Zakharkin, L. I. (Koord. Khim. **8** [1982] 1700/4; C.A. **98** [1983] No. 63625).

[20] Kalinin, V. N.; Izmailov, B. A.; Kazantsev, A. A.; Myakushev, V. D.; Zhdanov, A. A.; Zakharkin, L. I. (J. Organometal. Chem. **216** [1981] 295/320).

[21] Yanovskii, A. I.; Dubchak, I. L.; Shklover, V. E.; Struchkov, Yu. T.; Kalinin, V. N.; Izmailov, B. A.; Myakushev, V. D.; Zakharkin, L. I. (Zh. Strukt. Khim. **23** No. 5 [1982] 88/97; J. Struct. Chem. [USSR] **23** [1982] 728/36).

[22] Rabilloud, G.; Sillion, B. (J. Chem. Res. S **1981** 264/5).

[23] Korol'ko, V. V.; Vecherskaya, V. I.; Saratovkina, T. I. (Zh. Obshch. Khim. **50** [1980] 1580/3; J. Gen. Chem. [USSR] **50** [1980] 1284/7).

[24] Grishin, Yu. K.; Roznyatovskii, V. A.; Ustynyuk, Yu. A.; Kampel, V. Ts.; Bregadze, V. I. (Vestn. Mosk. Univ. Khim. **37** [1982] 488/91; Moscow Univ. Chem. Bull. **37** No. 5 [1982] 79/83; C.A. **98** [1983] No. 89540).

[25] Leites, L. A.; Vinogradova, L. E.; Bukalov, S. S.; Kampel, V. Ts.; Bregadze, V. I. (Izv. Akad. Nauk SSSR Ser. Khim. **1981** 2035/43; Bull. Acad. Sci. USSR Div. Chem. Sci. **30** [1981] 1670/6).

[26] Bochkarev, M. N.; Fedorova, E. A.; Razuvaev, G. A.; Bregadze, V. I.; Kampel, V. Ts. (J. Organometal. Chem. **265** [1984] 117/22).

[27] Yanovskii, A. I.; Antipin, M. Yu.; Struchkov, Yu. T.; Bregadze, V. I.; Usyatinskii, A. Ya.; Godovikov, N. N. (Izv. Akad. Nauk SSSR Ser. Khim. **1982** 293/8; Bull. Acad. Sci. USSR Div. Chem. Sci. **31** [1982] 266/70).

[28] Bregadze, V. I.; Usyatinskii, A. Ya.; Suleimanov, G. Z.; Godovnikov, N. N. (Izv. Akad. Nauk SSSR Ser. Khim. **1981** 1927/8; C.A. **95** [1981] No. 220084).

[29] Usyatinskii, A. Ya.; Bregadze, V. I.; Shcherbina, T. M.; Godovnikov, N. N. (Izv. Akad. Nauk SSSR Ser. Khim. **1981** 1428/31; Bull. Acad. Sci. USSR Div. Chem. Sci. **30** [1981] 1148/50).

[30] Kasymbekova, Z. K.; Prokof'ev, A. I.; Bubnov, N. N.; Solodovnikov, S. P.; Bregadze, V. I.; Kampel, V. Ts.; Petriashvili, M. V.; Godovnikov, N. N.; Kabachnik, M. I. (Izv. Akad. Nauk SSSR Ser. Khim. **1983** 316/24; Bull. Acad. Sci. USSR Div. Chem. Sci. **32** [1983] 282/9).

[31] Kovredov, A. I.; Meiramov, M. G.; Kazantsev, A. V.; Zakharkin, L. I. (Zh. Obshch. Khim. **51** [1981] 854/9; J. Gen. Chem. [USSR] **51** [1981] 708/12).

[32] Longato, B.; Bresadola, S. (Inorg. Chem. **21** [1982] 168/73).
[33] Morandini, F.; Longato, B.; Bresadola, S. (J. Organometal. Chem. **239** [1982] 377/84).
[34] Longato, B.; Morandini, F.; Bresadola, S. (Inorg. Chim. Acta **39** [1980] 27/34).
[35] Basato, M.; Longato, B.; Morandini, F.; Bresadola, S. (Inorg. Chem. **23** [1984] 3972/6).
[36] Zakharkin, L. I.; Pisareva, I. V. (Izv. Akad. Nauk SSSR Ser. Khim. **1983** 1158/61; Bull. Acad. Sci. USSR Div. Chem. Sci. **32** [1983] 1046/8).
[37] Plešek, J.; Heřmánek, S. (Collection Czech. Chem. Commun. **46** [1981] 687/92).
[38] Grushin, V. V.; Tolstaya, T. P.; Lisichkina, I. N.; Grishin, Yu. K.; Shcherbina, T. M.; Kampel, V. Ts.; Bregadze, V. I.; Godovikov, N. N. (Izv. Akad. Nauk SSSR Ser. Khim. **1983** 472/4; Bull. Acad. Sci. USSR Div. Chem. Sci. **32** [1983] 429/31).
[39] Merkushev, E. B.; Simakhina, N. D.; Grigor'ev, M. G. (Izv. Akad. Nauk SSSR Ser. Khim. **1980** 2649; C.A. **94** [1981] No. 103456).
[40] Grushin, V. V.; Tolstaya, T. P.; Lisichkina, I. N. (Dokl. Akad. Nauk SSSR **261** [1981] 99/102; Dokl. Chem. Proc. Acad. Sci. USSR **256/261** [1981] 456/9).

[41] Bregadze, V. I.; Usyatinskii, A. Ya.; Godovikov, N. N. (Izv. Akad. Nauk SSSR Ser. Khim. **1981** 398/401; Bull. Acad. Sci. USSR Div. Chem. Sci. **30** [1981] 315/8).
[42] Grushin, V. V.; Tolstaya, T. P.; Vanchikov, A. N. (Dokl. Akad. Nauk SSSR **264** [1982] 868/72; Dokl. Chem. Proc. Acad. Sci. USSR **262/267** [1982] 163/7).
[43] Zakharkin, L. I.; Pisareva, I. V. (J. Organometal. Chem. **267** [1984] 73/9).
[44] Zakharkin, L. I.; Pisareva, I. V. (Izv. Akad. Nauk SSSR Ser. Khim. **1981** 2794/6; Bull. Acad. Sci. USSR Div. Chem. Sci. **30** [1981] 2328/30).
[45] Zakharkin, L. I.; Pisareva, I. V. (Izv. Akad. Nauk SSSR Ser. Khim. **1984** 396/402; Bull. Acad. Sci. USSR Div. Chem. Sci. **1984** 355/61).
[46] Zakharkin, L. I.; Pisareva, I. V. (Izv. Akad. Nauk SSSR Ser. Khim. **1982** 718/9; Bull. Acad. Sci. USSR Div. Chem. Sci. **31** [1982] 644).
[47] Zakharkin, L. I.; Pisareva, I. V. (Izv. Akad. Nauk SSSR Ser. Khim. **1984** 472/3; Bull. Acad. Sci. USSR Div. Chem. Sci. **1984** 438).
[48] Kalinin, V. N.; Kobel'kova, N. I.; Krasnokutskaya, E. V.; Zakharkin, L. I. (Izv. Akad. Nauk SSSR Ser. Khim. **1983** 1200/2; Bull. Acad. Sci. USSR Div. Chem. Sci. **32** [1983] 1084/6).
[49] Kalinin, V. N.; Usatov, A. V.; Popello, I. A.; Zakharkin, L. I. (Izv. Akad. Nauk SSSR Ser. Khim. **1982** 1433; Bull. Acad. Sci. USSR Div. Chem. Sci. **31** [1982] 1281).
[50] Zakharkin, L. I.; Pisareva, I. V. (Zh. Obshch. Khim. **51** [1981] 1280/8; J. Gen. Chem. [USSR] **51** [1981] 1084/91).

[51] Zakharkin, L. I.; Kovredov, A. I.; Ol'shevskaya, V. A.; Antonovich, V. A. (J. Organometal. Chem. **267** [1984] 81/91).
[52] Zakharkin, L. I.; Kovredov, A. I.; Ol'shevskaya, V. A.; Shaugumbekova, Z. S. (J. Organometal. Chem. **226** [1982] 217/22).
[53] Zakharkin, L. I.; Kovredov, A. I.; Ol'shevskaya, V. A.; Shaugumbekova, Z. S. (Izv. Akad. Nauk SSSR Ser. Khim. **1980** 1691; C.A. **94** [1981] No. 65740).
[54] Zakharkin, L. I.; Kovredov, A. I.; Shaugumbekova, Z. S.; Vinogradova, L. E.; Leites, L. A. (Zh. Obshch. Khim. **51** [1981] 1575/82; J. Gen. Chem. [USSR] **51** [1981] 1337/42).
[55] Zakharkin, L. I. Kovredov, A. I.; Kazantsev, A. V.; Meiramov, M. G. (Zh. Obshch. Khim. **51** [1981] 357/61; J. Gen. Chem. [USSR] **51** [1981] 289/92).
[56] Churkina, L. A.; Zvereva, T. D.; Shingel, I. A.; Ol'dekop, Yu. A. (Vestsi Akad. Navuk Belarusk.SSR Ser. Khim. Navuk **1982** 59/66; C.A. **97** [1982] No. 6359).
[57] Churkina, L. A.; Zvereva, T. D.; Shingel, I. A.; Ol'dekop, Yu. A. (Vestsi Akad. Navuk Belarusk.SSR Ser. Khim. Navuk **1983** 56/64; C.A. **98** [1983] No. 160774).
[58] Drygina, O. V.; Dorofeenko, G. N.; Okhlobystin, O. Yu. (Zh. Obshch. Khim. **51** [1981] 868/75; J. Gen. Chem. [USSR] **51** [1981] 720/6).

[59] Sosin, S. L.; Alekseeva, V. P.; Litvinova, M. D.; Ezhova, T. M.; Shevchenko, Yu. V. (Dokl. Akad. Nauk SSSR **271** [1983] 871/4; Dokl. Chem. Proc. Acad. Sci. USSR **268/273** [1983] 249/52).
[60] Zakharkin, L. I.; Kalinin, V. N.; Shitikov, V. K.; Sergeev, V. A. (Zh. Obshch. Khim. **51** [1981] 593/5; J. Gen. Chem. [USSR] **51** [1981] 468/70).

[61] Kalinin, V. N.; Izmailov, B. A.; Kazantsev, A. A.; Zhdanov, A. A.; Zakharkin, L. I. (Zh. Obshch. Khim. **54** [1984] 1208/9; J. Gen. Chem. [USSR] **54** [1984] 1082).
[62] Vdovin, V. M.; Bespalova, N. B.; Kalinin, V. N.; Popov, A. V.; Zakharkin, L. I. (Zh. Obshch. Khim. **54** [1984] 1197/8; J. Gen. Chem. [USSR] **54** [1984] 1072/3).
[63] Zakharkin, L. I.; Kovredov, A. I.; Savel'eva, I. S.; Safronova, E. V. (Zh. Obshch. Khim. **51** [1981] 2383; J. Gen. Chem. [USSR] **51** [1981] 2056).
[64] Zakharkin, L. I.; Kovredov, A. I.; Ol'shevskaya, V. A. (Zh. Obshch. Khim. **52** [1982] 1911/8; J. Gen. Chem. [USSR] **52** [1982] 1694/700).
[65] Zakharkin, L. I.; Kovredov, A. I.; Shaugumbekova, Z. S.; Vinogradova, L. E.; Leites, L. A. (Zh. Obshch. Khim. **51** [1981] 1582/6; J. Gen. Chem. [USSR] **51** [1981] 1343/6).
[66] Zakharkin, L. I.; Kovredov, A. I.; Ol'shevskaya, V. A. (Zh. Obshch. Khim. **51** [1981] 2807/8; J. Gen. Chem. [USSR] **51** [1981] 2056).
[67] Zakharkin, L. I.; Kovredov, A. I.; Ol'shevskaya, V. A. (Izv. Akad. Nauk SSSR Ser. Khim. **1982** 673/5; Bull. Acad. Sci. USSR Div. Chem. Sci. **31** [1982] 599/602).
[68] Zakharkin, L. I.; Kovredov, A. I.; Ol'shevskaya, V. A. (Izv. Akad. Nauk SSSR Ser. Khim. **1981** 2159/61; Bull. Acad. Sci. USSR Div. Chem. Sci. **30** [1981] 1775/7).
[69] Sergeev, V. A.; Kalinin, V. N.; Shitikov, V. K.; Svetogorov, Y. E.; Chizhova, N. V.; Danilova, M. P.; Chimishkin, A. L.; Zakharkin, L. I. (Zh. Obshch. Khim. **51** [1981] 863/8; J. Gen. Chem. [USSR] **51** [1981] 716/20).
[70] Zakharkin, L. I.; Zhigareva, G. G. (Zh. Obshch. Khim. **53** [1983] 953/4; J. Gen. Chem. [USSR] **53** [1983] 841/2).

[71] Chari, S. L.; Chiang, S. H.; Jones, M. (J. Am. Chem. Soc. **104** [1982] 3138/45).
[72] Kazantsev, A. V.; Ibraev, M. I. (Zh. Obshch. Khim. **51** [1981] 2501/3; J. Gen. Chem. [USSR] **51** [1981] 2155/7).
[73] Zakharkin, L. I.; Kovredov, A. I.; Ol'shevskaya, V. A. (Zh. Obshch. Khim. **53** [1983] 1431/2; J. Gen. Chem. [USSR] **53** [1983] 1287).
[74] Korshak, V. V.; Bekasova, N. I.; Solomatina, A. I.; Frunze, T. M.; Sakharova, A. A.; Mel'nik, O. A. (Izv. Akad. Nauk SSSR Ser. Khim. **1982** 1904; Bull. Acad. Sci. USSR Div. Chem. Sci. **31** [1982] 1694/5).
[75] Zakharkin, L. I.; Pisareva, I. V.; Vasil'eva, N. S. (Zh. Obshch. Khim. **52** [1982] 711; J. Gen. Chem. [USSR] **52** [1982] 618).
[76] Usyatinskii, A. Ya.; Ryabov, A. D.; Bregadze, V. I.; Shcherbina, T. M.; Godovikov, N. N. (Izv. Akad. Nauk SSSR Ser. Khim. **1982** 1598/603; Bull. Acad. Sci. USSR Div. Chem. Sci. **31** [1982] 1420/4).
[77] Suleimanov, G. Z.; Bregadze, V. I.; Kampel, V. Ts.; Petriashvili, M. V.; Godovikov, N. N.; Sokolov, V. I. (Izv. Akad. Nauk SSSR Ser. Khim. **1982** 1606/7; Bull. Acad. Sci. USSR Div. Chem. Sci. **31** [1982] 1427/8).
[78] Bregadze, V. I.; Dzhashiashvili, T. K.; Sadzhaya, D. N.; Petriashvili, M. V.; Ponomareva, O. B.; Shcherbina, T. M.; Kampel, V. Ts.; Kukushkina, L. B.; Rochev, V. Ya.; Godovikov, N. N. (Izv. Akad. Nauk SSSR Ser. Khim. **32** [1983] 907/12; Bull. Acad. Sci. USSR Div. Chem. Sci. **1983** 824/7).

13.11.4 1,12-$C_2B_{10}H_{12}$ and Derivatives

Short-lived absorption spectra of a methanolic solution of **1,12-$C_2B_{10}H_{12}$**, under the action of a pulse of accelerated electrons, have been recorded over the whole spectral range from the UV to the IR spectra [1]. The disappearance of the Raman spectrum background for solid 1,12-$C_2B_{10}H_{12}$, just before melting, is accompanied by a reduction in the intensity of Rayleigh scattering [2]. For the boron atoms in 1,12-$C_2B_{10}H_{12}$ a cage "umbrella" angle of 116° is used in an empirical relationship, which includes the number of adjacent cage carbon atoms, to predict a $^1J(^{11}B$-$^1H)$ of 165 Hz which is exactly the measured value [3]. The ^{11}B quadrupolar coupling constant for *closo*-1,12-$C_2B_{10}H_{12}$ is found to be 887 kHz [4]. Two phase transitions were observed for 1,12-$C_2B_{10}H_{12}$ at 303 and 240 K by calorimetry, relaxation data, and ^{11}B NMR line shape analysis [5]. Aromatic solvent-induced ^1H NMR shifts are reported for 1,12-$C_2B_{10}H_{12}$ and correlated to PRDDO MO derived hydrogen charges [6]. The He(I) photoelectron spectrum of 1,12-$C_2B_{10}H_{12}$ exhibits bands at (in eV) 10.6, 11.2, 12.2, 13.6, 15.9, and 17.1 [7]. MNDO MO calculations on 1,12-$C_2B_{10}H_{12}$ as well as other carboranes are used to assess the accuracy of A. J. Stone's theory (1980) of their electronic structure and bonding [8]. The protonation of the 1,12-$C_2B_{10}H_{12}$ has been studied by the MNDO method; the calculated proton affinity for the compound is 119.3 kcal/mol and the calculations predict B-B-B face protonation [9].

A rapid high-yield conversion of 1,12-$C_2B_{10}H_{12}$ to [K(18-crown-6)][*nido*-2,9-$C_2B_9H_{12}$] in the presence of 18-crown-6 ether and KOH is presently the best method to prepare the **[*nido*-2,9-$C_2B_9H_{12}]^-$** ion [10].

[1,12-$C_2B_{10}H_{11}$-2-IC_6H_5][BF_4] is prepared in 80% yield by the reaction of 1,12-$C_2B_{10}H_{11}$-2-HgI with C_6H_6, $(CH_3CO)_2O$, H_2SO_4, and $K_2S_2O_8$. Reaction of [1,12-$C_2B_{10}H_{11}$-2-IC_6H_5][BF_4] with aqueous NaF produces 1,12-$C_2B_{10}H_{11}$-2-F, melting point 256°C. Reaction of [1,12-$C_2B_{10}H_{11}$-2-IC_6H_5][BF_4] with aqueous $NaNO_2$ produces 1,12-$C_2B_{10}H_{11}$-2-OH, melting point 301°C [11]. An electron-diffraction study of the structure of 1,12-$C_2B_{10}H_{10}$-1,12-I_2 shows the cage to have the expected near-icosahedral geometry with the following bond distances (in Å): 1.706(10) for B-C, 1.779(8) for B(2)-B(3), 1.778(13) for B(2)-B(7), 1.210(23) for B-H, 2.095(15) for C-I [12].

ESR data indicate that the 1,12-$C_2B_{10}H_{11}$-1-$P(CH_3)(OC_2H_5)[OC(CH_3)_3]$ radical has a trigonal bipyramid structure about the phosphorus atom with the carborane ring in the apical position [13]. An electron diffraction study of the molecular structure of 1,12-$C_2B_{10}H_{10}$-1,12-$(CH_3)_2$ shows the compound to have the expected icosahedral carbon-boron framework with the following bond distances (in Å): 1.716(13) for B-C, 1.777(7) for B(2)-B(3), 1.766(20) for B(2)-B(7), 1.533(19) for C-C, 1.216(19) for B-H, 1.088(33) for C-H [14]. Infrared and Raman spectra of mercury derivatives of 1,12-$C_2B_{10}H_{12}$ are reported [15].

In addition to those compounds mentioned above in this section, a supplemental list of 1,12-$C_2B_{10}H_{12}$ derivatives, along with reagents used to synthesize each one, and physical properties of each product, is given in Table 13/12.

Table 13/12

$1,12\text{-}C_2B_{10}H_{12}$ Derivatives; Synthesis and Physical Properties (m.p. = melting point in °C; b.p. = boiling point in °C; IR = infrared spectrum reported; NMR = nuclear magnetic resonance spectrum reported; MS = mass spectrum reported).

compound (product)	reagents used to synthesize compound listed in the first column	properties of product	Ref.
$1,12\text{-}C_2B_{10}H_{11}\text{-}1\text{-}SH$	$1,12\text{-}C_2B_{10}H_{12}$, NaH, S	^1H, ^{11}B NMR	[16]
$1,12\text{-}C_2B_{10}H_{11}\text{-}2\text{-}CH_3$	$1,12\text{-}C_2B_{10}H_{11}\text{-}2\text{-}I$, CH_3MgBr, $[P(C_6H_5)_3]_2PdCl_2$	m.p. 139 to 140	[17]
$1,12\text{-}C_2B_{10}H_{10}\text{-}1,12\text{-}(CH_2C_6H_4\text{-}4'\text{-}R)_2$	$M_2[1,12\text{-}C_2B_{10}H_{10}]$, $p\text{-}R\text{-}C_6H_4CH_2X$ (M = Li, Na; R = CH_3 when X = Br; R = CH_3O when X = Cl)	m.p. 167 to 168, R = CH_3; 140 to 141, R = CH_3O	[18]
$1,12\text{-}C_2B_{10}H_{10}\text{-}1,12\text{-}(CH_2C_6H_4\text{-}4'\text{-}COOH)_2$	$1,12\text{-}C_2B_{10}H_{10}\text{-}1,12\text{-}(CH_2C_6H_4\text{-}4'\text{-}CH_3)_2$, CrO_3, CH_3COOH, H_2SO_4	m.p. 390 to 395	[18]
$1,12\text{-}C_2B_{10}H_{10}\text{-}1,12\text{-}(CH_2C_6H_4\text{-}4'\text{-}OH)_2$	$1,12\text{-}C_2B_{10}H_{10}\text{-}1,12\text{-}(CH_2C_6H_4\text{-}4'\text{-}OCH_3)_2$, HI, 100°C	m.p. 285 to 286	[18]
$1,12\text{-}C_2B_{10}H_{11}\text{-}2\text{-}C_6H_4\text{-}3'\text{-}F$	$1,12\text{-}C_2B_{10}H_{11}\text{-}2\text{-}I$, $3\text{-}F\text{-}C_6H_4MgBr$, $[(C_6H_5)_3P]_2PdCl_2$, or $[(C_6H_5)_3P]_4Pd$	b.p. 120 at 1 Torr	[19]
$1,12\text{-}C_2B_{10}H_{11}\text{-}2\text{-}C_6H_4\text{-}4'\text{-}F$	$1,12\text{-}C_2B_{10}H_{11}\text{-}2\text{-}I$, $4\text{-}F\text{-}C_6H_4MgBr$, $[(C_6H_5)_3P]_2PdCl_2$, or $[(C_6H_5)_3P]_4Pd$	m.p. 50 to 51	[19, 20]
$1,12\text{-}C_2B_{10}H_{11}\text{-}2\text{-}COOH$	$1,12\text{-}C_2B_{10}H_{11}\text{-}2\text{-}CH_3$, CrO_3, H_2SO_4, CH_3COOH	m.p. 130 to 131	[17, 21]
$1,12\text{-}C_2B_{10}H_{10}\text{-}1,12\text{-}(COOCH_2CH{=}CH_2)_2$	$1,12\text{-}C_2B_{10}H_{10}\text{-}1,12\text{-}(COCl)_2$, $H_2C{=}CHCH_2OH$, $N(C_2H_5)_3$	m.p. 39 to 39.5	[22]
$1,12\text{-}C_2B_{10}H_{11}\text{-}2\text{-}Tl(OOCCF_3)_2$	$1,12\text{-}C_2B_{10}H_{11}\text{-}2\text{-}HgOOCCF_3$, $Tl(OOCCF_3)_3$	m.p. 193 to 194	[23]
$1,12\text{-}C_2B_{10}H_{11}\text{-}2\text{-}TlX_2$	$1,12\text{-}C_2B_{10}H_{11}\text{-}2\text{-}Tl(OOCCF_3)_2$, NaX (X = Cl, Br, SCN)		[23]

References for 13.11.4 on p. 230

References for 13.11.4:

[1] Zimina, G. M.; Vannikov, A. V.; Stanko, V. I. (Khim. Vysokikh Energ. **16** [1982] 335/9; High Energy Chem. [USSR] **16** [1982] 263/6; C.A. **97** [1982] No. 101576).
[2] Bukalov, S. S.; Leites, L. A. (Opt. Spektrosk. **56** [1984] 10/2; Opt. Spectrosc. [USSR] **56** [1984] 6/7).
[3] Jarvis, W.; Abdou, Z. J.; Onak, T. (Polyhedron **2** [1983] 1067/70).
[4] Leffler, A. J. (J. Chem. Phys. **81** [1984] 2574/6).
[5] Blumenfel'd, A. L.; Fedin, E. I. (Magn. Resonance Relat. Phenom. Proc. Congr. 20th AMPERE, Tallinn 1978 [1979], p. 454; C.A. **93** [1980] No. 185562).
[6] Jarvis, W.; Inman, W.; Powell, B.; DiStefano, E. W.; Onak, T. (J. Magn. Resonance **43** [1981] 302/15).
[7] Fehlner, T. P.; Wu, M.; Meneghelli, B. J.; Rudolph, R. W. (Inorg. Chem. **19** [1980] 49/54).
[8] Brint, P.; Cronin, J. P.; Seward, E.; Whelan, T. (J. Chem. Soc. Dalton Trans. **1983** 975/80).
[9] DeKock, R. L.; Jasperse, C. P. (Inorg. Chem. **22** [1983] 3843/8).
[10] Busby, D. C.; Hawthorne, M. F. (Inorg. Chem. **21** [1982] 4101/3).

[11] Grushin, V. V.; Tolstaya, T. P.; Lisichkina, I. N. (Izv. Akad. Nauk SSSR Ser. Khim. **1983** 2165/8; Bull. Acad. Sci. USSR Div. Chem. Sci. **1983** 1957/60).
[12] Dorofeeva, O. V.; Mastryukov, V. S.; Golubinskii, A. V.; Vilkov, L. V.; Almenningen, A. (Zh. Strukt. Khim. **22** No. 5 [1981] 51/6; J. Struct. Chem. [USSR] **22** [1981] 682/6).
[13] Tumanskii, B. L.; Degtyarev, A. N.; Bubnov, N. N.; Solodovnikov, S. P.; Bregadze, V. I.; Godovikov, N. N.; Kabachnik, M. I. (Izv. Akad. Nauk SSSR Ser. Khim. **1980** 2627/30; C.A. **94** [1981] No. 102438).
[14] Mastryukov, V. S.; Atavin, E. G.; Golubinskii, A. V.; Vilkov, L. V.; Stanko, V. I.; Gol'tyapin, Yu. V. (Zh. Strukt. Khim. **23** No. 1 [1982] 51/5; J. Struct. Chem. [USSR] **23** [1982] 41/5).
[15] Leites, L. A.; Vinogradova, L. E.; Bukalov, S. S.; Kampel, V. Ts.; Bregadze, V. I. (Izv. Akad. Nauk SSSR Ser. Khim. **1981** 2035/43; Bull. Acad. Sci. USSR Div. Chem. Sci. **30** [1981] 1670/6).
[16] Plešek, J.; Heřmánek, S. (Collection Czech. Chem. Commun. **46** [1981] 687/92).
[17] Zakharkin, L. I.; Kovredov, A. I.; Ol'shevskaya, V. A.; Antonovich, V. A. (J. Organometal. Chem. **267** [1984] 81/91).
[18] Zakharkin, L. I.; Kalinin, V. N.; Shitikov, V. K.; Sergeev, V. A. (Zh. Obshch. Khim. **51** [1981] 593/5; J. Gen. Chem. [USSR] **51** [1981] 468/70).
[19] Zakharkin, L. I.; Kovredov, A. I.; Ol'shevskaya, V. A.; Shaugumbekova, Z. S. (J. Organometal. Chem. **226** [1982] 217/22).
[20] Zakharkin, L. I.; Kovredov, A. I.; Ol'shevskaya, V. A. (Izv. Akad. Nauk SSSR Ser. Khim. **1981** 2159/61; Bull. Acad. Sci. USSR Div. Chem. Sci. **30** [1981] 1775/7).

[21] Zakharkin, L. I.; Kovredov, A. I.; Ol'shevskaya, V. A. (Zh. Obshch. Khim. **53** [1983] 1431/2; J. Gen. Chem. [USSR] **53** [1983] 1287).
[22] Korshak, V. V.; Bekasova, N. I.; Solomatina, A. I.; Frunze, T. M.; Sakharova, A. A.; Mel'nik, O. A. (Izv. Akad. Nauk SSSR Ser. Khim. **1982** 1904; Bull. Acad. Sci. USSR Div. Chem. Sci. **31** [1982] 1694/5).
[23] Bregadze, V. I.; Usyatinskii, A. Ya.; Godovikov, N. N. (Izv. Akad. Nauk SSSR Ser. Khim. **1983** 1405/7; Bull. Acad. Sci. USSR Div. Chem. Sci. **32** [1983] 1274/6).

13.11.5 Polymers Containing $C_2B_{10}H_x$ Groups

A polymer with repeating units of -$SC_2B_{10}H_{10}SCH_2C_6H_4CH_2$-, where the $C_2B_{10}H_{10}$ unit contains adjacent carbons in the cage and where the benzene ring is *ortho* substituted, is prepared by the action of $K_2[1,2\text{-}C_2B_{10}H_{10}\text{-}1,2\text{-}(S)_2]$ with C_6H_4-1,2-$(CH_2Br)_2$ [1]. Polymerization of the carborane epithiopropane, 1,2-$C_2B_{10}H_{10}$-1-C_6H_5-2-CH_2CHCH_2S by $(C_2H_5)_2O \cdot BF_3$ gives a carborane-substituted polythioethylene [2]. The addition of 3% finely dispersed red phosphorus to films of *p,p'*-biphenylenediamine-*m*-carboranedicarboxylic acid polyamide (*m*-carborane = 1,7-$C_2B_{10}H_{10}$ derivative) leads to the formation of thermally stable structures at relatively low temperatures [3].

Liquid polymers with hydroxyl or carboxyl end groups are obtained by copolymerization of 1,2-$C_2B_{10}H_{11}$-1-CH=CH_2, 1,2-$C_2B_{10}H_{11}$-1-C(CH_3)=CH_2, and 1,2-$C_2B_{10}H_{11}$-1-CH_2OC(O)CH=CH_2, with butadiene, isoprene, or chloroprene in dioxane and with 4,4'-azobis(4-cyano-1-pentanol) or 4,4'-azobis(4-cyanovaleric acid) as initiator [4]. A study of the polymerization of 1,2-$C_2B_{10}H_{11}$-1-(C_6H_4-4'-CH=CH_2) under various conditions shows that the compound can undergo polymerization by thermal, radical, or anionic mechanisms but not by a cationic mechanism [5].

Thermal degradation studies have been carried out on polyesters derived from bishydroxymethyl compounds of 1,2- or 1,7-$C_2B_{10}H_{12}$ [6]. Polymers derived from an N,N'-bis(hydroxymethyl)aminophenyl derivative of 1,2-$C_2B_{10}H_{12}$ have been reported [7].

$[N(C_2H_5)_4]OH$ degrades the carborane cages in polyamines which contain 1,7-$C_2B_{10}H_x$ moieties in the polymeric chain; the product is a polyamine which contains $C_2B_9H_x$ units [8]. Polyisocyanurates containing 1,2-C_2B_{10} carborane units have been studied with regard to high thermal and thermal oxidative stability [9].

Polyphenylene copolymers containing mono- or bis(4-acetylbenzyl) derivatives of $C_2B_{10}H_{12}$ were studied [10].

NMR relaxation processes were investigated in polyesters prepared from carboxyphenyl derivatives of 1,2- and 1,7-$C_2B_{10}H_{12}$ [11]. The temperature dependency of the dielectric constant of polyarylates containing 1,2-$C_2B_{10}H_{10}$-1,2- units in the chain has been examined [12]. The spin-spin and spin-lattice relaxation times for polyarylates consisting of dihydroxydiphenyl ether/1,7-$C_2B_{10}H_{10}$-1,7- copolymers, both in solution and in the solid state, have been studied over a range of temperatures [12]. Molecular motions in polyesters containing carborane groups have been studied [13]. Carborane-containing cellulose esters are obtained by acylation of cellulose with 1,*x*-$C_2B_{10}H_{11}$-1-COCl (*x* = 2, 7, 12) [14].

Structural changes in carborane-containing C_6H_5OH/CH_2O resins upon thermal treatment at 200 to 1200°C in vacuum and air have been examined by X-ray diffraction [15].

1,2-$C_2B_{10}H_{10}$-1,2-$(CH_2OH)_2$ reacts with organospirobicyclosilazasiloxanes to give siloxane polymers [16]. Oligomeric carboranylmethoxy polyorganosilazanes have been prepared by the amminolysis of 1,2-$C_2B_{10}H_{11}$-1-CH_2SiCl_2(OR) (R = CH_3, C_6H_5, CH=CH_2) [17]. Carboranylmethyl polyorganosilazanes were prepared via ammonolysis of 1,2-$C_2B_{10}H_{11}$-1-CH_2SiR(OR')Cl (R = Cl, CH_3; R' = CH_3, C_2H_5) or coammonolysis of 1,2-$C_2B_{10}H_{11}$-1-CH_2SiR(OR')Cl with 1,2-$C_2B_{10}H_{11}$-1-CH_2SiRCl_2 or with R"$SiCl_3$ (R" = CH_3, C_2H_5) [18]. The hydrolytic polycondensation of $C_2B_{10}H_{11}$-CH_2SiRR'Cl (R, R' = CH_3, Cl) occurs with the formation of oligomers with a cyclic silasesquioxane structure [19]. Carborane exocyclic vinylsilanes of the type 1,2-$C_2B_{10}H_{10}$-1,2-(-CH_2O-Si(R)(CH=CH_2)OCH_2-) (R = CH_3, OCH_3, Cl) have been homopolymerized and copolymerized with methyl methacrylate [20]. Hydrolytic copolycondensation of 1,2-$C_2B_{10}H_{11}$-1-$CH_2Si(CH_3)_2$-OCH_3 and 1,2-$C_2B_{10}H_{10}$-1,2-$[CH_2Si(CH_3)_2OCH_3]_2$ with $ClSi(CH_3)_3$ or $(ClSi(CH_3)_2)_2O$ gives the corresponding siloxanes containing bis(dimethylsilanylmethyl)carborane or $Si(CH_3)_2O$ units and (carboranylmethyl)dimethylsilanyl or $Si(CH_3)_3$ end groups [21]. Similarly, hydrolytic copoly-

References for 13.11.5 on pp. 239/41

condensation of 1,7-$C_2B_{10}H_{11}$-1-$CH_2Si(CH_3)_2OCH_3$ and 1,7-$C_2B_{10}H_{10}$-1,7-$[CH_2Si(CH_3)_2OCH_3]_2$ with $ClSi(CH_3)_3$ or $(ClSi(CH_3)_2)_2O$ gives the corresponding siloxanes containing bis(dimethylsilanylmethyl)carborane or $Si(CH_3)_2O$ units and (carboranylmethyl)dimethylsilanyl or $Si(CH_3)_3$ end groups [21]. Hydrolytic copolycondensation of 1,2-$C_2B_{10}H_{11}$-1-$CH_2Si(CH_3)(OCH_3)_2$, 1,2-$C_2B_{10}H_{11}$-1-$CH_2Si(OCH_3)_3$, and 1,2-$C_2B_{10}H_{11}$-1-$CH_2Si(OCH_2CH_3)_3$ with $ClSi(CH_3)_3$ gives the corresponding linear siloxanes containing (carboranylmethyl)methylsiloxane or (carboranylmethyl)-silasesquioxane units and $Si(CH_3)_3$, HOSi, or CH_3CH_2O end groups. Hydrolytic copolycondensation of 1,2-$C_2B_{10}H_{11}$-1-$CH_2Si(CH_3)(OCH_3)_2$ with $Cl_2Si(CH_3)_2$ gives cyclic and linear siloxanes having (carboranylmethyl)methylsiloxane and $Si(CH_3)_2O$ units, and SiOH end groups [22]. The ammonolysis reaction of (carboranylmethyl)chloroorganosilanes (carboranyl = $C_2B_{10}H_{11}$) and their coammonolysis reactions with $ClSi(CH_3)_3$ result in low-molecular weight (carboranylmethyl)organosilazane oligomers with terminal aminosilane or trimethylsilanyl groups [23]. Carborane siloxane oligomers, in which the carborane is either 1,2- or 1,7-$C_2B_{10}H_x$, are prepared from the hydrolysis of $C_2B_{10}H_{11}$-1-$CH_2Si(CH_3)(OCH_3)_2$ and $C_2B_{10}H_{11}$-1-CH_2SiCl_3 [24]. A pyrolysis gas chromatogram of a carborane-silicone polymer (the carborane unit appears to be the 1,7-$C_2B_{10}H_{12}$ isomer) is reported [25]. Pyrolysis of carborane-siloxane polymers based on 1,7-$C_2B_{10}H_{10}$-1,7- units produces materials with the composition SiC and B_4C [26]. Pyrolysis of a 1,7-C_2B_{10}-containing siloxane polymer produces the parent 1,7-$C_2B_{10}H_{12}$ [27]. Carborane-silicone polymers with C_2B_{10} carborane frameworks have been used in connection with capillary chromatography columns [28].

For additional data, see Table 13/13.

Table 13/13
Polymers Containing $C_2B_{10}H_x$ Groups; C—C (with $B_{10}H_{10}$) = 1,2-$C_2B_{10}H_{10}$; -$CB_{10}H_{10}C$- = 1,7-$C_2B_{10}H_{10}$; -$CB_{10}H_{10}C$- (bracketed) = 1,12-$C_2B_{10}H_{10}$.

polymer		Ref.
$[-N{=}P(OCH_2CF_3)_2-]$ $[-N{=}P(OCH_2CF_3)(C_2B_{10}H_{10}C_6H_5)-]_n$		[29]
$[(-N{=}P(OCH_2CF_3)_2-)_2(-N{=}P(OCH_2CF_3)(C_2B_{10}H_{10}R)-)_1]_n$	R = CH_3, C_6H_5	[30]

Table 13/13 (continued)

polymer		Ref.
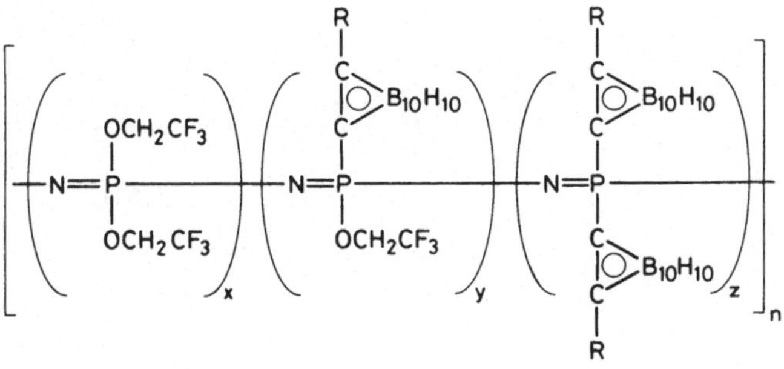	$R = CH_3, C_6H_5$	[30]
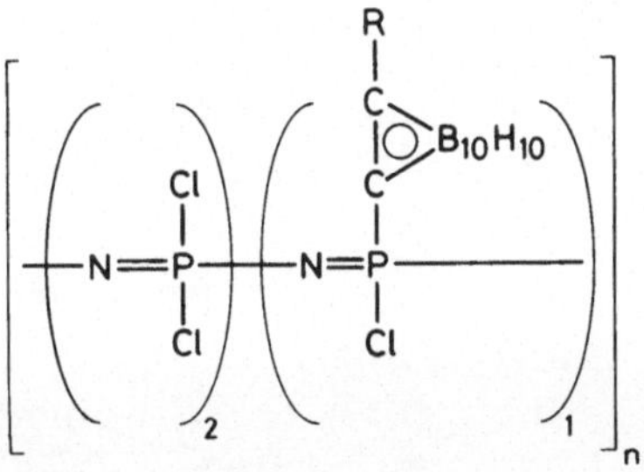	$R = CH_3, C_6H_5$	[30]
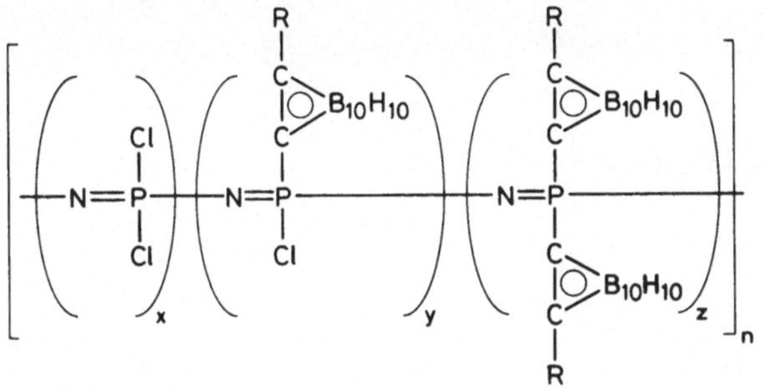	$R = CH_3, C_6H_5$	[30]
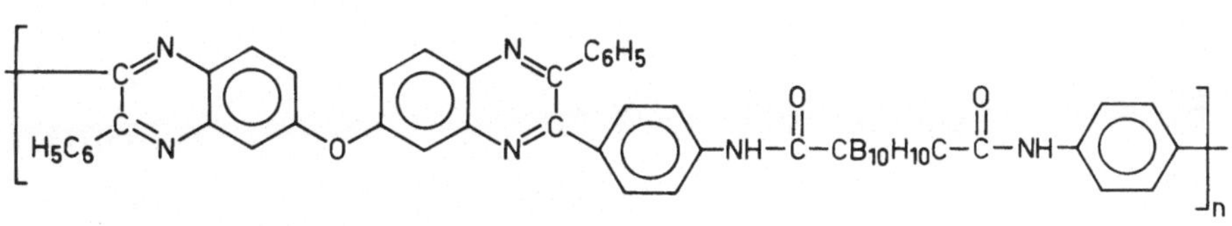		[31]
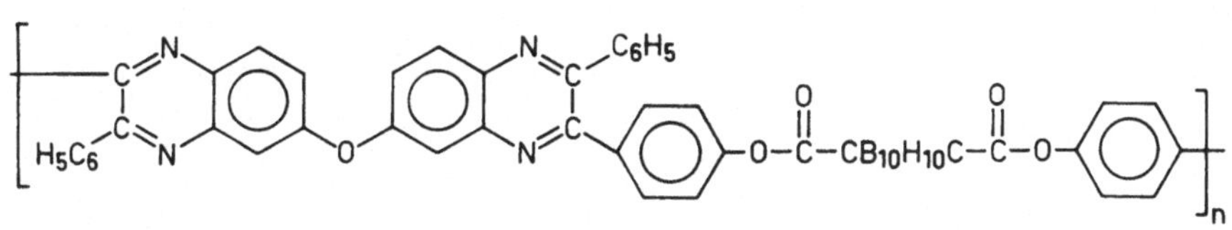		[31]

References for 13.11.5 on pp. 239/41

Table 13/13 (continued)

polymer	Ref.
[quinoxaline units with H_5C_6, C_6H_5, N, O; $-O-\overset{O}{\overset{\Vert}{C}}-CB_{10}H_{10}C-\overset{O}{\overset{\Vert}{C}}-O-$]$_n$	[31]
[quinoxaline units with H_5C_6, C_6H_5, N; $-O-\overset{O}{\overset{\Vert}{C}}-CB_{10}H_{10}C-\overset{O}{\overset{\Vert}{C}}-O-$]$_n$	[31]
$-[-\underset{\Vert}{\underset{O}{C}}-CB_{10}H_{10}C-\underset{\Vert}{\underset{O}{C}}-NH-C_6H_4-C_6H_4-NH-]_n-$	[32 to 37]
$-[-OCCB_{10}H_{10}CCON(CH_3)-C_6H_4-CH_2-C_6H_4-N(CH_3)-]_n-$	[32]
$-[-HN-C_6H_3(HO)-CH_2-C_6H_3(OH)-NH-OCCB_{10}H_{10}CCO-]_n-$	[38]
$-[-HN-C_6H_3(HO)-CH_2-C_6H_3(OOCCB_{10}H_{10}CCO-)-NH-OCCB_{10}H_{10}CCO-]_n-$	[38]
$-[-HN-C_6H_3(HO)-X-C_6H_3(OOC-R-CO-)-NHOC-R-CO-]_n-$ R = $CB_{10}H_{10}C$, X = CH_2	[38]
$[(-HN)_2C_6(NH-\overset{O}{\overset{\Vert}{C}}-CB_{10}H_{10}C-\overset{O}{\overset{\Vert}{C}}-)_2]_n$	[39]

Table 13/13 (continued)

polymer	Ref.
$[-CO-C_6H_4-CB_{10}H_{10}C-C_6H_4-CO-NH-C_6H_4-O-C_6H_4-NH-]_n$	[40, 41]
$[-CO-C_6H_4-C_2B_{10}H_{10}-C_6H_4-CO-NH-C_6H_4-O-C_6H_4-NH-]_n$	[40, 41]
imide polymer: $-N(CO)_2C_6H_3-C_2B_{10}H_{10}-C_6H_3(CO)_2N-C_6H_4-O-C_6H_4-$, repeating unit $[\ ]_n$	[41, 42]
imide polymer: $-N(CO)_2C_6H_3-CB_{10}H_{10}C-C_6H_3(CO)_2N-C_6H_4-O-C_6H_4-$, repeating unit $[\ ]_n$	[41, 42]
imide polymer: $-N(CO)_2C_6H_3-C_2B_{10}H_{10}-C_6H_3(CO)_2N-R-$, repeating unit $[\ ]_n$; R = $-C_6H_4-C(C_6H_5)_2-C_6H_4-$	[42]
imide polymer: $-N(CO)_2C_6H_3-CB_{10}H_{10}C-C_6H_3(CO)_2N-R-$, repeating unit $[\ ]_n$	[42]

References for 13.11.5 on pp. 239/41

Table 13/13 (continued)

polymer	Ref.
$[-O-C_6H_4-C\overset{\frown}{-}C(B_{10}H_{10})-C_6H_4-O-C(=O)-C_6H_4-O-C_6H_4-C(=O)-]_n$	[41]
$[-O-C_6H_4-C\overset{\frown}{-}C(B_{10}H_{10})-C_6H_4-OCOC_6H_4OCO-]_n$	[43]
$[-O-C_6H_4-CH_2-C\overset{\frown}{-}C(B_{10}H_{10})-CH_2-C_6H_4-OCOC_6H_4-O-C_6H_4-CO-]_n$	[41, 43]
$[-C(=O)-C_6H_4-C\overset{\frown}{-}C(B_{10}H_{10})-C_6H_4-C(=O)-O-C_6H_4-O-C_6H_4-O-]_n$	[41, 43, 44]
$[-C(=O)-C_6H_4-C\overset{\frown}{-}C(B_{10}H_{10})-C_6H_4-C(=O)-O-C_6H_4-C(CH_3)_2-C_6H_4-O-]_n$	[44]
$[-C(=O)-C_6H_4-C\overset{\frown}{-}C(B_{10}H_{10})-C_6H_4-C(=O)-O-C_6H_4-C(\text{fluorenylidene})-C_6H_4-O-]_n$	[44]
$[-C(=O)-C_6H_4-C\overset{\frown}{-}C(B_{10}H_{10})-C_6H_4-C(=O)-O-C_6H_4-C(\text{phthalide: }C_6H_4-C(=O)-O)-C_6H_4-O-]_n$	[44]
$[-C(=O)-C_6H_4-CB_{10}H_{10}C-C_6H_4-C(=O)-O-C_6H_4-O-C_6H_4-O-]_n$	[41, 43, 44]
$[-C(=O)-C_6H_4-CB_{10}H_{10}C-C_6H_4-C(=O)-O-C_6H_4-C(CH_3)_2-C_6H_4-O-]_n$	[44]

Table 13/13 (continued)

polymer	Ref.
$[-C(=O)-C_6H_4-CB_{10}H_{10}C-C_6H_4-C(=O)-O-C_6H_4-C(C_{12}H_8)-C_6H_4-O-]_n$ (C = fluorene-9,9-diyl)	[44]
$[-C(=O)-C_6H_4-CB_{10}H_{10}C-C_6H_4-C(=O)-O-C_6H_4-C(C_6H_4C(=O)O)-C_6H_4-O-]_n$ (C = phthalide-3,3-diyl)	[44]
$[-C_6H_4-O-C\overline{}C(B_{10}H_{10})-C_6H_4-O-C(=O)-C_6H_4-C\overline{}C(B_{10}H_{10})-C_6H_4-C(=O)-]_n$	[45]
$CH_3C(OC_2H_5)_2-[-C_6H_4-CH_2-CB_{10}H_{10}C-CH_2-C_6H_4-R-]_n-C(OC_2H_5)_2CH_3$ where R = biphenyl-3,5-diyl, $-C_6H_4-CO-CH=C(CH_3)-C_6H_4-$, $(-CH=C(C_6H_5)-)_m-$	[46]
$[-CH_2-C(CH_3)(C(=O)-OCH_3)-]_n-[CH_2-CH(C_6H_4-C\overline{}CH(B_{10}H_{10}))-]_m$	[47]
$[-OCH_2CH{=}CHCH_2OOCCB_{10}H_{10}CCO-]_n$	[48]
$[-OCH_2C{\equiv}CCH_2OOCCB_{10}H_{10}CCO-]_n$	[48]

References for 13.11.5 on pp. 239/41

Table 13/13 (continued)

polymer	Ref.
$[\text{-O(CH}_2)_6\text{OOCCB}_{10}\text{H}_{10}\text{CCOO(CH}_2)_6\text{OOCCH=CHCO-}]_n$	[48]
$\{[\text{-O(CH}_2)_6\text{OOCCH=CHCO-}]_k\text{O(CH}_2)_6\text{OOCCB}_{10}\text{H}_{10}\text{CCO-}\}_n$	[48]
$\{[\text{-O(CH}_2)_2\text{OOCCH=CHCO}]_k\text{O(CH}_2)_2\text{OOCCB}_{10}\text{H}_{10}\text{CCO-}\}_n$	[48]
$\{[\text{-O(C}_2\text{H}_4)_2\text{O}_2\text{OCCH=CHCO}]_k\text{O(C}_2\text{H}_4)_2\text{OCCB}_{10}\text{H}_{10}\text{CCO-}\}_n$	[48]
$[\text{-OCH}_2\text{CH=CHCH}_2\text{OOCCB}_{10}\text{H}_{10}\text{CCO-}]_n$	[48]
$[\text{-OCH}_2\text{C}\equiv\text{CCH}_2\text{OOCCB}_{10}\text{H}_{10}\text{CCO-}]_n$	[48]
$[\text{-O(CH}_2)_6\text{OOCCB}_{10}\text{H}_{10}\text{CCOO(CH}_2)_6\text{OOCCH=CHCO-}]_n$	[48]
$\{[\text{-O(CH}_2)_6\text{OOCCH=CHCO}]_k\text{O(CH}_2)_6\text{OOCB}_{10}\text{H}_{10}\text{CCO-}\}_n$	[48]
$\{[\text{-O(CH}_2)_6\text{OOCCH=CHCO}]_k\text{O(CH}_2)_6\text{OOCCB}_{10}\text{H}_{10}\text{CCO-}\}_n$	[48]
$\{[\text{-O(C}_2\text{H}_4)_2\text{O}_2\text{OCCH=CHCO}]_k\text{O(C}_2\text{H}_4)\text{OOCCB}_{10}\text{H}_{10}\text{CCO-}]_n$	[48]
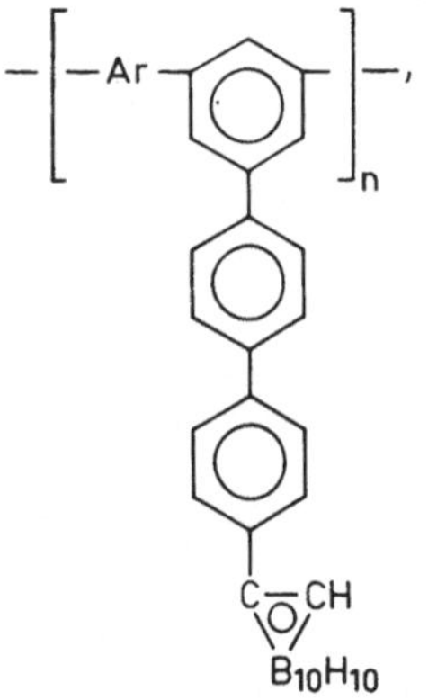 $-[-\text{Ar}-\text{C}_6\text{H}_3(\text{C}_6\text{H}_4\text{-C}_6\text{H}_4\text{-C(-CH)B}_{10}\text{H}_{10})-]_n-$, where Ar = $\text{-C}_6\text{H}_4\text{-}$, $\text{-C}_6\text{H}_4\text{-C}_6\text{H}_4\text{-}$, $\text{-C}_6\text{H}_4\text{-O-C}_6\text{H}_4\text{-}$, $\text{-C}_6\text{H}_4\text{-S-C}_6\text{H}_4\text{-}$, $\text{-C}_6\text{H}_4\text{-CH}_2\text{-C}_6\text{H}_4\text{-}$	[49]
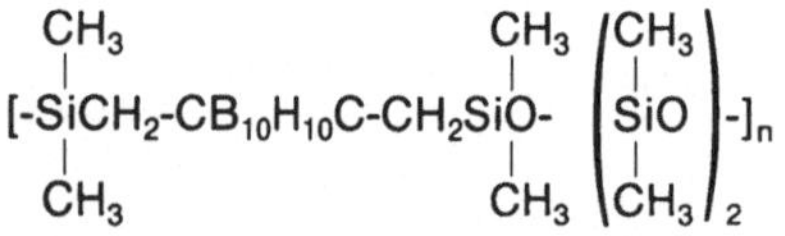 $[\text{-Si(CH}_3)_2\text{CH}_2\text{-CB}_{10}\text{H}_{10}\text{C-CH}_2\text{Si(CH}_3)_2\text{O-}(\text{Si(CH}_3)_2\text{O})_2\text{-}]_n$	[50]
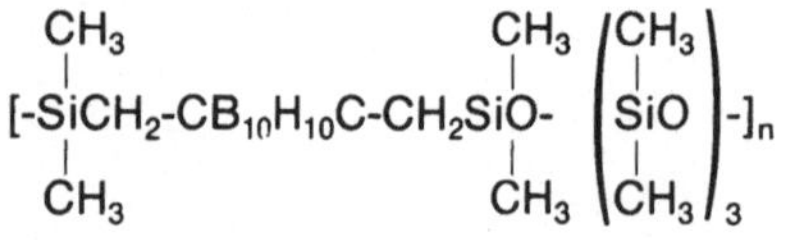 $[\text{-Si(CH}_3)_2\text{CH}_2\text{-CB}_{10}\text{H}_{10}\text{C-CH}_2\text{Si(CH}_3)_2\text{O-}(\text{Si(CH}_3)_2\text{O})_3\text{-}]_n$	[50]
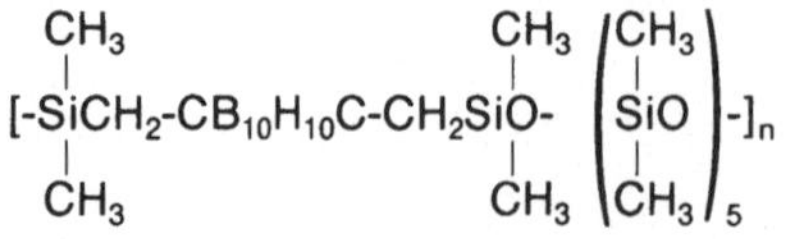 $[\text{-Si(CH}_3)_2\text{CH}_2\text{-CB}_{10}\text{H}_{10}\text{C-CH}_2\text{Si(CH}_3)_2\text{O-}(\text{Si(CH}_3)_2\text{O})_5\text{-}]_n$	[50]
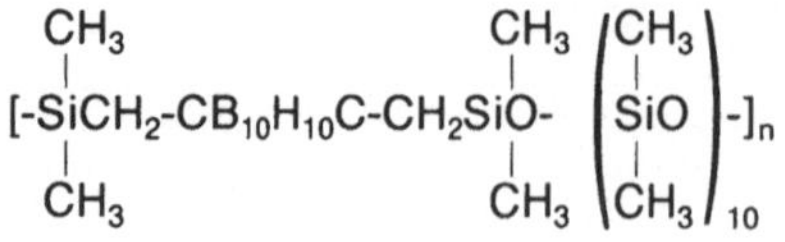 $[\text{-Si(CH}_3)_2\text{CH}_2\text{-CB}_{10}\text{H}_{10}\text{C-CH}_2\text{Si(CH}_3)_2\text{O-}(\text{Si(CH}_3)_2\text{O})_{10}\text{-}]_n$	[50]

Table 13/13 (continued)

polymer	Ref.
$[\text{-Si(CH}_3)_2\text{CH}_2\text{-CB}_{10}\text{H}_{10}\text{C-CH}_2\text{Si(CH}_3)_2\text{O-}(\text{Si(CH}_3)(\text{CH}_3)\text{O})_{80}\text{-}]_n$	[50]
$[\text{-Si(CH}_3)_2\text{CH}_2\text{-CB}_{10}\text{H}_{10}\text{C-CH}_2\text{Si(CH}_3)_2\text{O-}(\text{Si(CH}_3)(\text{C}_6\text{H}_5)\text{O})_2\text{-}]_n$	[50]
$[\text{-Si(CH}_3)_2\text{CH}_2\text{-CB}_{10}\text{H}_{10}\text{C-CH}_2\text{Si(CH}_3)_2\text{O-}(\text{Si(CH}_3)_2\text{O-Si(C}_6\text{H}_5)_2\text{O-Si(CH}_3)_2\text{O})\text{-}]_n$	[50]
$[\text{-Si(CH}_3)_2\text{CH}_2\text{-CB}_{10}\text{H}_{10}\text{C-CH}_2\text{Si(CH}_3)_2\text{O-}(\text{Si(CH}_3)(\text{C}_6\text{H}_5)\text{O})_3\text{-}]_n$	[50]

References for 13.11.5:

[1] Teixidor, F.; Rudolph, R. W. (J. Organometal. Chem. **241** [1983] 301/12).

[2] Sergeev, V. A.; Nedel'kin, V. I.; Shustova, T. V.; Zakharkin, L. I. (Izv. Akad. Nauk SSSR Ser. Khim. **1983** 1905/7; Bull. Acad. Sci. USSR Div. Chem. Sci. **1983** 1729/31).

[3] Korshak, V. V.; Bekasova, N. I.; Komarova, L. G.; Kats, G. A.; Komarova, L. I. (Vysokomol. Soedin. A **26** [1984] 86/8; Polym. Sci. [USSR] A **26** [1984] 95/8).

[4] Reed, S. F. (J. Polym. Sci. Polym. Chem. Ed. **19** [1981] 1863/6).

[5] Korshak, V. V.; Sosin, S. L.; Zakharkin, L. I.; Kovredov, A. I.; Alekseeva, V. P.; Antipova, B. A.; Shaugumbekova, Z. S. (Vysokomol. Soedin. B **23** [1981] 219/21; C.A. **95** [1981] No. 7840).

[6] Sobolevskii, M. V.; Zhigach, A. F.; Sarishvili, I. G.; Sytova, I. M.; Soleva, N. M.; Kochneva, L. N.; Vishnevskii, F. N. (Plasticheskie Massy **1980** 12/3; C.A. **92** [1980] No. 181795).

[7] Sergeev, V. A.; Shitikov, V. K.; Koloskova, G. N.; Pakhomov, V. I.; Batenina, N. V. (Plasticheskie Massy **1980** 24/5; C.A. **92** [1980] No. 182024).

[8] Bekasova, N. I.; Komarova, L. G.; Kats, G. A.; Korshak, V. V. (Makromol. Chem. **185** [1984] 2313/8).

[9] Sergeev, V. A.; Shitikov, V. K.; Chizhova, N. V.; Kalinin, V. N.; Zakharkin, L. I.; Abdrakhmanov, I. S.; Rudenko, T. M.; Chimishkyan, A. L. (U.S.S.R. 787422 [1980]; C.A. **94** [1981] No. 104360).

[10] Korshak, V. V.; Teplyakov, M. M.; Gelashvili, T. L.; Kalinin, V. N.; Zakharkin, L. I. (Vysokomol. Soedin. A **22** [1980] 262/8; C.A. **92** [1980] No. 181934).

[11] Aliguliev, R. M.; Zelenev, Y. V.; Aliev, G. M.; Khiteeva, D. M.; Matochkin, V. S. (Acta Polym. **31** [1980] 400/1; C.A. **93** [1980] No. 240101).

[12] Tregub, A. I.; Issajew, K. S.; Achrijew, A. S.; Iwanow, P. I.; Matotschkin, W. S.; Selenew, J. V. (Plaste Kautsch. **28** [1981] 137/9).

[13] Tregub, A. I.; Issajew, K. S.; Iwanow, P. I.; Selenew, J. V. (Plaste Kautsch. **29** [1982] 336/8).

[14] Korshak, V. V.; Bekasova, N. I.; Komarova, L. G.; Kats, G. A.; Babchinitser, T. M. (Vysokomol. Soedin. A **26** [1984] 43/7; C.A. **100** [1984] No. 123020).
[15] Strel'chenko, L. S.; Yurkovskii, I. M.; Valetskii, P. M.; Pshenichkin, P. A.; Kuchinskaya, T. K.; Vinogradova, S. V.; Korshak, V. V. (Vysokomol. Soedin. B **22** [1980] 72/5; C.A. **92** [1980] No. 164600).
[16] Kotrelev, G. V.; Zhdanova, E. A. (Vysokomol. Soedin. B **23** [1981] 289/92; C.A. **95** [1981] No. 62782).
[17] Kalinin, V. N.; Izmailov, B. A.; Zakharkin, L. I.; Zhdanov, A. A.; Petrushkina, E. A.; Myakushev, V. D. (U.S.S.R. 715588 [1980]; C.A. **92** [1980] No. 199358).
[18] Izmailov, B. A.; Kalinin, V. N.; Zhdanov, A. A.; Zakharkin, L. I.; Teplyakov, M. M.; Dmitrienko, A. V. (U.S.S.R. 715589 [1980]; C.A. **92** [1980] No. 182134).
[19] Izmailov, B. A.; Kalinin, V. N.; Myakushev, V. D.; Zhdanov, A. A.; Zakharkin, L. I. (Zh. Obshch. Khim. **50** [1980] 1558/61; J. Gen. Chem. [USSR] **50** [1980] 1263/6).
[20] Frunze, T. M.; Sakharova, A. A.; Mel'nik, O. A.; Izmailov, B. A.; Kalinin, V. N. (Vysokomol. Soedin. A **23** [1981] 2077/82; C.A. **96** [1982] No. 69519).

[21] Izmailov, B. A.; Kalinin, V. N.; Myakushev, V. D.; Zhdanov, A. A.; Zakharkin, L. I. (Zh. Obshch. Khim. **53** [1983] 1813/9; J. Gen. Chem. [USSR] **53** [1983] 1632/7).
[22] Izmailov, B. A.; Kalinin, V. N.; Myakushev, V. D.; Zhdanov, A. A.; Zakharkin, L. I. (Zh. Obshch. Khim. **53** [1983] 1807/13; J. Gen. Chem. [USSR] **53** [1983] 1627/32).
[23] Kalinin, V. N.; Izmailov, B. A.; Kazantsev, A. A.; Zhdanov, A. A.; Zakharkin, L. I. (Zh. Obshch. Khim. **51** [1981] 859/63; J. Gen. Chem. [USSR] **51** [1981] 713/6).
[24] Kalinin, V. N.; Izmailov, B. A.; Kazantsev, A. A.; Myakushev, V. D.; Zhdanov, A. A.; Zakharkin, L. I. (J. Organometal. Chem. **216** [1981] 295/320).
[25] Riska, G. D.; Estes, S. A.; Beyer, J. O.; Uden, P. C. (Spectrochim. Acta B **38** [1983] 407/17).
[26] Walker, B. E.; Rice, R. W.; Becher, P. F.; Bender, B. A.; Coblenz, W. S. (Am. Ceram. Soc. Bull. **62** [1983] 916/23).
[27] Sarto, L. G.; Estes, S. A.; Uden, P. C.; Siggia, S.; Barnes, R. M. (Anal. Letters **14** [1981] 205/18).
[28] Uden, P. C.; Henderson, D. E.; DiSanzo, F. P.; Lloyd, R. J.; Tetu, T. (J. Chromatogr. **196** [1980] 403/14).
[29] Fewell, L. L.; Basi, R. J.; Parker, J. A. (J. Appl. Poplym. Sci. **28** [1983] 2659/71).
[30] Allcock, H. R.; Scopelianos, A. G.; O'Brien, J. P.; Bernheim, M. Y. (J. Am. Chem. Soc. **103** [1981] 350/7).

[31] Raubach, H.; Schumann, A.; Oelert, H.; Korshak, V. V.; Krongauz, E. S.; Bekasova, N. I. (Izv. Akad. Nauk Kaz.SSR Ser. Khim. **1981** 14/9; C.A. **96** [1982] No. 104865).
[32] Korshak, V. V.; Bekasova, N. I.; Komarova, L. G. (Vysokomol. Soedin. A **24** [1982] 2424/8; Polym. Sci. [USSR] **24** [1982] 2788/93).
[33] Pavlova, S. A.; Gribkova, P. N.; Balykova, T. N.; Polina, T. V.; Komarova, L. G.; Bekasova, N. I.; Korshak, V. V. (Vysokomol. Soedin. A **25** [1983] 1270/6; Polym. Sci. [USSR] **25** [1983] 1475/83).
[34] Pavlova, S. A.; Gribkova, P. N.; Balykova, T. N.; Polina, T. V.; Komarova, L. G.; Bekasova, N. I.; Korshak, V. V. (Vysokomol. Soedin. A **24** [1982] 1712/7; Polym. Sci. [USSR] **24** [1982] 1953/9).
[35] Korshak, V. V.; Gribova, I. A.; Pavlova, S. A.; Nekrasov, Y. S.; Avetisyan, Y. L.; Gribkova, P. N.; Serov, I. K. (Vysokomol. Soedin. A **22** [1980] 515/9; Polym. Sci. [USSR] **22** [1980] 568/74).
[36] Korshak, V. V.; Pavlova, S. A.; Gribkova, P. N.; Balykova, T. N.; Polina, T. V.; Zakharkin, L. I.; Kalinin, V. N.; Bekasova, N. I.; Komarova, L. G. (Vysokomol. Soedin. A **26** [1984] 119/23; Polym. Sci. [USSR] **26** [1984] 132/7).

[37] Korshak, V. V.; Pavlova, S. A.; Gribkova, P. N.; Polina, T. V.; Babchinitser, T. M. (Vysokomol. Soedin. A **26** [1984] 104/10; Polym. Sci. [USSR] **26** [1984] 116/23).
[38] Bekasova, N. I.; Korshak, V. V.; Surikova, M. A.; Komarova, L. I.; Voloshina, I. Yu. (Vysokomol. Soedin. A **23** [1981] 138/44; Polym. Sci. [USSR] **23** [1981] 155/63).
[39] Korshak, V. V.; Bekasova, N. I.; Komarova, L. G.; Katz, G. A. (Vysokomol. Soedin. A **26** [1984] 48/52; Polym. Sci. [USSR] **26** [1984] 54/9).
[40] Voishchev, V. S.; Beloglazov, V. A.; Valetskii, P. M.; Sidorenko, V. I. (Vysokomol. Soedin. B **23** [1981] 123/7; C.A. **94** [1981] No. 192821).

[41] Voishchev, V. S.; Valetskii, P. M.; Sidorenko, V. I.; Voishcheva, O. V.; Vinogradova, S. V.; Korshak, V. V. (Vysokomol. Soedin. A **24** [1982] 2211/9; Polym. Sci. [USSR] **24** [1982] 2537/46).
[42] Aksenov, A. I.; Burtseva, T. A.; Valetskii, P. M.; Vinogradova, S. V.; Korshak, V. V. (Vysokomol. Soedin. B **22** [1980] 602/7; C.A. **94** [1981] No. 31159).
[43] Zhuravleva, I. V.; Vinogradova, N. K.; Pavlova, S. S. A. (Vysokomol. Soedin. A **23** [1981] 2351/9; Polym. Sci. [USSR] **23** [1981] 2555/64).
[44] Ivanov, P. I.; Zelenev, Yu. V. (Vysokomol. Soedin. B **22** [1980] 187/92; C.A. **92** [1980] No. 215887).
[45] Strel'chenko, L. S.; Sidorenko, V. I.; Genin, Ya. V.; Kalachev, A. I.; Pshenichkin, P. A.; Valetskii, P. M.; Vinogradova, S. V.; Korshak, V. V. (Vysokomol. Soedin. A **23** [1981] 1229/37; Polym. Sci. [USSR] **23** [1981] 1364/73).
[46] Zhuravleva, I. V.; Vinogradova, N. K.; Gelashvili, T. L.; Teplyakov, M. M.; Pavlova, S. S. A.; Korshak, V. V. (Vysokomol. Soedin. A **26** [1984] 22/6; Polym. Sci. [USSR] **26** [1984] 23/9).
[47] Korshak, V. V.; Askadskii, A. A.; Slonimskii, G. L.; Sosin, S. L.; Zakharkin, L. I.; Kovredov, A. I.; Bychko, K. A.; Antipova, B. A.; Shaugumbekova, Z. S. (Vysokomol. Soedin. A **23** [1981] 2051/63; C.A. **96** [1982] No. 69540).
[48] Korshak, V. V.; Bekasova, N. I.; Solomatina, A. I.; Vagina, Z. P.; Klimentova, N. V.; Suprun, A. P. (Vysokomol. Soedin. A **25** [1983] 989/96; Polym. Sci. [USSR] **25** [1983] 1143/52).
[49] Korshak, V. V.; Teplyakov, M. M.; Khotina, I. A.; Kalinin, V. N.; Tugov, I. I.; Elagina, V. P. (Vysokomol. Soedin. A **23** [1981] 1461/5; Polym. Sci. [USSR] **23** [1983] 1612/7).
[50] Izmailov, B. A.; Kalinin, V. N.; Zhdanov, A. A.; Zakharkin, L. I. (Vysokomol. Soedin. A **25** [1983] 1253/8; Polym. Sci. [USSR] **25** [1983] 1454/60).

13.11.6 Other Carborane Compounds Containing a Total of Ten Boron Atoms

The exopolyhedral $J(^{11}B-^{11}B)$ for **1:3'-[2,4-$C_2B_5H_6$]$_2$**, **Fig.** 13-9, p. 242, is found to be 119 to 124 Hz; 151 Hz for 3:3'-[2,4-$C_2B_5H_6$]$_2$; and ≧100 Hz for 3:5'-[2,4-$C_2B_5H_6$]$_2$. An intracage coupling constant of 9.5 Hz for B(1)B(5) of **3:3'-[2,4-$C_2B_5H_6$]$_2$** is found. The % s-character of the coupled boron atoms in 3:3'-[2,4-$C_2B_5H_6$]$_2$: 34% for B(1), 41% for B(3), and 39% for B(5) [1].

The molecular structure of 2-(*closo*-2',4'-$C_2B_5H_6$-3'-)-1,8-(η-C_5H_5)$_2$-*closo*-1,8,5,6-Co_2-$C_2B_5H_6$, a coupled-cage cobaltacarborane containing a boron-boron linkage, has been determined from single-crystal X-ray diffraction data. Lattice parameters of the triclinic crystals are a = 9.155(3), b = 15.659(11), c = 7.142(2) Å; α = 99.64(4)°; β = 101.11(3)°, γ = 100.69(5)°; Z = 2; space group $P\bar{1}$-C_i^1 (No. 2) [2]. 1,1-[(C_6H_5)$_3$P]$_2$-1-H-1,2,4-Rh$C_2B_{10}H_{12}$ is prepared by reaction of [N(CH_3)$_4$][$C_2B_{10}H_{13}$] with [(C_6H_5)$_3$P]$_3$RhCl and has been characterized by NMR and X-ray crystallography. The crystals are monoclinic and have lattice parameters a = 12.38(2), b = 21.26(2), c = 16.69(2) Å; β = 102.1(1)°; Z = 4; D_{calc} = 1.42 g/cm^3; space group $P2_1/n$-C_{2h}^5 (No. 14) [3].

References for 13.11.6 on p. 242

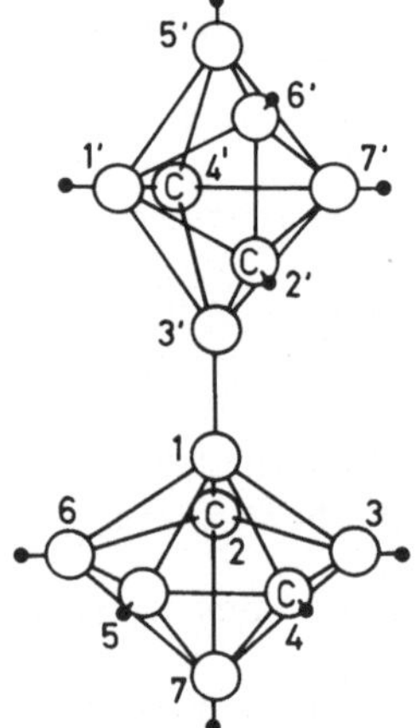

Fig. 13-9. Structure of $1{:}3'$-$[2,4\text{-}C_2B_5H_6]_2$ [1].

References for 13.11.6:

[1] Anderson, J. A.; Astheimer, R. J.; Odom, J. D.; Sneddon, L. G. (J. Am. Chem. Soc. **106** [1984] 2275/83).
[2] Micciche, R. P.; Plotkin, J. S.; Sneddon, L. G. (Inorg. Chem. **22** [1983] 1765/8).
[3] Hewes, J. D.; Knobler, C. B.; Hawthorne, M. F. (J. Chem. Soc. Chem. Commun. **1981** 206/7).

13.12 Carboranes Containing 11 Boron Atoms

For previous data on carboranes containing 11 boron atoms, see "Boron Compounds" 2nd Suppl. Vol. 2, 1982, pp. 326/7.

$1\text{-}CB_{11}H_{11}\text{-}1\text{-}(CH_3)_2NH$ is obtained by treating $7\text{-}CB_{10}H_{12}\text{-}7\text{-}N(CH_3)_3$ with $(CH_3)_3NBH_3$ at 180 to 200°C. The ^{11}B NMR spectrum of $1\text{-}CB_{11}H_{11}\text{-}1\text{-}(CH_3)_2NH$ exhibits signals at δ (in ppm) = −8.37, J(B-H) = 142 Hz (1B); = −13.65, J(B-H) = 136 Hz (5B); = −15.69, J(B-H) = 148 Hz (5B). Methylation of $1\text{-}CB_{11}H_{11}\text{-}1\text{-}(CH_3)_2NH$ produces **$1\text{-}CB_{11}H_{11}\text{-}1\text{-}N(CH_3)_3$**, which can be reduced to the parent **$[1\text{-}CB_{11}H_{12}]^-$** anion with Na in liquid NH_3. The ^{11}B NMR spectrum of $[1\text{-}CB_{11}H_{12}]^-$ exhibits signals at δ (in ppm) = −7.40, J(B-H) = 141 Hz (1B); = −13.65, J(B-H) = 140 Hz (5B); = −15.07, J(B-H) = 140 Hz (5B) [1]. Solubilities of carboranates in water (in g per 100 g H_2O): 58.52 for $Na[CB_{11}H_{12}]\cdot 8H_2O$, 120.3 for $K[CB_{11}H_{12}]\cdot 2H_2O$, 31.60 for $Rb[CB_{11}H_{12}]$, and 9.15 for $Cs[CB_{11}H_{12}]$. The infrared spectra of the $M[CB_{11}H_{12}]$ salts showed broad band features in the B-H stretching range 2430 to 2580 cm^{-1}, the bands attributed to stretch-deformations of the core lie in the range 1000 to 1100 cm^{-1}, and the bands attributed to lattice deformations in the range 720 to 730 cm^{-1}. The UV spectra of $M[CB_{11}H_{12}]$ show edge absorption in the range of 200 to 267 nm. The temperatures of dehydration for $Li[CB_{11}H_{12}]\cdot H_2O$ and for the Na and K salts are 150, 100, and 70°C, respectively. Temperatures for the start of decomposition are 190, 290, and 260°C for Li, Na, and K salts, respectively. The Rb and Cs salts are stable up to 600°C [2].

References for 13.12:

[1] Plešek, J.; Jelinek, T.; Drdakova, E.; Heřmánek, S.; Štíbr, B. (Collection Czech. Chem. Commun. **49** [1984] 1559/62).
[2] Myasoedov, S. F.; Tsimerinova, T. V.; Solntsev, K. A.; Kuznetsov, N. T. (Zh. Neorgan. Khim. **29** [1984] 1421/4; Russ. J. Inorg. Chem. **29** [1984] 817/8).

13.13 Carborane Derivatives Containing 12 Boron Atoms

$[2,3\text{-}C_2B_3H_5\text{-}2,3\text{-}(C_2H_5)_2]\text{-}5\text{-}Co[B_9H_{12}\text{-}1\text{-}thf]$ (thf = $O(CH_2)_4$ = tetrahydrofuran) is formed from $CoCl_2$, tetrahydrofuran, $Na[2,3\text{-}C_2B_4H_5\text{-}2,3\text{-}(C_2H_5)_2]$, and $Na[B_5H_9]$; and $[2,3\text{-}C_2B_3H_5\text{-}2,3\text{-}(C_2H_5)_2]\text{-}6\text{-}Co[B_9H_{12}\text{-}2\text{-}thf]$ is produced from $CoCl_2$, tetrahydrofuran, $Na[2,3\text{-}C_2B_4H_5\text{-}2,3\text{-}(C_2H_5)_2]$, and $Na[B_9H_{14}]$ [1]. The crystal and molecular structure of $[2,3\text{-}C_2B_3H_5\text{-}2,3\text{-}(CH_3)_2]\text{-}5\text{-}Co[B_9H_{12}\text{-}1\text{-}O(CH_2)_4]$ has been determined. Lattice parameters of the monoclinic crystals are a = 8.692(2), b = 9.658(4), c = 22.383(9) Å; β = 97.51(4)°; Z = 4; D_{calc} = 1.184 g/cm^3; space group $P2_1/n\text{-}C^5_{2h}$ (No. 14) [2].

References for 13.13:

[1] Borodinsky, L.; Grimes, R. N. (Inorg. Chem. **21** [1982] 1921/7).
[2] Borodinsky, L.; Sinn, E.; Grimes, R. N. (Inorg. Chem. **21** [1982] 1928/36).

13.14 Carborane Derivatives Containing 13 Boron Atoms

$[2,3\text{-}C_2B_4H_4\text{-}2,3\text{-}(C_2H_5)_2]\text{-}5\text{-}Co[B_9H_{12}\text{-}1\text{-}thf]$ (thf = $O(CH_2)_4$ = tetrahydrofuran) is produced from $CoCl_2$, tetrahydrofuran, $Na[2,3\text{-}C_2B_4H_5\text{-}2,3\text{-}(C_2H_5)_2]$, and $Na[B_5H_9]$ [1]. The crystal and molecular structure of $[2,3\text{-}C_2B_4H_4\text{-}2,3\text{-}(C_2H_5)_2]\text{-}5\text{-}Co[B_9H_{12}\text{-}1\text{-}O(CH_2)_4]$ have been determined. The monoclinic crystals have a = 14.175(4), b = 9.032(6), c = 17.771(3) Å; β = 112.12(3)°; D_{calc} = 1.165 g/cm^3; Z = 4; space group $P2_1/n\text{-}C^5_{2h}$ (No. 14) [2].

References for 13.14:

[1] Borodinsky, L.; Grimes, R. N. (Inorg. Chem. **21** [1982] 1921/7).
[2] Borodinsky, L.; Sinn, E.; Grimes, R. N. (Inorg. Chem. **21** [1982] 1928/36).

13.15 Carborane Derivatives Containing 16 Boron Atoms

For previous data on carboranes containing 16 boron atoms, see "Boron Compounds" 2nd Suppl. Vol. 2, 1982, pp. 327.

$[1,2\text{-}C_2B_7H_7\text{-}1,2\text{-}(C_2H_5)_2]\text{-}6\text{-}Co[B_9H_{12}\text{-}2\text{-}thf]$ (thf = $O(CH_2)_4$ = tetrahydrofuran) is formed from $CoCl_2$, tetrahydrofuran, $Na[2,3\text{-}C_2B_4H_5\text{-}2,3\text{-}(C_2H_5)_2]$, and $Na[B_9H_{14}]$ [1]. The crystal and molecular structure of $[1,2\text{-}C_2B_7H_7\text{-}1,2\text{-}(C_2H_5)_2]\text{-}6\text{-}Co[B_9H_{12}\text{-}2\text{-}O(CH_2)_4]$ have been determined. The monoclinic crystals have lattice parameters a = 13.166(7), b = 12.166(5), c = 14.786(4) Å; β = 105.99(5)°; D_{calc} = 1.182 g/cm^3; space group $P2_1/c\text{-}C^5_{2h}$ (No. 14) [2].

References for 13.15:

[1] Borodinsky, L.; Grimes, R. N. (Inorg. Chem. **21** [1982] 1921/7).
[2] Borodinsky, L.; Sinn, E.; Grimes, R. N. (Inorg. Chem. **21** [1982] 1928/36).

13.16 Carborane Compounds Containing 18 Boron Atoms

For previous data on carboranes containing 18 boron atoms, see "Boron Compounds" 2nd Suppl. Vol. 2, 1982, pp. 327/9.

3-(8′-*nido*-5′,6′-$C_2B_8H_{11}$)-1,2-$C_2B_{10}H_{11}$, prepared by heating *nido*-7,8-$C_2B_9H_{13}$ at 80°C, shows ^{11}B NMR resonances at δ (in ppm) = 8.5 for B(8′); = 6.9 for B(7′); = 5.3 for B(1′); = −2.4 for B(9′); = −2.7 for B(3′), B(9,12); = −7.9 for B(10′); = −8.8 for B(8) or B(10); = −11.6 for B(3); = −13.1 for B(4,11); = −13.6 for B(5,7); = −14.6 for B(6); = −27.0 for B(2′); = −37.4 for B(4′). ^{1}H NMR data: 6.54, 5.04 and 3.55 ppm. A pK_a = 6.18 has been measured for the compound [1]. The crystalline material is monoclinic, space group C_2/c-C^6_{2h} (No. 15), with a = 21.099(9), b = 7.081(2), c = 22.065(9) Å; β = 94.27(3)°; Z = 8. Atomic parameters, bond angles, and distances are given [2].

***nido-nido*-$[RCB_9H_{10}CN{=}NCB_9H_{10}CR]^{2-}$** (R = H, CH_3, C_6H_5) is formed by KOH degradation of $RCB_{10}H_{10}CN{=}NCB_{10}H_{10}CR$ [3].

$[Cr(CH_3NH_2)_6]Br[C_2B_9H_{12}]_2$ and $[Cr(RNH_2)_5X][C_2B_9H_{12}]_2$ (R = CH_3, X = Br, Cl; R = C_2H_5, X = Cl) are reported, and their thermolysis from 20 to 250°C investigated. IR spectra, UV/visible spectra of aqueous solutions of the four compounds, and the molar conductances of $[Cr(CH_3NH_2)_5X]$-$[C_2B_9H_{12}]_2$ (X = Cl, Br) were also measured [4]. $[(C_2B_9H_{11})_2Cr^{III}]^-$, and $[(C_2B_9H_{11})_2Co^{III}]^-$ have been used as examples of a simple generalization of the Mingos/Wade equation [5].

The conformation of $(8\text{-}CH_3O\text{-}C_2B_9H_{10})_2Ni$ has been studied by X-ray diffraction [6]. $[C_2B_9H_{10}\text{-}8\text{-}OCH_3]_2Ni$ is monoclinic, space group $P2_1/c$-C^5_{2h} (No. 14) with a = 15.208(3), b = 13.762(4), c = 10.363(1) Å; γ = 117.25(2)° [7].

Two reduction potentials, −0.62 and −1.93 V, are found for $(1,7\text{-}PCB_9H_{10}\text{-}1\text{-}CH_3)_2Co$ [8].

The $[8,8'\text{-}\mu\text{-}I\text{-}3\text{-}Co(1,2\text{-}C_2B_9H_{10})_2]$ complex with an iodonium bridge between both carborane ligands was prepared and its constitution was established by NMR methods [9].

Crystalline $[NH(C_2H_5)_3][Co(1,2\text{-}C_2B_9H_{11})_2]$ is orthorhombic; the Co is sandwiched between two $C_2B_9H_{11}$ units with the C_2B_3 faces of the two ligands nearly parallel. Lattice parameters are a = 29.048(9), b = 11.457(4), c = 7.100(2) Å; D_{calc} = 1.180 g/cm^3; Z = 4. Space group $Pna2_1$-C^9_{2v} (No. 33) [10]. Electrochemical properties of $[(1,2\text{-}C_2B_9H_{11})_2Co]^-$ [11 to 13] and $[(1,7\text{-}C_2B_9H_{11})_2Co]^-$ [13] have been studied. The magnetic susceptibility of $[Co(C_2B_9H_{11})_2]^-$ has been determined [14].

$(C_2B_9H_{12})_2[ML_3]X$ (M = Co, Cr; L = $H_2NCH_2CH_2NH_2$; X = Cl, Br) are prepared by the reaction of $KC_2B_9H_{12}$ with $[ML_3]X_3$ [15].

The use of capillary isotachophoresis for the separation of $[(C_2B_9H_{11})_2Co]^-$ and its chloro and bromo derivatives has been studied in H_2O/CH_3OH and H_2O/C_2H_5OH media; the order of migration of the derivatives is molecular weight-dependent [16]. $M[Co(C_2B_9H_{11})_2]$ (M = Cs, $N(CH_3)_4$) is formed by electrolysis of $K[7,8\text{-}C_2B_9H_{12}]$ in $(CH_3)_2SO$ on a cobalt anode. Similarly, electrolysis of $K[7,8\text{-}C_2B_9H_{12}]$ on a nickel anode gives a mixture of $[N(CH_3)_4][Ni(C_2B_9H_{11})_2]$ and $Ni(C_2B_9H_{11})_2$ [17]. The synthesis, properties, and chemical behavior of $[(C_2B_9H_{11})_2Co]^-$ anion derivatives, namely anions with -S- or -SS- bridges and of neutral species with -SR- (R = H, alkyl, aryl) or -SSR- bridges as well as with -$\overset{+}{S}(CH_3)_2$ groups, are described [6]. Reversed-phase ion-pair liquid chromatographic separations involving the heteroborane anions $[(1,2\text{-}C_2B_9H_{11})_2Co]^-$, $[(1,2\text{-}C_2B_9H_{10}I)_2Co]^-$, $[8,8'\text{-}S\text{-}(1,2\text{-}C_2B_9H_{10})_2Co]^-$, $[8,8'\text{-}C_6H_4\text{-}(1,2\text{-}C_2B_9H_{10})_2Co]^-$, $[(1,2\text{-}C_2B_9H_{11})_2Fe]^-$, and $[(1,2\text{-}C_2B_9H_{11})_2Ni]^-$ have been carried out [18].

The halogenation of $Cs[(C_2B_9H_{11})_2Co]$ by elemental halogens and γ-radiation-induced halogenation by $CHBr_3$, $CHCl_3$, or CCl_4 in polar solvents proceeds alternatively in both ligands yielding successively 8-; 8,8′-, 8,9,8′-, 8,9,8′,9′-, 8,9,12,8′,9′-, and 8,9,12,8′,9′,12′-halogen derivatives [19]. The bromination of the $Cs[Co(C_2B_9H_{11})_2]$ has been studied [20]. Stepwise bromination of bis(1,2-dicarbollyl)cobalt(III) by γ-irradiation in $CHBr_3/C_6H_5NO_2$ gives 8-monobromo and 8,8′-dibromo derivatives of this compound [21].

The crystal and molecular structure of methyltriethylammonium μ-8,8′-oxa-3,3′-commobis-(undecahydro-1,2-dicarba-3-cobalta-*closo*-dodecaborate)(1−), $[N(C_2H_5)_3CH_3][O(C_2B_9H_{10})_2Co]$, has been determined by the heavy-atom method. The crystals are orthorhombic with a = 12.508(7), b = 15.935(2), c = 12.618(2) Å; D_{calc} = 1.199 g/cm³; Z = 4; space group $Pna2_1$-C^9_{2v} (No. 33) [22].

The $[(C_2B_9H_{12})_2Co]^-$ anion has been used in the determination of ^{137}Cs in urine [23], in the extraction of ^{137}Cs from milk [24], in the extraction of ^{137}Cs from biological materials [25], for the extraction of ^{137}Cs [26, 27] and ^{90}Sr [27] from aqueous solutions, for the extraction of alkali metal cations [28], for the determination of strontium-90 in urine [29], for radiostrontium extraction [30], in the extraction of ^{89}Sr and ^{90}Sr from milk [31], for the extraction of Sr^{2+} and Ba^{2+} from aqueous solutions [32], for the extraction of ^{90}Sr and ^{137}Ba from aqueous solutions [33], in the extraction of Ba^{2+} [34], in the extraction of Ce^{3+} [35], for the extraction of Eu^+ from aqueous solutions [36], and the extraction of Pd^{II} [37]. A chlorinated derivative of $[(C_2B_9H_{11})_2Co]^-$ has been used for the extraction of ^{137}Cs from aqueous solutions [41].

The diffusion potentials of the hydrophobic anion $[(1,2\text{-}C_2B_9H_{11})_2Co]^-$ through bimolecular lipid membranes have been recorded [38]. Toxicity tests on $K[Co(C_2B_9H_{11})_2]$ have been conducted [39]. Accumulation of $K[Co(C_2B_9H_{11})_2]$ in animal tissues for neutron capture therapy has been studied [40].

$Cs[Fe(C_2B_9H_{11})_2]$ is prepared in 91% yield by the electrolysis of $K[C_2B_9H_{12}]$ in a 0.1 N NaBr solution in $(CH_3)_2SO$ with an Fe anode followed by treatment with CsCl [42]. Crystals of 8,8'-$CH_3O(C_2B_9H_{10})_2Fe$ are orthorhombic, space group $P2_12_12_1$-D^4_2 (No. 19), with a = 12.128(2), b = 21.521(3), c = 6.991(4) Å [43].

$[(C_2H_5)_3PRhC_2B_9H_{10}]_2$ is prepared by treating $LRh[P(C_2H_5)_3]Cl$ (L = cyclooctadiene) with $Cs_2[7\text{-}(7'\text{-}7',8'\text{-}C_2B_9H_{11})\text{-}7,8\text{-}C_2B_9H_{11}]$, and its crystal structure was determined [44].

3,3'-$Pt(1,2\text{-}C_2B_9H_{11})_2$ is formed by treating 7,8-$C_2B_9H_{13}$ with chloroplatinic acid [45].

References for 13.16:

[1] Janoušek, Z.; Plešek, J.; Štíbr, B.; Heřmánek, S. (Collection Czech. Chem. Commun. **48** [1983] 228/31).
[2] Subrtova, V.; Linek, A.; Hasek, J. (Acta Cryst. B **38** [1982] 3147/9).
[3] Aono, K.; Totani, T. (J. Chem. Soc. Dalton Trans. **1981** 1190/5).
[4] Shevchenko, Y. N.; Davidenko, N. K.; Svitsyn, R. A.; Yashina, N. I.; Egorova, N. V. (Zh. Neorgan. Khim. **25** [1980] 2180/6; Russ. J. Inorg. Chem. **25** [1980] 1207/12).
[5] Nishimura, E. K. (Inorg. Nucl. Chem. Letters **17** [1981] 269/71).
[6] Janoušek, Z.; Plešek, J.; Heřmánek, S.; Baše, K.; Todd, L. J.; Wright, W. F. (Collection Czech. Chem. Commun. **46** [1981] 2818/33).
[7] Petricek, V.; Maly, K.; Petrina, A.; Base, K.; Linek, A. (Z. Krist. **166** [1984] 1/10).
[8] Geiger, W. E.; Brennan, D. E.; Little, J. L. (Inorg. Chem. **21** [1982] 2529/31).
[9] Plešek, J.; Štíbr, B.; Heřmánek, S. (Collection Czech. Chem. Commun. **49** [1984] 1492/6).
[10] Borodinsky, L.; Sinn, E.; Grimes, R. N. (Inorg. Chem. **21** [1982] 1686/9).

[11] Hung, L. Q. (J. Electroanal. Chem. Interfacial Electrochem. **115** [1980] 159/74).
[12] Issaurat, B.; Amblard, G.; Gavach, C. (Bioelectrochem. Bioenerg. **11** [1983] 37/50; C.A. **100** [1984] No. 31168).
[13] Geiger, W. E.; Brennan, D. E. (Inorg. Chem. **21** [1982] 1963/6).
[14] Volkov, V. V.; Ikorskii, V. N.; Dvurechenskaya, S. Ya. (Izv. Akad. Nauk SSSR Ser. Khim. **1983** 252/4; Bull. Acad. Sci. USSR Div. Chem. Sci. **32** [1983] 219/22).
[15] Shevchenko, Yu. N.; Yashina, N. I.; Svitsyn, R. A.; Egorova, N. V. (Zh. Obshch. Khim. **51** [1981] 1258/63; J. Gen. Chem. [USSR] **51** [1981] 1065/70).
[16] Koval, M.; Kaniansky, D.; Matel, L.; Macasek, F. (J. Chromatogr. **243** [1982] 144/8).

[17] Erdman, A. A.; Zubreichuk, Z. P.; Shirokii, V. L.; Maier, N. A.; Ol'dekop, Yu. A. (Vestsi Akad. Navuk Belarusk.SSR Ser. Khim. Navuk **1984** 86/8; C.A. **101** [1984] No. 192180).
[18] Plzak, Z.; Plešek, J.; Štíbr, B. (J. Chromatogr. **212** [1981] 283/93).
[19] Matel, L.; Macasek, F.; Rajec, P.; Heřmánek, S.; Plešek, J. (Polyhedron **1** [1982] 511/9).
[20] Svec, A.; Rajec, P.; Matel, L. (Chem. Listy **75** [1981] 987/90; C.A. **96** [1982] No. 21677).

[21] Matel, L.; Macasek, F.; Kamenista, H. (Radiochem. Radioanal. Letters **46** [1981] 1/6).
[22] Petřina, A.; Petříček, V.; Malý, K.; Šubrtová, V.; Linek, A.; Hummel, L. (Z. Krist. **154** [1981] 217/26).
[23] Scasnar, V.; Koprda, V. (J. Radioanal. Chem. **59** [1980] 389/98).
[24] Koprda, V.; Scasnar, V. (Radiochem. Radioanal. Letters **44** [1980] 349/58).
[25] Koprda, V.; Scasnar, V. (Chem. Zvesti **34** [1980] 480/96; C.A. **94** [1981] No. 1590).
[26] Scasnar, V.; Koprda, V. (Radiochem. Radioanal. Letters **50** [1982] 333/43).
[27] Koprda, V.; Scasnar, V.; Galan, P. (J. Radioanal. Chem. **80** [1983] 55/62).
[28] Skarda, V.; Rais, J.; Kyrs, M. (J. Inorg. Nucl. Chem. **41** [1979] 1443/6).
[29] Scasnar, V. (Anal. Chem. **56** [1984] 605/8).
[30] Scasnar, V.; Koprda, V. (Chem. Zvesti **36** [1982] 379/87; C.A. **97** [1982] No. 68571).

[31] Koprda, V.; Scasnar, V. (J. Radioanal. Chem. **77** [1983] 71/8).
[32] Vanura, P.; Makrlik, E.; Rais, J.; Kyrs, M. (Collection Czech. Chem. Commun. **47** [1982] 1444/64).
[33] Selucky, P.; Vanura, P.; Rais, J.; Kyrs, M. (Radiochem. Radioanal. Letters **38** [1979] 297/302).
[34] Podzimek, I.; Kyrs, M.; Rais, J. (J. Inorg. Nucl. Chem. **42** [1980] 1481/6).
[35] Podzimek, I.; Kyrs, M.; Rais, J.; Vanura, P. (Polyhedron **2** [1983] 331/7).
[36] Vanura, P.; Benešova, M.; Makrlik, E.; Kyrs, M.; Rais, J. (Collection Czech. Chem. Commun. **49** [1984] 1367/81).
[37] Prihoda, J.; Kyrs, M. (J. Radioanal. Chem. **80** [1983] 49/54).
[38] Amblard, G.; Issaurat, B.; D'Epenoux, B.; Gavach, C. (J. Electroanal. Chem. Interfacial Electrochem. **144** [1983] 373/90).
[39] Spryshkova, R. A.; Karaseva, L. I.; Bratsev, V. A.; Serebryakov, N. G. (Med. Radiol. **26** [1981] 62/4; C.A. **95** [1981] No. 108411).
[40] Spryshkova, R. A.; Bratsev, V. A.; Sherman, T. L.; Stanko, V. I. (Med. Radiol. **26** [1981] 51/5; C.A. **97** [1982] No. 158872).

[41] Koprda, V.; Scasnar, V. (J. Radioanal. Chem. **51** [1979] 245/52).
[42] Shirokii, V. L.; Erdman, A. A.; Zubreichuk, Z. P.; Maier, N. A.; Ol'dekop, Yu. A. (Zh. Obshch. Khim. **53** [1983] 951/2; C.A. **99** [1983] No. 53923).
[43] Subrtova, V.; Maly, K.; Petricek, V.; Linek, A. (Acta Cryst. B **38** [1982] 2028/31).
[44] Behnken, P. E.; Knobler, C. B.; Hawthorne, M. F. (Angew. Chem. Intern. Ed. Engl. **22** [1983] 722/3).
[45] Chernyshev, E. A.; Knyazeva, L. K.; Belyakova, Z. V.; Kisin, A. V.; Kirillova, N. I.; Gusev, A. I.; Alekseev, N. V. (Zh. Obshch. Khim. **53** [1983] 1433/4; J. Gen. Chem. [USSR] **53** [1983] 1289/90).

13.17 Carboranes Containing 19 Boron Atoms

Action of KOH/C_2H_5OH on $(1,2\text{-}C_2B_{10}H_{11}\text{-}1\text{-}CH_2CH_2CH_2\text{-})_2$, followed by addition of $[N(CH_3)_4]Cl$, yields $\mathbf{[N(CH_3)_4][7,8\text{-}C_2B_9H_{11}\text{-}7\text{-}(1',2'\text{-}C_2B_{10}H_{10}\text{-}1'\text{-}(CH_2)_6\text{-})]}$ [1].

The crystal structure of the salt **$[(C_6H_5)_3P]_3Rh[$*nido*$-7,8-C_2B_9H_{11}R]$** [R = 1′-(*closo*-1′,2′-$C_2B_{10}H_{11}$)] has been determined. The monoclinic crystals have lattice parameters a = 14.819(4), b = 25.429(6), c = 17.578(4) Å; β = 98.65(2)°; Z = 4; D_{meas} = 1.257, D_{calc} = 1.267 g/cm³; space group $P2_1/c-C_{2h}^5$ (No. 14) [2]. ***closo-nido*-$[RCB_{10}H_{10}CN{=}NCB_9H_{10}CR]^-$** (R = H, CH_3, C_6H_5) is formed by KOH degradation of $RCB_{10}H_{10}CN{=}NCB_{10}H_{10}CR$ [3]. *closo-nido*-$[RCB_{10}H_{10}$-$CNHNHCB_9H_{10}CR]^{-1}$ (R = H, CH_3, C_6H_5) and $[RCB_{10}H_{10}CNR''NR'CB_9H_{10}CR]^{-1}$ (R = H, CH_3; R′ = R″ = C_2H_5) are formed from the action of piperidine base on the *closo-closo* compounds $RCB_{10}H_{10}CNHNHCB_{10}H_{10}CR$ and $RCB_{10}H_{10}CNR''NR'CB_{10}H_{10}CR$, respectively. Free energy barriers have been estimated for interconversion of the diastereo isomers of $[RCB_{10}H_{10}CN(CH_3)$-$N(CH_3)CB_9H_{10}CR]^-$ (R = H, CH_3) [4].

closo-2,2-$((C_6H_5)_3P)_2$-1-H-8-(1′-*closo*-1′,2′-$C_2B_{10}H_{11}$)-2,1,8-$IrC_2B_9H_{10}$ is prepared in 41% yield by heating $[(cod)Ir(PPh_3)_2]$[*nido*-7-(1′-*closo*-1′,2′-$C_2B_{10}H_{11}$)-7,8-$C_2B_9H_{11}$] (cod = cyclooctadiene) in cyclohexane at the reflux temperature [5].

References for 13.17:

1] Totani, T.; Aono, K.; Yamamoto, K.; Tawara, K. (J. Med. Chem. **24** [1981] 1492/9).
[2] Knobler, C. B.; Marder, T. B.; Mizusawa, E. A.; Teller, R. G.; Long, J. A.; Behnken, P. E.; Hawthorne, M. F. (J. Am. Chem. Soc. **106** [1984] 2990/3004).
[3] Aono, K.; Totani, T. (J. Chem. Soc. Dalton Trans. **1981** 1190/5).
[4] Aono, K.; Totani, T. (J. Chem. Soc. Dalton Trans. **1981** 1196/203).
[5] Doi, J. A.; Mizusawa, E. A.; Knobler, C. B.; Hawthorne, M. F. (Inorg. Chem. **23** [1984] 1482/4).

13.18 Carborane Derivatives Containing 20 Boron Atoms

For previous data on carboranes containing 20 boron atoms, see "Boron Compounds" 2nd Suppl. Vol. 2, 1982, pp. 330/4.

$(1,2-C_2B_{10}H_{11}-9-)_2S_2$, melting point 332 to 333°C, is obtained from the reaction of $1,2-C_2B_{10}H_{12}$ with $S_2Cl_2/AlCl_3$. Similarly, $(1,7-C_2B_{10}H_{11}-9-)_2S_2$, melting point 312 to 314°C, is formed from $1,7-C_2B_{10}H_{12}$ and $S_2Cl_2/AlCl_3$ [1]. The molecule $(1,2-C_2B_{10}H_{11}-9-S-)_2CH_2$ consists of two carborane icosahedra connected by an S-CH_2-S chain, as determined by X-ray crystallography; crystals of the compound are monoclinic, space group $P2_1-C_2^2$ (No. 4), with a = 12.491(7), b = 14.103(19), c = 11.974(9) Å; β = 92.52(5)°; Z = 2; atomic parameters are given [2].

$[1,2-C_2B_{10}H_{10}-1-R-2-CH_2CH(CH_2SO_2C_6H_5)S-]_2$ (R = $CH(CH_3)_2$, C_6H_5) is formed from the oxidation of $1,2-C_2B_{10}H_{10}-1-R-2-CH_2CH(SH)CH_2SC_6H_5$ with H_2O_2; the melting points of the products are 190 to 191°C for R = $CH(CH_3)_2$ and 199 to 200°C for R = C_6H_5 [3].

Reaction of $1,2-C_2B_{10}H_{10}-1,2-(SH)_2$ with $I_2/NaHCO_3$ produces **$[1,2-C_2B_{10}H_{10}-1,2-(S-)_2]_2$** (melting point 235°C) containing an eight-membered $S_2C_2S_2C_2$ ring. The ^{11}B NMR spectrum of $[1,2-C_2B_{10}H_{10}-1,2-(S-)_2]_2$ shows resonances at −3.67 ppm, J(B-H) = 141 Hz, and at −7.27 ppm, J(B-H) = 161 Hz, in an area ratio of 1:4, respectively [4]. $[1,2-C_2B_{10}H_{10}-1,2-(SCH_2-)_2]_2$, containing a 12-membered $C_2SC_2SC_2SC_2S$ ring, is prepared by the action of KOH on $1,2-C_2B_{10}H_{10}-1,2-(SH)_2$ followed by the addition of $BrCH_2CH_2Br$. The ^{11}B NMR spectrum of $[1,2-C_2B_{10}H_{10}-1,2-(SCH_2-)_2]_2$ shows resonances at −3.56 and −7.32 ppm in an area ratio of 1:4, respectively [4].

Electrophilic ring cleavage of $1,2-C_2B_{10}H_{10}-1-R-2-CH_2(CHSCH_2)$, in which $CHSCH_2$ is a three-membered C_2S ring, R = CH_3, $CH(CH_3)_2$, C_6H_5, with Cl_2, Br_2, or I_2 produces the disulfides **$[1,2-C_2B_{10}H_{10}-1-R-2-CH_2CH(S-)CH_2X]_2$**, X = Cl, Br, I; melting points (in °C) for the latter compounds: 99 to 100 for R = $CH(CH_3)_2$, X = Cl; 161 to 162 for R = C_6H_5, X = Cl; 142 to 143 for

References for 13.18 on pp. 250/2

R = CH_3, X = Br; 135 to 136 for R = $CH(CH_3)_2$, X = Br; 144 to 145 for R = C_6H_5, X = Br; 144 to 145 for R = CH_3, X = I; 124 to 125 for R = $CH(CH_3)_2$, X = I; 159 to 160 for R = C_6H_5, X = I [5].

(1,2-$C_2B_{10}H_{11}$-9-Se-)$_2$ is prepared by the reaction of 1,2-$C_2B_{10}H_{12}$ with Se/$AlCl_3$. In a similar fashion, (1,7-$C_2B_{10}H_{11}$-9-Se-)$_2$ is formed from the reaction of 1,7-$C_2B_{10}H_{12}$ with Se/$AlCl_3$ [6]. (1,2-$C_2B_{10}H_{10}$-1-R-2-)$_2Se_2$ is prepared from (a) the reaction of Na[1,2-$C_2B_{10}H_{10}$-1-R] with Se/NH_3 followed by oxidation with air or (b) the reaction of Li[1,2-$C_2B_{10}H_{10}$-1-R] with Se in $(C_2H_5)_2O$ or C_6H_6 followed by oxidation; melting points (in °C) 250 to 252 for R = H, 197 to 199 for R = C_6H_5, 163 to 164 for R = C_3H_7, 238 to 240 for R = CH_3 [7]. (1,7-$C_2B_{10}H_{11}$-7-)$_2Se_2$, melting point 243 to 244, is prepared from the reaction of Na[1,7-$C_2B_{10}H_{11}$] with Se/NH_3 followed by oxidation with air [7].

(1,*x*-$C_2B_{10}H_{11}$-9-)$_2Te_2$ ($x = 2$, 7) is formed from the reaction of 1,*x*-$C_2B_{10}H_{11}$-9-$TeCl_3$ with $Na_2S/C_2H_5OH/H_2O$; the melting point of the compound is 256 to 247°C for $x = 2$ and 219 to 220°C for $x = 7$ [8].

Treatment of (1,2-$C_2B_{10}H_{11}$-9-)$_2$Hg with Se in tetrahydrofuran or $CH_3OCH_2CH_2OCH_3$ yields (1,2-$C_2B_{10}H_{11}$-9-)HgSe(1,2-$C_2B_{10}H_{11}$-9-), melting point 273 to 274°C. Analogously, reaction of (1,2-$C_2B_{10}H_{11}$-9-)$_2$Hg with Te gives (1,2-$C_2B_{10}H_{11}$-9-)HgTe(1,2-$C_2B_{10}H_{11}$-9-), melting point 141 to 142°C. Fusion of (1,2-$C_2B_{10}H_{11}$-9-)$_2$Hg with Se at 270 to 300°C provides the diselenide (1,2-$C_2B_{10}H_{11}$-9-)$_2Se_2$, melting point 280 to 281°C. Analogously, treatment of (1,7-$C_2B_{10}H_{11}$-9-)$_2$Hg with Se in tetrahydrofuran or $CH_3OCH_2CH_2OCH_3$ yields (1,7-$C_2B_{10}H_{11}$-9-)HgSe(1,7-$C_2B_{10}H_{11}$-9-), melting point 224 to 226°C, and the fusion of (1,7-$C_2B_{10}H_{11}$-9-)$_2$Hg with Se at 270 to 300°C provides the diselenide (1,7-$C_2B_{10}H_{11}$-9-)$_2Se_2$, melting point 259 to 261°C [9].

Reaction of (1,2-$C_2B_{10}H_{11}$-3-NH_2-1-CH_2-)$_2$ with maleic anhydride and acetic anhydride yields **(1,2-$C_2B_{10}H_{11}$-3-N(-COCH=CHCO-)-1-CH_2)$_2$**, melting point 345 to 350°C [10].

Oxidation of the $[HCB_{10}H_{10}CCB_{10}H_{10}CNH_2]^{4-}$ anion by $KMnO_4$ in liquid NH_3 gives exclusively the hydrazo compound $HCB_{10}H_{10}CCB_{10}H_{10}CNHNHCB_{10}H_{10}CCB_{10}H_{10}CH$. The 1,1'-azo-1,2-carboranes **$RCB_{10}H_{10}CN{=}NCB_{10}H_{10}CR$** (melting points, in °C: 270 for R = H, 225 for R = CH_3, 243 to 244.5 for R = C_6H_5) are formed from the oxidation of $[1,2\text{-}C_2B_{10}H_{10}\text{-}1\text{-}R\text{-}2\text{-}NH_2]^{2-}$ anions by $KMnO_4$ in liquid NH_3. $RCB_{10}H_{10}CN{=}NCB_{10}H_{10}CR$ is reduced to $RCB_{10}H_{10}CNHNHCB_{10}H_{10}CR$, and degraded by KOH to ***closo-nido*-$[RCB_{10}H_{10}CN{=}NCB_9H_{10}CR]^-$** and ***nido-nido*-$[RCB_9H_{10}CN{=}NCB_9H_{10}CR]^{2-}$** [11]. $RCB_{10}H_{10}CN{=}NCB_{10}H_{10}CR$ (R = H, CH_3, C_6H_5) reacts with R'Li (R' = CH_3, C_2H_5, C_6H_5, 4-ClC_6H_4, 4-$(CH_3)_2N$-C_6H_4) to produce the N-monosubstituted derivatives **$RCB_{10}H_{10}CNHNR'CB_{10}H_{10}CR$**, along with small amounts of the hydrazocarboranes $RCB_{10}H_{10}$-$CNHNHCB_{10}H_{10}CR$. Further nitrogen substitution occurs when $RCB_{10}H_{10}CNHNR'CB_{10}H_{10}CR$ reacts with NaH and subsequently with R"X (R" = CH_3, C_2H_5, C_4H_9, CH_3SO_2; X = halogen) to yield **$RCB_{10}H_{10}CNR''NR'CB_{10}H_{10}CR$**. Variable-temperature ^{1}H NMR spectroscopy was used to estimate the free-energy barriers to conformational interconversion: 15.5 kcal/mol for $RCB_{10}H_{10}$-$CNHNR'CB_{10}H_{10}CR$, (R = CH_3, R' = C_2H_5) and >22 kcal/mol for $RCB_{10}H_{10}CNR''NR'CB_{10}H_{10}CR$ (R = H, CH_3; R' = R" = C_2H_5). The *closo-closo* compounds $RCB_{10}H_{10}CNHNHCB_{10}H_{10}CR$ and $RCB_{10}H_{10}CNR''NR'CB_{10}H_{10}CR$ can be converted into the corresponding *closo-nido* ones, $[RCB_{10}H_{10}CNHNHCB_9H_{10}CR]^-$ and $[RCB_{10}H_{10}CNR''NR'CB_9H_{10}CR]^-$, in the presence of piperidine base [12].

(1,2-$C_2B_{10}H_{11}$-3)-NHCONH-(1,2-$C_2B_{10}H_{11}$-3-), melting point 290°C, is obtained from 1,2-$C_2B_{10}H_{11}$-3-NH_2/$KMnO_4$/$(CH_3)_2CO$ [13].

(1,2-$C_2B_{10}H_{11}$-1-CH_2O-)$_2$P(R)X is formed from the reaction of $B_{10}H_{14}$ with $(HC{\equiv}CCH_2O)_2P(R)X$ in the presence of $C_6H_5N(CH_3)_2$; melting points, in °C, 160 to 161 for R = CH_3, X = S; 122 to 124 for R = CH_3, X = O; 125 to 127 for R = C_6H_5, X = O [14].

(1,7-$C_2B_{10}H_{10}$-1-CH_3-7-)$_2$AsCl, boiling point 200°C (1 Torr), is formed from 1,7-$C_2B_{10}H_{10}$-1-$CH(CH_3)_2$-7-Li/$AsCl_3$. (1,7-$C_2B_{10}H_{10}$-1-CH_3-7-)$_2$AsCl reacts with Cl_2/H_2O to form (1,7-$C_2B_{10}H_{10}$-

$1-CH_3-7-)_2AsO_2H$. $(1,2-C_2B_{10}H_{10}-1-CH(CH_3)_2-2-)_2AsCl$, melting point 151 to 152°C, is formed from $1,2-C_2B_{10}H_{10}-1-CH(CH_3)_2-2-Li/AsCl_3$. $(1,2-C_2B_{10}H_{10}-1-CH(CH_3)_2-2-)_2AsCl$ reacts with Cl_2/H_2O to yield $(1,2-C_2B_{10}H_{10}-1-CH(CH_3)_2-2-)_2AsO_2H$ [15].

$(1,2-C_2B_{10}H_{11}-1-)_2CH_2$, prepared by heating $1,2-C_2B_{10}H_{11}-1-CH_2C{\equiv}CH$ with $B_{10}H_{14}$ in the presence of CH_3CN, shows carborane C-H stretching frequencies in the infrared spectrum at 3055 and 3097 cm^{-1} in the solid phase. In contrast, $(1,2-C_2B_{10}H_{11}-1-CH_2-)_2$ has only one carborane C-H stretching frequency at 3061 cm^{-1} [16]. $(1,2-C_2B_{10}H_{10}-1-CH_3-2-CH_2-)_2CH_2$, melting point 258 to 260°C, is obtained from the reaction of $1,2-C_2B_{10}H_{10}-1-CH_3-2-Li$ and $BrCH_2CH_2CH_2Cl$. $(1,2-C_2B_{10}H_{11}-1-CH_2CH_2CH_2-)_2$, melting point 162 to 163°C, is produced from the reaction of $(CH_3CN)_2B_{10}H_{12}$ with 1,9-decadiyne [17].

Treatment of $1,2-C_2B_{10}H_{10}-1-Li-2-(1',2'-C_2B_{10}H_{11}-1'-)$ with $C_6H_5N_3$ produces **$1,2-C_2B_{10}H_{10}-1-N_3(H)C_6H_5-2-(1',2'-C_2B_{10}H_{11}-1'-)$**, melting point 141 to 143°C; the latter compound reacts with CH_3COOH to give **$1,2-C_2B_{10}H_{10}-1-NH_2-2-(1',2'-C_2B_{10}H_{11}-1'-)$**. The latter reacts with NOCl to form **$1,2-C_2B_{10}H_{10}-1-Cl-2-(1',2'-C_2B_{10}H_{11}-1'-)$** [17].

$1,4-(1',2'-C_2B_{10}H_{10}-1'-CH_3COOCH_2-2'-)_2-C_6H_4$, melting point 199 to 200°C, is formed from $B_{10}H_{14}/1,4-(CH_3COOCH_2C{\equiv}C-)_2-C_6H_4/C_6H_5N(CH_3)_2$; reaction of $1,4-(1',2'-C_2B_{10}H_{10}-1'-CH_3COOCH_2-2'-)_2-C_6H_4$ with CH_3OH/HCl gives **$1,4-(1',2'-C_2B_{10}H_{10}-1'-HOCH_2-2'-)_2-C_6H_4$**, melting point 314 to 315°C [18].

$[(1,7-C_2B_{10}H_{11}-1-)C(OH)(CH_3)C_5H_4]_2Fe$, melting point 219 to 222°C, is prepared from the reaction of $Li_2[1,7-C_2B_{10}H_{10}]$ with $(CH_3COC_5H_4)Fe(C_5H_5)$; dehydration of $[(1,7-C_2B_{10}H_{11}-1-)C(OH)(CH_3)C_5H_4]_2Fe$ yields $[(1,7-C_2B_{10}H_{11}-1-)C(=CH_2)C_5H_4]_2Fe$, melting point 126 to 128°C [19].

An X-ray structure determination of $Cl_2Si(1,2-C_2B_{10}H_{10})_2SiCl_2$ with a six-membered $-SiC_2SiC_2-$ ring shows the crystal to be monoclinic, space group $C2/c-C^6_{2h}$ (No. 15), with a = 17.352(3), b = 6.891, c = 19.096(4) Å; β = 90.36(1)°; Z = 4; all B-B distances are normal [20]. $[1,2-C_2B_{10}H_{10}-1-CH_3-2-CH_2Si(CH_3)_2-]_2O$, melting point 120 to 121°C, is formed by the action of LiC_4H_9 on $1,2-C_2B_{10}H_{10}-1-CH_3-2-CH_2Si(CH_3)_2OCH_3$. $1,2-C_2B_{10}H_{10}-1-CH_3-2-CH_2Si(CH_3)_2(1',2'-C_2B_{10}H_{10}-1'-CH_3-2'-)$ is produced from $1,2-C_2B_{10}H_{10}-1-CH_3-2-Li$ and $1,2-C_2B_{10}H_{10}-1-CH_3-2-CH_2Si(CH_3)_2Cl$ [21]. Crystals of $[((CH_3)_3SiO)_2(RCH_2)Si]_2O$, where $R = 1,2-C_2B_{10}H_{11}-1-$, are monoclinic with space group $P2_1/n-C^5_{2h}$ (No. 14), a = 10.521(7), b = 20.31(1), c = 11.270(7) Å; β = 107.45(5)°; Z = 2, D_{calc} = 1.075 g/cm³ [22]. **$(1,2-C_2B_{10}H_{10}-1-CH_3-x-)_2SiC_3H_6$ (x = 2, 7) is prepared by the reaction of $Li[1,x-C_2B_{10}H_{10}-1-CH_3]$ and $Cl_2SiC_3H_6$ (where SiC_3H_6 is silacyclobutane ring); the melting points of $(1,2-C_2B_{10}H_{10}-1-CH_3-x-)_2SiC_3H_6$ are 300 to 301°C for x = 2 and 150 to 151°C for x = 7 [23].**

Crystals of $(CH_3)_2Ge(1,2-C_2B_{10}H_{10}-1,2-)_2(CH_2)_2$ with a seven-member GeC_6 ring are monoclinic, a = 7.1383(4), b = 32.337(4), c = 28.943(4) Å; β = 91.59(1)°; Z = 12; space group $P2_1/c-C^5_{2h}$ (No. 14) [24]. Treatment of $HGe(C_6F_5)_2Ge(C_6F_5)_2H$ with $1,7-C_2B_{10}H_{11}-9-HgCH_3$ results in a compound with a B-Hg-Ge-Ge-Hg-B chain containing two carborane cages at the ends of the chain. The prepared compound undergoes oxidative insertion with $Pt(P(C_6H_5)_3)_n$ (n = 3, 4) to give the chain B-Hg-Pt-Ge-Ge-Pt-Hg-B [25].

Reaction of $(1,2-C_2B_{10}H_{11}-9-)_2Hg$ with naphthyllithium gives $[(1,2-C_2B_{10}H_{11}-9-)_2Hg]^{4-}$. Reaction of $(1,2-C_2B_{10}H_{11}-9-)_2Hg$ with LiC_4H_9 and subsequently with $(CH_3)_3SiCl$ gives $[1,2-C_2B_{10}H_9-1-Si(CH_3)_3-9-]_2Hg$; similarly, reaction of $(1,7-C_2B_{10}H_{11}-9-)_2Hg$ with LiC_4H_9 and subsequently with $(CH_3)_3SiCl$ gives $[1,7-C_2B_{10}H_9-1,7-[Si(CH_3)_3]_2-9-]_2Hg$. Treating $(1,2-C_2B_{10}H_{10}-1-CH_3-9-)_2Hg$ with LiC_4H_9 and subsequently with CH_3I yields $(1,2-C_2B_{10}H_9-1,2-(CH_3)_2-9-)_2Hg$ [26]. $(1,7-C_2B_{10}H_{10}-1-CH_3-7-CH_2-)_2Hg$, melting point 242 to 244°C, is prepared from the reaction of $1,7-C_2B_{10}H_{10}-1-CH_3-7-CH_2MgBr$ with $HgCl_2$ [27]. Treatment of $(1,7-C_2B_{10}H_{11}-9-)_2Hg$ with BrF_3 produces, after addition of $(C_2H_5)_2O \cdot BF_3$, the bromonium salt **$[(1,7-C_2B_{10}H_{11}-9-)_2Br][BF_4]$**. Treatment of

References for 13.18 on pp. 250/2

$(1,7\text{-}C_2B_{10}H_{11}\text{-}9\text{-})_2Hg$ with $C_6H_5BrF_2$ produces, after addition of $(C_2H_5)_2O \cdot BF_3$, the phenylbromonium salt **$[(1,7\text{-}C_2B_{10}H_{11}\text{-}9\text{-})BrC_6H_5][BF_4]$** [28].

$(1,2\text{-}C_2B_{10}H_{11}\text{-}9\text{-})HgSn(acac)_2\text{-}(\text{-}9\text{-}1,2\text{-}C_2B_{10}H_{11})$, melting point 171 to 173°C, is formed from the action of $Sn(acac)_2$ on $(1,2\text{-}C_2B_{10}H_{11}\text{-}9\text{-})_2Hg$ (acac = acetylacetonate = $C_5H_7O_2$). Similarly, $1,2\text{-}C_2B_{10}H_{11}\text{-}9\text{-}Sn(acac)_2HgCl$, melting point 186 to 188°C, is produced from the reaction of $Sn(acac)_2$ with $1,2\text{-}C_2B_{10}H_{11}\text{-}9\text{-}HgCl$ [29].

Reaction of $(1,2\text{-}C_2B_{10}H_{11})_2Hg$ with Se gives $(1,2\text{-}C_2B_{10}H_{11})HgSe(1,2\text{-}C_2B_{10}H_{11})$. Further reaction yields $(1,2\text{-}C_2B_{10}H_{11})_2Se_2$. Reaction of $(1,7\text{-}C_2B_{10}H_{11})_2Hg$ with Se or Te gives $(1,7\text{-}C_2B_{10}H_{11})HgM(1,7\text{-}C_2B_{10}H_{11})$ (M = Se or Te). Further substitution of mercury in the Se reaction takes place to a lesser extent to yield $(1,7\text{-}C_2B_{10}H_{11})_2Se_2$. $(1,7\text{-}C_2B_{10}H_{11})_2Te$ is obtained from this type of reaction as well as from the reaction of $1,7\text{-}C_2B_{10}H_{11}Tl(OOCCF_3)_2$ with tellurium, iodide ion, and dimethylformamide [30].

$(1,2\text{-}C_2B_{10}H_{10}\text{-}1\text{-}R\text{-})_2Hg$ reacts with Yb in tetrahydrofuran to give $(1,2\text{-}C_2B_{10}H_{10}\text{-}1\text{-}R\text{-})_2Yb$ (R = H, CH_3, C_6H_5). Similar reactions using La and Tm yield $(1,2\text{-}C_2B_{10}H_{10}\text{-}1\text{-}R\text{-})_3La$ and $(1,2\text{-}C_2B_{10}H_{10}\text{-}1\text{-}R\text{-})_3Tm$, respectively [31]. Treatment of $(1,2\text{-}C_2B_{10}H_{10}\text{-}x\text{-}R\text{-}2\text{-})_2Hg$ (in which x = 1, 9; R = CH_3, C_6H_5) with $LnCl_3$ (Ln = Tm, Yb) in tetrahydrofuran gives $1,2\text{-}C_2B_{10}H_{10}\text{-}x\text{-}R\text{-}2\text{-}LnCl_2$ as tetrahydrofuran adducts [32].

Bis(B-carboranyl)thallium acetates (where carboranyl = $C_2B_{10}H_{11}$) are prepared from the action of hot aqueous sodium acetate on carboranylthallium bis(trifluoroacetates). Bis-(B-carboranyl)thallium chlorides are similarly prepared using $LiAlH_4$ followed by NaCl as the reagents [33].

$(1,x\text{-}C_2B_{10}H_{10}\text{-}1\text{-}CH_3\text{-}2\text{-}CH_2\text{-})_2Ti(C_5H_5)$ (x = 2, 7) is obtained by reaction of the corresponding carboranylmethylmagnesium chloride with a halogen complex of Ti [34].

Oxidative insertion of tin(II) acetylacetonate (acetylacetonate = acac) into RHgR′ (R = *closo*-$1,2\text{-}C_2B_{10}H_{11}\text{-}9\text{-}$; R′ = R, CH_3, Cl) gives polymetallic $RHgSn(acac)_2R'$, $R'HgSn(acac)_2R$, and $RSn(acac)_2R'$, in which the acac ligands on the hexacoordinate Sn^{IV} are *cis*. In C_6H_6, $Sn(acac)_2$ inserted into $RHgCH_3$ at the B-Hg bond to give $RSn(acac)_2CH_3$, whereas in C_6H_6/tetrahydrofuran insertion occurs at both the B-Hg and C-Hg bonds to give $RSn(acac)_2CH_3$ and $RHgSn(acac)_2CH_3$ [35].

$(1,2\text{-}C_2B_{10}H_{11}\text{-}1\text{-})_2Ni \cdot L$ (L = bpy = 2,2′-bipyridyl) is formed from the action of $NiBr_2 \cdot L$ on $1,2\text{-}C_2B_{10}H_{11}\text{-}1\text{-}Li$. The preparation of $(1,2\text{-}C_2B_{10}H_{11}\text{-}1\text{-}CO_2)_2Ni \cdot bpy$ has also been reported [36]. Reaction of $Li[1,2\text{-}C_2B_{10}H_{11}]$ with $CoCl_2 \cdot bpy$ yields $(1,2\text{-}C_2B_{10}H_{11}\text{-}1\text{-})_2Co \cdot bpy$. The dilithio salt $Li_2[1,2\text{-}C_2B_{10}H_{10}]$ reacts with $CoCl_2 \cdot bpy$ to produce $(1,2\text{-}C_2B_{10}H_{10}\text{-}1,2\text{-})Co \cdot bpy$ which is proposed to have a CCCo three-membered ring [37]. $[1,2\text{-}C_2B_{10}H_{11}\text{-}1\text{-}P(C_6H_5)_2]_2MX_2$ (M = Co, Ni; X = Cl, Br, SCN) complexes have been prepared by treating a C_2H_5OH solution of MX_2 with a CH_2Cl_2 solution of $1,2\text{-}C_2B_{10}H_{11}\text{-}1\text{-}P(C_6H_5)_2$ [38].

$(1,2\text{-}C_2B_{10}H_{11}\text{-}1\text{-})_2Pd \cdot bpy$ is prepared from $1,2\text{-}C_2B_{10}H_{11}\text{-}1\text{-}Li$ and $PdCl_2 \cdot bpy$ in tetrahydrofuran [39]. $(1,x\text{-}C_2B_{10}H_{11}\text{-}1\text{-}C{\equiv}C)_2M[P(C_6H_5)_2]_2$ (x = 2, 7) is formed from the action of $Cl_2M[P(C_6H_5)_2]_2$ on $1,2\text{-}C_2B_{10}H_{11}\text{-}1\text{-}C{\equiv}CLi$, M = Pd, Pt [40]. Some chemistry of $[1,2\text{-}C_2B_{10}H_{11}\text{-}1\text{-}CH_2P(C_6H_5)_2]_2MCl_2$ (M = Pd, Pt) has been described which leads to intramolecular metallation through a carborane B-H bond [41].

References for 13.18:

[1] Zakharkin, L. I.; Pisareva, I. V. (Izv. Akad. Nauk SSSR Ser. Khim. **1981** 2794/6; Bull. Acad. Sci. USSR Div. Chem. Sci. **30** [1981] 2328/30).

[2] Novák, C.; Šubrtová, V.; Línek, A.; Hašek, J. (Acta Cryst. C **39** [1983] 1393/6).

[3] Kazantsev, A. V.; Shustova, T. V.; Zakharkin, L. I. (Izv. Akad. Nauk SSSR Ser. Khim. **1982** 2134/7; Bull. Acad. Sci. USSR Div. Chem. Sci. **31** [1982] 1884/7).

[4] Teixidor, F.; Rudolph, R. W. (J. Organometal. Chem. **241** [1983] 301/12).

[5] Zakharkin, L. I.; Shustova, T. V.; Kazantsev, A. V. (Zh. Obshch. Khim. **51** [1981] 1071/8; J. Gen. Chem. [USSR] **51** [1981] 893/9).

[6] Zakharkin, L. I.; Pisareva, I. V. (Izv. Akad. Nauk SSSR Ser. Khim. **1982** 718/9; Bull. Acad. Sci. USSR Div. Chem. Sci. **31** [1982] 644).

[7] Zakharkin, L. I.; Krainova, N. Y.; Zhigareva, G. G.; Pisareva, I. V. (Izv. Akad. Nauk SSSR Ser. Khim. **1982** 1650/1; Bull. Acad. Sci. USSR Div. Chem. Sci. **31** [1982] 1468/9).

[8] Zakharkin, L. I.; Pisareva, I. V. (Izv. Akad. Nauk SSSR Ser. Khim. **1984** 472/3; Bull. Acad. Sci. USSR Div. Chem. Sci. **33** [1984] 438).

[9] Bregadze, V. I.; Kampel, V. Ts.; Usayatinskii, A. Ya.; Ponomareva, O. B.; Godovikov, N. N. (Izv. Akad. Nauk SSSR Ser. Khim. **1982** 1434; Bull. Acad. Sci. USSR Div. Chem. Sci. **31** [1982] 1282).

[10] Sergeev, V. A.; Danilova, M. P.; Shitikov, V. K.; Kalinin, V. N.; Zakharkin, L. I. (Izv. Akad. Nauk SSSR Ser. Khim. **1981** 226/7; C.A. **95** [1981] No. 62286).

[11] Aono, K.; Totani, T. (J. Chem. Soc. Dalton Trans. **1981** 1190/5).

[12] Aono, K.; Totani, T. (J. Chem. Soc. Dalton Trans. **1981** 1196/203).

[13] Zakharkin, L. I.; Zhigareva, G. G.; Litonina, E. I. (Zh. Obshch. Khim. **52** [1982] 2367; J. Gen. Chem. [USSR] **52** [1982] 2106).

[14] Rys, E. G.; Godovikov, N. N.; Kabachnik, M. I. (Izv. Akad. Nauk SSSR Ser. Khim. **1983** 2640/4; Bull. Acad. Sci. USSR Div. Chem. Sci. **32** [1983] 2372/5).

[15] Zakharkin, L. I.; Pisareva, I. V. (Zh. Obshch. Khim. **51** [1981] 1280/8; J. Gen. Chem. [USSR] **51** [1981] 1084/91).

[16] Leites, L. A.; Vinogradova, L. E.; Kovredov, A. I.; Shaugumbekova, Z. S.; Zakharkin, L. I. (Izv. Akad. Nauk SSSR Ser. Khim. **1982** 2170/2; Bull. Acad. Sci. USSR Div. Chem. Sci. **31** [1982] 1920/2).

[17] Totani, T.; Aono, K.; Yamamoto, K.; Tawara, K. (J. Med. Chem. **24** [1981] 1492/9).

[18] Kovredov, A. I.; Shaugumbekova, Z. S.; Kazantsev, A. V.; Zakharkin, L. I. (Zh. Obshch. Khim. **54** [1984] 577/9; J. Gen. Chem. [USSR] **54** [1984] 509/10).

[19] Sosin, S. L.; Alekseeva, V. P.; Litvinova, M. D.; Ezhova, T. M.; Shevchenko, Y. V. (Dokl. Akad. Nauk SSSR **271** [1983] 871/4; Dokl. Chem. Proc. Acad. Sci. USSR **268/273** [1983] 249/52).

[20] Ryan, R. R.; Schaeffer, R. (Cryst. Struct. Commun. **10** [1981] 133/5).

[21] Kalinin, V. N.; Izmailov, B. A.; Kazantsev, A. A.; Myakushev, V. D.; Zhdanov, A. A.; Zakharkin, L. I. (J. Organometal. Chem. **216** [1981] 295/320).

[22] Yanovskii, A. I.; Dubchak, I. L.; Shklover, V. E.; Struchkov, Yu. T.; Kalinin, V. N.; Izmailov, B. A.; Myakushev, V. D.; Zakharkin, L. I. (Zh. Strukt. Khim. **23** No. 8 [1982] 88/97; J. Struct. Chem. [USSR] **23** [1982] 728/36).

[23] Vdovin, V. M.; Bespalova, N. B.; Kalinin, V. N.; Popov, A. V.; Zakharkin, L. I. (Zh. Obshch. Khim. **54** [1984] 1197/8; J. Gen. Chem. [USSR] **54** [1984] 1072/3).

[24] Yanovskii, A. I.; Timofeeva, T. V.; Struchkov, Yu. T.; Shemyakin, N. F.; Zakharkin, L. I. (Izv. Akad. Nauk SSSR Ser. Khim. **1980** 1150/2; C.A. **93** [1980] No. 204783).

[25] Bochkarev, M. N.; Fedorova, E. A.; Razuvaev, G. A.; Bregadze, V. I.; Kampel, V. Ts. (J. Organometal. Chem. **265** [1984] 117/22).

[26] Bregadze, V. I.; Kampel, V. Ts.; Godovikov, N. N. (Izv. Akad. Nauk SSSR Ser. Khim. **1981** 2594/6; Bull. Acad. Sci. USSR Div. Chem. Sci. **30** [1981] 2154/6).

[27] Zakharkin, L. I.; Kovredov, A. I.; Kazantsev, A. V.; Meiramov, M. G. (Zh. Obshch. Khim. **51** [1981] 357/61; J. Gen. Chem. [USSR] **51** [1981] 289/92).

[28] Grushin, V. V.; Tolstaya, T. P.; Lisichkina, I. N. (Izv. Akad. Nauk SSSR Ser. Khim. **1982** 2412; Bull. Acad. Sci. USSR Div. Chem. Sci. **31** [1982] 2127).
[29] Suleimanov, G. Z.; Bregadze, V. I.; Kampel, V. T.; Petriashvili, M. V.; Godovikov, N. N.; Sokolov, V. I. (Izv. Akad. Nauk SSSR Ser. Khim. **1982** 1606/7; Bull. Acad. Sci. USSR Div. Chem. Sci. **1983** 1427/8).
[30] Bregadze, V. I.; Kampel, V. Ts.; Usyatinskii, A. Ya.; Ponomareva, O. B.; Godovikov, N. N. (J. Organometal. Chem. **233** [1982] C33/C34).

[31] Suleimanov, G. Z.; Bregadze, V. I.; Koval'chuk, N. A.; Beletskaya, I. P. (J. Organometal. Chem. **235** [1982] C17/C18).
[32] Suleimanov, G. Z.; Koval'chuk, N. A.; Bregadze, V. I.; Kurbanov, T. K.; Agaeva, R. A.; Godovikov, N. N.; Beletskaya, I. P. (Zh. Org. Khim. **19** [1983] 2258/63; Russ. J. Org. Chem. **19** [1983] 1969/73).
[33] Bregadze, V. I.; Usyatinskii, A. Ya.; Godovikov, N. N. (Izv. Akad. Nauk SSSR Ser. Khim. **1981** 1426/8; Bull. Acad. Sci. USSR Div. Chem. Sci. **30** [1981] 1146/8).
[34] Kovredov, A. I.; Meiramov, M. G.; Kazantsev, A. V.; Zakharkin, L. I. (Zh. Obshch. Khim. **51** [1981] 854/9; J. Gen. Chem. [USSR] **51** [1981] 708/12).
[35] Bregadze, V. I.; Suleimanov, G. Z.; Kampel, V. Ts.; Petriashvili, M. V.; Mamedova, S. G.; Petrovskii, P. V.; Godovikov, N. N. (J. Organometal. Chem. **263** [1984] 131/8).
[36] Ol'dekop, Yu. A.; Maier, N. A.; Erdman, A. A.; Prokopovich, V. P. (Vestsi Akad. Navuk Belarusk.SSR Ser. Khim. Navuk **1983** 114/6; C.A. **98** [1983] No. 160775).
[37] Ol'dekop, Yu. A.; Maier, N. A.; Erdman, A. A.; Prokopovich, V. P. (Zh. Obshch. Khim. **52** [1982] 2256/9; J. Gen. Chem. [USSR] **52** [1982] 2008/10).
[38] Pena, J. P.; Trivino, L. M. S.; Contreras, J. G. (Bol. Soc. Chilena Quim. **27** [1982] 40/2).
[39] Ol'dekop, Yu. A.; Maier, N. A.; Erdman, A. A.; Prokopovich, V. P. (Vestsi Akad. Navuk Belarusk.SSR Ser. Khim. Navuk **1982** 72/6; C.A. **98** [1983] No. 198405).
[40] Zakharkin, L. I.; Kovredov, A. I.; Ol'shevskaya, V. A. (Izv. Akad. Nauk SSSR Ser. Khim. **1982** 673/5; Bull. Acad. Sci. USSR Div. Chem. Sci. **31** [1982] 599/602).

[41] Kalinin, V. N.; Usatov, A. V.; Zakharkin, L. I. (J. Organometal. Chem. **254** [1983] 127/30).

13.19 Carborane Derivatives Containing 26 Boron Atoms

$Cs_2[(C_2B_9H_{11})_2Co_2(C_2B_8H_{10})]$ is obtained by treating $[NH(CH_3)_3][C_2B_9H_{12}]$ with 40% NaOH, heating to remove $N(CH_3)_3$, and subsequent addition of $CoCl_2$ and then Cs_2SO_4 [1]. The magnetic susceptibility of $[Co_2(C_2B_8H_{10})(C_2B_9H_{11})_2]^{2-}$ has been determined [2].

References for 13.19:

[1] Volkov, V. V.; Dvurechenskaya, S. Ya. (Izv. Akad. Nauk SSSR Ser. Khim. **1981** 2356/9; Bull. Acad. Sci. USSR Div. Chem. Sci. **30** [1981] 1940/3).
[2] Volkov, V. V.; Ikorskii, V. N.; Dvurechenskaya, S. Ya. (Izv. Akad. Nauk SSSR Ser. Khim. **1983** 252/4; Bull. Acad. Sci. USSR Div. Chem. Sci. **32** [1983] 219/22).

13.20 Carborane Derivatives Containing 30 Boron Atoms

For previous data on carboranes containing 30 boron atoms, see "Boron Compounds" 1st Suppl. Vol. 3, 1981, p. 255.

$(1,2\text{-}C_2B_{10}H_{11}\text{-}1\text{-}CH_2O\text{-})_3PO$, melting point 247 to 248°C, is prepared from the action of PCl_5 on $1,2\text{-}C_2B_{10}H_{11}\text{-}1\text{-}CH_2OH$ [1]. $(1,2\text{-}C_2B_{10}H_{11}\text{-}1\text{-}CH_2O\text{-})_3PO$, melting point 255 to 257°C, is also formed from the reaction of $B_{10}H_{14}$ with $(HC{\equiv}CCH_2O)_3PO$ in the presence of $C_6H_5N(CH_3)_2$ [2].

$(1,7\text{-}C_2B_{10}H_{11}\text{-}1\text{-}CH_2O)_3PO$, melting point 246 to 247°C, is produced from the reaction of $1,7\text{-}C_2B_{10}H_{11}\text{-}1\text{-}CH_2OH$ with PCl_5 [3].

$(1,7\text{-}C_2B_{10}H_{10}\text{-}1\text{-}CH_3\text{-}7\text{-}CH_2)_3As$, melting point 297 to 298°C, is prepared from the action of $AsCl_3$ on $1,7\text{-}C_2B_{10}H_{10}\text{-}1\text{-}CH_3\text{-}7\text{-}CH_2MgBr$ [3].

Crystals of **$1,3,5\text{-}(1,2\text{-}C_2B_{10}H_{11}\text{-}1\text{-}CH_2C_6H_5\text{-}4'\text{-})_3\text{-}C_6H_3$** are triclinic, a = 15.7528(2), b = 12.7004(6), c = 15.460(2) Å; α = 109.97(1)°, β = 111.92(1)°, γ = 100.69(2)°; space group $P\bar{1}\text{-}C_i^1$ (No. 2); Z = 2. All three carboranylmethylphenyl fragments have the same conformation; the CH group of the carborane skeleton "hangs" over the benzene ring plane [4].

$(1,2\text{-}C_2B_{10}H_{10}\text{-}1\text{-}R\text{-})_3La$ and $(1,2\text{-}C_2B_{10}H_{10}\text{-}1\text{-}R\text{-})_3Tm$ (R = H, CH_3, C_6H_5) are formed from the reactions of $(1,2\text{-}C_2B_{10}H_{10}\text{-}1\text{-}R\text{-})_2Hg$ with La and Tm, respectively [5]. $(1,2\text{-}C_2B_{10}H_{10}\text{-}1\text{-}C_6H_5\text{-}2\text{-})_3M$ (M = La, Tm, Yb) is formed from $1,2\text{-}C_2B_{10}H_{10}\text{-}1\text{-}C_6H_5\text{-}2\text{-}Li/MCl_3$ [6]; $(1,2\text{-}C_2B_{10}H_{10}\text{-}1\text{-}R\text{-}2\text{-})_3M$ and $(1,2\text{-}C_2B_{10}H_{10}\text{-}1\text{-}R\text{-}9\text{-})_3M$ (R = H, CH_3, C_6H_5; M = La, Sm, Tm, Yb) are prepared by treating $(1,2\text{-}C_2B_{10}H_{10}\text{-}1\text{-}R\text{-}2\text{-})_3Hg$ and $(1,2\text{-}C_2B_{10}H_{10}\text{-}1\text{-}R\text{-}9\text{-})_2Hg$ with metal amalgam M/Hg [7].

References for 13.20:

[1] Kazantsev, A. V.; Meiramov, M. G.; Kovredov, A. I.; Zakharkin, L. I. (Izv. Akad. Nauk SSSR Ser. Khim. **1982** 1603/5; Bull. Acad. Sci. USSR Div. Chem. Sci. **1982** 1425/7).

[2] Rys, E. G.; Godovikov, N. N.; Kabachnik, M. I. (Izv. Akad. Nauk SSSR Ser. Khim. **1983** 2640/4; Bull. Acad. Sci. USSR Div. Chem. Sci. **1983** 2372/5).

[3] Zakharkin, L. I.; Kovredov, A. I.; Kazantsev, A. V.; Meiramov, M. G. (Zh. Obshch. Khim. **51** [1981] 357/61; J. Gen. Chem. [USSR] **51** [1981] 289/92).

[4] Lindeman, S. V.; Shklover, V. E.; Struchkov, Yu. T.; Khotina, I. A.; Salykhova, T. M.; Teplyakov, M. M.; Korshak, V. V. (Makromol. Chem. **185** [1984] 417/27).

[5] Suleimanov, G. Z.; Bregadze, V. I.; Koval'chuk, N. A.; Beletskaya, I. P. (J. Organometal. Chem. **235** [1982] C17/C18).

[6] Bregadze, V. I.; Koval'chuk, N. A.; Godovikov, N. N.; Suleimanov, G. Z.; Beletskaya, I. P. (J. Organometal. Chem. **241** [1983] C13/C15).

[7] Suleimanov, G. Z.; Bregadze, V. I.; Koval'chuk, N. A.; Godovikov, N. N.; Beletskaya, I. P. (Dokl. Akad. Nauk SSSR **270** [1983] 343/7; Dokl. Chem. Proc. Acad. Sci. USSR **268/273** [1983] 171/5).

13.21 Carborane Derivatives Containing 34 Boron Atoms

$Cs_3[(C_2B_9H_{11})_2Co_3(C_2B_8H_{10})_2]$ is obtained by treating $[NH(CH_3)_3][C_2B_9H_{12}]$ with 40% NaOH, heating to remove $N(CH_3)_3$, and subsequent addition of $CoCl_2$ and then Cs_2SO_4 [1]. The magnetic susceptibility of $Cs_3[Co_3(C_2B_8H_{10})_2(C_2B_9H_{11})_2]$ at 20°C is $\chi_{mol} = 666 \times 10^{-6}$ cm³/mol. The susceptibility of the anion is 573×10^{-6} cm³/mol [2].

References for 13.21:

[1] Volkov, V. V.; Dvurechenskaya, S. Ya. (Izv. Akad. Nauk SSSR Ser. Khim. **1981** 2356/9; Bull. Acad. Sci. USSR Div. Chem. Sci. **1981** 1940/3).

[2] Volkov, V. V.; Ikorskii, V. N.; Dvurechenskaya, S. Ya. (Izv. Akad. Nauk SSSR Ser. Khim. **1983** 252/4; Bull. Acad. Sci. USSR Div. Chem. Sci. **1983** 219/22).

13.22 Carboranes Containing 40 Boron Atoms

For previous data on carboranes containing 40 boron atoms, see "Boron Compounds" 1st Suppl. Vol. 3, 1981, p. 255.

Formation of **$[1,2\text{-}C_2B_{10}H_{10}\text{-}1\text{-}(1',2'\text{-}C_2B_{10}H_{11}\text{-}1'\text{-})\text{-}2\text{-}NH]_2$**, melting point above 300°C, is accomplished by treatment of $1,2\text{-}C_2B_{10}H_{10}\text{-}1\text{-}(1',2'\text{-}C_2B_{10}H_{11}\text{-}1'\text{-})\text{-}2\text{-}NH_2$ with Na/NH_3 followed by $KMnO_4$ [1, 2].

References for 13.22:

[1] Aono, K.; Totani, T. (J. Chem. Soc. Dalton Trans. **1981** 1190/5).
[2] Totani, T.; Aono, K.; Yamamoto, K.; Tawara, K. (J. Med. Chem. **24** [1981] 1492/9).

13.23 Carborane Derivatives Containing 42 Boron Atoms

$Cs_4[Co_4(C_2B_9H_{11})_2(C_2B_8H_{10})_3]$ is obtained by treating $[NH(CH_3)_3][C_2B_9H_{12}]$ with 40% NaOH, heating to remove $N(CH_3)_3$, and subsequent addition of $CoCl_2$ and then Cs_2SO_4 [1, 2]. The magnetic susceptibility of $Cs_4[Co_4(C_2B_9H_{11})_2(C_2B_8H_{10})_3]$ at 20°C has been determined to be 837×10^{-6} cm^3/mol. The susceptibility of the anion is 713×10^{-6} cm^3/mol [3].

References for 13.23:

[1] Volkov, V. V.; Dvurechenskaya, S. Ya. (Izv. Akad. Nauk SSSR Ser. Khim. **1981** 2356/9; Bull. Acad. Sci. USSR Div. Chem. Sci. **1981** 1940/3).
[2] Volkov, V. V.; Dvurechenskaya, S. Ya. (Koord. Khim. **8** [1982] 263/4; C.A. **97** [1982] No. 23968).
[3] Volkov, V. V.; Ikorskii, V. N.; Dvurechenskaya, S. Ya. (Izv. Akad. Nauk SSSR Ser. Khim. **1983** 252/4; Bull. Acad. Sci. USSR Div. Chem. Sci. **32** [1983] 219/22).

Physical Constants and Conversion Factors

Avogadro constant N_A (or L) $= 6.02214 \times 10^{23}$ mol^{-1}
Faraday constant $F = 9.64853 \times 10^{4}$ C/mol
molar gas constant $R = 8.31451$ J·mol^{-1}·K^{-1}
molar volume (ideal gas) $V_m = 2.24141 \times 10^{1}$ L/mol
(273.15 K, 101325 Pa)

Planck constant $h = 6.62608 \times 10^{-34}$ J·s
elementary charge $e = 1.60218 \times 10^{-19}$ C
electron mass $m_e = 9.10939 \times 10^{-31}$ kg
proton mass $m_p = 1.67262 \times 10^{-27}$ kg

1 kg = 2.205 pounds
1 m = 3.937×10^{1} inches = 3.281 feet
1 m^3 = 2.642×10^{2} gallons (U.S.)
1 m^3 = 2.200×10^{2} gallons (Imperial)

Force	N	dyn	kp
1 N	1	10^{5}	1.019716×10^{-1}
1 dyn	10^{-5}	1	1.019716×10^{-6}
1 kp	9.80665	9.80665×10^{5}	1

Pressure	Pa	bar	kp/m^2	at	atm	Torr	lb/in^2
1 Pa = 1 N/m^2	1	10^{-5}	1.019716×10^{-1}	1.019716×10^{-5}	9.86923×10^{-6}	7.50062×10^{-3}	1.450378×10^{-4}
1 bar = 10^6 dyn/cm^2	10^{5}	1	1.019716×10^{4}	1.019716	9.86923×10^{-1}	7.50062×10^{2}	1.450378×10^{1}
1 kp/m^2 = 1 mm H_2O	9.80665	9.80665×10^{-5}	1	10^{-4}	9.67841×10^{-5}	7.35559×10^{-2}	1.422335×10^{-3}
1 at (technical)	9.80665×10^{4}	9.80665×10^{-1}	10^{4}	1	9.67841×10^{-1}	7.35559×10^{2}	1.422335×10^{1}
1 atm = 760 Torr	1.01325×10^{5}	1.01325	1.033227×10^{4}	1.033227	1	7.60×10^{2}	1.469595×10^{1}
1 Torr = 1 mmHg	1.333224×10^{2}	1.333224×10^{-3}	1.359510×10^{1}	1.359510×10^{-3}	1.315789×10^{-3}	1	1.933678×10^{-2}
1 lb/in^2 = 1 psi	6.89476×10^{3}	6.89476×10^{-2}	7.03069×10^{2}	7.03069×10^{-2}	6.80460×10^{-2}	5.17149×10^{1}	1

Work, Energy, Heat	J	kW·h	kcal	Btu	eV
1 J = 1 W·s = 1 N·m = 10^7 erg	1	2.778×10^{-7}	2.39006×10^{-4}	9.4781×10^{-4}	6.242×10^{18}
1 kW·h	3.6×10^6	1	8.604×10^2	3.41214×10^3	2.247×10^{25}
1 kcal	4.1840×10^3	1.1622×10^{-3}	1	3.96566	2.6117×10^{22}
1 Btu (British thermal unit)	1.05506×10^3	2.93071×10^{-4}	2.5164×10^{-1}	1	6.5858×10^{21}
1 eV	1.602×10^{-7}	4.450×10^{-14}	3.8289×10^{-11}	1.51840×10^{-10}	1

$1\ cm^{-1} = 1.239842 \times 10^{-4}$ eV

1 hartree = 27.2114 eV

1 Hz = 4.135669×10^{-15} eV

1 eV ≙ 23.0578 kcal/mol

Power	kW	hp	kp·m·s^{-1}	kcal/s
1 kW = 10^3 J	1	1.35962	1.01972×10^2	2.39006×10^{-1}
1 hp (horsepower, metric)	7.3550×10^{-1}	1	7.5×10^1	1.7579×10^{-1}
1 kp·m·s^{-1}	9.80665×10^{-3}	1.333×10^{-2}	1	2.34384×10^{-3}
1 kcal/s	4.1840	5.6886	4.26650×10^2	1

References:

International Union of Pure and Applied Chemistry, Manual of Symbols and Terminology for Physicochemical Quantities and Units, Pergamon, London 1979; Pure Appl. Chem. **51** [1979] 1/41.

The International System of Units (SI), National Bureau of Standards Spec. Publ. 330 [1972].

Landolt-Börnstein, 6th Ed., Vol. II, Pt. 1, 1971, pp. 1/14.

ISO Standards Handbook 2, Units of Measurement, 2nd Ed., Geneva 1982.

Cohen, E. R., Taylor, B. N., Codata Bulletin No. 63, Pergamon, Oxford 1986.

Key to the Gmelin System of Elements and Compounds

System Number	Symbol	Element
1		Noble Gases
2	H	Hydrogen
3	O	Oxygen
4	N	Nitrogen
5	F	Fluorine
6	**Cl**	**Chlorine**
7	Br	Bromine
8	I	Iodine
8a	At	Astatine
9	S	Sulfur
10	Se	Selenium
11	Te	Tellurium
12	Po	Polonium
13	B	Boron
14	C	Carbon
15	Si	Silicon
16	P	Phosphorus
17	As	Arsenic
18	Sb	Antimony
19	Bi	Bismuth
20	Li	Lithium
21	Na	Sodium
22	K	Potassium
23	NH_4	Ammonium
24	Rb	Rubidium
25	Cs	Caesium
25a	Fr	Francium
26	Be	Beryllium
27	Mg	Magnesium
28	Ca	Calcium
29	Sr	Strontium
30	Ba	Barium
31	Ra	Radium
32	**Zn**	**Zinc**
33	Cd	Cadmium
34	Hg	Mercury
35	Al	Aluminium
36	Ga	Gallium

HCl

$ZnCl_2$

$CrCl_2$

$ZnCrO_4$

System Number	Symbol	Element
37	In	Indium
38	Tl	Thallium
39	Sc, Y La–Lu	Rare Earth Elements
40	Ac	Actinium
41	Ti	Titanium
42	Zr	Zirconium
43	Hf	Hafnium
44	Th	Thorium
45	Ge	Germanium
46	Sn	Tin
47	Pb	Lead
48	V	Vanadium
49	Nb	Niobium
50	Ta	Tantalum
51	Pa	Protactinium
52	**Cr**	**Chromium**
53	Mo	Molybdenum
54	W	Tungsten
55	U	Uranium
56	Mn	Manganese
57	Ni	Nickel
58	Co	Cobalt
59	Fe	Iron
60	Cu	Copper
61	Ag	Silver
62	Au	Gold
63	Ru	Ruthenium
64	Rh	Rhodium
65	Pd	Palladium
66	Os	Osmium
67	Ir	Iridium
68	Pt	Platinum
69	Tc	Technetium[1]
70	Re	Rhenium
71	Np,Pu . . .	Transuranium Elements

Material presented under each Gmelin System Number includes all information concerning the element(s) listed for that number plus the compounds with elements of lower System Number.

For example, zinc (System Number 32) as well as all zinc compounds with elements numbered from 1 to 31 are classified under number 32.

[1] A Gmelin volume titled "Masurium" was published with this System Number in 1941.

A Periodic Table of the Elements with the Gmelin System Numbers is given on the Inside Front Cover